NETWORK FLOWS AND MONOTROPIC OPTIMIZATION

NETWORK FLOWS AND MONOTROPIC OPTIMIZATION

R. T. ROCKAFELLAR

Department of Mathematics
University of Washington
Seattle, Washington

A WILEY-INTERSCIENCE PUBLICATION

JOHN WILEY & SONS

New York · Chichester · Brisbane · Toronto · Singapore

Library of Congress Cataloging in Publication Data:

Rockafellar, R. Tyrrell, 1935–
  Network flows and monotropic optimization.

  (Pure and applied mathematics)
  "A Wiley-Interscience publication."
  Includes bibliographical references and index.
  1. Network analysis (Planning) 2. Duality theory
(Mathematics) 3. Convex programming. 4. Linear program-
ming. I. Title. II. Series: Pure and applied mathematics
(John Wiley & Sons)
T57.85.R63 1984   519.7   83-23478
ISBN 0-471-88078-7

Printed in the United States of America

10 9 8 7 6 5 4 3 2 1

277236

PREFACE

This book is aimed at the kinds of optimization problems in which duality is as important a tool in computation as it is in theory and interpretation. These problems are characterized by a very rich interplay between combinatorial structure and convexity properties. They fill the spectrum between integer programming, on the one hand, and general convex programming on the other. Network programming, linear programming, and a broader subject that we call monotropic programming all are included.

Problems concerning flows and potentials in networks start us out at one end of the spectrum. The study of such problems abounds in results of combinatorial nature about paths, cuts, trees, and other objects, and such results are essential to the design of almost every algorithm. General linear programming problems also exhibit a combinatorial substructure. This is sometimes regarded as centering on the geometry of convex polyhedra, but tableau representations of linear systems of variables are an even more important part. A whole school of thought has grown up around the possible or desirable patterns of signs among the coefficients in such tableaus and the techniques for achieving them by a sequence of pivoting transformations.

In both network programming and linear programming, conditions for feasibility or optimality typically concern the relationships between a primal problem and a dual problem, and these relationships have a deep practical significance. One is not so much involved with constraint functions and their associated Lagrange multipliers, which are the prime focus in convex and nonconvex programming more generally, as one is with pairs of primal and dual variables whose values must fall into a certain pattern with respect to each other. Such a pattern can be anything from a complementary slackness condition to Ohm's law.

What we have attempted to do in this book is to take these common ideas, develop them in such a way as to illuminate the strong connections between linear and network programming and thereby enhance both subjects, and finally use this as a springboard in treating a much larger class of problems where the ideas find their full realization. The larger class encompasses all

v

problems where a preseparable convex function is minimized subject to linear constraints. (A convex function is *preseparable* if it is the sum of linear functions composed with convex functions of a single variable, as is any quadratic convex function, for instance.) These we call *monotropic* programming problems, because of the characteristic relationships they require between primal and dual variables; "monotropic" means "changing or varying in one direction only" and is a term which we propose as pertaining to the monotonicity properties of convex functions of a single variable.

Monotropic programming problems enjoy a remarkably complete and symmetric duality theory, almost every bit as constructively useful as the one for linear programming problems. But they also have inherent combinatorial properties. The latter are related to the possible tableau representations for the underlying linear systems of variables. On an abstract level these properties are the substance of the theory of oriented real matroids. They lead to a development of pivoting algorithms of such a form that specialization to combinatorial subroutines, like the generation of appropriate paths or cuts in the network case instead of the algebraic manipulation of coefficient matrices, is readily accomplished.

In line with our overall purpose, we emphasize the aspects of network programming that lend themselves to generalization and, at the other end of the spectrum, forgo discussion of topics that do not reside in the special nature of monotropic programming. Thus we omit combinatorial results about networks, even ones involving duality, if they are not ultimately tied in with network flows or useful for later purposes of analogy. On the other hand, we ignore computational methods of general convex programming that might in particular be applied in monotropic programming, and we never even come to the definition of the conjugate of a convex function of *several* variables, despite the fact that such functions in the case of a single variable form the foundation for our theory of duality.

Nevertheless, in order to flesh out the book as an introduction and make it more widely useful, we include many examples and details that would not be truly essential. We also provide an extensive list of exercises. The exercises serve a double purpose. They should be useful to anyone trying to learn the subject, but they also act as a repository for numerous results and observations that ought to be recorded somewhere and yet do not belong in the chapters proper, because they would take too much space, obscure the main points, or break the continuity. These exercises, in cases where they contain facts that are important enough to be cited at various places in the text, are supplied with extensive hints that amount to an outline of proof.

Each chapter begins with some remarks intended to orient the reader and ends with a section of supplementary notes and references to the literature.

A few words should be said about our treatment of algorithms. Although we try to indicate the various modes in which a procedure can be implemented, and to suggest advantages or disadvantages associated with different options, we hold back from stating algorithms in immediately programmable form, and we spend little time in discussing computational complexity. Some readers may feel this to be a serious lack; the emphasis on computers nowadays is all-pervasive. Our approach is partly just a matter of personal inclination, but there are some sound reasons behind it too.

Our objective is to forge theoretical links between problems and procedures that at first might seem quite different. We want to use these links to enrich, by way of analogy, the possible approaches that can be used in a given case, as well as to identify seemingly different approaches as being essentially the same. This requires us to view each algorithm first in its most basic form and only then consider how it might be elaborated with tricks and numerical shortcuts.

We hope that readers will recognize the long-run value of this conceptual method, as a complement to other ways of proceeding, and not be disappointed that we have not simply presented a "best" way of carrying out each computational task. In fact for much of what we discuss, there is no one "best" algorithm, nor is there as yet a theory of computational complexity that is capable of making meaningful comparisons. Even for the entirely combinatorial subroutines in network programming, since we are interested in them mainly as components in other procedures, we cannot pass judgment on them in isolation. We must take into account various features of their structure and behavior that might seem irrelevant to the narrow purpose at hand.

In truth, many of the coded procedures now available, such as shortest path algorithms, schemes for executing pivoting transformations and the like, do not seem well adapted to the intrinsic needs of our larger subject. Rather than trying to squeeze the subject into a preconceived mold, it makes more sense to develop it along its own natural lines. One can hope that this will provide stimulus for further work on the algorithms in question.

Besides the development of a general framework for a branch of optimization theory that has not hitherto been treated as a unit, this book offers some more specific contributions. In network programming it brings out the full role of potentials in duality with that of flows. This has long been perceived in electrical theory, but the ideas have not yet found their way to the hearts of economists and mathematical programmers. The book also contains the first comprehensive treatment of network programming problems with nonlinear costs. It presents the theory of conjugate convex functions of a single variable in a constructive manner with numerous examples and demonstrates its applications. This may help to popularize ideas that could be put to use much more widely than they have been.

In the area of monotropic programming, the text expounds duality results that have not previously been available in book form. It puts matroidal concepts in a form that is appropriate for tracing their role and that of duality in the design of algorithms. Among the by-products of this are extensions of the simplex method of linear programming and the out-of-kilter algorithm of network programming to general piecewise linear programming and beyond.

Although the first nine chapters deal with networks and only the last two exclusively with monotropic programming, this division of space is more apparent than real. The network chapters gradually build up the concepts to where the transition to the broader domain is very well motivated and ripe with analogies. Many of the proofs then carry over with little change. However, readers interested mainly in finding out about monotropic programming should be able to dive right into Chapter 10 and refer back to earlier material only as needed.

For readers who wish to approach the whole subject but would appreciate some guidance and shortcuts, we have designated with an asterisk * those sections that can most easily be skipped over on the first round.

Any book of this length is a long time in the making, if not in the writing, and in the present case it has been both. Many of the ideas have fascinated the author since the early 1960s but have not previously been put into print. Lecture notes on network programming from a course given at the University of Grenoble in 1973–74 formed the written nucleus out of which the book finally grew. The main job of writing extended from March 1976 to July 1979, with gaps, of course, for other activities. A final effort in the summer of 1982 went into updating the references and expanding the material on monotropic programming in the last two chapters in order to make it more accessible and self-contained.

During all that time there were many students and colleagues who helped by going through portions of the text and providing criticisms. Most notable among these was Jonathan Spingarn, who spent months at the task. The faithful and conscientious typist almost from the beginning has been Patricia Monohon, and it was she also who in fine style executed all the figures. The Air Force Office of Scientific Research, through the guidance of Dr. Joseph Bram, provided under grants AF-AFOSR-77-3204 and F4960-82-K-0012 many months of salary support without which this enormous project could never have come to fruition. The Air Force Office of Scientific Research and the University of Washington are to be commended for fostering the kind of circumstances in which such a long-term effort can be made.

R. T. ROCKAFELLAR

Seattle, Washington
April 1984

CONTENTS

NETWORK FLOWS
AND MONOTROPIC
OPTIMIZATION

1

NETWORKS

Many interesting and important problems of optimization arise in the study of transportation networks, electrical networks, and networks representing various kinds of interactions of a mechanical or economic nature. Other problems arise in areas that might seem quite far afield, but abstract models based on networks have been found to be valuable in their analysis. A celebrated case is that of "matching" problems, where objects in one set must be paired off, as many as possible, with "compatible" objects in another set. It turns out that these can be solved by maximizing the amount of material flowing through a certain transportation network.

The general theory of networks, which attempts to strengthen and unify the conceptual framework for handling problems in such diverse contexts, concerns itself with the relationship between two types of mathematical structure. First, there is a purely combinatorial foundation provided by a "directed graph," with nodes joined by oriented arcs. Besides being open to exploitation by graph-theoretical methods and algorithms, this has the valuable property of being easy to represent schematically. Second, there is the structure of the dual systems of variables corresponding to "flows" and "potentials" in the network. This is developed in terms of linear algebra and elementary convex analysis. Of particular interest for optimization are constraints and costs of the separable convex type. Introducing these corresponds in a profound way to treating each arc of the network as if it had a monotonic "characteristic curve," expressing the relationship between possible flows and potential differences in the arc much as if it were a sort of generalized electrical "black box." In economic problems potentials are prices that influence flows of goods.

Duality appears at all levels and dominates much of the subject. It serves to draw attention to many aspects of parallelism between different parts of the theory, thereby simplifying ideas and frequently suggesting how an approach in one context may be carried over to another. It leads to computational techniques that, in taking advantage also of the favorable data-processing possibilities associated with graphs, are often highly effective. In this way

1

problems of much larger scale can be solved than would be the case if they were treated as instances of general linear or convex optimization.

There is still another feature of great importance that emphasizes the need for special treatment of optimization for problems of the network type: often they can be demonstrated to have solutions purely in integers (i.e., combinatorial solutions). The tediousness of techniques such as integer programming is then rendered unnecessary.

1A. DEFINITION OF A NETWORK

The notion of a network must be put on a firm foundation before any detailed analysis is possible. As an aid to intuition a network is often shown pictorially as in Figure 1.1. There are two classes of objects here: the *nodes* (also called *vertexes* or *points*), which are represented as small circles, and the *arcs* (also called *edges*, *lines*, *branches*, or *links*), which are represented by arrows. The direction of the arrow furnishes the arc with an *orientation*, a feature that is useful in many situations. Note from the diagram that two nodes can be joined by more than one arc with the same orientation. It is generally convenient, however, to exclude *loops* (i.e., arcs that go from a node to itself).

One is led in this way to define a *network* formally as a triple consisting of two abstract sets A and N and a function that assigns to each $j \in A$ a pair $(i, i') \in N \times N$ such that $i \neq i'$. The elements of A are called *arcs*, and those of N *nodes*; it is assumed $N \neq \varnothing$. Instead of introducing a symbol for the function in the definition, we shall just write $j \sim (i, i')$ and call i the *initial* node of j and i' the *terminal* node. (The symbol $\sim$ can be read in this context as "corresponds to.") The arc j is said to be *incident* to i and i', whereas these nodes, by virtue of the existence of such an arc, are said to be *adjacent* to each other.

In some treatments of the subject multiple arcs in the same direction (arcs *in parallel*) are excluded, that is, for each (i, i') one allows at most one arc j with $j \sim (i, i')$. Then one can simply write $j = (i, i')$, and A can be identified with a subset of $N \times N$. In this case the network is called a *directed graph*, or a *digraph*. However, terminology differs; some writers also call the more general

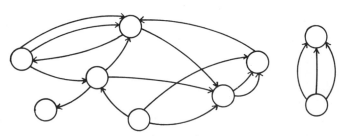

Figure 1.1

object a digraph. Others exclude parallel arcs but allow loops. (Of course then
A is an arbitrary subset of $N \times N$, i.e., any binary *relation* among the elements
of N.) The term "graph," by itself, is used for a similar structure where the arcs
are not oriented. There are also models in which some arcs of a graph are
oriented but others are not.

For present purposes it seems preferable to use the term "network" as
defined here, although in deference to the close relationship to graph theory, a
network will usually be denoted by the letter G. The choice of properties in the
definition is dictated by the desire to make as close and natural as possible the
connection with the flows and potentials to be introduced in this chapter. As
already mentioned, the emphasis on such concepts is an essential and distinguishing feature of "network theory," in contrast to "graph theory." A
network will typically have associated with its nodes and arcs various numbers,
intervals, functions, and so forth, related to conditions on flows and potentials.

The reason for excluding loops is that they cause a technical nuisance with
hardly anything to compensate for it. As for arcs in parallel, they cause no
theoretical harm at all and are even convenient in certain models (e.g., in
Chapter 7, where the complicated characteristics of a single arc are represented
by replacing it by a set of parallel arcs with simpler characteristics). Moreover,
in forcing one to think of arcs as fundamental objects in their own right, not
just as ordered pairs of nodes, they bring about a healthy notational freedom
that lends itself better to generalizations toward other areas of optimization, as
will be seen.

Nevertheless, parallel arcs are not strictly needed; they can usually be
avoided by the ruse indicated in Figure 1.2. This may well be desirable in many
cases when it comes to computation, for then one can apply certain highly
efficient methods that are known for storing and manipulating directed graphs
in a computer.

The fact that the arcs in a network are all "oriented" should not be taken to
mean that they necessarily all represent one-way links in some sense. Actually
some writers do adopt this interpretation of "oriented" arcs and are therefore
led also to work with "unoriented" arcs representing "two-way" links. Although this point of view may be justified in some graph-theoretic contexts, it
can be a serious impediment to the full understanding of models and algorithms in network optimization.

Figure 1.2

The direction of an arc should be seen mainly as a sort of sign convention. Whether in a given situation an arc is to be regarded as one-way forward, one-way backward, two-way, or even no-way will depend on other considerations and, for example, may well change during the execution of an algorithm from one iteration to the next. This may be true even if in the underlying model giving rise to the network the arcs do have an intrinsic one-way interpretation.

Problems of finding a path through a network under such changing conditions are very fundamental and will be taken up in the next chapter.

Obviously, any network can be represented as in Figure 1.1, but the appearance of the diagram depends heavily on the placement of the circles and the particular way the arrows are drawn. (The places where arcs seem to intersect in the diagram, other than at nodes, are of no significance.) It must be borne in mind that it is the abstract network G which is the subject of inquiry, not its somewhat accidental pictorial representation.

Under the general definition a network can consist of more than one "component" (two components in Figure 1.1), as will be discussed more fully in Chapter 2. It may even have "isolated nodes" which are not adjacent to any other nodes. No one is really interested in these possibilities in putting models together. But they must be allowed theoretically, since they can occur if, in the course of some development, a number of arcs are removed from the network.

1B. EXAMPLES OF NETWORKS

The following models will help to delineate, in a preliminary way, the kinds of things the theory of network seeks to encompass.

EXAMPLE 1. (*Logical Connections*)

The elements of N are mathematical properties of some kind, whereas the elements of A are mathematical arguments: $j \sim (i, i')$ means that j demonstrates that i implies i'. This rather artificial model lies more on the combinatorial end of things, yet one can imagine the relevance of some of the ideas developed later. For instance, the arguments might be assigned various weights (degrees of difficulty), and one might search for the "easiest" way to prove a certain property by a sequence of arguments starting from another property.

EXAMPLE 2. (*States of a Discrete System*)

Consider a system, mechanism, or game having only a finite number of states, which are assumed one after another, although the succession is not uniquely determined. The set N consists of the states, and A is the subset of $N \times N$ formed by the pairs (i, i') such that the state i' is one of the possible direct successors of the state i. Here G is a digraph. Again, notions of "flow" or "potential" do not on the surface seem to have much relevance, although it is clear that the study of paths from one state to another might be valuable.

One might be tempted to allow loops in this model, on the grounds that a state could succeed itself in the next time period. However, to some extent this would reflect a confusion between the network just described and a certain space-time version of it called its dynamic representation. *Dynamic networks* are of great importance and will be discussed later in this chapter.

The next model is much more fertile as a source of goals and concepts in general network optimization.

EXAMPLE 3.   (*Transportation Network*)

The elements of N are certain "places" (cities, warehouses, factories, retail outlets, ports, etc.), whereas the elements of A are transportation links (roads, railroads, shipping routes, and services, etc.): $j \sim (i, i')$ means that j is one of the direct links between i and i'.

In this case it is clear why one might want to allow parallel arcs since there may be two different means of transportation between the same locations that may not be easily lumped together (e.g., because of different nonlinear rate structures). On the other hand, the distinction between having $j \sim (i, i')$ or $j \sim (i', i)$ may not be apparent, unless there is only one direction of transport of interest, or unless each two-way link can validly be regarded as a pair of one-way links, one in each direction. However, an arbitrary choice of one of the two possible orientations for each link is convenient mathematically in representing a flow of material through the link.

For $j \sim (i, i')$, a quantity of material passing through j can be denoted by a positive number if it is moving from i to i', and by a negative number if it is moving in the opposite direction. Such quantities may be subject to various constraints and costs, and they may be influenced by prices prevalent at various locations ("potentials" at the nodes).

EXAMPLE 4.   (*Communications Network*)

This is very similar to the preceding, except for interpretation: the nodes are again places (telephone exchanges, transmission facilities, satellites, etc.), whereas the arcs are links (cables, microwave relay links, etc.). The material being transported consists of "calls" or "message units." It must be said, though, that many of the applications in this area lie beyond the scope of this book because they are also "probabilistic" in their formulation. For instance, an important concern is how to estimate how much capacity can be expected to be available between two locations, given that the network has to cope simultaneously with random demands (having known statistical distributions) at other locations. Another problem is that of analyzing the "reliability" of the network (equipment being subject to random breakdown) or its " vulnerability" to disaster.

EXAMPLE 5.   (*Hydraulic Network*)

Water is transported between reservoirs, cities, farms, and neighborhoods by means of tunnels, irrigation channels, pipes, and so forth. Many of the same features as can be found in transportation networks may be relevant, but there may also be stochastic aspects such as those mentioned for communications networks. Flows encounter "resistance" which requires the expenditure of "energy" (an associated cost); this aspect relates to optimization. Of a like nature are *power transmission networks*.

EXAMPLE 6.   (*Mechanical Network*)

The nodes are "joints" between arcs representing linear mechanical elements such as rods, beams, and springs. This time it is forces that are in some sense transmitted, whereas the positions of the joints play the role of "potentials." However, both are generally vector valued.

EXAMPLE 7.   (*Electrical Network*)

The nodes represent electrical junctions, whereas the arcs are not only "wires" but components such as resistors, batteries, generators, and diodes. Electricity moves in the network, subject to laws relating current to voltage (potential difference). The networks occurring in modern electronic systems have many features too sophisticated to fall within the bounds of the theory expounded here. Nevertheless, this example is extremely interesting as a source of analogies.

The diversity of the preceding examples makes clear the need for developing the mathematical theory of networks in a reasonably abstract manner that clarifies the fundamental ideas common to all the applications and places them in an efficient and convenient frame of reference. Such has been the motivation of many mathematicians who have worked on this subject, as has been the story in other areas of science. Indeed, in this respect network theory can serve students as an excellent example of the role and place of mathematics as a discipline, the way conceptual progress depends on inputs at many different levels. Sharp distinctions between "pure" and "applied" are impossible. Relevance, simplicity, economy of thought, and flexibility of outlook are the guidelines for all contributions of lasting value.

1C.   INCIDENCES

Let us turn to the question of how a network G and its associated flows and potentials may be represented numerically. The central idea is that of the node-arc *incidence function* of G, which is defined by

$$e(i, j) = \begin{cases} +1 & \text{if } i \text{ is the initial node of the arc } j, \\ -1 & \text{if } i \text{ is the terminal node of the arc } j, \\ 0 & \text{in all other cases.} \end{cases}$$

(A good device for remembering the signs in this definition is that "an arrow always goes from where it is to where it isn't.") This function is often expressed in terms of a node-arc incidence matrix: let the nodes and arcs be numbered in a certain order, $N = \{i_1, i_2, \ldots, i_m\}$, $A = \{j_1, j_2, \ldots, j_n\}$, and construct the array shown in Figure 1.3.

Observe that the incidence matrix E has in each column exactly one $+1$ and one -1. Conversely, if E is any $m \times n$ matrix with this property, it can be interpreted as the incidence matrix for a certain uniquely determined network G with

$$|N| = m \quad \text{and} \quad |A| = n.$$

(For a set S, we denote by $|S|$ the *cardinality* of S, i.e., the number of elements in S.) As an example, the matrix

$$E = \begin{bmatrix} 1 & 1 & 0 & 0 & 0 & 0 & 0 & 0 \\ -1 & 0 & -1 & 1 & 0 & 0 & 0 & -1 \\ 0 & -1 & 1 & 0 & -1 & 1 & 0 & 0 \\ 0 & 0 & 0 & -1 & 1 & 0 & 1 & 0 \\ 0 & 0 & 0 & 0 & 0 & -1 & -1 & 1 \end{bmatrix}$$

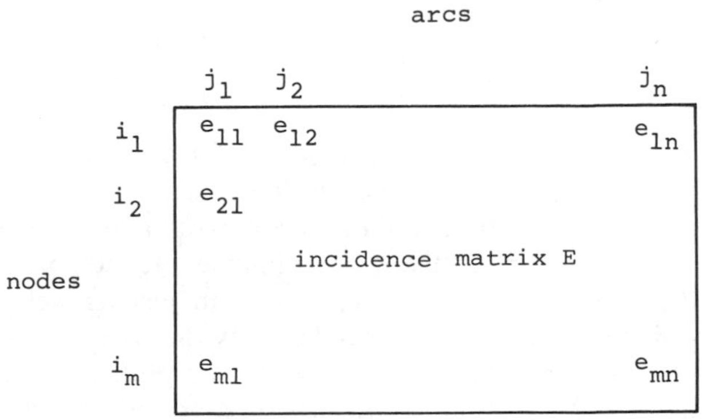

Figure 1.3

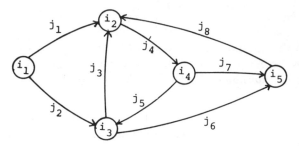

Figure 1.4

corresponds to thc nctwork in Figuie 1.4. Of course, if loops were permitted, this correspondence would break down.

Many writers go so far as to denote the nodes of the network simply by $i = 1,\ldots,m$ and the arcs by $j = 1,\ldots,n$, so that one has the equivalent notation

$$e(i, j) = e_{ij}.$$

Strictly speaking, this doesn't make sense. (How can the number 1 be a node and an arc as well as serve all its other functions?) But in practice such notation rarely causes any confusion and acts merely as a shorthand for the more formal expression in Figure 1.3. It is of obvious value for manipulations in a computer: for such purposes nodes and arcs have to be given numerical labels sooner or later anyway.

For purposes of theory, on the other hand, there is no reason for having to think of nodes and arcs as indexed in some fixed order. This could even be awkward. Suppose one wants to speak of a sequence of arcs generated in some construction. If the arcs already have been subjected to an underlying index system, it will presumably be necessary to refer to the sequence as $j_{k_1},\ldots,j_{k_r}$. The extraction of a subsequence could lead to a monstrosity like

$$j_{k_{s_1}},\ldots,j_{k_{s_p}}.$$

Such nested subscripts ought to be avoided as far as possible, not only as a kindness to humanity but because they are unjustifiably costly to set in print.

Therefore a slight broadening of outlook and terminology is desirable. After all, what is an $m \times n$ matrix really, if not a function defined on the product of the sets $\{1,2,\ldots,m\}$ and $\{1,2,\ldots,n\}$? One can just as well think of the incidence function of G as a matrix E indexed by the abstract sets N and A. It is then easy to pass back and forth between the "function" and "matrix" points of view according to the dictates of a situation. For purposes of computation a fixed order can always be introduced at the last minute.

Besides the incidence function (matrix) E it is sometimes useful to represent a network by its *adjacency function (matrix)* $\hat{E}$, which is defined on $N \times N$ by

$$\hat{e}(i, i') = \begin{cases} 1 \text{ if } (i, i') \text{ is an arc,} \\ 0 \text{ if } (i, i') \text{ is not an arc.} \end{cases}$$

Of course this presupposes that G is a digraph (i.e., $A \subset N \times N$). The network in Figure 1.4, for example, has the adjacency matrix

$$\hat{E} = \begin{bmatrix} 0 & 1 & 1 & 0 & 0 \\ 0 & 0 & 0 & 1 & 0 \\ 0 & 1 & 0 & 0 & 1 \\ 0 & 0 & 1 & 0 & 1 \\ 0 & 1 & 0 & 0 & 0 \end{bmatrix}$$

Clearly any $m \times m$ matrix can be the adjacency matrix of a network with m nodes, provided only that it consist entirely of 0's and 1's, with just 0's along the diagonal.

The way networks are typically represented in a computer is not in terms of either an incidence matrix or adjacency matrix but a "linked list structure," which in compressed form stores the information contained in the rows and columns of the adjacency matrix (see Section 1L). It is the incidence representation, however, that yields the most insights in the treatment of "flows" and "potentials" in networks, and therefore it is the most important for the problems in this book. Of course all calculations involving incidences can ultimately be carried out in a computer in terms of the "linked list" representation.

1D. FLOWS

By a *flow* in a network G, we shall in general mean nothing more than an arbitrary function $x: A \rightarrow R$. The value $x(j)$, called the *flux* in the arc j, is interpreted in most applications as the quantity of material flowing in the arc j under the sign convention already mentioned.

The kind of material is the same for all arcs. For an initial understanding of the concepts, it is helpful to think of j as a "canal" and $x(j)$ as the number of liters of water per second passing any point of j in a steady flow from the initial node i to the terminal node i'. The amount entering at i agrees at all times with the amount leaving at i', but this may be positive, negative, or zero depending on the physical direction of flow. Possibly water is being pumped in or withdrawn from the network at the various nodes; equations of conservation will be considered shortly.

For a fixed ordering of the arcs, $A = \{j_1, \ldots, j_n\}$, a flow x can be regarded as a vector $(x_1, \ldots, x_n)$. Thus a flow in the network of Figure 1.4 can be

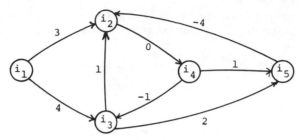

Figure 1.5

indicated as in Figure 1.5; this would correspond to $x = (3, 4, 1, 0, -1, 2, 1, -4)$. In function terms one would have $x(j_2) = 4$, and so forth. Of course much of what was said about "incidence function" versus "incidence matrix" also applies here. One could denote the arcs directly by $j = 1, \ldots, n$, so that $x(j)$ is just an equivalent notation for x_j:

$$x = (x(1), x(2), \ldots, x(n)) = (x_1, x_2, \ldots, x_n).$$

Or, as is more suited to present theoretical needs, one can recall that a vector in R^n is really no different from a function on the set $\{1, \ldots, n\}$ and thus treat x as a "vector" indexed by the abstract set A (i.e., an element of the space R^A). In fact the two approaches are equivalent, so one can use them both with perfect freedom.

There is another notation for flows which is kin to the representation of a network by its adjacency matrix. Like the latter it requires that the network be a digraph, so that each arc j can be *identified* with a pair of nodes (i, i'). The idea is simply to write $x(i, i')$ in this case in place of $x(j)$, taking $x(i, i')$ to be 0 for pairs (i, i') that do not correspond to an arc. The flows in G are then regarded as the functions on $N \times N$ which vanish at all pairs (i, i') having adjacency 0. In such terms the flow shown in Figure 1.5 receives a matrix representation:

$$x = \begin{bmatrix} 0 & 3 & 4 & 0 & 0 \\ 0 & 0 & 0 & 0 & 0 \\ 0 & 1 & 0 & 0 & 2 \\ 0 & 0 & -1 & 0 & 1 \\ 0 & -4 & 0 & 0 & 0 \end{bmatrix}$$

This notation is less compact and to that degree less attractive, except in cases where the network is close to being a complete digraph. (A digraph is *complete* if $A = N \times N$.)

1E. DIVERGENCE

Incidences enter the picture when we try to analyze what happens to a flow x at a node i, particularly the question of inputs and "leakage." The local structure of G at i is represented by row i of the incidence matrix E. Quantities of material heading away from i are associated with arcs j such that $x(j) > 0$ and $e(i, j) = 1$, or such that $x(j) < 0$ and $e(i, j) = -1$. Thus the total amount physically departing from i is the sum of all terms of the form $e(i, j)x(j)$ that are positive.

Similarly, quantities arriving at i correspond to the cases where $x(j) > 0$ and $e(i, j) = -1$, or $x(j) < 0$ and $e(i, j) = 1$. The sum of all terms of the form $e(i, j)x(j)$ which happen to be negative is thus the negative of the total amount physically arriving at i. For all other arcs j, one has $e(i, j)x(j) = 0$. Therefore the sum of $e(i, j)x(j)$ over *all* arcs j gives us the total departing from i minus the total arriving at i.

This quantity is the *divergence* of the flow at i; it will be denoted by $y(i)$. Thus

$$y(i) = \sum_{j \in A} e(i, j)x(j) = [\text{divergence of } x \text{ at } i].$$

For example, considering the flow x in Figure 1.5 at the node i_3, one sees that four units arrive on $j_2 \sim (i_1, i_3)$, while two units depart on $j_6 \sim (i_3, i_5)$, and one unit each on $j_5 \sim (i_4, i_3)$ and $j_3 \sim (i_3, i_2)$. This node thus has the property that

$$[\text{total arriving}] = [\text{total departing}],$$

so that $y(i_3) = 0$. At i_5, however, one has $y(i_5) = -7$. In other words, seven more units arrive at i_5 than leave it, meaning that seven units are withdrawn from the flow at this node.

In general, a node i is said to be a *source* for the flow x if $y(i) > 0$ and a *sink* if $y(i) < 0$. If $y(i) = 0$, the flow is *conserved* at i.

One calls y the *divergence function* (*vector*) associated with x. The definition is summarized by

$$y = Ex = \text{div } x.$$

Notice that matrix notation is convenient and appropriate here, even though the implied summation may be over all $j \in A$ instead of $j = 1, \ldots, n$.

In the case of the flow in Figure 1.5 one could use the incidence matrix determined earlier for the network in question and calculate

$$y = \begin{bmatrix} 1 & 1 & 0 & 0 & 0 & 0 & 0 & 0 \\ -1 & 0 & -1 & 1 & 0 & 0 & 0 & -1 \\ 0 & -1 & 1 & 0 & -1 & 1 & 0 & 0 \\ 0 & 0 & 0 & -1 & 1 & 0 & 1 & 0 \\ 0 & 0 & 0 & 0 & 0 & -1 & -1 & 1 \end{bmatrix} \cdot \begin{bmatrix} 3 \\ 4 \\ 1 \\ 0 \\ -1 \\ 2 \\ 1 \\ -4 \end{bmatrix} = \begin{bmatrix} 7 \\ 0 \\ 0 \\ 0 \\ -7 \end{bmatrix}$$

Thus i_1 is a source, i_5 is a sink, and the flow is conserved at the other nodes (i_2, i_3, and i_4).

The fact that the seven units created at the sole source i_1 in this example are exactly matched by seven units destroyed at the sole sink i_5 is no accident. Physical intuition suggests, and algebra confirms, that the total amount created at the sources always equals the total amount destroyed at the sinks. This is expressed by the *total divergence principle*:

$$\sum_{i \in N} y(i) = 0 \quad \text{for} \quad y = \operatorname{div} x.$$

To verify the principle, one need only insert the formula defining $y(i)$ and interchange sums:

$$\sum_{i \in N} y(i) = \sum_{i \in N} \sum_{j \in A} e(i, j) x(j) = \sum_{j \in A} \sum_{i \in N} e(i, j) x(j).$$

The last sum vanishes because

$$\sum_{i \in N} e(i, j) = 0 \quad \text{for all } j \in A.$$

Indeed, each column of the incidence matrix contains exactly one 1 and one -1 and hence adds up to 0.

1F.   VECTOR OPERATIONS

Two flows x and x' in G can be added together ("superimposed") to produce a resultant flow x'': $x''(j) = x(j) + x'(j)$ for all $j \in A$. Likewise a flow can be multiplied by a scalar: $x' = \lambda x$ means that $x'(j) = \lambda x(j)$ for all $j \in A$. The properties of these operations are completely obvious in vector terms, but it is important to gain an early understanding of what they might mean physically.

As an illustration, let us add to the flow x in Figure 1.5 the flow x' depicted in Figure 1.6, which represents one unit passing from i_1 to i_5 by way of i_2 and i_3. (There is a source of one unit at i_1 and a sink of one unit at i_5, whereas for x

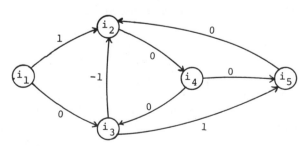

Figure 1.6

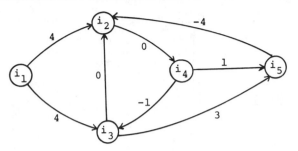

Figure 1.7

it was seven units in each case.) Figure 1.7 shows the flow $x'' = x + x'$, which has a source of eight units at i_1 and a sink of eight units at i_5.

The special thing to notice is the cancellation which takes place in the arc $j_3 \sim (i_3, i_2)$: this arc is "used" by both x and x', but not by x''. In physical terms the superposition of the two flows has entailed a "diversion" of a unit of x, which used to go from i_1 to i_2 by way of i_3, into one accomplishing the same thing directly, and a new unit has been added going from i_1 to i_3 to i_5. (It is not always the case that the superposition of two flows has a unique interpretation along such lines; the analysis of flows will be studied in more detail in Chapter 4.)

Obviously, the flow $2x'$, say, would represent two units moving in the same pattern as x' from i_1 to i_2 to i_3 to i_5, whereas $-x'$ would represent one unit moving in the reverse pattern.

Trivially, one has the rules:

$$\text{div}(x + x') = \text{div}\,x + \text{div}\,x',$$

$$\text{div}(\lambda x) = \lambda \,\text{div}\,x.$$

1G.   CIRCULATIONS AND THE AUGMENTED NETWORK

A special role is played by the flows x in G such that $\text{div}\,x = 0$ (i.e., x is conserved at *every* node). Such flows are called *circulations*. Sums and scalar multiples of circulation are again circulations. Thus the set of all circulations forms a linear subspace of R^A: the *circulation space* $\mathcal{C}$. Clearly $\mathcal{C}$ is the "row null space" of the incidence matrix E, the kernel of the linear transformation $x \to Ex = \text{div}\,x$ from R^A to R^N.

One of the reasons why circulations are important is that theoretical discussions can often be simplified in terms of them. This is due to the fact that every flow in G can be identified with a circulation in a certain larger network. In the case of the flow x in Figure 1.5, the idea is illustrated by Figure 1.8.

In general, one forms from G a new network $\overline{G}$ by adding a new node $\bar{i}$ (the *distribution node*) and an arc $j_i \sim (\bar{i}, i)$ (a *distribution arc*) for each of the old nodes i. This is the *augmented network*; its node and arc sets are $\overline{N}$ and $\overline{A}$. (Notice right here, incidentally, how awkward it would be for this discussion if

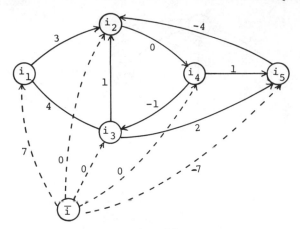

Figure 1.8

the convention had been embraced that the arcs of a network are always numbered $1,\dots,n$.) To each flow x in G there corresponds the flow $\bar{x}$ in $\bar{G}$, defined by

$$\bar{x}(j) = x(j) \quad \text{for all old arcs,}$$

$$\bar{x}(j_i) = y(i) \quad \text{for all distribution arcs.}$$

The fact that $\bar{x}$ is conserved at all old nodes i, regarded as nodes of $\bar{G}$, is immediate from the definition of $\bar{x}(j_i)$, whereas the fact that $\bar{x}$ is conserved at the distribution node $\bar{i}$ is equivalent to the total divergence principle already enunciated. Thus $\bar{x}$ is a circulation in $\bar{G}$. Conversely, every circulation in $\bar{G}$ corresponds in this way to a flow in G.

Sometimes one can pass to a setting of circulations by a simpler device, specifically in cases where there is just a fixed pair of nodes permitted to act as source or sink. Such cases arise very frequently. For the flow in Figure 1.5 the corresponding representation as a circulation is in Figure 1.9. The new arc $\bar{j}$ is called a *feedback arc*.

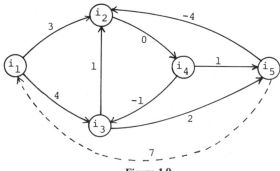

Figure 1.9

1H. DYNAMIC VERSION OF A NETWORK

For some models the "static" notion of a flow, which has been stressed up until now, is not appropriate. It is necessary instead to think of the material as starting out at various nodes and making definite progress in time through other nodes, undergoing interactions along the way.

For example, suppose one wanted to rush a large quantity of material through a transportation network from a certain "supply point" to a "demand point" in the shortest time possible. The capacities of the many alternate transportation links may be limited, and there may be potential bottlenecks at some intermediate handling points. The flow would have to be organized in the time scale: one would have to specify how much should be entered into each

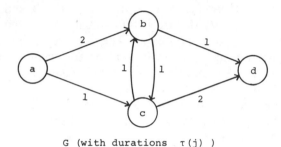

G (with durations $\tau(j)$)

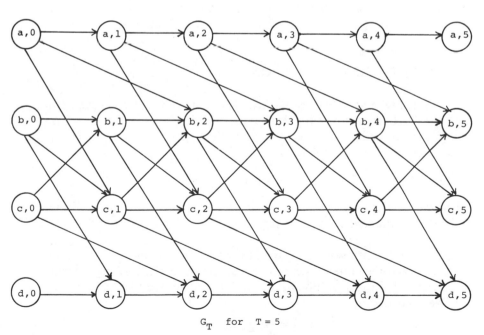

G_T for $T = 5$

Figure 1.10

transportation link and when, how much should temporarily be held over at certain intermediate points so as not to block the progress of other crucial parts of the flow, and so forth.

Surprisingly enough, there is a general way of reducing such complicated dynamic situations to static ones in a "space-time" framework. This does depend, however, on the legitimacy of regarding each arc in G as "one-way" in the direction of its orientation.

Let each arc of G have associated with it a non-negative integer $\tau(j)$ called its *duration* (the time taken to traverse it, measured in discrete units). A network G_T, corresponding to any specified time interval $0 \leq t \leq T$, is then constructed as follows (see Figure 1.10). The node set is

$$N_T = N \times \{0, 1, \ldots, T\}.$$

To obtain the arc set A_T, form for each $j \in A$ with $j \sim (i, i')$, and each integer $t \geq 0$ such that $t + \tau(j) \leq T$, the arc j_t with initial node given by the pair $(i, t) \in N_T$ and terminal node $(i', t + \tau(j))$. This arc j_t represents the fact that one can pass via j from being at i at time t to being at i' at time $t + \tau(j)$. Also include in A_T, for each $i \in N$ and time t, $0 \leq t < T$, a *holdover arc* with initial node (i, t) and terminal node $(i, t + 1)$.

Non-negative flows in G_T correspond to non-negative "dynamic" flows in G, complete with a specification of how much material enters an arc j or passes through a node i at each time t. Fluxes in the holdover arcs of G_T represent material kept stationary at nodes of G during certain time periods.

The dynamic model may also be useful where flows, as such, are not present. For instance, the network of Example 2 in Section 1B may be extended in time by regarding each arc as having duration 1.

1I. POTENTIALS AND TENSION

Duality in the study of flows is often closely tied to the following notion. A *potential* in G is an arbitrary real-valued function u on the node set N. The value $u(i)$ is called the potential at node i. With an arc $j \sim (i, i')$, one associates the potential difference

$$v(j) = u(i') - u(i) = [\text{tension across } j].$$

(The sign of the difference depends of course on the orientation of the arc.) This defines the *tension function* (or *vector*) v on A which is the *differential* of the potential u.

As example is shown in Figure 1.11 where the numbers at the nodes are potentials and the numbers next to the arcs are the corresponding tensions.

Observe that the definition of the tension v corresponding to the potential u can be expressed by

$$v(j) = - \sum_{i \in N} u(i) e(i, j).$$

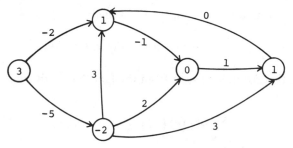

Figure 1.11

Therefore in compact notation

$$v = -uE = \Delta u.$$

The relationships between all the variables that have been introduced are summarized by the tableau in Figure 1.12, which is said to define a pair of *dual linear systems*. Rows correspond to relationships in the flow-divergence system, whereas columns correspond to those in the potential-tension system.

Potentials and their corresponding tensions can be added together or multiplied by scalars. Two different potentials can give rise to the same tension. In particular, if $u'(i) = u(i) + $ const. for all $i \in N$, then $\Delta u' = \Delta u$. The full possibilities in this direction will be analyzed in Section 6B.

In general, one calls v a *differential* in G if $v = \Delta u$ for *some* potential u (which, as we have just seen, can never be unique). The set of all differentials is preserved under addition and scalar multiplication and thus, like the circulation space $\mathscr{C}$, forms a linear subspace of R^A. It is called the *differential space* and denoted by $\mathscr{D}$. Thus $\mathscr{D}$ is the "row space" of the incidence matrix E, the range of the linear transformation $u \mapsto -uE$ from R^N to R^A (which is the "negative adjoint" of the linear transformation $x \mapsto Ex$).

It is natural to use the following inner product notation for elements of R^N and R^A:

$$u \cdot y = \sum_{i \in N} u(i) \cdot y(i), \qquad v \cdot x = \sum_{j \in A} v(j) \cdot x(j).$$

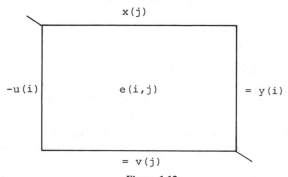

Figure 1.12

One then has the fundamental *conversion formula*:

$$v \cdot x = -u \cdot y \quad \text{if } y = \operatorname{div} x, \, v = \Delta u.$$

The validity of this formula is seen from the fact that both sides reduce to the expression

$$- \sum_{\substack{i \in N \\ j \in A}} u(i) e(i, j) x(j) = -uEx.$$

An immediate consequence of the conversion formula is the fact that

$$v \cdot x = 0 \quad \text{for all } v \in \mathcal{D}, x \in \mathcal{C}.$$

Less immediate perhaps is the stronger fact that

$$\mathcal{D} = \mathcal{C}^{\perp} = \{ v \in R^A | v \cdot x = 0 \quad \text{for all } x \in \mathcal{C} \},$$

$$\mathcal{C} = \mathcal{D}^{\perp} = \{ x \in R^A | v \cdot x = 0 \quad \text{for all } v \in \mathcal{D} \},$$

or in other words, that the circulation space $\mathcal{C}$ and differential space $\mathcal{D}$ are *orthogonally complementary* to each other. However, this follows by elementary linear algebra from the definitions of $\mathcal{C}$ and $\mathcal{D}$ in terms of E (see Exercises 1.7 and 1.8 at the end of the chapter). The same would hold even if E had nothing to do with incidences in a network.

Much as any flow in G can be regarded as part of a circulation in the augmented network $\overline{G}$, so can every potential in G be regarded as "part of" a differential in $\overline{G}$; see Figure 1.13, which demonstrates this for the potential in Figure 1.11. Indeed, given any potential u in G with associated tension v, define the potential $\overline{u}$ in $\overline{G}$ by

$$\overline{u}(i) = \begin{cases} u(i) & \text{for all old nodes } i, \\ u(\overline{i}) = 0 & \text{for the distribution node } \overline{i}. \end{cases}$$

The tension $\overline{v} = \Delta \overline{u}$ in $\overline{G}$ then satisfies

$$\overline{v}(j) = v(j) \quad \text{for all old arcs } j,$$

$$\overline{v}(j_i) = u(i) \quad \text{for all distribution arcs } j_i.$$

Observe that the orthogonality of the circulation space $\overline{\mathcal{C}}$ and differential space $\overline{\mathcal{D}}$ for $\overline{G}$ corresponds to the conversion formula for G, written in the form $u \cdot y + v \cdot x = 0.$

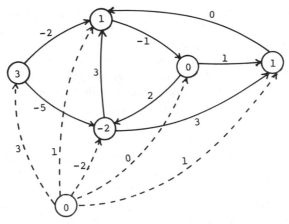

Figure 1.13

1J.*   PREVIEW OF OPTIMAL FLOWS AND POTENTIALS

To get some idea of the general kinds of network optimization problems to which the developments in this book are ultimately directed, consider the following situation. For each arc $j \in A$ there is an interval $C(j) \subset R$ and a function f_j: $C(j) \mapsto R$; likewise for each node $i \in N$ there is an interval $C(i) \subset R$ and a function f_i: $C(i) \mapsto R$. The *optimal flow* problem in broadest form is this:

$$\text{minimize } \sum_{j \in A} f_j(x(j)) + \sum_{i \in N} f_i(y(i)) \text{ over all flows}$$
$x \in R^A$ satisfying $x(j) \in C(j)$ for all $j \in A$ and
$y(i) \in C(i)$ for all $i \in N$, where $y = \text{div } x$.

In analogous terms there is also a general *optimal potential* problem:

$$\text{minimize } \sum_{i \in N} g_i(u(i)) + \sum_{j \in A} g_j(v(j)) \text{ over all}$$
potentials $u \in R^N$ satisfying $u(i) \in D(i)$ for all $i \in N$
and $v(j) \in D(j)$ for all $j \in A$, where $v = \Delta u$.

Under certain continuity and convexity assumptions about the functions involved, these two types of problems turn out to be "dual" to each other. They then possess joint optimality conditions of the form

$$(x(j), v(j)) \in \Gamma(j) \quad \text{and} \quad (y(i), u(i)) \in \Gamma(i),$$

where $\Gamma(j)$ and $\Gamma(i)$ are "characteristic curves" of a special kind associated with the arcs and nodes of the network. The relation $(x(j), v(j)) \in \Gamma(j)$ can be viewed as a generalization of Ohm's law (where $\Gamma(j)$ is merely a straight line through the origin, the slope of the line being the "resistance" of the arc j).

The theory of optimal flows and potentials will not be handled in such grand formulation until Chapters 8 and 9, although the case of linear and piecewise linear cost functions will be covered in Chapter 7. Much of the material studied along the way, however, in Chapters 2 through 6, either lays the foundation for solving optimal flow and potential problems or works out applications that actually correspond to special instances of such problems. In the latter category, for example, are the results in Chapter 5 on matching problems, which are flow problems of a very particular sort, as well as the optimal path theory in Chapter 6, which is concerned in effect with potentials. Chapters 3 and 6 are devoted in large measure to questions of the existence and determination of flows or potentials that are *feasible* (i.e., satisfy constraint systems of the kind specified in the problem given here). The answers to these questions are important in themselves, but they also enter into the general optimization algorithms developed in Chapters 7 and 9, as do results in Chapters 2 and 4 that bear on the representations of flows and potentials in terms of the combinatorial substructures of a network (paths, circuits, cuts, trees, etc.).

1K.* SOME GENERALIZATIONS

One can consider flows that are *vector valued* rather than merely real valued. This is natural, for instance, in the case of a transportation network serving simultaneously for movement of several different kinds of material: then each flux $x(i)$ is itself a vector of components $(x_1(i), \ldots, x_N(i))$, where $x_k(i)$ is the flux of the kth kind of material in the arc i. Because of such an interpretation, vector-valued flows are often referred to as *multicommodity flows*. However, as already seen, they also arise in entirely different contexts where no "commodities" are present, such as in the mechanical network of Example 6 in Section 1B.

Vector-valued flows are also associated with situations where only one kind of material is involved, but the kind of cancellation that may occur as described earlier, when two flows are superimposed, is not appropriate. Thus in a model of street traffic, the flow may all be in terms of a homogeneous material ("vehicles") moving between various points of "supply" and "demand." To make matters simple, let us suppose these points can be represented by "north," "south," "east," and "west." A certain volume of traffic must pass though the network from "north" to "south," another volume from "east" to "west," and so forth. The trouble is that these different kinds of traffic need to preserve their identities: the demand at "south" for traffic originating at "north" cannot be met by shunting to "south" some of the traffic originating at "east," with a compensating diversion of some of the "north"-originating vehicles to end up at "west." These different kinds of traffic therefore need to be treated as different "commodities" that interact by sharing the same network.

Another direction of generalization concerns the notion that the flux entering an arc may be multiplied by some factor before emerging at the other end. One then has a *network with gains*. (The "gain" factors need not be greater than unity.) This corresponds to replacing the -1's in the incidence matrix E by arbitrary negative numbers.

Many of the important results about ordinary networks break down when applied to vector-valued flow or flows in networks with gains. But some valuable and interesting facts remain. In exploring what can be accomplished in such directions, one reaches the idea of trying to mimic the theory of network optimization, including its combinatorial features, as far as possible in the case of a *general* pair of dual linear systems (i.e., two systems of real variables related as in Figure 1.12 but with an arbitrary real matrix E. Even "multicommodity" problems can be reduced to this case.)

Remarkably much turns out to be possible. A theory can be put together that spans the gap between network optimization and general convex optimization, passing by way of such subjects as linear programming which are intermediate in computational amenability and retain an important combinatorial flavor. The presentation of this theory is the goal of the latter part of this book. It is what we call monotropic programming. The primal and dual monotropic programming problems treated in Chapter 10 correspond to the general optimal flow and optimal potential problems described in the preceding section, except that the relations $y = \operatorname{div} x$ and $v = \Delta u$ are replaced by $y = Ex$ and $v = -uE$ for arbitrary E, not necessarily the incidence matrix of any network.

1L.*   EXERCISES

1.1. (*Incidence*). Draw a network whose incidence matrix is the following:

$$E = \begin{bmatrix} -1 & 0 & 0 & 1 & 0 & 0 \\ 0 & 1 & -1 & 0 & 1 & 0 \\ 0 & 0 & 1 & 0 & 0 & -1 \\ 0 & 0 & 0 & 0 & 0 & 0 \\ 1 & 0 & 0 & -1 & 0 & 0 \\ 0 & -1 & 0 & 0 & -1 & 1 \end{bmatrix}$$

1.2. (*Incidence*). Determine the incidence matrix for the network shown in Figure 1.14.

1.3. (*Divergence*). Figure 1.15 indicates a certain flow x and potential u for the same network as in Figure 1.14. Determine the divergence $y = \operatorname{div} x$ and tension $v = \Delta u$. Where are the sources and sinks of the flow?

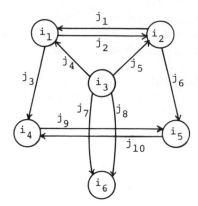

Figure 1.14

1.4. (*Divergence*). Let x be a flow in a network G with the property that x is conserved at every node except for a certain pair of nodes s and s'. Show that $y = \text{div } x$ satisfies $y(s) = -y(s')$, and give a physical interpretation of this relation.

1.5. (*Differentials*). The values shown next to the arcs in Figure 1.16 define a certain $v \in R^A$. Prove that v is a differential by displaying a particular potential u such that $\Delta u = v$.

1.6. (*Divergence*). The values shown at the nodes in Figure 1.17 define a certain $y \in R^N$. Does there exist a flow x such that $\text{div } x = y$?

1.7. (*Circulation Space*). Prove that the circulation space $\mathscr{C}$ and differential space $\mathscr{D}$ of a network G are always related in dimension by $\dim \mathscr{C} + \dim \mathscr{D} = |A|$.

(*Hint.*   Work with standard facts about the rank of a matrix E.)

1.8. (*Circulation Space*). Prove that the spaces $\mathscr{C}$ and $\mathscr{D}$ are orthogonally complementary: $\mathscr{C}^{\perp} = \mathscr{D}$ and $\mathscr{D}^{\perp} = \mathscr{C}$.

(*Hint.*   Build on Exercise 1.7.)

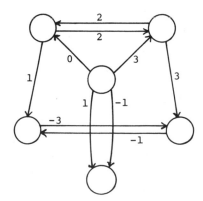
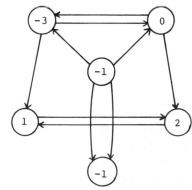

Figure 1.15

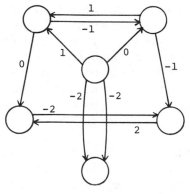

Figure 1.16

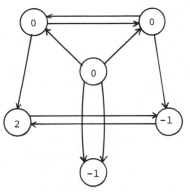

Figure 1.17

1M.*   COMMENTS AND REFERENCES

Although electrical networks have been studied for a long time, the usefulness of network flows and potentials in the modeling of problems in economics and operations research was not generally recognized until the 1950s. The book of L. R. Ford and D. R. Fulkerson [1962] has played an especially significant role in stimulating the growth and applications of network theory in these new areas. The notation and terminology in that book are natural for people familiar with linear programming and have widely been accepted in operations research. However, there are some drawbacks that need to be pointed out because they explain why a different approach is used here.

Most noteworthy is the convention of Ford and Fulkerson that flux values $x(j)$ must always be non-negative, that is, that the material can only move in the directions in which the arcs are oriented. Although this does not entail a real loss of generality, in the sense that any problem modeled by a network can be reformulated so as to bring it into compliance with the convention, there is a serious effect on theoretical aspects such as duality. The simplicity and symmetry of relationships between flows and potentials is obscured, particularly in the case of problems with nonlinear costs. In the description of algorithms, repeated reformulations are sometimes necessary in the subproblems that are generated. The connections with electrical theory, where no sign restrictions are imposed on flows, are harder to see.

In an article written earlier than the Ford and Fulkerson book but not nearly so well known, G. J. Minty [1960] developed a framework for network optimization that was more in harmony with the electrical viewpoint and made clear that the conditions characterizing optimal flows and potentials could always be interpreted as arising from generalized (possibly nonlinear) "resistance" relations associated with the arcs of the network. Minty's approach was adopted subsequently in the book of C. Berge and A. Ghouila-Houri [1962], and it has been used more recently by Iri [1969], although not in its full scope in either case.

In order to place duality in the brightest light, Minty reduced everything about flows and potentials to circulations and differentials (i.e., to the complementary subspaces $\mathscr{C}$ and $\mathscr{D}$). This is perfectly possible in terms of passing to the augmented network, as has been explained in Sections 1G and 1I. But a certain flexibility of outlook seems to be lost thereby in operations research models, and analogies with related areas like linear and convex programming tend to suffer. For these reasons we try in this book to proceed somewhat more broadly in a setting that readily encompasses Minty's as well as the one of Ford and Fulkerson.

The insistence on every arc having an orientation is dictated by our aim of concentrating on the features of network theory that ultimately have some bearing on flows and potentials. Certainly there are many interesting problems of a purely combinatorial nature where orientations are irrelevant. When orientations are put aside, one speaks simply of a *graph* (undirected) instead of a network. Of course a graph can also be viewed as a special kind of network in which every arc occurs paired with its reverse. Among the many texts treating graph theory and its applications, we mention Berge [1962], Bondy and Murty [1976], Busacker and Saaty [1965], Christophides [1975], Deo [1974], Harary [1969], and Wilson [1972].

The text of Lawler [1976] elaborates the more combinatorial aspects of the kind of network theory in the present work, whereas that of Helgason and Kennington [1980] fills in the connections with linear programming and recent developments in the design of algorithms for optimization problems having a networklike substructure. For the classical theory of electrical networks, see Chen [1971] and Mayeda [1972]. For probabilistic communications networks and related models, see Frank and Frisch [1971].

Computer representation of networks is covered in various books such as Deo [1974] and Helgason and Kennington [1980], but the article of Dial, Glover, Karney, and Klingman [1979] is particularly illuminating and accessible. This explains the use of a linked list structure and discusses the dramatic effect that such improvements in computer implementation technology have had on network calculations. A more complete description of some of the network codes in question is provided by Ali, Helgason, Kennington, and Hall [1977]. The success of this approach has been attributed to the fact that it corresponds to specializations of the simplex method of linear programming, that is, to a combinatorial representation of "pivoting" techniques. Pivoting will be covered in Chapter 4 and will play an important role thereafter, particularly in Chapters 7, 9, and 10. Although only some of the algorithms based on pivoting have not as yet received such close attention in terms of computer implementation as the ones just mentioned, it is good to keep in mind that much improvement could lie in that direction.

The notion of dual linear systems of variables stems from linear programming. It has been developed in many interesting directions by A. W. Tucker [1960], [1963], and proves especially useful in making the generalization from optimization problems in networks to other kinds of separable programming.

The terms "flux" and "divergence" have not previously been used in network theory as they are here and may not be entirely to the liking of electrical engineers because of other usage in connection with magnetism. However, they are simple, natural, and fill a definite need. The "distribution node" in the augmented network corresponds to "ground" in electrical theory.

Flows x belonging to the circulation space $\mathscr{C}$ are also said to satisfy *Kirchhoff's current law*, whereas tensions v in the differential space $\mathscr{D}$ satisfy *Kirchhoff's voltage law* (see the equivalent criterion in Exercise 2.12; see also Section 6A). This terminology honors the pioneering work of G. Kirchhoff in 1847. A discussion of these conditions in the terminology of combinatorial topology is furnished by Slepian [1968]. The fact that $\mathscr{C}$ and $\mathscr{D}$ are orthogonal is sometimes called *Tellegen's theorem*.

2

PATHS AND CUTS

The special character of network theory, as a branch of optimization, is due to its combinatorial foundations. Of utmost importance is the fact that certain purely combinatorial notions, like that of a "path" from one node to another, are intimately connected with concepts of a different order, like that of a flow of material from one node to another. At the heart of most of the computational methods for optimizing flows and potentials, and as the guarantee of their particular efficiency, are basic procedures involving only the manipulation of arcs and nodes. These procedures construct, or test the existence of, configurations such as paths with prescribed properties.

The most fundamental of the combinatorial aspects are described in this chapter. Paths are discussed along with different kinds of connectivity of a network. A dual notion of "cut" is introduced and used in formulating a central result for problems concerning the existence of paths: the painted network theorem. This theorem is proved constructively in terms of an algorithm that serves as a conceptual or practical component in a great many computational procedures that will be developed later.

2A. PATHS

A *path* P in a network G is a finite sequence $i_0, j_1, i_1, j_2, \ldots, j_r, i_r$ $(r > 0)$, where each i_k is a node, j_k is an arc, and either $j_k \sim (i_{k-1}, i_k)$ or $j_k \sim (i_k, i_{k-1})$. This formalizes an idea that is very natural in terms of the geometric representation of a network, that of a way of passing from one node to another through a succession of arcs but not necessarily keeping to the directions of the arrows. The *initial* node of P is i_0, and the *terminal* node is i_r. This is sometimes indicated by the notation $P: i_0 \rightarrow i_r$. If $i_0 = i_r$, P is called a *circuit*; however, in this case there is little point in distinguishing a particular node as the start and finish, so two paths which differ only in this respect are regarded as constituting the same circuit.

The arc j_k in P is said to be traversed *positively* or *negatively* according to whether $j_k \sim (i_{k-1}, i_k)$ or $j_k \sim (i_k, i_{k-1})$. If all arcs are traversed positively, P

is called a *positive* path, or positive circuit, and similarly *negative* path, or negative circuit, if all arcs are traversed negatively.

In many situations merely the sequence of nodes, or the list of arcs encountered, would be enough to describe a path unambiguously. Then one should not hesitate to exploit the simplification. It should be understood, though, that the full specification of the alternating sequence of nodes and arcs is necessary sometimes to make certain just which arcs are traversed and in which direction, as is important for our purposes.

For example, let i and i' be distinct nodes, and let j and j' be arcs such that $j \sim (i, i')$ and $j' \sim (i', i)$ (see Figure 2.1). The closed path P: i, j, i', j', i cannot be summarized by giving only i, i', i, since that would equally well describe the path P': i, j', i', j, i. Nor can P be summarized by j, j', since that could also refer to P'': i', j, i, j', i'. Such ambiguities would leave us in doubt as to whether j is traversed positively or negatively, and this is unacceptable.

An abbreviated notation that is often convenient in describing paths in G, if G is a digraph, is based on the property that, in passing from i to an adjacent node i', there are at most two possibilities: a "forward" arc or a "backward" arc. Symbolism like

$$P: i_0 \rightarrow i_1 \leftarrow i_2 \leftarrow i_3 \rightarrow i_4$$

is then almost self-explanatory: the corresponding arcs are $j_1 = (i_0, i_1)$, $j_2 = (i_2, i_1)$, $j_3 = (i_3, i_2)$, and $j_4 = (i_3, i_4)$. (In a computer the signs $\rightarrow$ and $\leftarrow$ could be replaced by $+$ and $-$ prefixed to the nodes, giving the sequential representation $+i_0, +i_1, -i_2, -i_3, +i_4$.)

A path P may traverse an arc more than once, maybe sometimes positively and sometimes negatively. Then it is a *path with multiplicities*. An *elementary* (or *simple*) path is a path without multiplicities which in fact uses no node more than once, except of course for the initial and terminal nodes when the path is a circuit. Figure 2.2 illustrates a path without multiplicities that is *not* elementary.

Given any path P: $s \rightarrow s'$ (i.e., with $i_0 = s$ and $i_r = s'$), one can construct a corresponding elementary path from s to s' simply by deleting superfluous portions of P. In algorithmic terms one proceeds along P until a node is encountered (other than s') that coincides with a previous node, say, $i_{k+p} = i_k$. Deleting $j_{k+1}, i_{k+1}, \ldots, j_{k+p}, i_{k+p}$ still leaves a path from s to s', so one can

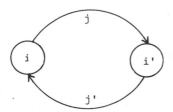

Figure 2.1

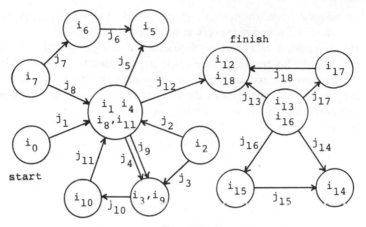

Figure 2.2

proceed further. This continues until s' is reached; then everything afterward is lopped off. The resulting path from s to s' is clearly elementary. (If $s' = s$, there is the degenerate possibility that no arcs at all are left at termination, so that one does not have what can be called a path. This is remote from the typical cases where the procedure might actually be applied.)

Almost everything of interest about paths can be reduced in this fashion to the case of elementary paths.

2B.   INCIDENCES FOR PATHS

For a path P without multiplicities, the set of all arcs traversed positively (the *positive* or *forward* arcs) is denoted by P^+, and the set of all arcs traversed negatively (the *negative* or *backward* arcs) is denoted by P^-. Arc-path incidences are accordingly defined by

$$e_P(j) = e(j, P) = \begin{cases} 1 & \text{if } j \in P^+ \\ -1 & \text{if } j \in P^-, \\ 0 & \text{otherwise.} \end{cases}$$

(This could be ambiguous if P had multiplicities, since P^+ and P^- might overlap, and anyway one might then want to replace ± 1 by other integers indicating the number of times the arc was traversed.) It is convenient to write $j \in P$ to mean $e_P(j) \neq 0$, even though P really denotes more than just the union of P^+ and P^-.

Observe that e_P, as a function on the set of arcs, can be regarded as a special flow. Physically it represents one unit flowing along the path P. If P is not a circuit, the function $y = \text{div } e_P$ has the value $+1$ only at the initial node

i_0 of P and -1 only at the terminal node i_r; thus i_0 is the unique source for the flow and i_r the unique sink. If P is a circuit, there is no source or sink, and e_P is a circulation.

An elementary path P can be completely reconstructed from its incidence function (vector) e_P. Consider, first, the case where P is not a circuit. The initial node i_0 is determined as the sole node which is a source of the flow e_P. Next, there is exactly one arc j incident to i_0 such that $e_P(j) \neq 0$; designate it by j_1 and its other node by i_1. Now j_2 can be determined as the only other arc j besides j_1 which is incident to i_1 and has $e_P(j) \neq 0$, and so forth. Almost the same process works if P is a circuit, except that initially an arbitrary arc of P must be designated as j_1 and its nodes denoted by i_0 and i_1, with the order chosen so that $j_1 \sim (i_0, i_1)$ if $e_P(j) = 1$ but $j_1 \sim (i_1, i_0)$ if $e_P(j) = -1$.

As a matter of fact the process just described shows that an elementary path P can be reconstructed from knowledge of the set $P^+ \cup P^-$ alone, except for determining which of the two possible directions it is to have.

Circumstances are not much different for a nonelementary path P without multiplicities. Knowing only P^+ and P^-, one could reconstruct P except for possible ambiguities about the order in which certain side circuits, such as shown in Figure 2.2, are to be traversed. However, this order is really of little interest; what *is* important is that the directions the individual arcs are traversed be beyond doubt.

For these reasons a path without multiplicities can almost be regarded as just a special kind of *signed set* of arcs (i.e., a subset P of A supplied with a partitioning into a "positive" part P^+ and "negative" part P^-, either of which might be empty). Although the formal definition of "path" given earlier will be maintained, both for the sake of concreteness and its closer ties to computation, the more abstract point of view will guide much of the theory.

2C.   CONNECTEDNESS

A network G is said to be *connected* if for every pair of different nodes s and s', there is a path $P: s \to s'$.

If G is not connected, its node set N can be partitioned into disjoint subsets N_k such that two different nodes can be joined by a path if and only if they belong to the same N_k. The sets N_k can be defined as the equivalence classes of nodes induced by the following binary relation: s' *is connected to* s if either $s' = s$ or there exists a path with initial node s and terminal node s'; see Exercise 2.6. Then there is a corresponding partition of the arc set A into disjoint subsets A_k, where A_k consists of the arcs whose initial node and terminal node both lie in N_k. (Since any arc, together with its initial and terminal nodes, forms a path, these two nodes must belong to the same equivalence class, i.e., the same set N_k.)

Each pair N_k, A_k, then constitutes a connected network G_k called a *component* of G. Obviously, G is connected if and only if there is just one component:

G itself. Note that the augmented network $\overline{G}$ is always connected, even if G is not.

A more restrictive property than connectedness is *strong connectedness*. In this case one asks that there exist, for every pair of different nodes s and s', a *positive* path P: $s \rightarrow s'$ (as defined in Section 2A).

It is not true in general that a network that fails to be strongly connected can be decomposed into *disjoint* pieces that are. Nevertheless, a useful decomposition of sorts is possible. Again one may proceed in terms of an equivalence relation: s' *is strongly connected to* s if either $s' = s$ or there is both a positive path from s to s' and a positive path from s' to s (see Exercise 2.6). Let the equivalence classes in N be indexed N_k, and as before, let A_k be the set of all arcs having both nodes in N_k. The network G_k determined by N_k and A_k is called a *strong* component of G. This time, however, the sets A_k, though disjoint, do not necessarily exhaust A and therefore may not form a partition. Thus there may be various arcs joining the strong components G_k. But it can be shown that all the arcs joining one strong component to another must "go in the same direction" (Exercise 2.7).

2D. FINDING A PATH FROM ONE PLACE TO ANOTHER

How can one test efficiently whether a given network is connected or strongly connected? More generally, if two different nodes s and s' are given, how can one construct a path from s to s' that avoids certain "forbidden" arcs and traverses certain other arcs, if at all, only in a specified direction?

In a general problem of this type, four categories of arcs may be considered in order to gain the greatest flexibility: arcs traversible in either direction (*two-way*), arcs traversible only positively (*one-way forward*), arcs traversible only negatively (*one-way backward*), and forbidden arcs (*no-way*). The practical virtue of allowing all four possibilities, rather than trying to "simplify" matters by some sort of modification of the underlying network, will be thoroughly apparent later, when the study of flows gets under way.

To formulate the problem, one should therefore specify, besides s and s', a partition of the arc set A into four disjoint subsets, some of which might be empty. Thanks to G. Minty, there is a happy way of describing such a partition in which, instead of referring to the four categories directly, one speaks of each of the arcs as having been "painted" one of four possible colors. The partition is then called a *painting* of A.

The colors green, white, black, and red will be used, respectively, to correspond to the four categories mentioned. (Green and red are easily remembered in terms of "go" and "stop," and likewise white and black have a natural duality which is convenient for our purposes.) In this terminology, for instance, a "green arc" is an arc in the "two-way" category of usability with respect to the paths in question.

The fundamental problem is then stated as follows, with a slight generalization concerning the initial and terminal nodes of the path.

Painted Path Problem.   *In the network G, two nonempty disjoint node sets N^+ and N^- are given, as well as a painting of the arcs by the colors green, white, black, and red. The problem is to determine a path*

$$P: N^+ \rightarrow N^-$$

(i.e., with initial node in N^+ and terminal node in N^-) such that every arc in P^+ is green or white, whereas every arc in P^- is green or black.

Any path with the specified color properties is said to be *compatible* with the given painting. Thus a *solution* to the problem is a compatible path P from N^+ to N^-. Observe that there is no real loss of generality in requiring further that P be an *elementary path, none of whose intermediate nodes belongs to N^+ or N^-.* This follows from our discussion of elementary paths in Section 2A; the extra properties can always be achieved through a constructive process whereby certain segments are deleted from a path that solves the problem in the more general sense. As a matter of fact the algorithm to be given will always furnish a solution with these extra properties, if a solution exists at all. Therefore in speaking of a solution to the painted path problem, one with these properties will always be meant, unless something to the contrary is mentioned.

It deserves to be emphasized that, in a painting, each arc of the network is given exactly one of the four colors listed, but some of the colors can remain unused. For example, in testing for connectedness all the arcs would be painted green, whereas for strong connectedness they would all be painted white or black (see Section 2I).

At this stage we do not place any additional burdens on a solution to the painted path problem, although it is easy to imagine some that might be worthwhile. Often there will be many solutions, and one could ask for the path among them with the fewest arcs or, given further data, the path that is shortest, longest, quickest, cheapest, easiest, or whatever. Such refinements of the problem will be considered in Chapter 6.

2E.   CUTS

In the algorithm about to be considered, as well as in almost every branch of the theory of networks, the concept of a "cut" is useful and plays a role dual to that of "path."

To formulate it, start by defining for arbitrary node sets S and S' (not necessarily disjoint) the arc sets

$$[S, S']^+ = \{ j \in A | j \sim (i, i') \quad \text{with } i \in S, i' \in S' \}.$$
$$[S, S']^- = \{ j \in A | j \sim (i', i) \quad \text{with } i \in S, i' \in S' \}.$$

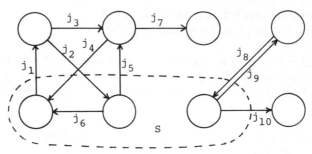

Figure 2.3

For the present, attention is to be focused on the case where $S' = N \setminus S$ (the *complement* of S in N). The sets $[S, N \setminus S]^+$ and $[S, N \setminus S]^-$ are disjoint, and one may therefore speak of the "signed set" of arcs having these as its positive and negative parts, respectively; this is denoted simply by $[S, N \setminus S]$. (Recall that a "signed set" is merely a set partitioned into two subsets, not necessarily nonempty, one designated "positive" and the other "negative.")

Any signed set of the form $[S, N \setminus S]$ will be called a *cut* in the network G. More exactly, a cut in G is defined to be a signed arc set Q (with positive and negative parts denoted by Q^+ and Q^-) such that, for *some* node set S (maybe not unique), one has $Q^+ = [S, N \setminus S]^+$ and $Q^- = [S, N \setminus S]^-$. A cut is illustrated in Figure 2.3.

The word "cut" for $Q = [S, N \setminus S]$ arises from the idea that any path P with initial node in S and terminal node in $N \setminus S$ must, at some stage, traverse one of the arcs in Q; the deletion of the arcs in Q would thus "cut" all such paths. The fact that P must use an arc of Q is formally established as follows. Let i be the first of the nodes of P not in S; such a node exists, because P goes from S to $N \setminus S$. The arc of P immediately preceding i, call it j, then joins a node of S with a node of $N \setminus S$, and hence it belongs to Q. Indeed either $j \in P^+ \cap Q^+$ or $j \in P^- \cap Q^-$.

Some writers use "cut" in a more general sense to mean any set of arcs (not signed) whose deletion would disrupt some class of paths or other, depending on the context. But this usage is unsatisfactory for the treatment to be given here.

The node set S is not always uniquely determined by the corresponding cut. In particular both $S = \emptyset$ and $S = N$ give rise to the *empty cut*. But if the network is connected and the cut Q is nonempty, there really is only one S such that $[S, N \setminus S] = Q$, and it can be determined by a simple construction (Exercise 2.14). At all events the *reverse* of a cut Q can be defined unambiguously. It consists of the same arcs, but with opposite orientations.

Incidences for arcs and cuts are defined by

$$
e_Q(j) = e(j, Q) = \begin{cases} 1 & \text{if } j \in Q^+ \\ -1 & \text{if } j \in Q^- \\ 0 & \text{in all other cases.} \end{cases}
$$

It is interesting to observe that e_Q, as a function on the set of all arcs, can be regarded as a differential. Indeed, it is the tension corresponding to the *negative* of the potential

$$e_S(i) = \begin{cases} 1 & \text{if } i \in S, \\ 0 & \text{if } i \notin S, \end{cases}$$

where S is any node set such that $Q = [S, N \setminus S]$. In symbols, $e_Q = -\Delta e_S = \Delta e_{N \setminus S}$. The situation is analogous to that for paths (e_P, for a path P, is a certain flow), and it hints strongly at the important role paths and cuts are to play in the analysis and synthesis of flows and tensions.

Following our earlier abuse of notation with paths (crime becomes a habit!), we shall write $j \in Q$ to mean that the arc j belongs to either Q^+ or Q^-, or in other words, $e_Q(j) \neq 0$, even though Q is more than just the union of Q^+ and Q^- (i.e., a *signed* set). The arcs of Q^+ are called the *positive* or *forward* arcs of Q, whereas those of Q^- are the *negative* or *backward* arcs.

2F.   PAINTED NETWORK ALGORITHM

A procedure will now be described that in q iterations or less, where $q = |N| - |N^+| - |N^-| + 1$, either constructs a solution to the painted path problem or establishes that none exist by producing a certain kind of cut (a solution to the "painted cut problem" which will be introduced in Section 2H). This is a *conceptual* algorithm, in the sense that its statement aims at clarifying fundamental ideas and leaving the possible modes of implementation as open and flexible as possible. Details of implementation can of course make a big difference in practice, but they raise other issues and are best relegated to a separate discussion (see Section 2G).

In the general step of the algorithm there is a node set $S \supset N^+$ and also a function $\theta: S \setminus N^+ \mapsto A$ which, by labeling each node $i \in S \setminus N^+$ with an arc $j \in A$, will serve to represent paths that have so far been constructed in order to reach the nodes of S from N^+ without passing outside of S. The exact requirements are the following; we shall call θ a *routing of S with base N^+* when these are fulfilled:

1. For each $i \in S \setminus N^+$, $\theta(i)$ is an arc joining i to another of the nodes in S.
2. Whenever a sequence is generated of the form $i, \theta(i), i', \theta(i'), i'', \theta(i''), \ldots$, where i' is the other node of $\theta(i)$, i'' is the other node of $\theta(i')$, and so forth, a node in N^+ is eventually reached.

A sequence as in 2 must stop when it reaches N^+, since θ is only defined on $S \setminus N^+$. The reverse of the sequence is then a path from N^+ to i that does not use any nodes outside of S. It will be called the path to i *associated* with θ, or the θ-path from N^+ to i.

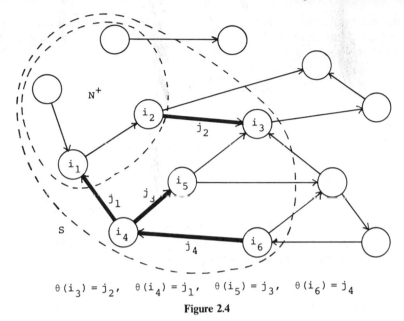

$$\theta(i_3) = j_2, \quad \theta(i_4) = j_1, \quad \theta(i_5) = j_3, \quad \theta(i_6) = j_4$$

Figure 2.4

Figure 2.4 illustrates a particular routing and brings out the fact that in many situations there may be more convenient ways of indicating a routing θ than the full description of its domain and values. In particular, if the network has no more than one arc joining any pair of nodes, a labeling of nodes by nodes rather than arcs would suffice. The general concept of a "routing" as employed here is closely related to that of a "rooted tree," which will be discussed in Section 4E.

In the algorithm we shall be interested only in routings θ which are *compatible with the given painting*. This means that the arc $\theta(i)$ must be green or white if i is its terminal node, whereas in the opposite case it must be green or black. Obviously, all the paths associated with θ are then themselves compatible with the painting: their positive arcs are all green or white, whereas their negative arcs are all green or black.

Statement of the Algorithm

Initially, let $S = N^+$; the routing θ is then "empty." The general step is as follows. There is a set $S \supset N^+$ with $S \cap N^- = \varnothing$ and a routing of S with base N^+. Inspecting the cut $Q = [S, N \setminus S]$, determine if there is an arc in Q^+ that is green or white, or an arc in Q^- that is green or black.

If there is not, the algorithm halts; in this case the painted path problem has no solution. (Indeed, as seen in the previous discussion of cuts, if the problem had a solution P, there would have to be at least one arc in $P^+ \cap Q^+$ or in $P^- \cap Q^-$, and this is impossible under the circumstances without violating the color requirements.)

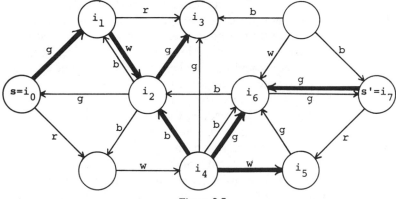

Figure 2.5

On the other hand, if there is such an arc j, let $\theta(i) = j$, where i is the node of j not in S, and redefine S as $S \cup \{i\}$. The extended θ is then still a routing compatible with the painting. If $i \in N^-$, the algorithm halts: the θ-chain P from N^+ to i solves the painted path problem. If $i \notin N^-$, then again $S \cap N^- = \varnothing$, and the general step is repeated.

EXAMPLE 1

A network is shown in Figure 2.5 with two distinguished nodes s and s' and a painting of the arcs; here $N^+ = \{s\}$ and $N^- = \{s'\}$. The notation indicates how the corresponding painted path problem was solved by the algorithm. The node i_k is the one added to S at the kth iteration, and $\theta(i_k)$ is the unique "heavy" arc joining i_k to a node of lower index. The solution is the path

$$P: s \to i_1 \to i_2 \leftarrow i_4 \to i_6 \leftarrow s'.$$

Constructing a Maximal Routing

The painted network algorithm can also be applied usefully with $N^- = \varnothing$. Termination then comes only with a cut (possibly the empty cut). At this point the set S consists of *all* possible nodes that can be reached from N^+ by paths compatible with the painting, and the routing θ embodies a system of paths to all these nodes. Then θ is called a *maximal (compatible) routing with base* N^+, relative to the given painting.

2G.*   PRIORITY RULES AND MULTIROUTINGS

In some of its details the painted network algorithm could be implemented in different ways, and this could affect its performance on a computer and its qualities as a subroutine in other algorithms. Supplementary procedures for handling "ties" are the chief source of flexibility. At each step a node $i \notin S$ is reached by way of an arc j from a node $i' \in S$, and then i is added to S. But there may be more than one such combination meeting the requirements, so additional criteria can be brought into play in making a choice, or several combinations can be processed in the same iteration.

An optional rule that will later be seen to have important theoretical consequences is the following. In selecting the arc j to be used, always prefer a green arc if available, rather than a white or black arc. This case will be referred to as the painted network algorithm with *arc discrimination*.

Another approach is to set up a "priority" ordering for nodes in S and then look to the *priority* of i' in choosing the combination to be used. There are two common ways of defining priorities, as will be explained, but both refer to the following pattern of implementation.

In the initial iteration (where $S = N^+$) the priority ordering of the nodes is introduced arbitrarily, and it is subsequently extended to new nodes as they are added to S. As the algorithm progresses, certain of the nodes of S achieve a special but permanent status; they are said to have been *thoroughly scanned*. (Initially, no nodes are in this category.) In a general iteration with $N^+ \subset S \subset N \setminus N^-$, one takes i' to be the node of highest priority among those in S that have not yet been thoroughly scanned. The arcs incident to i' are then inspected one by one to see if any meets the prescription for crossing the cut $Q = [S, N \setminus S]$ to some new node i (i.e., a green or white arc $j \sim (i', i)$ with $i \notin S$, or alternatively a green or black arc $j \sim (i, i')$ with $i \notin S$). If such an arc j is detected, one proceeds as already described (defining $\theta(i) = j$ and adding i to S with a certain priority). Otherwise the node i' is deemed henceforth to have been thoroughly scanned. Then one passes to the node of next highest priority among those in S that have not yet been thoroughly scanned, and so on. In the case where there are no nodes in S that have not yet been thoroughly scanned, the algorithm terminates with the cut Q, there being no solution to the painted path problem.

To get the most out of this implementation, bookkeeping details can be added to avoid duplication of effort from iteration to iteration in the inspection of the arcs emanating from a particular node.

The two most important ways of generating the priorities that are invoked in this procedure are the rule of *breadth-first search* and the rule of *depth-first search*. In breadth-first search each new node that is added to S is regarded as having a lower priority than any of the nodes already in S. (It is sent to the end of the scanning queue.) In depth-first search it receives a higher priority than any of the others.

The effect of breadth-first search is to keep the arc inspection procedure focused on a particular node i' until i' has been thoroughly scanned. This simplifies some aspects of the bookkeeping. It also has the consequence that the solution path which is obtained uses as few arcs as possible (Exercise 2.20). Breadth-first search is closely related to the "multipath" version of the painted network algorithm that will be explained later in this section.

The rule of depth-first search amounts to the following. If, having arrived at some node i', one finds it possible to pass to a new node i (not previously encountered), one always does so. Otherwise i' is permanently removed from the list of nodes worth looking at, and one backtracks (following the routing θ) to the node from which i' was reached.

Somewhat surprisingly perhaps, depth-first search can lead to dramatic improvements in efficiency in certain graph-theoretic applications (see Section 2M for references). The success lies in taking advantage of particular ways that networks can be represented and manipulated in a computer. But some reservations are in order, lest one think that depth-first search should therefore be used in all implementations of the painted network algorithm. When an algorithm is intended, like this one, as a building block in other computational procedures covering a wide variety of situations, its efficiency cannot always be engineered or assessed in isolation. Certain features may take on an unanticipated significance in a new context.

For example, depth-first search may well conflict with the rule of arc discrimination mentioned earlier. It is therefore unusable in circumstances requiring the special results produced by arc discrimination (e.g., see Exercises 4.11 and 4.42). It is also inferior in certain fundamental applications to flow problems where breadth-first search, in effect, turns out to be crucial (see Section 3J). For theoretical purposes, then, there is obviously wisdom in trying to concentrate on the essence of a procedure like the painted network algorithm and otherwise trying to leave it relatively flexible and open to adaptation.

Last among the procedures for handling "ties" that deserve mention here is one where a choice is not made, but instead all the tied combinations are processed together on an equal basis. This means in particular that more than one arc may be eligible for routing designation as $\theta(i)$ when a node i is reached, so it is necessary to deal with a set $\Theta(i)$ of such arcs.

The concept of a *multirouting* Θ of S with base N^+ becomes useful. This is just like that of a routing, except that $\Theta(i)$ is a set of arcs, any one of which can be followed in tracing back from nodes of $S \setminus N^+$ to S. In other words, Θ is a multirouting of S with base N^+ if it associates with each node $i \in S \setminus N^+$ a nonempty arc set $\Theta(i)$ such that no matter what element $\theta(i)$ might be selected from $\Theta(i)$ for each i, the resulting mapping θ would be a routing of S with base N^+.

Thus a multirouting Θ, in contrast to an ordinary routing, generally represents a whole family of paths from N^+ to each node i of $S \setminus N^+$. These are spoken of as Θ-*paths*, and Θ is said to be compatible with a given painting if all these paths are.

The *multipath version* of the painted network algorithm will now be described. In the general step the data on hand are as in the basic algorithm, except that there is a multirouting Θ rather than a routing θ. We determine the set consisting of *all* the nodes of $N \setminus S$ that can be reached from S by green or white arcs in Q^+, or by green or black arcs in Q^- (where Q is the cut $[S, N \setminus S]$), and for each such node i we denote the corresponding arc set by

$\Theta(i)$. All these nodes are then added to S, and unless N^- has been reached, the step is repeated. (If no nodes can be added to S in this way, the algorithm terminates as before with the cut Q, whose nature indicates that the painted path problem has no solution.) Clearly this procedure constructs a multirouting with base N^+ that is compatible with the given painting and (upon favorable termination) yields a possible multiplicity of compatible paths P: $N^+ \to N^-$. **Claim**: *These Θ-paths then constitute all the solutions to the painted path problem that use r arcs, where r is the number of iterations (of the preceding type) taken to reach N^-, and moreover r is the smallest number of arcs possible for a solution path.* Indeed, if we denote by N_k the set of nodes reached on the kth iteration ($k = 1, \ldots, r$), it is easy to see by induction that N_k consists of all nodes reachable from N^+ by compatible paths containing k arcs (but not reachable by compatible paths containing fewer arcs). For each $i \in N_k$, the arcs in $\Theta(i)$ give all the possible ways i can be reached from nodes in N_{k-1} (where $N_0 = N^+$). Any compatible path P: $N^+ \to N^-$ that uses only r arcs must have its consecutive nodes in $N_0, N_1, \ldots, N_r$ and therefore be one of the Θ-paths in question.

2H.   THEORETICAL IMPLICATIONS OF THE ALGORITHM

The fundamental information furnished by the painted network algorithm can be summarized elegantly in terms of a complementary problem. Let us say that Q is a cut that *separates N^+ from N^-* if Q is of the form $[S, N \setminus S]$ for some node set S such that $S \supset N^+$ and $S \cap N^- = \emptyset$. The notation for this is Q: $N^+ \downarrow N^-$.

Painted Cut Problem.   *In the network G, two nonempty disjoint node sets N^+ and N^- are given, as well as a painting of the arcs by the colors green, white, black, and red. The problem is to determine a cut*

$$Q: N^+ \downarrow N^-$$

(i.e., *separating N^+ from N^-*) *such that every arc in Q^+ is red or black, whereas every arc in Q^- is red or white.*

COLOR	PATHS	CUTS
green	two-way	no-way
white	one-way forward	one-way backward
black	one-way backward	one-way forward
red	no-way	two-way

Figure 2.6

A cut meeting the color requirements is said to be *compatible with the painting*, and if it also separates N^+ and N^-, it is a *solution* to the painted cut problem. Note the duality between the color code for paths and the one for cuts, as tabulated in Figure 2.6.

The painted network algorithm furnishes a constructive proof of the following fact.

Painted Network Theorem. *Let N^+ and N^- be two nonempty disjoint subsets of the node set N. Then for each painting of the arcs of G by the colors green, white, black, and red, one and only one of the following assertions is true:*

1. *The painted path problem has a solution P.*
2. *The painted cut problem has a solution Q.*

Indeed, Alternatives 1 and 2 correspond to the two possible outcomes of the algorithm when it terminates, as it ultimately must. If it terminates without a path $P: N^+ \rightarrow N^-$, one has a cut $Q: N^+ \downarrow N^-$ with the property that no arc in Q^+ is green or white, whereas no arc in Q^- is green or black. But then Q is a solution to the painted cut problem. The existence of such a cut precludes the existence of a path solving the painted path problem, so Alternatives 1 and 2 are mutually exclusive as claimed.

It is possible in Alternative 2 for Q to be the empty cut, if G is not connected and N^+ and N^- are contained in different components. But this is not a case of real interest.

There is another result closely related to the painted network theorem that we shall often want to use in a constructive manner. To formulate it in the best way, we need, parallel to the concept of an elementary path, that of an "elementary cut."

A cut Q is said to be *elementary* if the deletion of its arcs would increase the number of components of the network by exactly 1. For G connected, this is equivalent to the property that Q is of the form $[S, N \setminus S]$, where $\emptyset \neq S \neq N$ and every pair of nodes in S can be joined by a path using only nodes of S, and likewise every pair of nodes in $N \setminus S$ can be joined by a path using only nodes of $N \setminus S$. Some of the possibilities for nonelementary cuts are illustrated by Figures 2.3 and 2.7.

A fact that needs to be recorded theoretically, but for which we will not actually have much practical use, is that *every nonempty cut Q can be expressed as the disjoint union of elementary cuts*, in the special sense that there exist elementary cuts $Q_1, \ldots, Q_q$ having no arcs in common, such that

$$Q^+ = Q_1^+ \cup \cdots \cup Q_q^+ \quad \text{and} \quad Q^- = Q_1^- \cup \cdots \cup Q_q^-.$$

(see Exercise 2.9). This is analogous to the fact that a path without multiplicities, such as illustrated in Figure 2.2, can be expressed as the (arc-)disjoint union of a set of elementary paths.

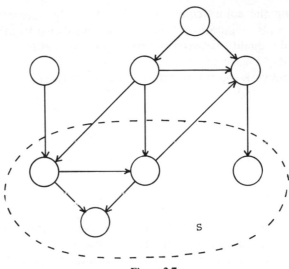

Figure 2.7

Minty's Lemma.   *Given any painting of the network G by the colors green, white, black, and red, and any arc j that is white or black, one and only one of the following assertions is true:*

    1.   *There exists an [elementary] circuit P such that P uses $\bar{j}$ and is compatible with the painting.*

    2.   *There exists an [elementary] cut Q such that Q uses $\bar{j}$ and is compatible with the painting.*

The word "elementary" has been placed in brackets in both Alternatives 1 and 2 to indicate that its presence or absence has no effect on the validity of the result. This has already been recognized in the case of paths: if there exists a path meeting the conditions, it can be reduced to an elementary path meeting the same conditions by a simple procedure. The corresponding fact about cuts follows from the decomposition principle just cited: given a cut Q meeting the conditions, it can be replaced by whichever of the "subcuts" Q_k in the decomposition is the one containing the arc j.

Minty's lemma can be derived from the painted network theorem in the following manner (this furnishes, of course, a *constructive* method, to be referred to as *Minty's algorithm*, for verifying which of Alternatives 1 and 2 holds in a given case). Designate the two nodes of a given arc j by s and s', with the notation chosen, so that $j \sim (s', s)$ in the case where j is white but $j \sim (s, s')$ in the case where j is black. Apply the painted network algorithm with $N^+ = \{s\}$ and $N^- = \{s'\}$. If a solution $Q = [S, N \setminus S]$ to the painted cut problem is obtained, it necessarily contains $\bar{j}$ (because $s \in S$, $s' \notin S$) and therefore satisfies Alternative 2 of Minty's lemma; conversely, any cut with the latter properties solves the painted cut problem. On the other hand, a solution

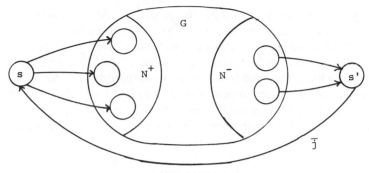

Figure 2.8

to the painted path problem, if it is elementary (as can be supposed without loss of generality), is a compatible path P': $s \to s'$ which does not use $\bar{j}$. (The reason why it cannot use $\bar{j}$ is that there is only one elementary path from s to s' that does so, i.e., s, $\bar{j}$, s', and this does not meet the color requirements because of the way s and s' were chosen.) Such paths P' correspond one-to-one with elementary circuits P satisfying Alternative 1 of Minty's lemma: $P = P'$, $\bar{j}$, s.

A fact not as evident, but nevertheless of some theoretical interest in motivating some generalizations to be made later, is that the painted network theorem can in turn be derived from Minty's lemma. The two results are therefore *equivalent* from a theoretical point of view. This is seen from Figure 2.8. The setting of the painted network theorem is extended, as shown, by the addition of certain new arcs and nodes, including a special arc $\bar{j}$. All the new arcs are painted white, and Minty's lemma is applied. Alternatives 1 and 2 of the two results are found to be in direct correspondence. (A compatible cut in the extended network, if it contains $\bar{j}$, cannot contain any other of the added arcs, due to their all being white.)

2I.*   APPLICATION TO CONNECTEDNESS

The painted network algorithm provides an efficient means of testing whether a network is connected and, if not, determining its components. Such a test sometimes plays a part in computational schemes where a number of arcs are tentatively removed from a network.

Select an arbitrary node s, and apply the algorithm with $N^+ = \{s\}$, $N^- = \varnothing$, and all arcs painted green. On termination, there will be a maximal routing θ with base s (i.e., base $\{s\}$) corresponding to a certain set S containing s. From the nature of the painting it is clear that S consists of all nodes reachable by paths starting at s (the nodes *connected to* s, as defined in Section 2C), and θ describes a particular system of paths accomplishing the task. If $S = N$, G is connected. If not, then the nodes in S, along with the arcs incident to them, form the component of G containing s. To determine another component, choose any node not in S and repeat the procedure. After a finite sequence of such calculations all the components will have been identified.

The corresponding *test for strong connectedness* needs just twice the effort. Apply the algorithm as before, starting from an arbitrarily chosen node s, but with all arcs white. This yields a maximal compatible routing θ_w associated with a node set S_w. Then repeat the application with the arcs all black, obtaining θ_b and S_b. Obviously, the nodes in S_w (besides s) are those that can be reached by a positive path from s, whereas those in S_b are the ones *from*

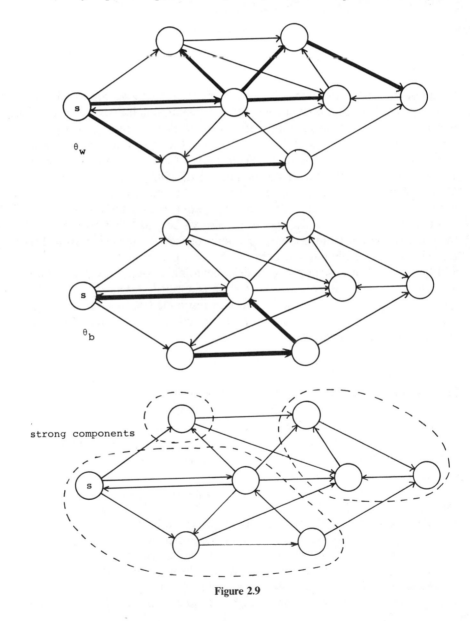

Figure 2.9

which s can be reached by a positive path. Therefore the set $S = S_w \cap S_b$ is comprised of all the nodes strongly connected to s. If $S = N$, G is strongly connected. If not, S furnishes the strong component containing s; see Figure 2.9.

Sometimes a mixed sort of connectedness may be in question. In the case of a city street network one can paint all two-way streets green, all one-way streets white, and all streets presently to be closed for repairs red. Will it be possible to go from any node to any other node?

2J.*   ACYCLIC NETWORKS

The property "opposite" to G being connected is presumably for G to be completely disconnected, that is, to be a network whose components are all isolated nodes. What then is opposite to G being strongly connected? A case can be made for assigning this role to the property that G is *acyclic*, which means G possesses no positive circuits. The duality between the two properties is explored more fully in the exercises at the end of this chapter (see Exercises 2.26 through 2.29); for instance, G is acyclic if and only if each of its strong components consists of a single node.

The class of acyclic networks is of great importance. Such networks often appear in models where an arc $j \sim (i, i')$ represents an irreversible causal or temporal relationship between i and i'. For example, the dynamic version of a network is by nature acyclic if the "durations" are all positive.

There is also a close connection between acyclic networks and *strict partial orderings*. Recall that a strict partial ordering on a set is a binary relation " $<$ " satisfying the axioms:

1. Antisymmetry, that is, $i < i'$ precludes $i' < i$.
2. Transitivity, that is, $i < i'$, $i' < i''$, implies $i < i''$.

In such a setting the elements of the set can be regarded as the nodes of a network G whose arcs are the pairs $j = (i, i')$ such that $i < i'$. Obviously, G is acyclic, for the existence of a positive circuit, amounting to a string of "inequalities" $i < i' < ,\ldots, < i$, would present a conflict between transitivity and antisymmetry. Also G is without parallel arcs.

Conversely, any acyclic network G without parallel arcs corresponds to a certain relation $<$ on the node set N: $i < i'$ if and only if there is an arc j (necessarily unique) such that $j \sim (i, i')$. Since G has no positive circuits, the relation is antisymmetric ($i < i' < i$ is impossible). But without something further it might *not* be transitive and therefore *not* a strict partial ordering. The "something further" consists of passing to an induced relation: $i <^\circ i'$ if and only if there is a positive path in G from i to i' (or equivalently, a string $i < \cdots < i'$). This relation is still antisymmetric, though also transitive, and therefore it *does* represent a strict partial ordering. It is called the *transitive closure of* < , and the network that is obtained by adding an arc $j \sim (i, i')$ for each pair such that $i <°i'$, but not $i < i'$, is called the *transitive closure of G*.

A common case where G is the network associated with a strict partial ordering is that where N is a collection of subsets of some set and $i < i'$ means that $i \subset i'$ and $i \neq i'$.

Testing whether a given relation < , or its transitive closure <°, is a strict partial ordering therefore amounts to *testing whether a network is acyclic*. How can this be accomplished? One way would be to check whether the strong components of the network are just the individual nodes, but this would involve much duplication in the calculations from node to node. A more efficient procedure is furnished in Exercise 2.31. If part of the interest lies in detecting some specific positive circuit that prevents the network from being acyclic, then the algorithms of Chapter 6 can be specialized to this purpose (see Exercise 6.30).

2K.* PLANAR NETWORKS AND DUALITY

The duality between paths and cuts is brought to its fullest realization in the case of *planar* networks. These are networks that can be represented pictorially in a plane (or equivalently, on the surface of a sphere) without any arcs crossing each other away from the nodes.

Consider any such representation of a planar network G that is connected and moreover cannot be disconnected by the removal of any single arc. A *dual network G^** may then be constructed, following the pattern in Figure 2.10. The nodes of G^* are the regions into which the plane (or sphere) is partitioned by the representation of G. These are called the *faces* of G (relative to the particular representation). Each arc of G borders two different faces (due to the connectedness assumptions). Each such pair of faces is joined by an arc of G^* oriented so as to cross the corresponding arc of G "from right to left." (The determination of which is "right" and which is "left" depends of course on which side of the plane the network is viewed from; assume that a particular side has been designated as part of the specification of the representation.)

Observe that G^* is a network satisfying the same assumptions as G. The faces of G^* correspond to the nodes of G. If G^* is thought of formally as viewed from the other side of the plane, then its dual G^{**} can be identified with G (rather than just the "reverse" of G).

Circuits in G correspond to cuts in G^*, whereas cuts in G correspond to circuits in G^*. Moreover circulations in G correspond to differentials in G^*, whereas differentials in G correspond to circulations in G^*.

This thorough form of duality is interesting and illuminating, but of course it works only for special networks and depends on the particular geometric representations of them. For this reason it does not usually serve as an important vehicle in practical applications. In contrast, the general duality

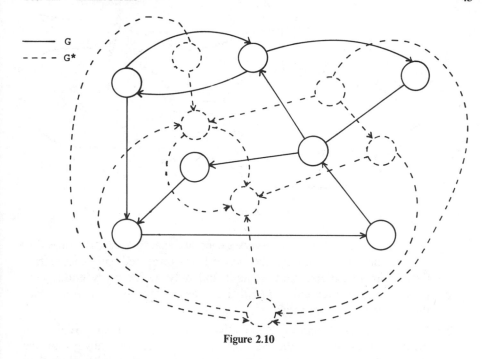

Figure 2.10

between circuits and cuts, circulations and differentials, which is valid for all networks, is very significant in both the organization of the overall theory and the design of algorithms for solving many problems.

2L.* EXERCISES

2.1. Is a path necessarily without multiplicities if it uses no node more than once, except that its initial and terminal nodes might coincide? Show that this is true except for one degenerate case.

2.2. (*Minimality of Elementary Circuits*). Prove that a circuit P without multiplicities is elementary if and only if it is *minimal* in the sense that there is no circuit without multiplicities that uses only arcs of P but at least one arc fewer. In fact if P is elementary and P' is a circuit without multiplicities which uses only arcs of P, then P' must be either P or the reverse of P.

2.3. What are the facts, analogous to those in Exercise 2.2, in the case of paths that are not circuits?

2.4. (*Euler Circuits*). One of the origins of graph theory was Euler's problem of the seven bridges of Königsberg. The bridges are shown in Figure 2.11. Is it possible to take a walk that starts and ends at the same place, having crossed each bridge exactly once? In general an *Euler circuit* in a

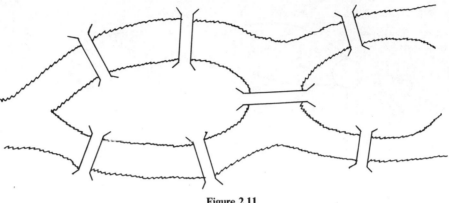

Figure 2.11

network G is a circuit that traverses each arc exactly once (the direction doesn't matter). Assuming the network is connected, prove constructively that an Euler circuit exists if and only if, for every node i, the number of arcs incident to i (called the *degree* of i) is even. What does this say about Königsberg?

2.5. (*Positive Euler Circuits*). A positive Euler circuit is a circuit that traverses each arc of the network exactly once and always in the positive direction. Assuming the network is connected (with more than one node) prove constructively that a positive Euler circuit exists if and only if for each node i the number of arcs with i as initial node (called the *outward demidegree* of i) equals the number of arcs with i as terminal node (called the *inward demidegree* of i).

(Notice that the problem of finding a circuit that traverses each arc twice, exactly once in each direction, can be reduced to that of determining a positive Euler circuit: replace each arc by a pair of oppositely directed arcs that can be traversed only positively.)

2.6. (*Connectedness*). Verify that the relations "connected to" and "strongly connected to" (for pairs of nodes, as defined in Section 2C) satisfy the axioms of an equivalence relation—reflexivity, symmetry, and transitivity.

2.7. (*Acyclic Metanetwork of Strong Components*). Let us form a network G' whose nodes are the strong components of the network G; there is an arc in G' from G_1 to G_2 if and only if there is an arc in G whose initial node lies in G_1 and whose terminal node lies in G_2. (Described a different way, G' is obtained by coalescing each strong component of G into a single node and then coalescing each bundle of arcs in parallel into a single arc.) Prove that, in G', there are no positive circuits (i.e., G' is acyclic as defined in Section 2J).

2.8. (*Intersections of Circuits and Cuts*). Prove that if P is an elementary circuit and Q is a cut, then the number of arcs common to P and Q is even. In fact give two proofs: one based on a direct "geometric" argument and another in terms of the inner product $e_P \cdot e_Q$.

2.9. (*Decomposition into Elementary Cuts*). Let Q be a nonempty cut. Prove there exist elementary cuts $Q_1, \ldots, Q_r$ (definition in Section 2H) that are disjoint (with respect to arcs, of course) and satisfy $Q^+ = Q_1^+ \cup \cdots \cup Q_r^+$ and $Q^- = Q_1^- \cup \cdots \cup Q_r^-$.

(*Hint.* First, write $Q = [S, N \setminus S]$ and partition S into the subsets S_k corresponding to components that would be obtained if the arcs of Q were deleted from the network. This will yield a decomposition, in terms of the cuts $Q_k' = [S_k, N \setminus S_k]$, serving as a half-way achievement.)

2.10. (*Cut Criterion for Circulations*). Prove that a flow x is a circulation if and only if, for every cut Q, one has $e_Q \cdot x = 0$. (e_Q = incidence function/vector for Q.)

(*Hint.* Observe that each row of the incidence matrix of the network is the incidence vector for a certain cut.)

2.11. (*Cut Criterion for Differentials*). Prove that $v \in R^A$ is a differential if and only if v can be expressed as a linear combination of incidences e_Q for various cuts Q. (Invoking Exercise 2.7, one could restrict the cuts to being elementary.)

2.12. (*Circuit Criterion for Differentials*). Prove that $v \in R^A$ is a differential if and only if, for every elementary circuit P, one has $e_P \cdot v = 0$.

(*Hint.* Reduce to the case of a connected network G and argue by induction, building G up one node at a time.)

2.13. (*Circuit Criterion for Circulations*). Prove that x is a circulation if and only if x can be expressed as a linear combination of incidences e_P for various elementary circuits P.

(*Hint.* Use Exercises 1.8 and 2.12. A more constructive approach will be available in Chapter 4.)

2.14. (*Initial Side of a Cut*). Let Q be a nonempty cut in a connected network. Prove there is a *unique* node set S such that $[S, N \setminus S] = Q$.

(*Hint.* Let N^+ denote the set of all initial nodes of arcs in Q^+ and terminal nodes of arcs in Q^-; the set S will consist of N^+ and all nodes reachable from N^+ by paths not meeting Q. Notice that this furnishes a constructive method of determining S: apply the painted network algorithm with the arcs of Q painted red and all other arcs green.)

2.15. Demonstrate by counterexample that the assertion of Exercise 2.14 can fail to be true when the network is not connected.

2.16. (*Minimality of Elementary Cuts*). Prove that a nonempty cut Q is elementary if and only if it is *minimal* in the sense that there is no nonempty cut using only arcs of Q but at least one arc fewer. In fact if Q is elementary and Q' is a nonempty cut using only arcs of Q, then Q' is either Q or the reverse of Q.

(*Hint.* Reduce to the case of a connected network and make use of the preceding exercise.)

2.17. (*Problem of the Missionaries and Cannibals; Berge*). A total of *M* missionaries and *C* cannibals must cross a river using a boat that can take only *B* of them at a time. The cannibals must never outnumber the missionaries—on either bank of the river or in the boat—or disaster will ensue (from the *missionaries'* point of view!). The boat cannot cross the river by itself. Initially the boat and all the missionaries and cannibals are on the left bank.

Set up the problem as one of finding a path from a node *s* to another node *s'* in a certain network.

(*Hint.* Think of this as a discrete system in the manner of Example 2 in Section 1B, the states being the possible "nondisasterous" configurations of missionaries, cannibals, and boat. These configurations need to be described precisely, along with the criteria for which ones can directly follow which others.)

2.18. (*Painted Network Algorithm*). Apply the painted network algorithm to the example in Figure 2.12 with $N^+ = \{s\}$, $N^- = \{s'\}$. Describe what happens in the fashion of Example 1 in Section 2F.

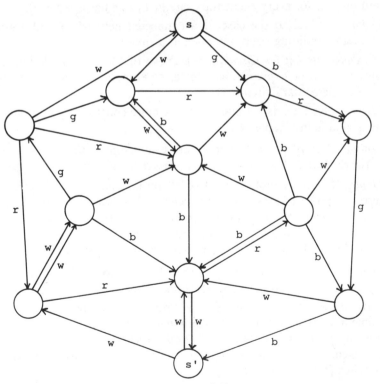

Figure 2.12

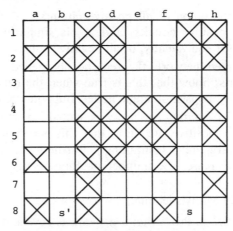

Figure 2.13

2.19. (*The Knight's Gambol*; *Berge*). For the chessboard shown in Figure 2.13, is it possible for a knight to pass from square s to square s' by some succession of moves without landing on any of the crossed squares, and if so, how? Apply the painted network algorithm.

(*Hint*. View the squares as the nodes of a certain network, but do not actually draw a representation of it. Execute the algorithm in adapted form directly on the chessboard diagram, *representing the routing in a suitable way* as it is calculated.)

2.20. (*Least Solution to the Painted Path Problem*). Show that if the painted network algorithm is executed with breadth-first search (see Section 2G), then the solution (if any) that is obtained to the painted path problem is one using the fewest arcs possible. (In fact there is a close connection with the multipath version of the painted network algorithm and the multirouting Θ that it generates. Let $N_0(= N^+)$, $N_1, \dots, N_r$ be the sets defined at the end of Section 2G. Demonstrate that under breadth-first search the remaining nodes in S are precisely those of N_k at the stage when the last of the nodes in N_{k-1} has been thoroughly scanned. Moreover the routing θ satisfies $\theta(i) \in \Theta(i)$ for all i, and the path it yields therefore has the minimality property in question.)

2.21. Are there any cases where the painted network algorithm *must* take exactly $q + 1$ iterations (q = number of nodes not in N^+ or N^-)?

2.22. What does the painted network theorem say if all arcs are painted white? Show that this special case is actually *equivalent* to the general case.

(*Hint*. Apply the special result to an appropriately modified network.)

2.23. (*Arc Dichotomy*). Prove that the arc set A of an arbitrary network G satisfies

$$A = A_P \cup A_Q, \qquad A_P \cap A_Q = \varnothing,$$

where A_P is the union of all positive circuits (i.e., the set of all arcs belonging to positive circuits) and A_Q is similarly the union of all positive cuts (or equivalently, of all negative cuts).

2.24. Can the painted network theorem be sharpened to speak only of *elementary* cuts? Show that this is true when the network is connected and N^+ and N^- consist of single nodes, but sometimes false otherwise.

2.25. (*Existence of Circuits*). An arc $\bar{j}$ is called an *isthmus* if, by itself, it constitutes a positive cut. Show that this is true if and only if there is no elementary circuit containing $\bar{j}$. Furthermore removal of an isthmus always increases the number of components of the network by exactly 1.

(*Hint.*   Apply Minty's lemma.)

2.26. (*Acyclic Networks*). Prove that a network is acyclic if and only if every arc belongs to some positive cut.

2.27. (*Strong Connectedness*). Prove that a connected network is strongly connected if and only if every arc belongs to some positive circuit. Also it is strongly connected if and only if it contains no nonempty positive cut.

2.28. Compare Exercises 2.26 and 2.27 with Exercises 2.23 and 2.7. What are the relationships?

2.29. (*Acyclic Networks*). Prove that a network is acyclic if and only if each of its strong components consists of a single node.

2.30. (*Predecessors and Successors*). A node i is said to be an *immediate predecessor* of a node i' if there is an arc $j \sim (i, i')$ and similarly an *immediate successor* in the case of $j \sim (i', i)$. Prove that if every node has an immediate predecessor, then every node also has an immediate successor. Moreover in this event the network cannot be acyclic. Must it be strongly connected?

2.31. (*Test for Acyclicity*). Justify the following algorithm. Let N^+ denote the set of all nodes not having an immediate predecessor (see Exercise 2.30). If $N^+ = \emptyset$, the network is not acyclic. If $N^+ \neq \emptyset$, apply the painted network algorithm with all arcs painted white (and N^- taken empty), but with one extra condition: never consider for addition to the set S at any iteration any node having an immediate predecessor not in S. Stop when no further progress is possible under this restriction. If at that point $S = N$, the network is acyclic, whereas if $S \neq N$, it is not.

2.32. (*Finding all Paths*). Modify the painted network algorithm so that the information it generates furnishes, in some suitable sense, *all* solutions to the painted path problem.

(*Note.*   Not only with the calculation be more tedious, but very often there will be such an enormous number of solutions that it would be impractical to record them all.)

2.33. (*Number of paths*). Let M be the adjacency matrix of a network (see Section 1C) and M^k its kth power. Prove that the entry of M^k in the row and column corresponding to the nodes i and i', respectively, is the number of positive paths from i to i' containing exactly k arcs.

2.34. (*Planar Networks*). Let G be a connected planar network that cannot be disconnected by the removal of any one arc, and let F denote the set of faces of G relative to a particular planar representation. For each face $k \in F$ and arc $j \in A$, let

$$e^*(k, j) = \begin{cases} 1 & \text{if the face on the right side of } j \text{ is } k \\ -1 & \text{if the face on the left side of } j \text{ is } k \\ 0 & \text{if } j \text{ is not one of the boundary arcs of } k. \end{cases}$$

Associate with each flow $x \in R^A$ the function $z = \text{rot } x$ on F defined by

$$z(k) = (\text{rot } x)(k) = \sum_{j \in A} e^*(k, j)x(j).$$

(This is the sum of the fluxes around the boundary of the face k, with signs corresponding to the counterclockwise direction.)

Show that the array of values $e^*(k, j)$ can be identified with the incidence matrix of the dual network G^*. Moreover a flow x in G satisfies $\text{rot } x = 0$ if and only if there is a potential u on N such that $x = \Delta u$. On the other hand, a flow x is a circulation in G if and only if there is a function w on F such that

$$x(j) = -\sum_{k \in F} w(k)e^*(k, j) \quad \text{for all } j \in A.$$

2M.* COMMENTS AND REFERENCES

The theory of graphs and networks has long suffered from a lack of standard terminology, with various authors using different words for the same thing or the same word in conflicting senses. In part this is due to the many areas of application and the different viewpoints they suggest as to which concepts are the important and fundamental ones.

In the case of "paths" and the like, it is hard to find two texts that agree, but what is worse, none of the proposed terminologies is adequate for our purposes. One difficulty with these terminologies lies in the tradition of stressing a dichotomy between graphs (unoriented) and digraphs or networks (oriented). Many authors choose a term like "path," "chain," "walk," or "arc progression" for the undirected path concept, where the arcs and even the path have no particular orientation or direction of transversal, and a different term for the extreme case of the corresponding directed concept, where *every* arc *must* be traversed positively. The case that is critical for our framework, where orientation is important but the directions in which various arcs may be traversed can depend on various circumstances, is often ignored.

The same unsatisfactory dichotomy is seen in connection with "cuts" (more commonly called "cut-sets") and "circuits." Here another difficulty arises: the terms that are adopted are usually reserved for what we call *elementary* cuts and circuits (with no words furnished for the "nonelementary" cases). Although, in principle, most everything about cuts and circuits can be reduced to the "elementary" case, this can be a nuisance in practice.

For example, the cut produced by the painted network algorithm (when the painted path problem has no solution) will not always be elementary. To get around this, the painted network theorem would have to be replaced systematically by Minty's lemma, and the corresponding algorithm would have to be augmented by a decomposition routine based on Exercise 2.9. In many respects this would be tedious theoretically as well as unnecessarily burdensome in computations.

The terms adopted in this book allow a simple and very flexible treatment that in the context of orientation requirements ("paintings") can depend on various configurations of data and may be modified repeatedly. Such is the natural setting for most of the algorithms of network optimization, as will be seen in later chapters. It is also the motivation for a number of other changes we have made in existing approaches, for instance in developing the painted network theorem as a substitute for the original lemma of Minty [1960] and in expressing it with sets N^+ and N^- instead of just a pair of nodes s and s'.

Incidentally, Minty used paintings of three colors only. He did not admit the category we refer to as "black," preferring to appeal to a reversal of such arcs when necessary so as to make them all "white." However, this is a "simplification" that can get in one's way and actually force matters to come out in a more complicated form than necessary.

The notion of a "routing" and its compatibility with a given painting is dictated by the same goals. Other authors speak of "rooted trees" or "arborescences" in similar circumstances where the question of orientation is not so subtle. "Trees" will be discussed in Chapter 4.

The virtues of depth-first search in streamlining a number of graph-theoretic computer algorithms are surveyed by Aho, Hopcroft, and Ullman [1974] and Deo [1974]. See these texts in particular for refined tests for connectedness, strong components, acyclicity, and planarity. For Kuratowski's famous characterization of when a graph (or network) is planar, see Wilson [1972, Chap. 5] or Bondy and Murty [1975, Chap. 9].

Although the existence of an Euler circuit is easy to characterize (see Exercise 2.4), the opposite holds for a *Hamiltonian* circuit, which is a circuit that passes through each *node* exactly once. The theory of Hamiltonian circuits is important for a number of applications but is not nearly so simple or complete; see Christophides [1975, Chap. 10].

3

FLOWS AND CAPACITIES

In most problems involving flows there are restrictions in at least some of the arcs of the possible flux values. There may also be restrictions on the divergence values allowed at various nodes. A fundamental question is whether there exists a flow meeting such requirements, and if so, how it might be determined. The constructive answer to this question turns out to be crucial, in more ways than might be expected, in various phases of the optimization of both flows and potentials.

A particular optimization problem, more basic than the rest, is studied in this chapter before the problem of existence is tackled: the celebrated max flow problem of L. R. Ford and D. R. Fulkerson. Besides its strong intuitive appeal and many important applications, it has the interesting feature of engendering a dual problem which is entirely combinatorial in character: the min cut problem. It helps prepare the way, theoretically and computationally, for the subsequent treatment of other pairs of optimization problems in duality.

The climax of the development comes with the results on the feasible distribution problem and the main algorithm for solving it. Everything in the chapter up through Section 3I is crucial for what happens later in the book.

3A. CAPACITY INTERVALS

Although more general kinds of restrictions on flux values must sometimes be considered, the basic case is that where the flux in an arc j is allowed to range over a nonempty closed real interval $C(j)$, called the *capacity interval* for j. It is supposed in what follows that such an interval has been assigned to every arc of the network G. A flow x in G is called *feasible with respect to capacities* if $x(j) \in C(j)$ for all $j \in A$.

To fix notation, we shall always express the capacity intervals by

$$C(j) = [c^-(j), c^+(j)],$$

where $c^+(j)$ is called the *upper capacity* for the arc j and $c^-(j)$ the *lower capacity*. The only conditions placed on $c^+(j)$ and $c^-(j)$ are those corresponding to $C(j)$ being a nonempty real interval, namely,

$$c^-(j) \leq c^+(j), c^+(j) > -\infty, c^-(j) < +\infty.$$

In particular, $c^+(j)$ could be $+\infty$ and $c^-(j)$ could be $-\infty$, in which case the generic closed bracket notation for $C(j)$ is not apt: one really has $C(j) = (-\infty, +\infty)$, not $C(j) = [-\infty, +\infty]$. Always remember, whatever the exigencies of generic notation might suggest, that $C(j)$ is a *real* interval and never contains $+\infty$ or $-\infty$, which are *not* real numbers and can *never be flux values for any flow*. Nevertheless, $+\infty$ and $-\infty$ will often be referred to as possible "endpoints" of $C(j)$, and they will enter into certain arithmetic calculations where their presence is convenient and harmless.

The following examples of capacity intervals should especially be kept in mind:

1. $C(j) = [-c, c]$ with $0 \leq c < +\infty$; the flow x can use the arc j in either direction, but the flux must satisfy $|x(j)| \leq c$.
2. $C(j) = [0, c]$ with $0 \leq c < +\infty$; same as the preceding example, but the arc can only be used in the positive direction.
3. $C(j) = [0, +\infty)$; the arc can only be used in the positive direction, but there is no upper bound on the flux.
4. $C(j) = (-\infty, +\infty)$; in effect there is no constraint whatever on the flux in the arc j.
5. $C(j) = [c, c]$ with $-\infty < c < +\infty$; there is an exact requirement, $x(j) = c$.

The usefulness of other types of intervals is less apparent at this stage, but they enter the picture when certain problems are converted from one form to another, as well as in subroutines used by a number of algorithms.

3B. FLUX ACROSS A CUT

The presence of capacity intervals places a direct constraint on the flux in each arc, but there are other constraints on a flow which then follow indirectly. The most important of these is associated with cuts.

For a cut Q and flow x, the quantity

$$e_Q \cdot x = \sum_{j \in Q^+} x(j) - \sum_{j \in Q^-} x(j)$$

(where e_Q is the incidence function for Q) is called, naturally enough, the *flux of x across Q.* It may be interpreted as the net amount of material flowing across Q in the direction of the orientation of Q.

To reinforce this interpretation, write $Q = [S, N \setminus S]$ for a node set S, and define the *divergence of x from S* by

$$y(S) = \sum_{i \in S} y(i), \quad \text{where } y = \text{div } x.$$

This quantity represents the net amount of material originating in S (i.e., the total amount of source minus the total amount of sink). It is related to the first quantity by the fundamental *divergence principle*:

$$[\textit{divergence of x from S}] = [\textit{flux of x across Q}],$$

or in symbols,

$$y(S) = e_Q \cdot x \quad \text{for } y = \text{div } x,\ Q = [S, N \setminus S].$$

The proof of the divergence principle is very elementary: using the definitions of y and e_Q, the two sides can be reduced to

$$\sum_{i \in S} \sum_{j \in A} e(i, j) x(j) = \sum_{j \in A} \sum_{i \in S} e(i, j) x(j)$$

where $e(i, j)$ gives the incidence of node i with arc j. The principle is illustrated in Figure 3.1.

It is interesting to note that the "total divergence rule" of Section 1E, written $y(N) = 0$, can be construed as the special case of the divergence principle where $S = N$ (and Q is therefore the empty cut).

What constraint on the flux of x across a cut Q is implied by the flux constraints for the individual arcs of Q? For arcs $j \in Q^+$, one has $c^-(j) \le$

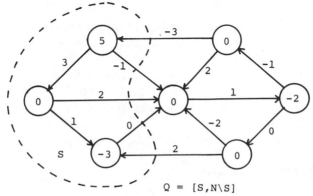

$$Q = [S, N \setminus S]$$

$$[\text{divergence of x from S}] = 2 = [\text{flux of x across Q}]$$

Figure 3.1

$x(j) \le c^+(j)$, whereas for $j \in Q^-$ the corresponding limitation, expressed in the direction of flow given by the orientation of Q, is $-c^+(j) \le -x(j) \le -c^-(j)$. Adding over all the arcs of Q, one gets

$$c^-(Q) \le [\text{flux of } x \text{ across } Q] \le c^+(Q),$$

where

$$c^+(Q) = \sum_{j \in Q^+} c^+(j) - \sum_{j \in Q^-} c^-(j),$$

$$c^-(Q) = \sum_{j \in Q^+} c^-(j) - \sum_{j \in Q^-} c^+(j).$$

(In these sums, $+\infty$ and $-\infty$ are treated in the obvious way; it is impossible for the problematical expression $(+\infty) + (-\infty)$ to arise.)

The number $c^+(Q)$ is called the *upper capacity* for the cut Q, and $c^-(Q)$ the *lower capacity*. Obviously, these are the endpoints of the nonempty closed real interval

$$C(Q) = \sum_{j \in Q^+} C(j) - \sum_{j \in Q^-} C(j)$$

that is, the set of all real numbers that can be represented in the form

$$\sum_{j \in Q^+} x(j) - \sum_{j \in Q^-} x(j)$$

as $x(j)$ ranges over $C(j)$. The latter is the *capacity interval associated with* Q. The notation

$$C(Q) = [c^-(Q), c^+(Q)]$$

will be used, subject to the same warning issued previously: always remember that $C(Q)$ is a real interval and never contains $+\infty$ or $-\infty$, even though these values may appear as $c^+(Q)$ or $c^-(Q)$, respectively.

The definitions of $c^+(Q)$ and $c^-(Q)$ give occasion for mentioning a convention that will be used throughout this book, namely, that *an empty sum of numbers is to be interpreted as* 0. Then cases where $Q^+ = \emptyset$ or $Q^- = \emptyset$ do not require special treatment.

3C. MAX FLOW PROBLEM

Let G be a network with capacity intervals, and let N^+ and N^- be nonempty disjoint sets of nodes of G.

Consider any flow x that is conserved at all nodes outside of N^+ and N^-, or in other words, has $y(i) = 0$ for all $i \notin (N^+ \cup N^-)$, where $y = \operatorname{div} x$. The total divergence rule of Section 1E says that

$$0 = \sum_{i \in N} y(i) = \sum_{i \in N^+} y(i) + \sum_{i \in N^-} y(i),$$

or in the notation of the preceding section,

$$y(N^+) = -y(N^-).$$

Here $y(N^+)$ represents the net amount of source in N^+, whereas $-y(N^-)$ is the net amount of sink in N^-. The common value is thus to be interpreted as the net amount flowing from N^+ to N^-. It will be called the *flux of x from N^+ to N^-*. Of course this notion makes sense only for a flow that is conserved at all nodes outside of N^+ and N^-.

The following problem is the object of the present investigation.

Max Flow Problem. *Maximize the flux from N^+ to N^- over the set of all flows x that are conserved at all nodes outside N^+ and N^- as well as feasible with respect to capacities.*

Figure 3.2 displays a max flow problem which will be solved later in this chapter. Capacity intervals are shown next to the arcs.

For the time being, the existence of at least one flow satisfying all the constraints (i.e., the conditions of conservation and capacity) is taken for granted. (The question will be reopened in Section 3H.) As x ranges over the set of such flows, a corresponding (nonempty) set of flux values from N^+ to N^- is produced. The least upper bound of this set of flux values is called the *supremum in the max flow problem*. Possibly it is $+\infty$. If it is finite, there is the question of whether it is attained by some flow x. (Attainment is clearly

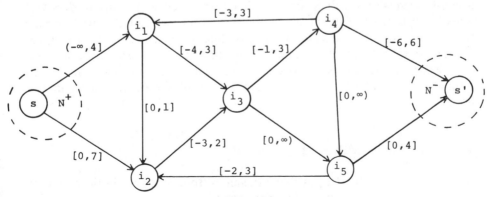

Figure 3.2

impossible if the supremum is $+\infty$, since no flow is allowed to have infinite flux values in any arcs.) A flow for which the supremum is attained is called a *solution* to the max flow problem.

In general, in treating problems of optimization, one uses the neutral terms *supremum* (least upper bound) and *infimum* (greatest lower bound) when attainment may be in question—*or merely if it is not convenient at the moment to make a positive assertion in this respect*. The stronger terms *maximum* and *minimum* are definitely to be understood as implying attainment. To avoid misunderstandings, note that these terms always refer to certain *numbers*, not to the actual *solutions* to the problems (i.e., the flows, potentials, or whatever, for which these values are attained). Often the four terms are abbreviated as sup, inf, max, and min.

Some of the special character of the max flow problem may be seen from writing out the constraint system in full:

$$\sum_{j \in A} e(i, j)x(j) = 0 \quad \text{for every } i \notin N^+ \cup N^-,$$

$$x(j) \le c^+(j) \quad \text{for every } j \in A,$$

$$x(j) \ge c^-(j) \quad \text{for every } j \in A.$$

(Inequalities where $c^+(j) = +\infty$ or $c^-(j) = -\infty$ are superfluous and can be omitted.) Subject to this, one wants to maximize the value $y(N^+) = -y(N^-)$, or in other words, the expression

$$\sum_{i \in N^+} \sum_{j \in A} e(i, j)x(j) = \sum_{j \in A} \left(\sum_{i \in N^+} e(i, j) \right) x(j).$$

This is a linear programming problem in the variables $x(j), j \in A$.

A *linear programming problem* is by definition an optimization problem involving a finite number of real variables as unknowns, in which a linear function is maximized or minimized subject to a constraint system comprised of a finite number of linear equations or weak linear inequalities. ("Weak" refers to inequalities with $\le$ or $\ge$ rather than $<$ or $>$.) Many readers will already have some acquaintance with such problems, and this will provide them with additional motivation and insights. Some reference to linear programming will therefore be made from time to time. However, no real knowledge of the theory is required, and indeed, all the basic facts will ultimately be established in the natural course of the developments in this book.

3D.   MAX FLOW MIN CUT

A simple condition on the possible flux values from N^+ to N^- in the max flow problem may be derived from the relations in the preceding section, as applied

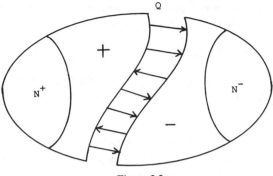

Figure 3.3

to the situation schematized in Figure 3.3. *If x is any flow satisfying the constraints of the problem, and Q is any cut separating N^+ from N^- (recall the notation $Q: N^+ \downarrow N^-$), then*

$$[\,\textit{flux of x from } N^+ \textit{ to } N^-\,] \le c^+(Q).$$

Indeed, to say that Q separates N^+ from N^- is to say that Q is of the form $[S, N \setminus S]$, where $S \supset N^+$, $S \cap N^- = \varnothing$. Then $y(i) = 0$ for all $i \in S \setminus N^+$, so that $y(N^+) = y(S)$. But by the divergence principle, one has $y(S) = e_Q \cdot x$ (the flux of x across Q), and an upper bound for the latter quantity is $c^+(Q)$.

Inasmuch as each cut $Q: N^+ \downarrow N^-$ furnishes an upper bound to the supremum in the max flow problem, the *least* of these upper bounds is of interest.

Min Cut Problem.   *Minimize $c^+(Q)$ over all cuts Q separating N^+ from N^-.*

An immediate consequence of the preceding discussion is the inequality

$$[\text{sup in max flow problem}] \le [\text{min in min cut problem}].$$

Observe that the use of "min" instead of "inf" follows our general guidelines; the minimization is over a *finite* set of possibilities and therefore is sure to be attained. As a matter of fact the equation sup = min holds between the two problems. This is the main result to be obtained in what follows.

The full statement of the result will use the notion of a *path P*: $N^+ \to N^-$ of *unlimited capacity*, which means that

$$c^+(j) = +\infty \quad \text{for all } j \in P^+,$$

$$c^-(j) = -\infty \quad \text{for all } j \in P^-.$$

If a path of such type exists at all, then of course there is an elementary one which has no intermediate nodes in N^+ or N^-.

Max Flow Min Cut Theorem (Ford and Fulkerson). *Assume there is at least one flow satisfying the constraints of the max flow problem. Then*

$$[\textit{sup in max flow problem}] = [\textit{min in min cut problem}].$$

This common value is $+\infty$ if there is an elementary path $P: N^+ \to N^-$ of unlimited capacity, whereas if there is no such path, it is finite, and the max flow problem has a solution.

A proof of the theorem will be given in Section 3G, in conjunction with the justification of an algorithm for solving both problems simultaneously. The algorithm is quite different from what would correspond to solving the max flow problem by general techniques of linear programming, and it provides a surprising theoretical bonus.

3E. NATURE OF THE MIN CUT PROBLEM

The min cut problem is interesting for several reasons, and its character merits further discussion. Most instructively, it can be viewed as a natural refinement of the painted cut problem, although this is not the way it was arrived at.

This interpretation is obtained by associating with the given system of capacity intervals the following painting of the arcs j of the network:

$$\begin{aligned} &\text{red} \quad \text{if } c^+(j) < +\infty, c^-(j) > -\infty \\ &\text{black if } c^+(j) < +\infty, c^-(j) = -\infty \\ &\text{white if } c^+(j) = +\infty, c^-(j) > -\infty \\ &\text{green if } c^+(j) = +\infty, c^-(j) = -\infty \end{aligned}$$

Observe that *a cut Q is compatible with this painting if and only if $c^+(Q) < +\infty$.* Namely both conditions are equivalent to having $c^+(j) < +\infty$ for all $j \in Q^+$ and $-c^-(j) < +\infty$ for all $j \in Q^-$. Assuming the existence of at least one such cut (as the problem is otherwise trivial), one sees that *there is no loss of generality in restricting the min cut problem to the class of all such cuts.* Indeed, cuts with $c^+(Q) = +\infty$ are of no interest in the minimization if there are cuts with $c^+(Q) < +\infty$. Thus the min cut problem consists essentially of minimizing the quantity $c^+(Q)$ over all cuts $Q: N^+ \downarrow N^-$ that are compatible with a certain painting.

In this setting an alternative interpretation of $c^+(Q)$, independent of "capacities," is helpful. The value $c^+(j)$ may be thought of as the *cost* of incorporating the arc j into Q in the positive direction, and $-c^-(j)$ as the cost of using j in the negative direction. Then $c^+(Q)$ is the *total cost* of the cut Q (refer to the formula), and the min cut problem consists of determining the *cheapest* cut separating N^+ from N^- that is compatible with a certain painting. Arcs j which the painting excludes from Q^+ have an infinite forward cost

$(c^+(j) = +\infty)$, whereas those it forbids from Q^- have an infinite backward cost $(-c^-(j) = +\infty)$; *thus cuts that violate the compatibility condition are deemed infinitely costly.*

This interpretation shows that the min cut problem could well arise in its own right. One might start with some painting and a structure of "forward" and "backward" costs (the two quite possibly different) for incorporating an arc into a cut $Q: N^+ \downarrow N^-$ and then look for the cheapest compatible cut.

What is remarkable is that in this context one would not dream of any connection with "flows." Yet if the forward and backward costs are written as $c^+(j)$ and $-c^-(j)$, the problem falls right into the domain of the max flow min cut theorem (assuming of course that $c^-(j) \le c^+(j)$).

An entirely analogous situation will be faced in Chapter 6. There interest will center on finding a "cheapest path" $P: N^+ \to N^-$ that is compatible with a given painting, and the dual problem will be to find a kind of potential function yielding the "greatest tension" from N^+ to N^-. The idea of constraints being represented by infinite costs is also an important omen for future developments.

Another thing to discuss about the min cut problem is its discrete nature. Since the collection of all cuts separating N^+ from N^- is finite, the problem may seem, to a beginner at any rate, almost trivial: all one has to do in principle is inspect the various cuts one by one to see which gives the lowest value of $c^+(Q)$.

To dispel this notion, a simple example suffices. Suppose the network is connected and has, outside of N^+ and N^-, at least 263 nodes, which is by no means a fantastic number for the kinds of models handled nowadays. According to Exercise 2.14, the cuts $Q: N^+ \downarrow N^-$ are in one-to-one correspondence with the node sets S such that $S \supset N^+$ and $S \cap N^- = \varnothing$, and hence the number of them is at least 2^{263}. *But this is greater than the estimated total number of all atoms in the observable universe!*[†] Surely, to speak of minimization "by inspection" for a set of this size is hardly different from speaking of minimizing a function over R^n by looking at all the points individually to see which one gives the lowest value. Nevertheless, the algorithm to be given in the next section is capable of solving such a min cut problem quite practically and efficiently.

Of course this is not the only case where a sharp distinction between "discrete" and "continuous" problems is unwarranted. In science one is accustomed to treating naturally discrete systems (matter, populations, etc.) as if they could be represented by a continuum. On the other hand, continuous models are often discretized. What is truly essential from the mathematical point of view is that there be enough structure to produce pleasing and

[†] There are an estimated 10^{56} grams of matter in the observable universe (Fred Hoyle, *Galaxies, Nuclei, and Quasars*, Harper and Row, New York, 1965, p. 112). The approximate number of atoms in one gram of some element is obtained by dividing Avogadro's number, 10^{23}, by the atomic weight. Thus an upper bound for the number of atoms in the universe is $10^{56} \times 10^{23} = 10^{79} \approx 2^{262.4}$, at least by this reckoning.

significant insights, as well as effective numerical approaches. The max flow
min cut theorem and its associated algorithm confirm, rather dramatically, that
this is true for the min cut problem.

3F. MAX FLOW ALGORITHM

The algorithm about to be given will simultaneously solve both the max flow
problem and the min cut problem, subject to some minor qualifications needed
to ensure its termination.

The key to being able to solve the problems simultaneously lies in the
following observation. (The notation of Sections 3C and 3D is assumed.)
Suppose one has determined a flow x satisfying the constraints of the max flow
problem, along with a cut $Q: N^+ \downarrow N^-$ such that

$$[\text{flux of } x \text{ across } Q] = c^+(Q).$$

*Then x must be a solution to the max flow problem and Q a solution to the min
cut problem* (with $c^+(Q) < +\infty$). For, as noted prior to the statement of the
min cut problem in Section 3D, one has

$$[\text{flux of } x \text{ across } Q] = [\text{flux of } x \text{ from } N^+ \text{ to } N^-],$$

and $c^+(Q)$ is an upper bound for this quantity. Equality with $c^+(Q)$ thus
implies there cannot be a flow satisfying the constraints and yielding a higher
flux from N^+ to N^- than x does. Hence x is a solution to the max flow
problem. At the same time, the flux of x from N^+ to N^- is a lower bound to
the minimum in the min cut problem, so that equality with $c^+(Q)$ implies Q
must be a solution to that problem.

The algorithm is based on a simple device for improving a given flow x
(which satisfies the constraints of the max flow problem) by superimposing on
it an additional flux along some path. An elementary path $P: N^+ \rightarrow N^-$, having
no intermediate nodes in N^+ or N^-, is called *flow augmenting* for x if

$$x(j) < c^+(j) \quad \text{for all } j \in P^+,$$

$$x(j) > c^-(j) \quad \text{for all } j \in P^-.$$

Then there is a number $\alpha > 0$ such that the flow

$$x'(j) = x(j) + \alpha e_P(j) = \begin{cases} x(j) + \alpha & \text{for } j \in P^+, \\ x(j) - \alpha & \text{for } j \in P^-, \\ x(j) & \text{for all other arcs,} \end{cases}$$

will again satisfy the constraints of the max flow problem. As a matter of fact α

can be any positive real number such that

$$\alpha \leq c^+(j) - x(j) \quad \text{for all } j \in P^+,$$

$$\alpha \leq x(j) - c^-(j) \quad \text{for all } j \in P^-,$$

and then it will be true that

$$c^-(j) \leq x'(j) \leq c^+(j) \quad \text{for all } j \in A.$$

Moreover x' will be conserved, like x, at all nodes outside of N^+ and N^-, and one will have

$$[\text{flux of } x' \text{ from } N^+ \text{ to } N^-] = [\text{flux of } x \text{ from } N^+ \text{ to } N^-] + \alpha.$$

This is immediate from the fact that

$$\text{div } x' = \text{div } x + \alpha \, \text{div } e_P,$$

where the divergence function $\text{div } e_P$ vanishes except at the initial node of P in N^+ (where the divergence is $+1$) and the terminal node of P in N^- (where the divergence is -1). Thus x' will be "better" than x, from the standpoint of the max flow problem.

Notice that a path of unlimited capacity from N^+ to N^-, as defined in Section 3D, turns out to be a flow-augmenting path such that α could be chosen arbitrarily large. Then clearly the supremum in the max flow problem would have to be $+\infty$.

Actually a path $P: N^+ \to N^-$ of unlimited capacity is precisely a solution to the painted path problem for the painting discussed in Section 3E. A solution to the corresponding painted cut problem is, as seen in Section 3E, precisely a cut Q such that $c^+(Q) < +\infty$. Therefore, by the painted network theorem, the existence of such a cut is *equivalent* to the nonexistence of a path of unlimited capacity. Moreover the state of affairs could be determined, if in doubt, by the painted network algorithm.

Algorithm

It is assumed that there does not exist a path of unlimited capacity from N^+ to N^- (see preceding paragraph). A flow x is given that satisfies the constraints of the max flow problem. The arcs j of the network are then painted as follows:

$$\text{green if } c^-(j) < x(j) < c^+(j)$$
$$\text{white if } c^-(j) = x(j) < c^+(j)$$
$$\text{black if } c^-(j) < x(j) = c^+(j)$$
$$\text{red} \quad \text{if } c^-(j) = x(j) = c^+(j)$$

The painted network algorithm is applied. If a solution $Q: N^+ \downarrow N^-$ to the painted cut problem is obtained, it satisfies

$$c^+(j) = x(j) \quad \text{for all } j \in Q^+,$$

$$c^-(j) = x(j) \quad \text{for all } j \in Q^-,$$

and consequently

$$[\text{flux of } x \text{ across } Q] = \sum_{j \in Q^+} x(j) - \sum_{j \in Q^-} x(j) = c^+(Q).$$

In this case x and Q are solutions to the respective problems by virtue of the argument at the beginning of this section. The algorithm then terminates. Otherwise a solution $P: N^+ \rightarrow N^-$ to the painted path problem is obtained. This is a flow-augmenting path relative to x, and the number

$$\alpha = \min \begin{cases} c^+(j) - x(j) & \text{for } j \in P^+, \\ x(j) - c^-(j) & \text{for } j \in P^-, \end{cases}$$

is positive and finite (the latter because of the exclusion of paths of unlimited capacity). The flow $x' = x + \alpha e_P$ satisfies the constraints of the max flow problem and has flux from N^+ to N^- which is greater by the amount α. The procedure is now repeated for x'.

EXAMPLE 1

The max flow algorithm can be applied to the network in Figure 3.2 starting with the *zero flow* ($x(j) = 0$ for all $j \in A$). This is possible because all the capacity intervals contain 0. One outcome is shown in Figure 3.4; the numbers

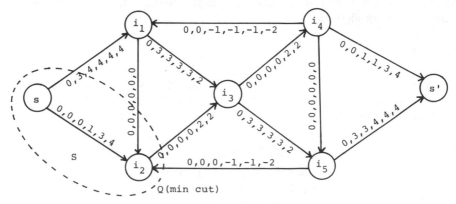

Figure 3.4

displayed next to each arc in the figure are the flux values initially and after iterations 1 to 5. The corresponding flow-augmenting paths are

$$P_1: s \rightarrow i_1 \rightarrow i_3 \rightarrow i_5 \rightarrow s', \qquad \alpha_1 = 3,$$

$$P_2: s \rightarrow i_1 \leftarrow i_4 \rightarrow s', \qquad \alpha_2 = 1,$$

$$P_3: s \rightarrow i_2 \leftarrow i_5 \rightarrow s', \qquad \alpha_3 = 1,$$

$$P_4: s \rightarrow i_2 \rightarrow i_3 \rightarrow i_4 \rightarrow s', \qquad \alpha_4 = 2,$$

$$P_5: s \rightarrow i_2 \leftarrow i_5 \leftarrow i_3 \leftarrow i_1 \leftarrow i_4 \rightarrow s', \qquad \alpha_5 = 1.$$

Iteration 6 yields not another flow augmenting path, but the cut $Q = [S, N \setminus S]$ corresponding to $S = \{s, i_2\}$. Thus the flow x at this stage solves the max flow problem, whereas Q solves the min cut problem. The flux of x from s to s' is 8, and this is of course also the value of $c^+(Q)$.

Note the cancellation of flow that takes place in this example at iteration 5 in the arcs (i_1, i_3) and (i_3, i_5), where the paths P_1 and P_5 run counter to each other. Such cancellation corresponds to a "rerouting" of an earlier flow, and the algorithm would not work if it were not allowed.

3G.   COMMENSURABILITY AND TERMINATION

How can one generally be sure that the improvement process in the max flow algorithm will not be repeated indefinitely without a solution being found? Although the flux from N^+ to N^- is increased at every iteration, it might be by ever-diminishing amounts. Conceivably, the supremum might never be achieved, or even worse, the flux values might approach a limit short of the supremum. Without some further condition the worst is indeed possible (Exercise 3.12).

A very mild condition suffices for many purposes. Recall that a collection of numbers is said to be *commensurable* if they can all be expressed as *whole* multiples of a certain "quantum" $\delta > 0$. Certainly any set of integers is commensurable ($\delta = 1$), but more generally, any finite set of rational numbers is commensurable, since it can be expressed in terms of a common denominator.

Suppose that the capacities $c^+(j)$, $c^-(j)$ and the flux values $x(j)$ at the start of the algorithm are commensurable. In this, $+\infty$ and $-\infty$ are deemed commensurable with all other numbers.) They are then all multiples of a certain δ, and hence so will be the numbers α and $x'(j)$ calculated in the improvement process, as is readily seen from the formulas. The situation is therefore self-perpetuating, and it follows that *at every iteration the flux from*

N^+ to N^- is increased by at least δ. There is no danger then of it increasing by amounts that dwindle indefinitely to zero. If there were infinitely many iterations, this could only mean that the supremum in the max flow problem is $+\infty$. But the latter is excluded by the assumption at the outset of the algorithm, which implies there is some finite upper bound $c^+(Q)$ to the supremum.

Therefore the algorithm, as it has been stated, must terminate after a finite number of iterations, if the foregoing commensurability condition is satisfied. Of course the condition does not present much limitation in practice. For computations, numbers are always rounded off to something rational anyway.

Nevertheless, for some theoretical purposes it is helpful to know that there is a slight refinement of the procedure which ensures termination of the algorithm without the expense of extra assumptions on the problem, such as commensurability of the data. The welcome fact in this vein is that the *max flow algorithm must terminate after a finite number of iterations, even without "commensurability" if the painted network algorithm at each stage is executed with arc discrimination* (as defined in Section 2G).

This is seen as follows. Each time a flow-augmenting path consisting entirely of green arcs is used, at least one of the arcs in the path turns white or black for the next iteration (i.e., one yielding the minimum in the calculation of α), whereas all the arcs that were already white or black remain so. Therefore a stage must come after finitely many iterations when, to make further progress toward a flow-augmenting path (if the algorithm has not in fact terminated), one must resort to white or black arcs.

Due to arc discrimination the painted network subroutine at that moment of its execution has a set S corresponding to a cut Q containing no green arcs. This property means that in every arc j of Q, the current flow x satisfies $x(j) = c^+(j)$ or $x(j) = c^-(j)$, or both. Hence the quantity

$$\sum_{j \in Q^+} x(j) - \sum_{j \in Q^-} x(j) = [\text{flux of } x \text{ across } Q] = [\text{flux of } x \text{ from } N^+ \text{ to } N^-]$$

is a certain sum of numbers $\pm c^+(j)$ and $\pm c^-(j)$. There are only finitely many sums of such a special form, since there are only finitely many arcs, so it follows that there are only finitely many different values that can possibly be assumed by the flux from N^+ to N^- at such a critical stage. None can ever be repeated, since the flux is increased at each iteration. Consequently the iterations must come to an end sooner or later.

Proof of the Max Flow Min Cut Theorem

From the observation made immediately before the statement of the algorithm, it is known that the following are equivalent: (1) there is a path $P: N^+ \rightarrow N^-$ of unlimited capacity, and (2) the minimum in the min cut problem is $+\infty$. Moreover in this event the supremum in the max flow problem is $+\infty$.

Suppose now that (1) and (2) are false. The max flow algorithm may be applied, and it will terminate (at least if arc discrimination is enforced in the painted network subroutine, as just discussed), yielding solutions x and Q to the respective problems such that (as already seen)

$$[\text{flux of } x \text{ from } N^+ \text{ to } N^-] = c^+(Q).$$

In conjunction with the general "sup $\leq$ min" inequality already derived for the problems at the time the theorem was stated, this equation says that "max = min."

Efficient Implementation

Some labor can be saved in each cycle of the max flow algorithm by carrying on calculations that will ultimately yield the flux augmentation value α, as the painted network algorithm is executed. This involves generating a certain node function a, which initially is defined only on N^+ and has the value $+\infty$ there at every node. At each iteration of the painted network algorithm (as applied in the context of the max flow algorithm), a new node i' is reached from an old node i by some arc j fitting one of two cases:

1. $j \sim (i, i')$ and $x(j) < c^+(j)$.
2. $j \sim (i', i)$ and $c^-(j) < x(j)$.

In Case 1 define

$$a(i') = \min\{a(i), c^+(j) - x(j)\},$$

whereas in Case 2

$$a(i') = \min\{a(i), x(j) - c^-(j)\}.$$

When N^- is finally reached, say, at node i^*, one has $a(i^*) = \alpha$. This is not hard to verify.

The point of this procedure, for implementing the algorithm on a computer, is that the differences $c^+(j) - x(j)$ and $x(j) - c^-(j)$ must in effect be calculated anyway as the painted network algorithm proceeds, in order to determine the colors of the arcs. By "storing" them in this fashion, one saves the trouble of recalculation when the flow-augmenting path is traced backward.

However, for real efficiency in the max flow algorithm the best approach involves using the multipath version of the painted network algorithm in a certain way at each iteration to produce a bigger improvement than would be possible one path at a time. This idea is explained in Section 3J in connection with a closely related algorithm, from which the approach can readily be adapted (see Exercise 3.31).

3H. FEASIBLE FLOWS

How can a flow satisfying the constraints of the max flow problem be determined, if one is not apparent? In Example 1 in Section 3F, the zero flow obviously met the conditions and could therefore be used to initiate the max flow algorithm. But the situation might not always be so simple.

The problem of determining such a flow fits into a general model in which the capacity intervals for the arcs are supplemented by intervals constraining the divergence values at the nodes. Conservation at node i corresponds to the interval $[0, 0]$ at i. However, a seemingly more restricted problem will turn out to be adequate as the cornerstone of the theory.

Feasible Distribution Problem. *Given capacity intervals $C(j) = [c^-(j), c^+(j)]$ for all arcs j and supply values $b(i)$ for all nodes i, find a flow x such that*

$$c^-(j) \leq x(j) \leq c^+(j) \quad \text{for all } j \in A,$$

$$y(i) = b(i) \quad \text{for all } i \in N \ (y = \text{div} \, x).$$

Here b is called the *supply* function (vector). If $b = 0$, one has the *feasible circulation problem*. The supply constraint can be written simply as $\text{div} \, x = b$.

A flow satisfying the supply and capacity constraints is called a *solution* to the feasible distribution problem. Necessary and sufficient conditions for the existence of a solution will be given later, and the obvious necessary condition, stemming from the total divergence principle in Section 1E is

$$0 = \sum_{i \in N} b(i) = b(N).$$

Although $b(i)$ is called the "supply" at node i, a negative value of $b(i)$ corresponds really to a *demand*, since the constraint $y(i) = b(i)$ says then that i is to be a sink of a certain magnitude. Thus the condition $b(N) = 0$ can be interpreted as asserting that the total of the positive supplies must equal the total of the positive demands.

EXAMPLE 2.   (*Commercial Shipments*)

A certain product must be sent from warehouses to customers by way of a transportation network G with capacity intervals. The warehouses comprise a node set S^+ with supplies $b(i) > 0$, whereas the customers form S^- with demands $a(i) > 0$. At the rest of the nodes the flow must be conserved. Thus one seeks a flow x that is feasible with respect to capacities and satisfies

$$0 \le y(i) \le b(i) \quad \text{for all } i \in S^+,$$

$$-y(i) = a(i) \quad \text{for all } i \in S^-,$$

$$y(i) = 0 \quad \text{for all other nodes.}$$

This does not, on the surface, fit the mold of the feasible distribution problem, but a conversion is easily made. Let

$$w = \sum_{i \in S^+} b(i) - \sum_{i \in S^-} a(i)$$

$$= [\text{total positive supply}] - [\text{total positive demand}].$$

Assuming $w \ge 0$, one can express the situation *equivalently* as the feasible distribution problem for the extended network G' shown in Figure 3.5. The possible slack in the inequality constraints is taken care of by consigning all surplus of the product to an abstract "storage" location. Actually, if $w = 0$, no "storage" is necessary, and the network need not be extended; the inequality constraints in this case *imply* equality for the given supplies and demands (due to $y(N) = 0$), and hence they can be written without loss of generality as $y = b$, where $b(i) = -a(i)$ for $i \in S^-$ and $b(i) = 0$ for $i \notin (S^+ \cup S^-)$.

Even if $w < 0$ in this example, some sense can be made out of the situation by construing it as the feasible distribution problem for the network in Figure 3.6. Later the question of *costs* of flows in such a context will be taken up (Chapter 7). For the fluxes in the added arcs in Figure 3.5, these would be *storage costs* associated with the various warehouses $i \in S^+$, whereas in the case of Figure 3.6 they would be *costs of default* associated with the customers $i \in S^-$.

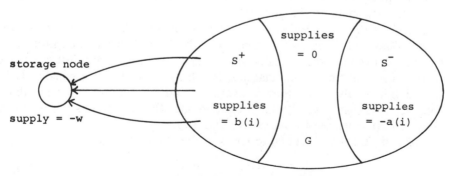

Figure 3.5

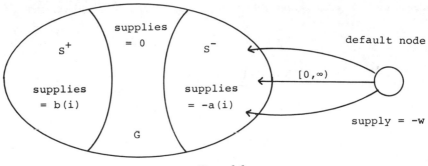

Figure 3.6

General Feasible Flow Problem

Example 2 demonstrates that a supply constraint of the form div $x = b$ is much less restrictive in practical terms than might be imagined. It will now be shown that even the generalized problem alluded in the preceding example may be brought under the same model. The problem in question consists of finding a flow x such that

$$x(j) \in C(j) \quad \text{for all } j \in A,$$

$$y(i) \in C(i) \quad \text{for all } i \in N \ (y = \text{div } x),$$

where $C(j)$ is the capacity interval associated with the arc j, as before, and $C(i)$ is now the *supply interval* associated with the node i. This is called the *feasible flow problem*.

Finding a flow that satisfies the constraints of the max flow problem corresponds to solving this problem in the case of $C(i) = (-\infty, +\infty)$ for all nodes in N^+ or N^-, and $C(i) = [0,0]$ for all other nodes. As another example, the feasible distribution problem is the case where $C(i) = [b(i), b(i)]$ for all nodes i, and the *feasible circulation problem* is the still more special case where $b(i) = 0$. However, the feasible flow problem can in turn be represented as a feasible circulation problem. All one has to do is pass to the usual augmented network in the manner of Figure 3.7.

Therefore *the three problems—feasible distribution, feasible circulation, and feasible flow—are equivalent*. Among them the feasible distribution problem is the most convenient theoretically, since it is quite intuitive and just as easy to treat as the problem for circulations, to which the facts can immediately be specialized. At the same time it avoids various notational and algorithmic complications that arise with the broader model. (For extensions to the case of *nonclosed* intervals, see Exercises 3.13 and 3.14.)

The main theorem is stated in terms of the quantity

$$b(S) = \sum_{i \in S} b(i),$$

which for any node set S represents the (net) *supply in S*.

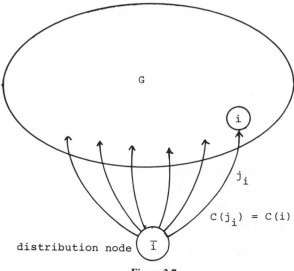

Figure 3.7

Feasible Distribution Theorem (Gale and Hoffman). *The feasible distribution problem has a solution if and only if $b(N) = 0$ and*

$$b(S) \le c^+(Q) \quad \text{for all cuts } Q = [S, N \setminus S].$$

A constructive proof of the theorem will be derived from a corresponding algorithm in the next section. But some general comments are appropriate here.

First the elementary nature of the *necessity* of the condition should be noted. The need for $b(N) = 0$ has already been observed. More generally, if $\text{div } x = b$, then

$$b(S) = [\text{divergence of } x \text{ from } S] = [\text{flux of } x \text{ across } Q]$$

by the divergence rule. Feasibility of x with respect to capacities then implies

$$b(S) \in C(Q) = [c^-(Q), c^+(Q)],$$

and hence in particular $b(S) \le c^+(Q)$.

Of course the inequality $b(S) \ge c^-(Q)$ is also necessary, but this condition does not need to be listed separately for the following reason. The property $0 = b(N) = b(S) + b(N \setminus S)$ entails $b(N \setminus S) = -b(S)$. At the same time the cut $Q' = [N \setminus S, S]$, which is the *reverse* of Q, has $c^+(Q') = -c^-(Q)$. Thus the inequality $b(S) \ge c^-(Q)$ is equivalent to $b(N \setminus S) \le c^+(Q')$.

3I. FEASIBLE DISTRIBUTION ALGORITHM

The feasible distribution problem can be solved in different ways. The method now to be described is appealing because of its resemblance to the max flow algorithm. (In fact it is equivalent to applying the max flow algorithm to a certain modified network that depends on the initial flow; see Exercise 3.22.) An alternative procedure will be given in Section 3K. In both cases, subject to a minor commensurability condition (or resorting to arc discrimination, or other restrictions such as will be described in Section 3J), either a solution to the problem is obtained or a cut is found that violates the condition in the feasible distribution theorem.

Algorithm

It is assumed that $b(N) = 0$. A flow x is given that is feasible with respect to the arc capacities. (Initially, one could choose $x(j)$ to be any number in the interval $C(j)$, e.g., the number nearest 0.) The node sets N^+ and N^- are then defined, depending on the values of $y = \text{div } x$, by

$$N^+ = \{i | b(i) > y(i)\}, \qquad N^- = \{i | b(i) < y(i)\}.$$

(The nodes in N^+ have a *surplus*, the supply being greater than what is used by x, while the nodes in N^- have a *deficit*.) If $N^+ = \varnothing = N^-$, then x is a solution to the feasible distribution problem, and the algorithm terminates. If not, then both N^+ and N^- are nonempty (because $b(i) - y(i)$ must sum to 0 as i ranges over all of N, due to $b(N) = 0 = y(N)$). The painted network algorithm is applied with the same painting of the arcs j as in the max flow algorithm:

$$\text{green if } c^-(j) < x(j) < c^+(j)$$
$$\text{white if } c^-(j) = x(j) < c^+(j)$$
$$\text{black if } c^-(j) < x(j) = c^+(j)$$
$$\text{red} \quad \text{if } c^-(j) = x(j) = c^+(j)$$

If the outcome is a solution $Q = [S, N \setminus S]$ to the painted cut problem, then $b(S) > c^+(Q)$, and it follows that the feasible distribution problem has no solution. If the outcome is a solution P to the painted path problem, then the quantity

$$\alpha = \min \begin{cases} c^+(j) - x(j) & \text{for } j \in P^+, \\ x(j) - c^-(j) & \text{for } j \in P^-, \\ b(i) - y(i) & \text{for the initial node of } P \text{ (in } N^+), \\ y(i) - b(i) & \text{for the terminal node of } P \text{ (in } N^-), \end{cases}$$

is positive and finite. The flow $x' = x + \alpha e_P$ is formed, and the procedure is repeated.

Obviously, the labeling refinement in the max flow algorithm can be utilized here in slightly modified form, to save effort in the calculation of α. However, an even more efficient approach to implementation will be discussed in Section 3J.

Justification of the Algorithm

Much of the reasoning is almost identical to that on which the max flow algorithm is based, and it need not be given again in detail. If $Q: N^+ \downarrow N^-$ solves the painted cut problem and corresponds to the set S (where $S \supset N^+$ and $S \cap N^- = \emptyset$), then

$$c^+(Q) = [\text{flux of } x \text{ across } Q] = [\text{divergence of } x \text{ from } S] = y(S)$$

$$= b(S) - [b(S) - y(S)] = b(S) - [b(N^+) - y(N^+)] < b(S),$$

where the fact has been used that $b(i) - y(i) = 0$ for $i \in S \setminus N^+$ but $b(i) - y(i) > 0$ for $i \in N^+ \neq \emptyset$. On the other hand, if $P: N^+ \to N^-$ solves the painted path problem, the quantities whose minimum defines α are all positive, and the last two, at any rate, are finite. Hence α is positive and finite. The flow $x' = x + \alpha e_P$ again satisfies the capacity constraints. For its divergence y', one has $b(i) - y'(i) = b(i) - y(i)$ for all nodes, except that

$$0 \leq b(i) - y'(i) = b(i) - y(i) - \alpha \quad \text{for the initial node of } P,$$

$$0 \geq b(i) - y'(i) = b(i) - y(i) + \alpha \quad \text{for the terminal node of } P.$$

In this sense x' is an improvement over x, since $b - y'$ is nearer to 0 than $b - y$ was.

Termination

The algorithm must terminate after a finite sequence of iterations, if the capacity bounds $c^+(j), c^-(j)$, the supplies $b(i)$, and the initial flow values $x(j)$, are *commensurable*. In this case all these values, as well as all subsequent flux and divergence values and flow alteration values α, are whole multiples of a certain $\delta > 0$. Since each of the nonzero differences $b(i) - y(i)$ progresses monotonically toward 0, changing by at least δ if at all, and two of them are affected by each iteration; the iterations cannot continue endlessly.

As in the case of max flow the algorithm must terminate *even without commensurability*, if arc discrimination is used in the painted network subroutine (Exercise 3.22).

Proof of the Feasible Distribution Theorem

The necessity of the condition has already been seen in Section 3H. The sufficiency follows from the fact that the algorithm must terminate, at least if

the painted network algorithm with arc discrimination is utilized at each
iteration.

EXAMPLE 3

A feasible distribution problem is indicated in Figure 3.8, where the supplies
$b(i)$ are shown at the nodes and the capacity intervals $C(j)$ next to the arcs.
The sequence of paths that is generated by the feasible distribution algorithm
is not necessarily unique, of course, because of arbitrary choices in the painted
network subroutine. The following sequence gives rise to the flows shown in

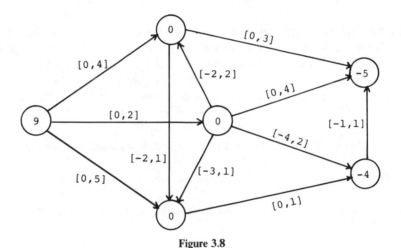

Figure 3.8

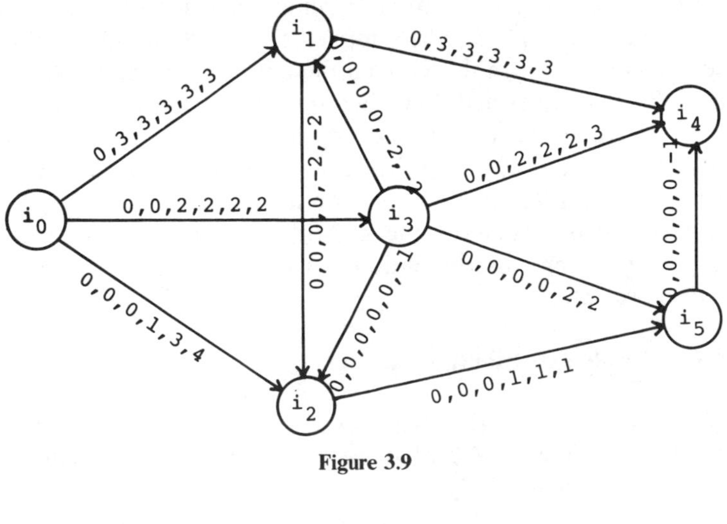

Figure 3.9

Figure 3.9, the last of which solves the problem in question:

N^+	N^-	P	α
i_0	i_4, i_5	$i_0 \rightarrow i_1 \rightarrow i_4$	3
i_0	i_4, i_5	$i_0 \rightarrow i_3 \rightarrow i_4$	2
i_0	i_5	$i_0 \rightarrow i_2 \rightarrow i_5$	1
i_0	i_5	$i_0 \rightarrow i_2 \leftarrow i_1 \leftarrow i_3 \rightarrow i_5$	2
i_0	i_5	$i_0 \rightarrow i_2 \leftarrow i_3 \leftarrow i_4 \leftarrow i_5$	1

EXAMPLE 4

Another feasible distribution problem is displayed in Figure 3.10, and the feasible distribution algorithm is applied to it in Figure 3.11. This time, however, the outcome is the detection of a cut $Q = [S, N \setminus S]$ with $1 = c^+(Q)$

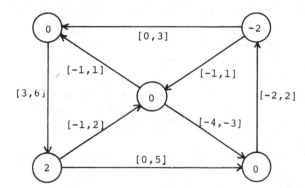

Figure 3.10

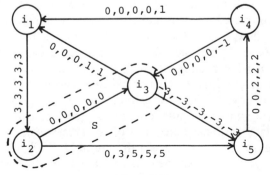

Figure 3.11

$< b(S) = 2$. The sequence of paths is as follows:

N^+	N^-	P	α
i_2, i_3	i_1, i_4, i_5	$i_2 \rightarrow i_5$	3
i_2, i_3	i_1, i_4	$i_2 \rightarrow i_5 \rightarrow i_4$	2
i_3	i_1	$i_3 \rightarrow i_1$	1
i_3	i_1	$i_3 \leftarrow i_4 \rightarrow i_1$	1

3J.*   MULTIPATH IMPLEMENTATION

An efficient approach to executing the feasible distribution algorithm is based on using the multipath version of the painted network algorithm as a subroutine. As explained near the end of Section 2G, this subroutine yields a multirouting Θ that represents all the compatible paths from N^+ to N^- containing r arcs, where r is the smallest number possible. Let us call these the *minimal improvement paths for x*.

The multirouting is generated along with a sequence of sets $N_0, N_1, \ldots, N_r$, where N_k consists of all the nodes reachable from N^+ by compatible paths using k arcs (but not by compatible paths using fewer arcs). For $i \in N_k$ the arcs in $\Theta(i)$ are all those by which i can be reached (compatible) from nodes in N_{k-1}. Of course $N^+ = N_0$, whereas N^- is disjoint from all the sets N_k except N_r. This is schematized in Figure 3.12.

Since the multirouting furnishes *all* the minimal improvement paths for x, any of which could serve in the flow modification procedure in the algorithm, the idea arises of using several of them at once to make an especially big improvement. However, matters are not quite so simple. Suppose we begin with a particular Θ^- path P_1, obtained by tracing back from an arbitrary node in $N^- \cap N_r$ to N^+, with an arbitrary arc in $\Theta(i)$ chosen in departing from each node i that is encountered. We can make the corresponding improvement in x, changing it to a flow ξ, and then without reinvoking the painted network algorithm, look for another Θ-path P_2 that can be used for an improvement of ξ. But the fact that x has been replaced by ξ means that certain arcs may now have a different color, and the surplus or deficit at certain nodes in N^+ and N^- may have been removed. Thus we cannot be sure that another arbitrarily chosen Θ-path will still be suitable, and extra precautions become necessary.

In the first place we must take note of all arcs whose colors have been changed and regard them as henceforth unusable. Each change is in fact always of the following type, in view of the nature of the flow improvement

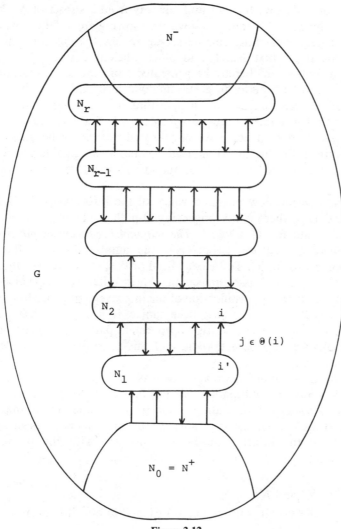

Figure 3.12

process:

> a green or white arc oriented from some N_{k-1} to N_k has
> become black, or alternatively, a green or black arc   (1)
> oriented from N_k to N_{k-1} has become white.

(Such an arc has become "saturated," as far as sending additional flux from
N_{k-1} to N_k is concerned.) In "tracing back," we must also avoid starting at a
node in N^- that is no longer a deficit node, or arriving at a node in N^+ that is
no longer a surplus node, or for that matter, arriving at any node i where the
arcs in $\Theta(i)$ have all become unusable.

The upshot is that we have to find our way from a subset of N^- back to a subset of N^+ by way of Θ much as if we were solving an auxiliary painted path problem. Indeed, everything can be set up in exactly this form and implemented using depth-first search and other efficient devices. A sequence of improvement paths can thereby be generated from the same multirouting Θ. This turns out to be superior to generating one such path in each of the basic iterations of the feasible distribution algorithm, due to the greatly reduced size and very special structure of the "auxiliary network" associated with Θ.

In a moment an even better way to take advantage of the multirouting Θ will be described. But first it is important to observe a valuable property of the approach just explained that will be found to carry over to the subsequent situation.

Let x' denote the flow on hand when all the information in Θ has been exhausted; that is, there is no longer any Θ-path that happens also to be an improvement path for x'. **Claim.** *The minimal improvement paths for x' (if any) use more than r arcs* (where r is the number used by the minimal improvement paths for x). To be convinced of this, start with the fact that the node sets N'^+ and N'^- associated with x' are contained, respectively, in N^+ and N^- (as seen from the justification of the improvement procedure in Section 3I). Since only the arcs of Θ have been subject to changes in color, and each change is in the pattern of (1), any improvement path for x' using only r arcs would actually have to be a Θ-path, and this possibility is excluded by the assumption on x'.

Let us speak of a *grand iteration* of the feasible distribution algorithm as one in which we apply the multipath version of the pointed network subroutine to a given flow x to get a multirouting Θ and then by some "grand improvement procedure" (not necessarily the same as the one we have been sketching) construct from the information in Θ a flow x' with the properties just exploited, namely:

1. $N'^+ \subset N^+$ and $N'^- \subset N^-$.
2. x' is still feasible with respect to capacities and differs from x only in arcs used by Θ.
3. Each color change in the painting for x', relative to the painting for x, is in conformity with (1).
4. No Θ-path is an improvement path for x'.

Then by the argument just given, each grand iteration increases the number of arcs (or equivalently, the number of nodes) that is used by any minimal improvement path. *The number of grand iterations that can be performed before termination of the algorithm is therefore bounded by $|N| - 1$.*

This conclusion is important, because it gives us a grip on the computational complexity of the algorithm. We are about to describe a "grand improvement procedure" based on ideas of Dinic and Karzanov, for which the number of arc operations in a certain sense is bounded by $|N|^2 + 2|A|$. Using this as the crucial ingredient in the recipe, one obtains what we shall refer to as the *Dinic-Karzanov implementation* of the feasible distribution algorithm. This then has an overall bound of

$$[|N| - 1](|N|^2 + 3|A|)$$

(where an extra $[|N| - 1]|A|$ is included to take account of the multipath version of the painted network subroutine, which involves looking at each arc at most once in each "grand iteration"). In particular, the number of arc operations in the digraph case is less than $4|N|^3$ (since $|A| \leq |N|^2$ when G is a digraph).

Grand Improvement Procedure (Dinic-Karzanov Approach)

At intermediate stages before the improved flow x' has been produced from x, there will be a flow ξ whose divergence will be denoted by η. (Initially, $\xi = x, \eta = y$.) The surplus nodes for ξ (where $b(i) > \eta(i)$) will always be among those in N^+, but the set of deficit nodes

$$D = \{i \in N | b(i) < \eta(i)\}$$

will generally not be included in N^-, except at the very end when we will set $x' = \xi$. At all times ξ will have Properties 2 and 3 which are described for x', and termination will come when Property 4 has been achieved too. In this way we will obtain a flow x' meeting all the requirements.

As things progress, certain arcs will be declared "closed" (the remaining arcs are said to be "open"), and certain nodes will be declared "blocked." These designations are permanent and will turn out to mean that no Θ-path using such arcs or nodes can be an improvement path for ξ.

The arcs in each set $\Theta(i)$ are assumed to be listed in some particular order, so that one can speak of the "first" arc of a certain type, and so forth.

The procedure for modifying ξ involves two basic operations. The first will be called *transmitting demands* from one of the node sets N_k to N_{k-1}. It is said to be possible if $0 < k \leq r$ and there is a node $i \in N_k \cap D$ such that $\Theta(i)$ contains an open arc j whose other node i' (in N_{k-1}) is not blocked. The operation consists of taking the *first* of such arcs in $\Theta(i)$ (the choice of i, if there is more than one such node available in N_k, does not matter) and treating it as follows, according to its orientation.

CASE 1. $j \sim (i', i)$. Calculate

$$\alpha = \min\{c^+(j) - \xi(j), \eta(i) - b(i)\}.$$

This will satisfy $0 < \alpha < \infty$. Declare j *closed* if $\alpha = c^+(j) - \xi(j)$. Replace $\xi(j)$ by $\xi(j) + \alpha$.

CASE 2. $j \sim (i, i')$. Calculate

$$\alpha = \min\{\xi(j) - c^-(j), \eta(i) - b(i)\}.$$

This will satisfy $0 < \alpha < \infty$. Declare j *closed* if $\alpha = \xi(j) - c^-(j)$. Replace $\xi(j)$ by $\xi(j) - \alpha$.

Observe that in both cases an additional α units of flux are sent from i' to i by way of j, so that the deficit $\eta(i) - b(i)$ at i is decreased by α (possibly removing i from D) and the deficit $\eta(i') - b(i')$ at i' is increased by α (possibly adding i' to D).

For bookkeeping purposes, since this change may later have to be undone at least in part, we record the pair (α, j) at i' and refer to it as a *requisition*. (The records at i' are maintained in the order of arrival, so that we can immediately recall the "most recent" requisition when needed.)

The second operation in the procedure for modifying ξ is called *rejecting demands* in N_k (returning an excess of transmitted demands back to N_{k+1}). To perform the operation (with $0 \le k < r$), take any node $i' \in N_k \cap D$ and recall the most recent requisition (α, j) at i'. (The other node i of j will be in N_{k+1}.) Calculate

$$\beta = \min\{\alpha, \eta(i') - b(i')\} > 0.$$

Subtract β from $\xi(j)$ if $j \sim (i', i)$, but add it if $j \sim (i, i')$. (This will decrease the deficit $\eta(i') - b(i')$ at i' by β and increase the deficit $\eta(i) - b(i)$ at i by β.) Correspondingly, replace the requisition (α, j) at i' by $(\alpha - \beta, j)$ if $\beta < \alpha$, but simply delete it from the records if $\beta = \alpha$. If the new deficit $\eta(i') - b(i')$ is 0 (i.e., if β equaled the old value $\eta(i') - b(i')$), the operation is finished; regard the node i' as *blocked*. Otherwise the requisition (α, j) will have been deleted; repeat the step with the most recent of the remaining requisitions at i'. (There will always be at least one, as long as the deficit at i' remains positive.)

Algorithm

The overall procedure is composed of the two operations in the following pattern. In the general step, a set N_k is up for inspection. If demands can be transmitted from N_k, do so until the possibilities have been exhausted; then move down to inspect N_{k-1}. If demands cannot be transmitted from N_k, there are two cases. If $k = r$, terminate and set $x' = \xi$. Otherwise reject demands from N_k until this is no longer possible; then move back up to inspect N_{k+1}.

Discussion

Some properties of the procedure are obvious from the description. At all stages the requisition records at the nodes in $N_0, \ldots, N_{r-1}$ tell exactly how much, and along which arcs, additional flux should be superimposed on x to get the flow ξ. This additional flux is carried solely by arcs in the multirouting

Θ, and it moves only in the "upward" direction: one has

$$x(j) \leq \xi(j) \leq c^+(j) \quad \text{if } j \text{ goes from some } N_{k-1} \text{ to } N_k,$$

$$c^-(j) \leq \xi(j) \leq x(j) \quad \text{if } j \text{ goes from some } N_k \text{ to } N_{k-1}.$$

Thus ξ always does have two of the properties slated for x', namely, Properties 2 and 3. Observe that an arc is declared closed as soon as it changes color, and it keeps this designation even though it may revert to its original color later during a "rejection" process.

For nodes in $N_1, \ldots, N_{r-1}$ (such nodes having $b(i) = y(i)$), it is always true that $b(i) \leq \eta(i)$, the deficit $\eta(i) - b(i)$ being equal to the total of the requisitioned values α recorded at i. For nodes in N_0 (where $b(i) > y(i)$), the total of the requisitioned values is equal instead to

$$[\eta(i) - b(i)] + [b(i) - y(i)] = \eta(i) - y(i);$$

hence $\eta(i) > y(i)$, but perhaps $b(i) > \eta(i)$ or $b(i) \leq \eta(i)$. For N_r, only the nodes belonging also to N^- are ever affected; these always have $b(i) \leq \eta(i) \leq y(i)$, with $b(i) < y(i)$. In summary, the deficit node set $D = \{i | \eta(i) > b(i)\}$, after starting out equal to N^-, can lose or gain certain nodes in $N_r \cap N^-$ or other sets N_k (including $N_0 = N^+$) from time to time, but the surplus node set $\{i | \eta(i) < b(i)\}$ is always included in N^+. Therefore Property 1 will be valid for x' if, on termination of the procedure, D does not meet $N_0, \ldots, N_{r-1}$.

To establish this and Property 4 for x', let us observe what happens when the procedure is brought into action. At the beginning all arcs used by Θ are open, and no nodes are blocked. Demands are first transmitted from N_r successively down through $N_{r-1}, \ldots, N_1, N_0$. (Some do reach N_0, because every Θ-path is an improvement path for x.) Of course demands that arrive in N_0 are balanced off against existing surpluses; that is, nodes $i' \in N_0$ that have been affected may not actually join D, because $\eta(i') = y(i') < b(i')$ at such nodes initially.

If any node in N_0 has joined D, the next step is to reject the corresponding excess in demands back to N_1 and declare each such node blocked. (The surpluses at such nodes have been exhausted.) At this point D does not meet N_0. Another attempt is made to transmit demands from N_1 to N_0, this time avoiding arcs that have become closed and nodes that have become blocked, and again the excess demands are rejected from N_0 back to N_1. This continues back and forth until, after rejecting excess demands from N_0, one finds there are no more demands that can be transmitted from N_1.

Any excess demands in N_1, represented by the presence of deficit nodes $i' \in N_1 \cap D$, are then rejected back to N_2, the nodes in question becoming blocked. (Note that for every such node the arcs in $\Theta(i')$ must all be either closed or connected to nodes in N_0 that became blocked earlier, for otherwise it would have been possible to transmit some of the demands in question.) Next one tries to retransmit demands from N_2 to N_1 and on, if possible, to N_0 (avoiding closed arcs and blocked nodes). Adjustments are made as before, and eventually one returns to the inspection of N_2, at which point D does not meet N_0 or N_1. The same thing is repeated until no more demands can be transmitted from N_2. Then any excess demands in N_2 are rejected back to N_3, transmission is attempted again, and so forth. Thus the procedure may move up and down the sequence $N_0, N_1, \ldots, N_r$ in a number of phases of transmission and rejection. But since the most recent requisitions are always first in the operation of rejecting demand from a node $i' \in N_k$, there is no danger of this operation ever creating a deficit at some node $i \in N_{k+1}$ that was declared blocked in some previous phase.

This explanation makes apparent the fact that at any stage when one passes from N_k up to N_{k+1}, there are no remaining deficits in $N_0, N_1, \ldots, N_k$; that is, these sets do not meet D. It is true therefore that in the final iteration, when upon reaching N_r (necessarily from N_{r-1}) one finds that no more demands can be transmitted, the sets $N_0, N_1, \ldots, N_{r-1}$ are all disjoint from D. Hence Property 1 is indeed achieved.

An arc of altered color cannot of course belong to any Θ-path that is an improvement path for ξ, because it is "saturated" in the only usable direction. This state persists for an arc unless some of the flux is canceled later during a rejection operation. In that event the "lower" node of the arc is declared blocked, and nothing incident to it is ever touched again. By induction, then, any arc in $\Theta(i)$ that is closed is either of altered color or leads down to a blocked node i'. Recalling that all blocked nodes satisfy $\eta(i') = b(i')$ (the ones in N_0 having reached this condition through the exhaustion of whatever surplus they had originally been blessed with), we see that no Θ-path using a closed arc or a blocked node can be an improvement path for ξ.

When the procedure terminates, the arcs associated by Θ with the remaining deficit nodes in N_r (if any) are all either closed or lead to blocked nodes, for otherwise demands could indeed be transmitted from N_r. Hence there can be no improvement paths left at all. We conclude from this that x' does have Property 4.

Termination and Computational Complexity

It will now be demonstrated that the procedure does terminate eventually, and in fact the number of flux changes is bounded by $|N|^2 + 2|A|$. Here a "flux change" means a change affecting a single arc j, where a quantity α or β is added or subtracted from $\xi(j)$.

An arc can only once undergo a flux change that causes it to be closed. Following this, it can be affected at most once more, namely, if the latest requisition involving it enters the demand rejection process, after which it is an arc associated with a blocked node. (A demand rejection operation on a node $i' \in N_k$ cannot involve two requisitions on the same arc j, for the earlier requisition would have had to come during an earlier transmission of demands from N_{k+1} and between the two transmissions there must have been a stage of demand rejection in N_k that would have left i' without any deficit.) The number of changes of these kinds is therefore bounded by $2|A|$.

For the rest of the argument we can concentrate on flux changes that occur in an arc prior to its becoming closed (if it ever does). Such changes can only occur during transmission of demands. We need only show that their number is bounded by $|N|^2$.

Let us view the pattern of execution of the algorithm as divided into "phases," where each phase consists of a unidirectional series of moves up or down the sequence $N_0, N_1, \ldots, N_r$. Any time a rejection phase ends and is followed by a transmission phase, there must have been at least one node that was handled during the rejection phase and then declared blocked, never to be handled again. The number of transmission phases is therefore bounded by the number of blocked nodes, hence certainly by $|N|$.

Now in any one operation of transmitting demands from node i, a sequence of open arcs in $\Theta(i)$ may be used, but all except possibly the last one will become closed. Hence there can be at most one flux change of the kind we are trying to reckon with. It follows that the number of such changes during any transmission phase cannot exceed the number of nodes operated on. No node is operated on more than once during a transmission phase, so this yields a bound of $|N|$. Multiplying by the bound on the number of such phases, we get the asserted bound $|N|^2$. (This could be sharpened by a more careful argument; see Exercise 3.32.)

3K.*   FLOW RECTIFICATION ALGORITHM

The procedure to be described now solves the feasible distribution problem in a manner complementary to the algorithm developed in the last two sections. Instead of maintaining feasibility with respect to capacities, and improving feasibility with respect to supplies by an application of the painted network algorithm at each iteration, it works with flows satisfying the supply constraint and improves feasibility with respect to capacities by involving Minty's algorithm (i.e., the painted network algorithm as applied in the context of Minty's lemma in Section 2H) at each iteration.

Algorithm

A flow x is given such that $\operatorname{div} x = b$. Define the arc sets

$$A^+ = \{ j | x(j) > c^+(j) \}, \qquad A^- = \{ j | x(j) < c^-(j) \}.$$

These consist of the arcs that violate the capacity constraints. If $A^+ = \varnothing = A^-$, then x is a solution to the feasible distribution problem, and the algorithm terminates. If not, let j denote any arc in either A^+ or A^-. Minty's algorithm is applied to $\bar{j}$ and the following painting of the arcs of the network:

$$\text{green if } c^-(j) < x(j) < c^+(j)$$
$$\text{white if } x(j) \le c^-(j), x(j) < c^+(j)$$
$$\text{black if } x(j) > c^-(j), x(j) \ge c^+(j)$$
$$\text{red} \quad \text{if } c^-(j) = x(j) = c^+(j)$$

(The arcs in A^+ arc all black, whereas those in A^- are white; in particular, $\bar{j}$ is either black or white.) If the outcome is a compatible cut $Q = [S, N \setminus S]$ containing $\bar{j}$, then $b(S) > c^+(Q)$, (as we will see later). Thus the condition in the feasible distribution theorem is violated, and the problem has no solution. The algorithm then terminates. If a compatible elementary circuit P containing $\bar{j}$ is the outcome, define

$$\alpha = \min \begin{cases} c^+(j) - x(j) & \text{for } j \in P^+, \\ x(j) - c^-(j) & \text{for } j \in P^-, \\ \bar{\alpha}, \end{cases}$$

where

$$\bar{\alpha} = \text{dist}\left(x(\bar{j}), \left[c^-(\bar{j}), c^+(\bar{j})\right]\right)$$

$$= \begin{cases} c^-(j) - x(j) & \text{if } j \in A^+ \\ x(j) - c^+(j) & \text{if } j \in A^-. \end{cases}$$

Then $0 < \alpha < \infty$. The flow $x' = x + \alpha e_P$ again satisfies the supply constraint, $\text{div } x' = b$. It is "better" than x in the sense that the capacity constraint violations in the various arcs are not worsened, and at least one is improved or eliminated (justification follows). The step is then repeated for the new flow.

EXAMPLE 5

Resolution of the feasible distribution problem in Figure 3.8 (Example 3) by means of the flow rectification algorithm is summarized in Figure 3.13. The sequence of circuits used was the following:

$\bar{j}$	P	$\bar{\alpha}$	α
(i_0, i_3)	$i_0 \to i_1 \leftarrow i_3 \leftarrow i_0$	7	2
(i_0, i_3)	$i_0 \to i_2 \leftarrow i_3 \leftarrow i_0$	5	3
(i_0, i_3)	$i_0 \to i_1 \to i_4 \leftarrow i_3 \leftarrow i_0$	2	2
(i_3, i_5)	$i_3 \to i_4 \to i_5 \leftarrow i_3$	2	1
(i_3, i_5)	$i_3 \to i_2 \to i_5 \leftarrow i_3$	1	1

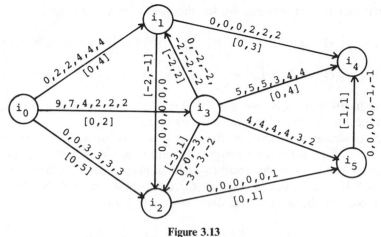

Figure 3.13

Justification

With a cut $Q = [S, N \setminus S]$ as the outcome of Minty's algorithm, one has

$$x(j) \geq c^+(j) \quad \text{for all } j \in Q^+$$

$$-x(j) \geq -c^-(j) \quad \text{for all } j \in Q^-,$$

with strict inequality for at least one arc, namely, $\bar{j}$. It follows that

$$c^+(Q) < \sum_{j \in Q^+} x(j) - \sum_{j \in Q^-} x(j)$$

$$= [\text{flux of } x \text{ across } Q]$$

$$= [\text{divergence of } x \text{ from } S]$$

$$= b(S),$$

so the condition in the feasible distribution theorem is violated, as claimed.
    With a circuit P as the outcome, one has

$$c^+(j) - x(j) > 0 \quad \text{for } j \in P^+,$$

$$x(j) - c^-(j) > 0 \quad \text{for } j \in P^-.$$

Since the arc $\bar{j}$ belongs to either A^- or A^+, the number $\bar{\alpha}$ is finite and positive, and so is α.

The choice of α clearly ensures for the flow $x' = x + \alpha e_P$ that

$$c^+(j) \geq x'(j) \quad \text{and} \quad c^-(j) - x'(j) = c^-(j) - x(j) - \alpha \quad \text{for all } j \in P^+,$$

$$c^-(j) \leq x'(j) \quad \text{and} \quad x'(j) - c^+(j) = x(j) - c^+(j) - \alpha \quad \text{for all } j \in P^-,$$

whereas in all other arcs one has $x'(j) = x(j)$, and the status of feasibility with respect to the capacity interval is unchanged.

The constraint violations by the arcs that were in $A^+ \cup A^-$ are thus no worse for x' than for x, and there is definite reduction for at least the arc $\bar{j}$. (The definition of α is designed to yield improvement in $\bar{j}$, without "overshooting." If α turns out to equal $\bar{\alpha}$, the infeasibility in $\bar{j}$ will be eliminated.) Since P is a circuit, one also has $\text{div } e_P = 0$, and hence

$$\text{div } x' = \text{div } x + \alpha \, \text{div } e_P = \text{div } x = b.$$

Termination

Assume that the capacity bounds $c^+(j)$, $c^-(j)$, and initial flux values $x(j)$ are commensurable. (This is no less stringent than the commensurability condition used to prove termination of the feasible distribution algorithm: since $\text{div } x = b$, the supplies $b(i)$ must belong to this same class of commensurability.) By the argument that is now quite familiar, all the flux values and flow augmentation values generated in the course of the algorithm will be whole multiples of a certain $\delta > 0$. Therefore when the capacity constraint violation corresponding to any arc in $A^+ \cup A^-$ is reduced, it is always reduced by at least the quantum δ. Only a finite amount of reduction is possible, and there are only finitely many arcs having need of it. Hence the algorithm must terminate. Without commensurability, termination can be assured by arc discrimination (Exercise 3.27).

Multipath Implementation

The procedures in Section 3J could be adapted to this method, since the step using Minty's algorithm amounts to a search for flow improvement paths from a certain node s to a certain s'.

Comparison with the Feasible Distribution Algorithm

Which of the two algorithms for the feasible distribution problem is "best" may depend on the context. The first algorithm has the advantage of "starting from almost nothing": there is no need for preliminary construction of a flow satisfying $\text{div } x = b$. Furthermore there is more flexibility and perhaps greater speed and efficiency in the application of the painted network algorithm at each iteration, due to the fact that *any* compatible path from a set N^+ to a set N^- is satisfactory. The flow rectification method, by employing Minty's algorithm, carries out this calculation only for paths from a particular node s to a node s'.

Nevertheless, it will be seen later that when the feasible distribution problem needs to be solved repeatedly as a subroutine in some more general algorithm of optimization, a flow satisfying div $x = b$ may be right at hand and only slightly infeasible. In such cases the second algorithm might be preferable. At any rate it is instructive as a prototype for the general "out-of-kilter algorithm" for finding optimal flows (Chapters 7 and 9). Its use has some theoretical ramifications that will be exposed later.

Yet, for solving a *single* problem, the flow rectification algorithm seems unlikely to be really advantageous, even starting from a slightly infeasible flow $\tilde{x}$ with div $\tilde{x} = b$. It is almost as easy to start the feasible distribution algorithm with the corresponding flow x such that $x(j)$ is the element of $[c^-(j), c^+(j)]$ nearest to $\tilde{x}(j)$. Then div x ought to differ only "slightly" from b.

3L.* NODE CAPACITIES AND DYNAMIC FLOWS

The model underlying the max flow problem and feasible distribution problem can easily be extended in a couple of ways.

First of all it may be desirable to place constraints on the "amounts flowing through" certain nodes. For instance, at certain junctions of a transportation network there may be limited "handling capacity." This is an idea distinct from that of supply constraints on the divergence permitted at the nodes.

To put the concept on a firm footing, let us write the formula for the divergence $y(i)$ associated with a flow x as a sort of conservation equation for all the flux "passing through" the node i,

$$-y(i) + \sum_{j \in A} e(i, j)x(j) = 0$$

In this equation there are in general some positive terms and some negative terms, and the total of the positive terms must equal the absolute value of the total of the negative terms. It is this quantity that is called the *flux of x through the node i*. It can be defined succinctly by the formula

$$[\text{flux of } x \text{ through } i] = \tfrac{1}{2}\left[|y(i)| + \sum_{j \in A} |e(i, j)x(j)|\right].$$

Figure 3.14 provides an example.

The question to be explored is how to incorporate in the max flow problem or feasible distribution problem constraints of the form

$$a^-(i) \le [\text{flux of } x \text{ through } i] \le a^+(i).$$

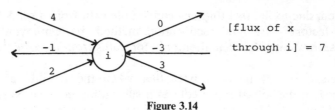

[flux of x

through i] = 7

Figure 3.14

The numbers $a^+(i)$ and $a^-(i)$ are called the upper and lower *node capacities* at i.

The case where this can be accomplished most simply is the one where the arc capacity constraints already in force imply $x(j) \geq 0$ in all the arcs incident to the node i, and x is required to be conserved at i. Then it is simply a matter of splitting i into an input node i^+ and output node i^- joined by an arc j^i, the *internal arc* of i, as in Figure 3.15. Conservation is required at both i^+ and i^-, and the flux of x through i becomes the flux through j^i, which is assigned $[a^-(i), a^+(i)]$ as capacity interval.

Little is changed if x is not required to be conserved at i, provided one of the two remaining possibilities of i being a source or a sink is eliminated in advance. If i could only be a source, conservation is enforced at i^- but not at i^+, and vice versa if i could only be a sink.

If the capacity constraints do not in themselves entail $x(j) \geq 0$ in the arcs, the treatment is more complicated and depends on reducing the model to the non-negative case. If the constraints imply $x(j) \leq 0$, there is no real problem: reverse the orientation of j, or more simply, attach j to i^- instead of i^+. Consider now the final possibility: an arc j with $c^-(j) < 0 < c^+(j)$. The trick to use is to replace j by a pair of opposed arcs j^+ and j^- as indicated in Figure 3.16. Given any flow x' in the modified network G' such that x' is feasible with respect to the capacities, there is a corresponding flow x in the original network G, obtained by

$$x(j) = x'(j^+) - x'(j^-),$$

and x is feasible with respect to the original capacities. Conversely, any x with

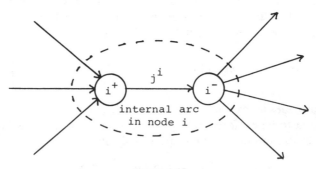

Figure 3.15

Figure 3.16

the latter properties corresponds in this way to some x' (not unique). Thus the modified network affords an equivalent formulation where the flux values are necessarily non-negative. This makes possible the representation of node capacities as before.

The case of nodes which can potentially be either sources or sinks is handled by passing first to the augmented network and representing the problem in terms of circulations.

Effect on the Min Cut Problem

The introduction in the max flow problem of internal arcs bearing node capacities naturally corresponds to a broadened outlook for the min cut problem. This is especially enlightening when the problem is interpreted, as in Section 3E, as one of finding a cheapest solution to a painted cut problem. There is no point in relating all the details here (see Exercise 3.28), but the upshot is that nodes, as well as arcs, can be incorporated into a sort of "generalized cut," and the numbers $a^+(i)$, $-a^-(i)$, are the costs for such incorporation. A dual situation, encountered in Chapter 6, is that of paths that must "pay a toll" to go through certain nodes (Exercise 6.8).

Dynamic Models

The dynamic version of a network, described in Section 1H, is the model for max flow problems or feasible distribution problem in which the "progress" of the flow is important. For a "nonhorizontal" arc of the type j_t (see Figure 1.10) a capacity interval $[c^-(j_t), c^+(j_t)]$ restricts the amount of flux which at time t can *enter* arc j of the underlying "static" network G. (Recall that the interpretation of the model requires that only non-negative flows be admitted: $c^-(j_t) \geq 0$.) Capacity intervals for the "horizontal" arcs restrict what can be *held over* at a node of G from one time period to the next (e.g., because of storage space or parking space). Supplies at the nodes of G_T correspond to supplies (positive or negative) at the nodes of G which in general may vary with time.

The dynamic version of a network is a complete success conceptually but sometimes less appealing practically because of its size. A relatively modest network, expanded through a number of time periods, may become enormous.

3M.* EXERCISES

3.1. (*Divergence Principle*). Show that the divergence principle is a special case of the conversion formula: $v \cdot x = -u \cdot y$ for $v = \Delta u$, $y = \operatorname{div} x$ (see Section 1I).

3.2. (*Max Flow Model*). Demonstrate that the max flow problem with sets N^+, N^-, is really no more general than its special case where N^+ and N^- consist of single nodes.

(*Hint.* Attach two new nodes s and s' to the network.)

(*Note.* Transformation to the "simpler" case might, however, retard computations.)

3.3. (*Max Flow Model*). Show that a max flow problem with general closed capacity intervals can always be converted into an equivalent max flow problem for which the intervals are all of the form $[c, +\infty)$, c finite.

(*Hint.* An arc with a bounded capacity interval can be replaced by a pair of arcs "in series," each having a half-bounded interval. An arc with interval $(-\infty, +\infty)$ can be replaced by two arcs with $[0, +\infty)$, as explained in Section 3L in connection with node capacities.)

(*Note.* Although the equivalence is of some interest theoretically, it seems to offer no significant computational advantage.)

3.4. (*Generalized Max Flow Min Cut Theorem; Minty*). Suppose in the max flow problem that there is at least one flow satisfying the constraints. *Then the set J of flux values from N^+ to N^- corresponding to such flows forms an interval and*

$$J = \bigcap_{Q:\, N^+ \downarrow N^-} C(Q).$$

Moreover this is true even if the capacity intervals associated with the arcs are not necessarily closed. Derive this result from the max flow min cut theorem as stated in Section 3D.

3.5. (*Max Flow Min Cut as Linear Programming*). Re-express the max flow problem as a linear programming problem in "standard form" and write down the corresponding dual problem. Show that a solution to the min cut problem furnishes a solution to this dual.

(*Hint.* Remember the correspondence between cuts and certain potentials.)

3.6. (*Least Solution to Painted Cut Problem*). Suppose that, in a given painted cut problem, one wants to find a solution Q having the least number of arcs ($|Q| \leq |Q'|$ for all other solutions Q'). Show how this can be formulated as a certain min cut problem, through the right choice of the numbers $c^+(j)$, $c^-(j)$.

(*Note.* It follows that the max flow algorithm can be used to determine such a Q.)

3.7. (*Painted Cuts*). Show that, starting from a painted cut problem, it is possible to choose $c^+(j)$ and $c^-(j)$ so that the following is true: Q solves the painted cut problem if and only if Q solves the min cut problem and $c^+(Q) < +\infty$. What does the max flow min cut theorem reduce to in this case?

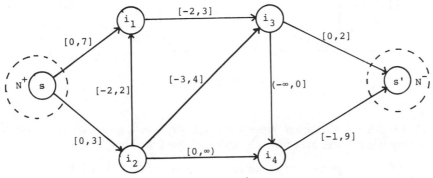

Figure 3.17

3.8. (*Max Flow Algorithm*). Solve the max flow problem and min cut problem for the network shown in Figure 3.17.

3.9. (*Max Flow Algorithm*). Solve the max flow problem and min cut problem for the network shown in Figure 3.18.

3.10. (*Max Flow Algorithm*). Solve the max flow problem and min cut problem for the network shown in Figure 3.19.

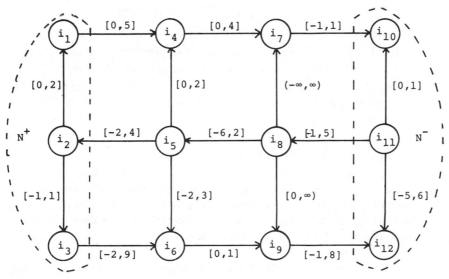

Figure 3.18

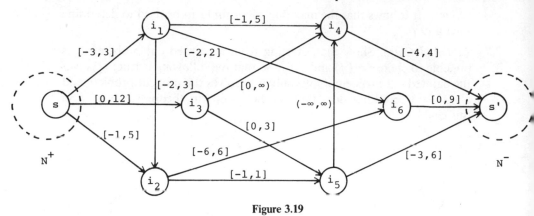

Figure 3.19

3.11. (*Tightened Max Flow Algorithm*). Fix any $\varepsilon > 0$, and modify the painting used in the max flow algorithm to the following:

> green if $x(j) \leq c^+(j) - \varepsilon$ and $x(j) \geq c^-(j) + \varepsilon$
> white if merely $x(j) \leq c^+(j) - \varepsilon$
> black if merely $x(j) \geq c^-(j) + \varepsilon$
> red   if neither is true

(Everything else in the algorithm stays the same.) Prove that termination is then inevitable (still assuming there are no paths of unlimited capacity from N^+ to N^-), and when it arrives, one has a flux x and a cut Q satisfying

$$c^+(Q) - [\text{flux of } x \text{ from } N^+ \text{ to } N^-] < \varepsilon|A|.$$

Furthermore show that the latter implies x comes within $\varepsilon' = \varepsilon|A|$ of furnishing the supremum in the max flow problem, whereas Q comes within ε' of furnishing the minimum in the min cut problem.

(*Note.* Since ε can be chosen arbitrarily small, this yields another constructive proof of the general fact that "sup = min" for the two problems.)

3.12. (*Max Flow Counterexample; Ford and Fulkerson*). Verify the details of the following example, which demonstrates the need for the commensurability condition for termination of the max flow algorithm, unless some refinement such as arc discrimination is used. The network is shown in Figure 3.20. All capacity intervals are $(-\infty, +\infty)$ except for the four horizontal arcs, and the supremum in the max flow problem is $+\infty$. Here $r = (-1 + \sqrt{5})/2$; this number has the property that $0 < r < 1$ and $r^k - r^{k+1} = r^{k+2}$ for all k.

The algorithm starts with the zero flow, and the first flow-augmenting path it utilizes is $s \to u_1 \to v_1 \to s'$. The "residual capacities" in the

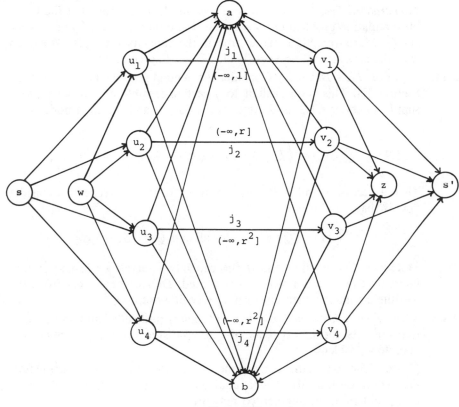

Figure 3.20

four special arcs are then $0, r, r^2, r^2$. Consider now more generally a situation where the current flow has, with respect to the four special arcs in some order or other, the "residual capacities" $0, r^k, r^{k+1}, r^{k+1}$. From the symmetry of the network it can harmlessly be supposed, for the sake of simpler notation, that this is true for the special arcs in their original order, j_1, j_2, j_3, j_4. On the next iteration there is a flow-augmenting path that has both j_2 and j_3 as forward arcs but does not use j_1 or j_4. On the succeeding iteration there is a flow-augmenting path that has j_2 as a forward arc and j_1 and j_3 as backward arcs but does not use j_4. The result of the two iterations is a situation where the special arcs, in a different order, have residual capacities $0, r^{k+1}, r^{k+2}, r^{k+2}$. This pattern can therefore be replicated interminably. Not only will the algorithm never come to a halt, but the flux from N^+ to N^- will converge to a finite value, despite the fact that the supremum is $+\infty$.

(*Note.* $\pm\infty$ could be replaced in this example by $\pm M$ for any real number M sufficiently large.)

3.13. (*Generalized Feasible Circulation Theorem*; *Minty*). Show that the feasible circulation problem has a solution if and only if $0 \in C(Q)$ for every cut Q. Moreover this remains true if the capacity intervals associated with the arcs are not necessarily closed (cf. Exercise 3.4).

3.14. (*Feasible Flow Theorem*). Consider the general feasible flow theorem in Section 3H. This is equivalent to a certain feasible flow problem with supply intervals $C(i)$ associated with nodes i, and for each node set S define

$$C(S) = \sum_{i \in S} C(i) = \left\{ \beta \mid \exists b(i) \in C(i) \quad \text{with} \sum_{i \in S} b(i) = \beta \right\}.$$

Then a necessary and sufficient condition for the existence of a feasible flow is that

$$C(S) \cap C(Q) \neq \varnothing \quad \text{for all cuts } Q = [S, N \setminus S].$$

Moreover this remains true if the capacity intervals $C(j)$ and supply intervals $C(i)$ are not necessarily closed. Derive these facts from the feasible circulation theorem in the preceding exercise.

3.15. (*Max Flow Constraints*). Derive a necessary and sufficient condition, in terms of cuts, for the existence of a flow satisfying the constraints of the max flow problem.

(*Hint.*   The constraints correspond to a case of the general feasible flow theorem in Section 3H. This is equivalent to a certain feasible distribution problem in the augmented network.)

3.16. (*Divergence Functions*). What properties of a node function b are necessary and sufficient for it to be the divergence of some flow x?

3.17. (*Feasible Distributions*). In Section 3A five special types of capacity intervals are described. State the five special cases of the feasible distribution theorem corresponding to all the arcs being of type 1, all of type 2, and so on.

3.18. (*Feasible Distribution*). Show that the feasible distribution problem with general closed intervals can always be restated equivalently as one in which all the capacity intervals are of the form $[0, +\infty)$. (Supplies have to be altered too.)

(*Hint.*   See Exercise 3.3.)

3.19. (*Circulations*). Show that there exists a circulation that has positive flux in every arc if and only if there does not exist any nonempty positive cut.

(*Note.*   For a connected network the latter condition is equivalent to strong connectedness; see Exercise 2.27.)

3.20. (*Feasible Distribution Algorithm*). Solve the feasible distribution problem in Figure 3.21.   Take as the initial flux values $x(j) = c^-(j)$ if $c^-(j) > 0$, $x(j) = c^+(j)$ if $c^+(j) < 0$, otherwise $x(j) = 0$.

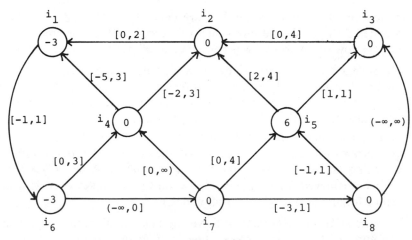

Figure 3.21

3.21. (*Feasible Distribution Algorithm*). Show that no feasible circulation exists for the network in Figure 3.22 by producing (algorithmically) a cut that violates the condition in the feasible distribution theorem.

3.22. (*Feasible Distribution Algorithm*). Prove that the feasible distribution algorithm must terminate, sooner or later, if arc discrimination is used in the painted network subroutine at each iteration (see discussion in Section 3G for the case of the max flow algorithm).

3.23. (*Feasible Distribution versus Max Flow*). Show that the feasible distribution algorithm with initial flow $\tilde{x}$ is equivalent to the max flow algorithm applied to a certain extended network depending on x.

(*Hint.* The idea is depicted roughly in Figure 3.23 where $\tilde{N}^+$ and $\tilde{N}^-$ denote the sets corresponding to $\tilde{x}$ which in general, during the course of

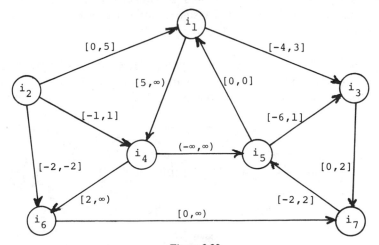

Figure 3.22

the feasible distribution algorithm, are denoted by N^+ and N^-. In the extended network shifted flux values and capacity intervals are used, so that the flow one works with is conserved at all nodes between s and s'.)

3.24. (*Feasible Flow Algorithm*). Devise an algorithm that is similar to the feasible distribution algorithm but solves a feasible flow problem with supply intervals $C(i) = [b^-(i), b^+(i)]$ *directly* (i.e., without requiring reformulation as a feasible circulation problem).

(*Hint.* The algorithm needs to be in two phases. The first phase concentrates on reducing violations of the form $y(i) < b^-(i)$ without worsening any violations of the form $y(i) > b^+(i)$. After $y(i) \geq b^-(i)$ has been achieved for all nodes, the second phase takes over with the goal of forcing $y(i) \leq b^+(i)$ for all nodes.)

3.25. (*Flow Rectification Algorithm*). Solve the feasible distribution problem in Figure 3.22 by means of the flow rectification algorithm (starting from any appropriate initial flow).

3.26. (*Flow Rectification Algorithm*). Determine by the flow rectification algorithm a cut Q in the network of Figure 3.22 such that $c^+(Q) < 0$.

3.27. (*Flow Rectification Algorithm*). Prove that the flow rectification algorithm must terminate sooner or later if arc discrimination is used in the painted network subroutine.

(*Hint.* Certain quantities of the form

$$\sum_{j \in A_0} \mathrm{dist}(x(j), C(j)) = \sum_{j \in A_0} \max\{0, x(j) - c^+(j), c^-(j) - x(j)\}$$

where A_0 is some subset of $A^+ \cup A^-$, have a special representation at critical stages.)

3.28. (*Node Capacities*). Generalize the max flow min cut theorem to the case where there are capacity intervals $[a^-(i), a^+(i)]$ associated with the nodes i as in Section 3L, as well as capacity intervals $[c^-(j), c^+(j)]$ associated with the arcs j. (Assume that $c^-(j) \geq 0$ for all $j \in A$. A notion of "generalized cut" involving both arcs and nodes will be needed.)

3.29. (*Dynamic Max Flow*). For the dynamic network of Figure 1.10 (see Section 1H), what is the maximum amount that can be sent from node a at time $t = 0$ so as to reach node d by $t = 5$, assuming that at most one

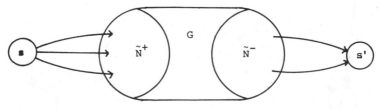

Figure 3.23

unit of flux can enter any arc of G during any unit time period? All flux values must be non-negative. There is no limit on storage, that is, on the amounts that can be held over at a node from one time period to the next. (After solving the problem in terms of the "space-time" representation of the network, translate the solution back into a "schedule" for how the flow is to be organized in terms of the original network.)

3.30.   (*Feasible Circulations*). Let G be a network with capacity intervals $[c^-(j), c^+(j)]$, and let $\bar{j}$ be any fixed arc. Let I be the set of all flux values $x(\bar{j})$ obtained as x ranges over all feasible circulations. Prove that I is a closed real interval, and explain how its endpoints could be calculated.

3.31.   (*Multipath Implementation*). Adapt the approach in Section 3J to the max flow algorithm.

(*Hint.*   Solving the max flow problem for given sets N^+ and N^- is like trying to solve a feasible distribution problem in which the supplies (in N^+) and demands (in N^-) are inexhaustible. Virtually the same procedure can be used, except that $b(i)$ is treated in the formulas as if it were $+\infty$ for $i \in N^+$ and $-\infty$ for $i \in N^-$, but of course 0 for all other nodes.)

3.32.   (*Multipath Implementation*). Sharpen the argument at the end of Section 3J to show that the number of "flux changes" in each application of the "grand improvement procedure" (Dinic-Karzanov approach) cannot exceed

$$\tfrac{1}{2}(|N| + 1)|N| + 2|A| - 1.$$

(*Hint.*   Only *unblocked* nodes in $N_1, \ldots, N_r$ are operated on in transmitting demands. In the first few rejection phases all the nodes that get blocked may be in N_0, but the number of such phases is limited, so that after a certain point each rejection phase must block at least one new node in $N_1 \cup \cdots \cup N_{r-1}$.)

3N.*   COMMENTS AND REFERENCES

More than any other event the discovery of the max flow min cut theorem by Ford and Fulkerson [1956] (independently also by Elias, Fernstein, and Shannon [1956]) marks the beginning of the era in network flow analysis when the emphasis shifted from purely electrical or mechanical problems toward applications in operations research and combinatorial optimization. The connection between the max flow min cut theorem and the duality theory for linear programming problems was recognized nearly from the start (Dantzig and Fulkerson [1956]).

In the common version of the theorem all capacity intervals are taken to be of the form $[0, c]$, and single nodes appear in place of N^+ and N^-. Such restrictions do not represent a significant shrinkage of generality, because of the tricks that can be used to reformulate a given problem. However, direct treatment of arbitrary intervals and multiple sources and sinks facilitates a number of applications. It helps to bring out analogies with dual results like the max tension min path theorem in Chapter 6, where it would not be sensible to focus on intervals $[0, c]$. It points the way to a direct and more natural, constructive proof of the feasible distribution theorem, a result that is essentially equivalent to the max flow min cut theorem, but tends to overshadow it in usefulness and therefore ought to be more in the center of the stage. Feasible distribution problems, as they arise in subroutines in many algorithms in network optimization (Chapters 7 and 9), definitely involve general capacity intervals.

The feasible distribution theorem was first published by Gale [1957] in the case where $c^-(j) \leq 0 \leq c^+(j)$ and by Hoffman [1960] in the case of arbitrary closed capacity intervals. Both results actually date from unpublished work of the authors in 1956, and both allow for divergence constraints more general than $y(i) = b(i)$, namely, $y(i) \geq b(i)$ in Gale's case and $b^-(i) \leq y(i) \leq b^+(i)$ in Hoffman's. Despite these differences the two results are equivalent to each other and to the feasible distribution theorem as we have stated it; see Section 3H.

Minty [1962] established the versions of the max flow min cut theorem and feasible circulation theorem that are valid for arbitrary capacity intervals, not necessarily even closed (see Exercises 3.4 and 3.13). A generalized statement of the max flow min cut theorem in the case of node capacities is furnished by Lawler [1976, p. 212].

In a network with capacity intervals, one could solve the max flow problem for *all* pairs of nodes s, s', and thereby generate a matrix of size $|N| \times |N|$. How can this be done efficiently, and just what characterizes the matrices that can be obtained in this manner? How could a network yielding a desired matrix be synthesized "cheaply"? For answers, see the book of T. C. Hu [1969] and its references.

The max flow algorithm (replete with labeling routine) was devised by Ford and Fulkerson [1957], who in this connection were the first to point out the theoretical importance of commensurability conditions and the like. Although data can be assumed in practice to be commensurable, the real implication of this termination peculiarity (and the counterexample of Ford and Fulkerson described in Exercise 3.12) is that the algorithm in its original, broad form could be erratic in speed and efficiency. In particular, one cannot give any bounds, in terms of the size of the network (rather than merely the capacities and initial flow values), on the number of steps that will be required. The same holds true even if the algorithm is supplemented by the rule of arc discrimination (a device suggested by work of Ellis Johnson [1966]), although that does at least guarantee finite termination for arbitrary data, and in a manner that is quite simple to explain and therefore useful in constructively proving the full max flow min cut theorem and related results.

In recent years much effort has gone into improving this state of affairs. J. Edmonds and R. M. Karp [1972] demonstrated, in effect, that if breadth-first search is used in the painted network subroutine, the algorithm surely will terminate (even without commensurability of data), and in fact the number of steps required is of order $0(|N||A|^2)$ (hence in the digraph case, no worse than $0(|N|^5)$). The effect of breadth-first search is to select in each iteration a flow-augmenting path with as few arcs as possible. The same thing could be accomplished of course by employing the multipath version of the painted network subroutine and any one of the paths it embodies.

The idea of getting more out of this approach by solving auxiliary painted network problems based on a multirouting is due to E. A. Dinic [1970], but his work, with details not available in English, remained unknown in the West for a long time. Dinic's method reduces the order of complexity in the digraph case to $0(|N|^4)$, and the refinement of A. V. Karzanov [1974] brings it down to $0(|N|^3)$. (The details of this approach are in the book of Adelson-Velsky, Dinic, and Karzanov [1975], in Russian. An exposition in English has been provided by S. Even [1976].) More recent work on implementing the Dinic-Karzanov approach has brought the order of complexity down as low as $0(|A||N|\log^2|N|)$, which is asymptotically better than $0(|N|^3)$ in networks that are even slightly sparse, in the sense of $A = 0(|N|^r)$ for some $r < 2$. See Galil and Naamad [1979] for such an implementation and additional references to the subject.

Any improvement in the max flow algorithm can be adapted to the feasible distribution algorithm, and, more easily, vice versa (see Exercises 3.23 and 3.31). We have taken this step in presenting the Dinic-Karzanov method in Section 3J, because the feasible distribution algorithm is more important to the procedures in later chapters. A change that has been made over the version that would be obtained by direct emulation of Dinic-Karzanov, is to have demands transmitted from N^- to N^+. This appears to take better advantage of the information generated by the multipath subroutine.

The flow rectification algorithm introduced in Section 3K can be regarded essentially as a simplification of a method of Minty [1962, p. 207].

For max flow problems in stochastic networks, consult J. R. Evans [1976]. Max flow problems in dynamic networks can often be reduced to min cost flow problems in the underlying "static" networks, as explained by Ford and Fulkerson [1962, pp. 142–150].

See Picard and Ratliff [1975] for a class of quadratic programming problems in 0–1 variables which can be reformulated as min cut problems for a certain network.

4

ANALYSIS OF FLOWS

Intuitively, the feasible distribution problem involves taking a quantity of material apportioned at various supply nodes and guiding it through the network to the demand nodes. In this sense the concept of a solution to the problem as a "flow" is somewhat unsatisfying, since only the flux in each arc is specified. It is desirable to know how the flow might be represented in more physical terms that respect the identity of the material in transit, for instance, as a superposition of "transport plans," each giving a path over which a certain amount of material is to move from a supply node to a demand node.

As a matter of fact the flows constructed by the algorithms of Chapter 3 do all appear as sums of an initial flow (perhaps identically zero) and amounts flowing along various paths. But these sums may entail cancellation in some arcs and thus may not furnish an appropriate realization from, say, an economic point of view.

This chapter explores the chief ways to realize or represent a flow, as well as properties of the special kinds of flows that are involved in such representations: integral, elementary, and extreme flows. Except for the integrality theorem in Section 4A, the results are not crucial to an understanding of the next two chapters and could be postponed on an initial reading. (They come into play in Chapter 7.)

4A. INTEGRAL FLOWS

A flow x is said to be *integral* if $x(j)$ is an integer for every arc j. Such flows are closely linked to combinatorial structures (e.g., arcs, paths, and special families of arcs and paths) particularly if the nonzero flux values are just ± 1. Their existence under certain restrictions is therefore of prime importance in applications of flow theory to a number of combinatorial problems. Several such applications will be described in Chapter 5.

The following result about integral flows can be drawn immediately from properties of the algorithms of Chapter 3.

Integrality Theorem for Flows

1. *If a max flow problem has all integral capacities $c^+(j), c^-(j)$, and has a solution at all, then it has at least one solution that is an integral flow. In fact any solution calculated by the max flow algorithm, starting from an integral flow, will be integral.*

2. *If a feasible distribution problem has integral supplies $b(i)$ and integral capacities $c^+(i), c^-(i)$, and if it has a solution at all, then it has at least one solution which is an integral flow. In fact any solution calculated by the feasible distribution algorithm or flow rectification algorithm, starting from an integral flow, will be integral.*

(*Note:* $+\infty$ *and* $-\infty$ *are considered "integral" in this context.*)

Proof. This is mainly a consequence of the commensurability arguments used to prove convergence of the algorithms in question. In each case it is known that if the initial data are all whole multiples of some $\delta > 0$, then the same will be true of the flux values on termination—and the algorithm will indeed terminate sooner or later with a solution to the problem. Integral flows correspond to $\delta = 1$. To conclude that the feasible distribution problem has an integral solution, it need only be observed further that if $c^+(j)$ and $c^-(j)$ are integral for all arcs j, then there does exist an integral flow with which the feasible distribution algorithm can be initiated. For example, the flow x having each flux $x(j)$ taken to be the element of $[c^-(j), c^+(j)]$ nearest to 0 fits the requirements.

For the max flow problem the task of finding a flow that is feasible with respect to capacities and conserved at all nodes outside N^+ and N^- is, as observed in Section 3H, a special case of the general feasible flow problem, which in turn is a feasible circulation problem in the augmented network. If $c^+(j)$ and $c^-(j)$ are integral, the argument already given shows there is an integral solution, if there is any solution at all. Thus the max flow algorithm can be initiated with an integral flow, and the proof of the theorem is complete.

It should be noted that when an integral solution exists, then unless it is the unique solution to the problem in question (as it well might not be), there will exist other solutions that are not integral. For if x_0 and x_1 are both integral solutions to the max flow problem, say, then the flow x defined by

$$x(j) = (1 - \lambda)x_0(j) + \lambda x_1(j) \quad \text{for all arcs } j,$$

where $\lambda \in (0, 1)$ will again satisfy the capacity constraints and will yield the same flux from N^+ to N^- as x_0 and x_1; therefore x will also be a solution to the problem, and generally a nonintegral one.

Application to Round-off

The integrality theorem will be seen to have many interesting consequences, and the following fact is a fine illustration. *For any circulation $\tilde{x}$ in a network G, there is an integral circulation $\bar{x}$ in G such that $|\tilde{x}(j) - \bar{x}(j)| < 1$ for all $j \in A$.*

To verify this in a constructive manner, consider the feasible circulation problem in which the capacity intervals $[c^-(j), c^+(j)]$ are defined by

$$c^+(j) = [\text{lowest integer} \geq \tilde{x}(j)],$$

$$c^-(j) = [\text{highest integer} \leq \tilde{x}(j)].$$

Obviously, $c^-(j) = \tilde{x}(j) = c^+(j)$ if $\tilde{x}(j)$ is an integer, but otherwise $c^-(j) < \tilde{x}(j) < c^+(j)$, $c^+(j) - c^-(j) = 1$. The feasible circulation problem in question has at least one solution, namely, $\tilde{x}$, and since $c^+(j)$ and $c^-(j)$ are integral, we may conclude from the integrality theorem (with $b(i) = 0$) that there is at least one integral solution $\bar{x}$. Such an $\bar{x}$ clearly meets the requirements and can be calculated by the feasible distribution algorithm.

EXAMPLE 1

For a given $m \times n$ real matrix $\tilde{X} = [\tilde{x}_{kl}]$, does there exist an $m \times n$ *integer* matrix $\bar{X} = [\bar{x}_{kl}]$ such that

$$|\bar{x}_{kl} - \tilde{x}_{kl}| < 1 \quad \text{for all } k, l,$$

$$\left| \sum_{l=1}^{n} \tilde{x}_{kl} - \sum_{l=1}^{n} \bar{x}_{kl} \right| < 1 \quad \text{for all } k,$$

$$\left| \sum_{k=1}^{m} \tilde{x}_{kl} - \sum_{k=1}^{m} \bar{x}_{kl} \right| < 1 \quad \text{for all } l,$$

$$\left| \sum_{k=1}^{m} \sum_{l=1}^{n} \tilde{x}_{kl} - \sum_{k=1}^{m} \sum_{l=1}^{n} \bar{x}_{kl} \right| < 1?$$

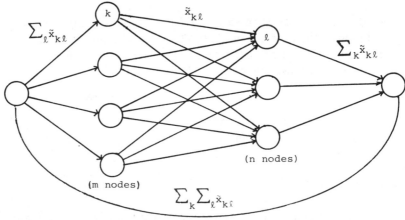

Figure 4.1

Yes, always. Simply associate $\tilde{X}$ with the circulation shown in Figure 4.1, and apply the round-off result just explained.

4B.   CONFORMAL REALIZATION OF FLOWS

An *elementary* flow is a flow of the form

$$x(j) = \alpha e_P(j) = \begin{cases} \alpha & \text{if } j \in P^+, \\ -\alpha & \text{if } j \in P^-, \\ 0 & \text{otherwise,} \end{cases}$$

where P is an elementary path and $\alpha > 0$. The question to be studied now is how a general flow can be expressed as a sum of elementary flows, but in this there is interest in restricting the class of paths to be admitted.

A path P is said to *conform* to a flow x if two conditions are met:

1. $x(j) > 0$ for all $j \in P^-$, and $x(j) < 0$ for all $j \in P^-$.
2. Either P is a circuit, or the initial and terminal nodes of P are a source and sink of x, respectively.

Condition 1 means that, as one moves along the path, one is always going "downstream" relative to the flow. Equivalently, the path is compatible with the painting where the arc j is as follows:

$$\begin{array}{ll} \text{white} & \text{if } x(j) > 0 \\ \text{black} & \text{if } x(j) < 0 \\ \text{red} & \text{if } x(j) = 0 \end{array}$$

This assignment of colors will be referred to as the *conformal painting* relative to x.

A flow having no circuit that conforms to it is called an *anticirculation*. In a sense this property is the "opposite" of being a circulation.

Proposition 1.   *The only flow that is both a circulation and an anticirculation is the zero flow.*

Proof.   Let x be a nonzero anticirculation, and give the network the corresponding conformal painting. No circuits are compatible with this painting, since x is an anticirculation. But there is at least one black or white arc $\bar{j}$, since x is nonzero. By Minty's lemma, there must be a cut Q containing $\bar{j}$ which is compatible with the painting and thus has

$$x(j) \leq 0 \quad \text{for all } j \in Q^+,$$
$$x(j) \geq 0 \quad \text{for all } j \in Q^-.$$

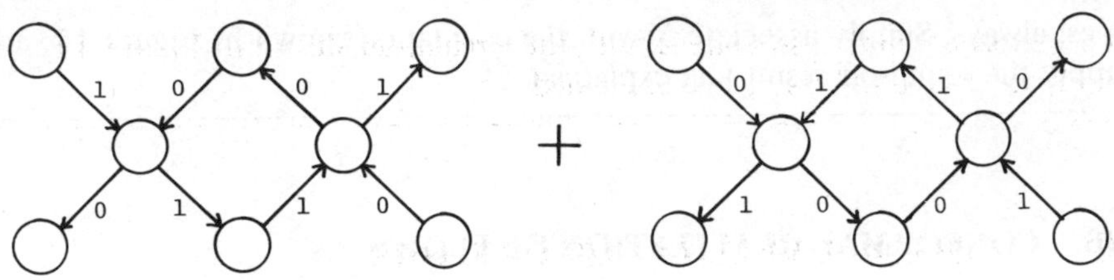

Figure 4.2

Then

$$[\text{flux of } x \text{ across } Q] = \sum_{j \in Q^+} x(j) - \sum_{j \in Q^-} x(j) \leq -|x(\bar{j})| < 0,$$

so x cannot be a circulation (see Exercise 2.10).

In contrast to circulations, which form a linear subspace of the space of flows, the sum of two anticirculations need not be an anticirculation. A counterexample is displayed in Figure 4.2.

The main result is the following.

Conformal Realization Theorem.   *Let x be any flow other than the zero flow. Then there exist elementary paths $P_1, \ldots, P_r$ that conform to x and positive numbers $\alpha_1, \ldots, \alpha_r$, which can be chosen integral if x is integral, such that*

$$x = \alpha_1 e_{P_1} + \cdots + \alpha_r e_{P_r}.$$

Moreover the paths (not necessarily unique) can be chosen so that if the circuits among them are indexed $P_1, \ldots, P_q$, then $\alpha_1 e_{P_1} + \cdots + \alpha_q e_{P_q}$ is a circulation whereas $\alpha_{q+1} e_{P_{q+1}} + \cdots + \alpha_r e_{P_r}$ is an anticirculation. Thus x is in particular represented as the sum of a circulation and an anticirculation.

The proof of this theorem will be represented in the next two sections in terms of a constructive procedure and its justification.

4C.*   REALIZATION ALGORITHM

In constructing a decomposition of the kind just described, it is helpful to use still another notion, that of the *support* of a flow x. This is the signed set formed by the arcs j such that $x(j) \neq 0$; the positive part consists of the arcs with $x(j) > 0$, and the negative part of those with $x(j) < 0$. (Recall that a *signed set* is a set S plus a partition of S into two subsets S^+ and S^-.) Two signed sets are said to be in *conformity* with each other if no arc belongs to the positive part of one set but to the negative part of the other.

Starting with an arbitrary flow $x \neq 0$, a sequence of elementary paths P_k that conform to x is to be constructed, along with numbers $\alpha_k > 0$ such that the flow

$$x_{k+1} = x - \left[\alpha_1 e_{P_1} + \cdots + \alpha_k e_{P_k} \right]$$

has support *strictly* included in the support of x_k and in conformity with it. (In this notation, $x_1 = x$.) Since the supports are strictly decreasing, the stage must come, say, at iteration r, when there is no support left at all, namely,

$$0 = x_{r+1} = x - \left[\alpha_1 e_{P_1} + \cdots + \alpha_r e_{P_r} \right].$$

Due to the way the paths are to be chosen, this will provide the desired representation of x.

In the first phase of the procedure only circuits are considered. The corresponding elementary flows $\alpha_k e_{P_k}$ are therefore circulations. This phase continues until an iteration is reached, call it q, when there does not exist a circuit that conforms to x_{q+1} (as must happen sooner or later, at worst when the zero flow is attained). Then x_{q+1} turns out to be an anticirculation, and the equation

$$x = \left[\alpha_1 e_{P_1} + \cdots + \alpha_q e_{P_q} \right] + x_{q+1}$$

therefore represents x as the sum of a circulation and an anticirculation. At that point the procedure switches over to dealing only with paths that are not circuits. (If x is a circulation, then $x_{q+1} = 0$ by Proposition 1 in Section 4B, so no switchover is necessary, and the algorithm simply terminates.)

Path-Calculating Procedure: First Phase

In the kth iteration, the flow x_k is on hand. Its support is in conformity with that of x and included in it. There is also an arc set $M \subset A$ (initially, $M = \varnothing$), consisting, as will be seen, of arcs ascertained not to belong to any circuit that conforms to x_k.

Give the network the conformal painting for x_k (see Section 4B), but with the modification that all arcs in M, whatever color they would have been, are to be painted red. If there are no black or white arcs, proceed to the second phase (outlined in the next subsection). Otherwise choose any arc $\bar{j}$ that is white or black, and apply Minty's lemma (and the corresponding algorithm).

If this yields a compatible cut containing $\bar{j}$, add all the arcs of the cut to M and try again (with a correspondingly altered painting and a different choice of $\bar{j}$). If it yields a compatible circuit containing $\bar{j}$, this is an elementary circuit that conforms to x_k (and hence with x). Denote it by P_k and calculate

$$\alpha_k = \text{min of } |x_k(j)| \quad \text{for arcs } j \text{ in } P_k.$$

Then $\alpha_k > 0$, and the flow $x_{k+1} = x_k - \alpha_k e_{P_k}$ has support strictly included in, and in conformity with, that of x_k (and hence that of x). Repeat the step with x_{k+1} (and the same M).

Second Phase

This begins when there are no black or white arcs as defined in the first phase. At that point (and in all later iterations) the flow x_k is an anticirculation (see Section 4D); the set M is dropped.

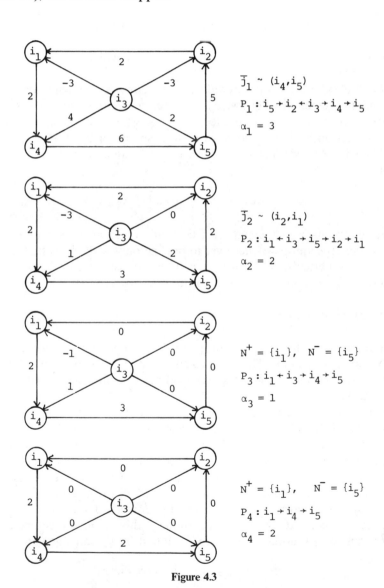

$$\bar{J}_1 \sim (i_4, i_5)$$
$$P_1 : i_5 \to i_2 \to i_3 \to i_4 \to i_5$$
$$\alpha_1 = 3$$

$$\bar{J}_2 \sim (i_2, i_1)$$
$$P_2 : i_1 \to i_3 \to i_5 \to i_2 \to i_1$$
$$\alpha_2 = 2$$

$$N^+ = \{i_1\}, \quad N^- = \{i_5\}$$
$$P_3 : i_1 \to i_3 \to i_4 \to i_5$$
$$\alpha_3 = 1$$

$$N^+ = \{i_1\}, \quad N^- = \{i_5\}$$
$$P_4 : i_1 \to i_4 \to i_5$$
$$\alpha_4 = 2$$

Figure 4.3

The arcs are simply given the conformal painting x_k. Let N^+ and N^-, respectively, denote the set of sources and set of sinks of x_k. If either N^+ or N^- is empty, then x_k must be the zero flow (see Section 4D), and the algorithm therefore terminates; the desired representation has been achieved. If not, apply the painted network algorithm. This necessarily yields an elementary path that conforms to x_k (and hence with x). Denote it by P_k and form α_k and x_{k+1} as before. Again x_{k+1} has support strictly included in, and in conformity with, that of x_k (and hence that of x). Repeat the step with x_{k+1}.

EXAMPLE 2

Figure 4.3 depicts the calculation of a conformal realization for a certain integral flow x that is not a circulation:

$$x = 3e_{P_1} + 2e_{P_2} + e_{P_3} + 2e_{P_4},$$

where P_1 and P_2 are circuits (i.e., $q = 2$).

4D.*   JUSTIFICATION OF THE ALGORITHM

The reason the supports of the flows x_k are strictly decreasing is that at each iteration x_{k+1} vanishes in the arcs j for which $|x_k(j)| = \alpha_k$, as well as in all arcs where x_k vanishes. Furthermore the choice of the painting ensures that $x_k(j) \geq \alpha_k$ for all $j \in P_k^+$ and $x_k(j) \leq \alpha_k$ for all $j \in P_k^-$, and hence x_{k+1} can only be positive in some of the arcs where x_k is positive, and negative in some where x_k is negative. Thus the support of x_{k+1} is again in conformity with that of x_k (and hence with that of x).

The assertion about the arcs in the set M in the first phase of the algorithm is established by an inductive argument. Suppose that at some stage of iteration k in the first phase all the arcs of M do have the status of not belonging to any circuit that conforms to x_k (as is true in the vacuous sense at the start of the algorithm, where $M = \varnothing$). Suppose a cut Q compatible with the painting is obtained. It must be shown that none of the arcs in Q belongs to any elementary circuit that conforms to x_k, so that when Q is added to M, the desired property of M is maintained. (The property is inherited by x_{k+1} in the next iteration, since any circuit that conforms to x_{k+1} also conforms to x_k.) The compatibility of Q with the given painting yields

$$x_k(j) < 0 \quad \text{for all } j \in Q^+ \backslash M,$$

$$x_k(j) > 0 \quad \text{for all } j \in Q^- \backslash M.$$

If P is an elementary circuit that conforms to x_k, then by our assumption on M

at this stage, P does not meet M. All the arcs in P^+ have $x_k(j) > 0$, whereas those in P^- have $x_k(j) < 0$, so any arc common to P and Q would have to be in $P^+ \cap Q^-$ or $P^- \cap Q^+$. Therefore

$$e_P \cdot e_Q = \sum_{j \in P^+ \cap Q^-} e_P(j) e_Q(j) + \sum_{j \in P^- \cap Q^+} e_P(j) e_Q(j)$$

$$= -|P^+ \cap Q^-| - |P^- \cap Q^+|.$$

But $e_P \cdot e_Q = 0$ (since e_P is a circulation and e_Q a differential). Hence $P^+ \cap Q^- = \varnothing$ and $P^- \cap Q^+ = \varnothing$, and it is true that none of the arcs of Q belongs to P.

Since none of the arcs of M at the end of the first phase (when *all* arcs have become red) belongs to any circuit that conforms to the flow x_{q+1} at that stage, x_{q+1} is indeed an anticirculation. (Any circuit that conforms to x_{q+1} would have to be compatible with the current painting, but this is impossible if there are only red arcs.) Thus the switchover to the second phase at this stage is justified, and all subsequent flows x_k are anticirculations.

In the second phase, if at any time N^+ or N^- is empty, then both must be empty (by the total divergence principle), and hence x_k is a circulation. But x_k is also an anticirculation, and hence by Proposition 1 in Section 4B one has $x_k = 0$, as claimed.

Finally, observe that if x is integral, then so are α_k and x_k at every iteration; this is evident from the formulas.

EXAMPLE 3

A flow can often be conformally realized in more than one way. Thus the flow x in Example 2 (Figure 4.3) can also be expressed as

$$x = 2e_{P_1} + 2e_{P_2} + e_{P_3} + 3e_{P_4},$$

where the paths are

$$P_1: i_1 \rightarrow i_4 \rightarrow i_5 \rightarrow i_2 \rightarrow i_1 \quad \text{(circuit)},$$

$$P_2: i_3 \rightarrow i_5 \rightarrow i_2 \leftarrow i_3 \quad \text{(circuit)},$$

$$P_3: i_4 \rightarrow i_5 \rightarrow i_2 \leftarrow i_3 \rightarrow i_4 \quad \text{(circuit)}$$

$$P_4: i_1 \leftarrow i_3 \rightarrow i_4 \rightarrow i_5 \quad \text{(noncircuit)}.$$

This also shows, incidentally, that even the decomposition into a circulation plus an anticirculation is not always unique.

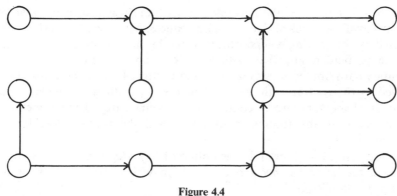

Figure 4.4

4E.   TREES

A number of questions about algebraic properties of flows and the space of circulations are tied to the theory of "trees" and "forests."

A network is called a *tree* if it is connected and contains no elementary circuits but has at least one arc; see Figure 4.4. The last provision simply excludes from treehood networks that consists of a single isolated node. An equivalent definition of a tree is that it is a network with at least two nodes, such that for any two distinct nodes i and i' there is a *unique* elementary path $P: i \rightarrow i'$ (Exercise 4.21). More generally, a set of arcs in a network G is said to *form a tree* if the subnetwork of G consisting of these arcs and the nodes incident to them is a tree.

Test of Whether a Network is a Tree

There is a simple way to determine algorithmically whether a given network (with $|N| \leq 2$) is a tree, and if so, to represent in convenient form the unique elementary paths joining its nodes. Select any node s, and apply the painted network algorithm with all arcs painted green to get a maximal routing θ with base $N^+ = \{s\}$. *If this is a routing of all of N* (i.e., $\theta(i)$ is defined for every $i \neq s$) *and uses every arc* (i.e., every arc j is designated as $\theta(i)$ for some node i, necessarily unique according to the definition of a "routing"), *then the network is a tree, but otherwise not.*

Indeed if the routing has these properties, then the network is connected, since there is a path from s to every other node. It cannot contain any elementary circuit, for if P were an elementary circuit, then every arc j of P would have to be $\theta(i)$ for one of the two nodes of P adjacent to it; since an elementary circuit has an equal number of nodes and arcs, this would imply for every node i of P that $\theta(i)$ is defined and is an arc of P, so that the routing could not possibly provide a path joining any node of P with s. Thus the network is in this case a tree.

On the other hand, if the network is a tree and this procedure is applied, every node will be reached due to connectedness. Moreover the nodes and the arcs used by the routing will constitute a tree by the argument just given, and hence in particular, any two distinct nodes i and i' can be joined by an elementary path limited to arcs used by the routing. If there were also an arc j not used by the routing with $j \sim (i, i')$ or $j \sim (i', i)$, then the network would contain an elementary circuit, contrary to its being a tree. Therefore every arc must be used by the routing in question, and the routing does have the properties asserted.

The test just given shows that the notion of a *routing with base s* is virtually equivalent to that of a *rooted tree*, a tree in which some node s has been designated as the "root."

A specific description of the unique elementary paths in a given tree provided by a routing θ of the node set of the tree with base s is the following. In the case of s and another node i, the two paths in question must be the θ-path from s to i and its reverse from i to s. In the case of two different nodes i and i' distinct from s, the unique elementary path from i to i' must be the one obtained by tracing the routing from i back to where it joins the θ-path from s to i', and then proceeding along the latter to i'.

Number of Arcs in a Tree

Another characterization emerges from the test just described, namely, that *a connected network is a tree if and only if*

$$[\textit{number of arcs}] = [\textit{number of nodes}] - 1 > 0.$$

For in a connected network a complete routing based on any node s assigns a different arc to each node $i \neq s$ and thus uses one less arc than the total number of nodes.

4F. FORESTS AND SPANNING TREES

An arc set $F \subset A$ is said to form a *forest* in the network G if every component of the subnetwork comprised of the arcs of F and the nodes incident to them is a tree. In other words, F is a forest if and only if no elementary circuits are included in F. See Figure 4.5 where the darkened arcs are the ones in F.

The test for a tree, using the painted network algorithm as prescribed in Section 4E, can readily be extended to a test of whether an arc set F forms a forest. As a matter of fact *F forms a forest in G if and only if it is the set of arcs used by some routing θ* (not necessarily of all of N, with a base $N^+ \subset N$) (Exercise 4.24). If the routing is based at a single node, the forest consists of a solitary tree.

The connection between forests and flows is the following. *An arc set F forms a forest if and only if there is no nonzero circulation whose support is*

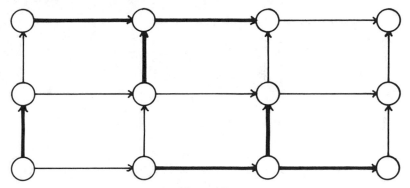

Figure 4.5

included in F, that is, no circulation x such that $x(j) = 0$ for all $j \notin F$ but $x(j) \neq 0$ for at least one $j \in F$ (Exercise 4.26).

It is worthwhile to explore what the latter condition implies in terms of the incidence matrix (function) E for the network. Circulations x are characterized as solutions to the system

$$\sum_{j \in A} e(i, j)x(j) = 0 \quad \text{for all } i \in N.$$

Let $e^j(i) = e(i, j)$; thus e^j denotes the "column" of E corresponding to the arc j. The system then becomes

$$\sum_{j \in A} x(j)e^j = 0.$$

In other words, a nonzero circulation can be interpreted as a choice of coefficients $x(j)$ that yields a relation of linear dependence among the corresponding columns of E. It follows that *an arc set F forms a forest in G if and only if the corresponding columns of the incidence matrix of G are linearly independent.*

It is natural to move from this fact to the study of sets of columns that form a *basis* for the column space of E (i.e., for the linear subspace of R^N generated by all the e^j, $j \in A$), since, among other things, this will yield a description of the rank of E.

A *maximal forest* in G is defined to be a forest that is not strictly contained in any other forest of G. (Any forest can be extended to a maximal forest.) A *spanning tree* for G is a tree meeting every node of G (see Figure 4.6).

Spanning Theorem.   *For any network G, the following properties of an arc set $F \subset A$ are equivalent*:

1.   *F forms a maximal forest for G.*
2.   *F forms a spanning tree for each component of G that is not just an isolated node.*

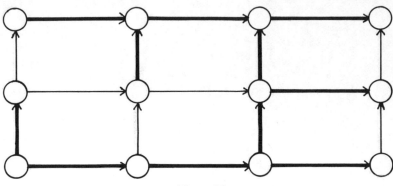

Figure 4.6

3. *F forms a forest such that $|F| = |N| - p$, where p is the number of components of G.*

4. *The columns e^j of the incidence matrix E corresponding to the arcs $j \in F$ form a basis for the column space of E.*

Proof. The equivalence of Properties 1 and 4 is clear from the preceding observations about linear independence.

To see that Property 1 implies Property 2, note first from the maximality of F that every component of G that is not just an isolated node must include an arc of F and hence one of the component trees of F. Let T be such a tree, and let S be its node set. The component of G that includes T cannot have nodes besides those in S. If it did, the cut $Q = [S, N \setminus S\}$ would be nonempty, and any arc in Q could be added to F without destroying the "forest" property. (This is true because any elementary circuit meeting Q at all must have at least two arcs in common with Q; see Exercise 2.8.) Thus each component tree of the forest corresponds to a component of G having the same node set as the tree, and every component of G that is not just an isolated node corresponds to a tree in this way. This is Property 2. The implication from Property 2 to Property 3 is immediate, in view of the fact, already noted in this section, that the number of arcs in a tree is always one less than the number of nodes. The latter generalizes to the rule that for any component G' of G,

$$\begin{bmatrix} \text{number of arcs} \\ \text{in a forest} \\ \text{that belong} \\ \text{to } G' \end{bmatrix} = \begin{bmatrix} \text{number of nodes} \\ \text{incident to the} \\ \text{forest that belong} \\ \text{to } G' \end{bmatrix} - \begin{bmatrix} \text{number of trees} \\ \text{of the forest} \\ \text{that are contained} \\ \text{in } G' \end{bmatrix}$$

$$\leq \begin{bmatrix} \text{number of nodes} \\ \text{in } G' \end{bmatrix} - 1.$$

Therefore any arc set F forming a forest in G satisfies

$$|F| \leq |N| - [\text{number of components of } G].$$

If equality holds, F must be maximal. Thus Property 3 implies Property 1.

Corollary.   *In a connected network (with at least one arc) a maximal forest is the same thing as a spanning tree.*

Those having some familiarity with linear programming will see the special significance of Property 4. In linear programming problems involving the constraint equation $Ex = b$, which may be written

$$\sum_{j \in A} x(j)e^j = b,$$

it is of interest to know which sets $F \subset A$ are such that the elements e^j for $j \in F$ form a basis for the column space (assumed to contain b). Each of these determines a unique x satisfying the equation and having $x(j) = 0$ for all $j \notin F$. The "simplex method" of computation depends on generating a sequence of such sets F. Geometrically, these sets can be identified with the spanning trees for the network, assuming that the network is connected.

4G.   TUCKER REPRESENTATIONS OF THE CIRCULATION SPACE

Spanning trees and forests are central in any discussion of how the circulation space $\mathscr{C}$ and differential space $\mathscr{D}$ of a network may be represented.

As already noted, $\mathscr{C}$ consists of the elements x of R^A such that

$$\sum_{j \in A} e(i, j)x(j) = 0 \quad \text{for all } i \in N. \tag{1}$$

This system of equations can be solved for some of the variables $x(j)$ in terms of the others and written *equivalently*, for various subsets $F \subset A$, as

$$\sum_{k \in F'} a(j, k)x(k) = x(j) \quad \text{for all } j \in F, \text{ where } F' = A \setminus F. \tag{2}$$

A system of the latter form is a *Tucker representation* of the subspace $\mathscr{C}$. If flux values for the arcs in F' are chosen arbitrarily, then (2) determines the unique flux values for the arcs in F which make the flow a circulation.

EXAMPLE 4

For the network in Figure 4.7, system (1) can be written in tableau notation as

$x(1)$	$x(2)$	$x(3)$	$x(4)$	$x(5)$	$x(6)$	
1	−1	−1	0	0	0	= 0
0	0	1	1	1	0	= 0
−1	0	0	−1	0	1	= 0
0	1	0	0	−1	−1	= 0

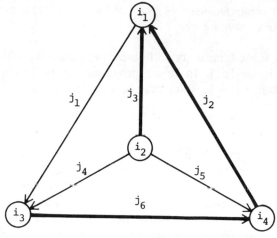

Figure 4.7

The set $F = \{j_2, j_3, j_6\}$ yields a Tucker representation (2) of $\mathscr{C}$, namely,

$x(1)$	$x(4)$	$x(5)$	
1	1	1	$= x(2)$
0	-1	-1	$= x(3)$
1	1	0	$= x(6)$

Proposition 2.   *An arc set $F \subset A$ corresponds to a Tucker representation of the circulation space $\mathscr{C}$ if and only if F forms a maximal forest for the network G.*

Proof.   Write the defining system (1) for $\mathscr{C}$ as

$$\sum_{j \in F} x(j)e^j = - \sum_{k \in F'} x(k)e^k. \tag{3}$$

The set F corresponds to a Tucker representation if and only if for every choice of numbers $x(k)$, $k \in F'$, there exist unique numbers $x(j)$, $j \in F$, such that (3) holds. This property will certainly hold if the vectors e^j, $j \in F$, form a basis for the column space of E. On the other hand, taking $x(k) = 0$ for all $k \in F'$, one sees from the uniqueness of the associated numbers $x(j)$ in (3) that the vectors e^j, $j \in F$, are linearly independent. Moreover (3) implies that every vector expressible as a linear combination of all the vectors e^j, $j \in F$, and e^k, $k \in F'$, is actually expressible as a linear combination of just e^j, $j \in F$. Thus $\{e^j | j \in F\}$ is a basis for the column space.

Dual Tucker Representations

Each Tucker representation (2) of $\mathscr{C}$ can immediately be translated into a Tucker representation of the differential space $\mathscr{D}$, namely,

$$- \sum_{j \in F} v(j)a(j, k) = v(k) \quad \text{for all } k \in F'. \tag{4}$$

This follows from the fact that $\mathscr{D} = \mathscr{C}^{\perp}$ (see Section 1H), that is, $v \in \mathscr{D}$ if and only if

$$0 = \sum_{j \in A} v(j)x(j) \quad \text{for all } x \in \mathscr{C}.$$

In terms of a Tucker representation (2), this property characterizes $\mathscr{D}$ as the set of all $v \in R^A$ satisfying

$$0 = \sum_{j \in F} v(j)\left[\sum_{k \in F'} a(j, k)x(k) \right] + \sum_{k \in F'} v(k)x(k)$$

$$= \sum_{k \in F'} x(k)\left[v(k) + \sum_{j \in F} v(j)a(j, k) \right]$$

for all choices of $x(k)$, $k \in F'$. This is obviously equivalent to (4). Conversely, each representation (4) of $\mathscr{D}$ yields a representation (2) of $\mathscr{C}$.

For instance, v is a differential for the network of Example 4 (Figure 4.7) if and only if it satisfies the system

$-v(2)$	1	1	1
$-v(3)$	0	-1	-1
$-v(6)$	1	1	0
	$= v(1)$	$= v(4)$	$= v(5)$

The Tucker representations of $\mathscr{C}$ and $\mathscr{D}$ therefore occur in dual pairs in one-to-one correspondence with the maximal forests F of G (or spanning trees, if G is connected). The two representations can be displayed jointly as in Figure 4.8.

It is interesting and useful to observe that the tableau defining the relationships between flows, divergences, potentials, and differentials in G (see Figure 4.9) can also be interpreted in terms of Tucker representations, namely, for the circulation and differential spaces of the usual augmented network $\bar{G}$ with "distribution arcs" j_i. This is seen by expressing the tableau in equivalent notation as in Figure 4.10 and noting that the distribution arcs j_i together constitute a particular spanning tree for $\bar{G}$. Tucker representations thus have a wider applicability than may have been apparent.

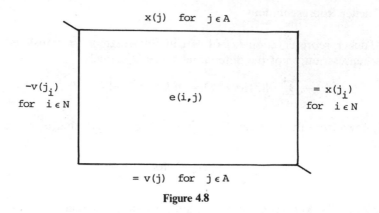

x(j) for j ∈ A

−v(jᵢ)
for i ∈ N

e(i,j)

= x(jᵢ)
for i ∈ N

= v(j) for j ∈ A

Figure 4.8

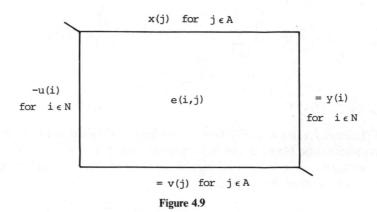

x(j) for j ∈ A

−u(i)
for i ∈ N

e(i,j)

= y(i)
for i ∈ N

= v(j) for j ∈ A

Figure 4.9

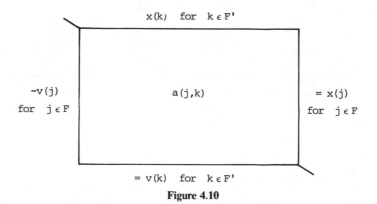

x(k) for k ∈ F'

−v(j)
for j ∈ F

a(j,k)

= x(j)
for j ∈ F

= v(k) for k ∈ F'

Figure 4.10

116

4H.*   BASIS THEOREM

The next goal is to relate the coefficients $a(j, k)$ in a Tucker representation to the combinatorial structure of the network. This involves associating with each maximal forest a special set of elementary circuits whose incidence vectors form a basis for $\mathscr{C}$, as well as a set of elementary cuts whose incidence vectors form a basis for $\mathscr{D}$.

Basis Theorem.   *Let $F \subset A$ be a maximal forest in G, and let the numbers $a(j, k)$ for $j \in F, k \in F' = A \setminus F$, be the coefficients in the corresponding Tucker representations of the circulation space $\mathscr{C}$ and differential space $\mathscr{D}$:*

1. *For each arc $k \in F'$ there is a unique elementary circuit P_k that contains k in its positive part and otherwise uses only arcs of F. The incidence vectors e_{P_k} form a basis for $\mathscr{C}$ and are given by*

$$e_{P_k}(j) = \begin{cases} a(j, k) & \text{for all } j \in F, \\ 0 & \text{for all } j \in F' \text{ except } j = k, \\ 1 & \text{for } j = k. \end{cases}$$

2. *For each arc $j \in F$ there is a unique elementary cut Q_j that contains j in its positive part and otherwise uses only arcs of F'. The incidence vectors e_{Q_j} form a basis for $\mathscr{D}$ and are given by*

$$e_{Q_j}(k) = \begin{cases} -a(j, k) & \text{for all } k \in F', \\ 0 & \text{for all } k \in F \text{ except } k = j, \\ 1 & \text{for } k = j. \end{cases}$$

Proof.   Only the argument for Part 1 will be furnished; the proof of Part 2 is left as Exercise 4.32. Let $k \in F', k \sim (i, i')$. Since F is a maximal forest, it includes a spanning tree T for the component of G containing i and i' (see the spanning theorem in Section 4F). There is a unique elementary path in T from i' to i, and this, together with the arc k, forms the circuit P_k. The incidence vector e_{P_k} may be regarded as a circulation having flux $+1$ in k but 0 in all other arcs of F'. Hence for any choice of coefficients λ_k the flow

$$x = \sum_{k \in F'} \lambda_k e_{P_k}$$

is a circulation satisfying $x(k) = \lambda_k$ for all $k \in F'$ and

$$x(j) = \sum_{k \in F'} \lambda_k e_{P_k}(j) \quad \text{for all } j \in F.$$

On the other hand, it is known from the Tucker representation that there is a *unique* circulation x satisfying $x(k) = \lambda_k$ for all $k \in F'$, namely, the one whose other flux values are given by

$$x(j) = \sum_{k \in F'} \lambda_k a(j, k) \quad \text{for all } j \in F.$$

Thus each element of $\mathcal{C}$ can be represented uniquely as a linear combination of the vectors e_{P_k}, and one has

$$\sum_{k \in F'} \lambda_k a(j, k) = \sum_{k \in F'} \lambda_k e_{P_k}(j)$$

for all $j \in F$ and all $\lambda_k \in R$. This shows that the vectors e_{P_k} form a basis for $\mathcal{C}$, and $e_{P_k}(j) = a(j, k)$. The proof is now finished.

The circuits P_k and cuts Q_j in the basis theorem are called the *fundamental* circuits and cuts associated with F.

Corollary. *In any Tucker representation of the spaces $\mathcal{C}$ and $\mathcal{D}$ all the coeffi cients $a(j, k)$ are $+1$, -1, or 0.*

The property in the corollary is important, since it makes the storage and manipulation of Tucker representations on a computer much easier than would be the case with similar representations of general subspaces not corresponding to circulations or tensions in a network.

The basis theorem indicates *how to determine the coefficients in a Tucker representation combinatorially* rather than by solving equations as at the beginning of Section 4G. For simplicity, suppose that the network is connected, so that the maximal forest F is a spanning tree, and let this be represented by a routing θ. As seen in Section 4E, the routing holds all the information needed to find the unique elementary path from any node of the network to any other node, using only arcs in the tree. It is therefore easy to generate from it a description of the circuit formed when any arc in F' is adjoined to F. Likewise it is easy to determine the two "connected" node sets S and $N \setminus S$ into which the tree splits if any of its arcs is deleted. Thus the bases of elementary circuits and cuts corresponding to F can be recovered from θ, and with them the coefficients $a(j, k)$.

EXAMPLE 5

In the network of Figure 4.11 the darkened arcs form a spanning tree F. The elementary circuits P_k corresponding to the arcs $k \in F' = A \setminus F$, as in the basis theorem, are summarized as follows:

arc $k \in F'$	P_k^+	P_k^-
j_3	j_2, j_3	j_1
j_4	j_1, j_4	j_2, j_6, j_8
j_5	j_1, j_5	j_2, j_6
j_9	j_6, j_9	j_7
j_{11}	j_6, j_8, j_{11}	j_7

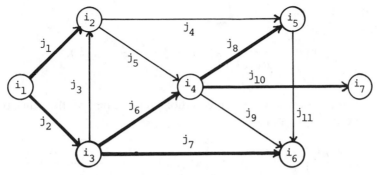

Figure 4.11

The incidences in this table yield the following Tucker representations for circulations x and differentials v:

	$x(3)$	$x(4)$	$x(5)$	$x(9)$	$x(11)$	
$-v(1)$	-1	1	1	0	0	$= x(1)$
$-v(2)$	1	-1	-1	0	0	$= x(2)$
$-v(6)$	0	-1	-1	1	1	$= x(6)$
$-v(7)$	0	0	0	-1	-1	$= x(7)$
$-v(8)$	0	-1	0	0	1	$= x(8)$
$-v(10)$	0	0	0	0	0	$= x(10)$
	$= v(3)$	$= v(4)$	$= v(5)$	$= v(9)$	$= v(11)$	

4I.*   PIVOTING

Sometimes it is necessary to pass from one Tucker representation to another. This can always be accomplished by a sequence of elementary transformations called pivoting steps. The simplex method in linear programming is based on pivoting, so the idea is obviously one of great importance for computations and will in fact be applied in Chapter 7 to the optimization of flows and potentials. The purpose of this section is to explain what pivoting is and how it can be expressed in the network case in terms of operations on maximal forests.

Let us start with a Tucker tableau, as in Figure 4.8, and consider any pair of arcs $\bar{j} \in F$ and $\bar{k} \notin F$ such that $a(\bar{j}, \bar{k}) \neq 0$. The equation represented by the $\bar{j}$ row of the tableau, namely,

$$\sum_{k \in F'} a(\bar{j}, k)x(k) = x(\bar{j}),$$

can be solved for $x(\bar{k})$:

$$x(\bar{k}) = \frac{1}{a(\bar{j},\bar{k})}x(\bar{j}) - \sum_{\substack{k \notin F \\ k \neq \bar{j}}} \frac{a(\bar{j},k)}{a(\bar{j},\bar{k})}x(k). \tag{1}$$

This can then be substituted in the equations represented by the other rows so as to obtain the expressions

$$x(j) = \frac{a(j,\bar{k})}{a(\bar{j},\bar{k})}x(\bar{j}) + \sum_{\substack{k \notin F \\ k \neq \bar{j}}} \left[a(j,k) - \frac{a(j,\bar{k})a(\bar{j},k)}{a(\bar{j},\bar{k})} \right] x(k)$$

$$\text{for each } k \in F, k \neq \bar{k}. \tag{2}$$

The point is that the systems (1) and (2) is equivalent to the row system in the given tableau, and it therefore constitutes a *new Tucker representation* of the circulation space. The new representation corresponds to the arc set

$$\tilde{F} = \left[F \backslash \bar{j} \right] \cup \bar{k}. \tag{3}$$

(*Note*: Strict usage of notation would require $\{\bar{j}\}$ and $\{\bar{k}\}$ in (3), but the braces are omitted for simplicity.) It follows that the coefficients in (1) and (2), when written in tableau form, yield at the same time a new Tucker representation for the differential space.

The transformation from the F tableau to the $\tilde{F}$ tableau is called *pivoting* on $(\bar{j}, \bar{k})$. It is summarized in Figure 4.12. It is possible if and only if $a(\bar{j}, \bar{k}) \neq 0$.

The following conclusion can be drawn immediately from the basis theorem in Section 4H.

Pivoting Theorem.   *Let F be a maximal forest, and let $\bar{j}$ and $\bar{k}$ be arcs such that $\bar{j} \in F$ and $\bar{k} \notin F$. Then the following conditions are equivalent:*

1.   *$a(\bar{j}, \bar{k}) \neq 0$ in the Tucker representation corresponding to F.*
2.   *$\bar{j}$ lies on the circuit $P_{\bar{k}}$ corresponding to $\bar{k}$ as in Part 1 of the basis theorem.*
3.   *$\bar{k}$ lies on the cut $Q_{\bar{j}}$ corresponding to $\bar{j}$ as in Part 2 of the basis theorem.*

When these conditions are satisfied, the set $\tilde{F}$ defined by Equation (3) is again a maximal forest.

The forests F and $\tilde{F}$ are said to be *adjacent*, and the same for the corresponding Tucker representations. The question now arises: Is it possible to pass from any maximal forest F to any other maximal forest F^* through a sequence of adjacent forests? Equivalently, is it possible to pass from any Tucker tableau to any other Tucker tableau (for the same dual linear systems) by a sequence of pivoting transformations? That the answer is yes can easily be seen in the following constructive manner.

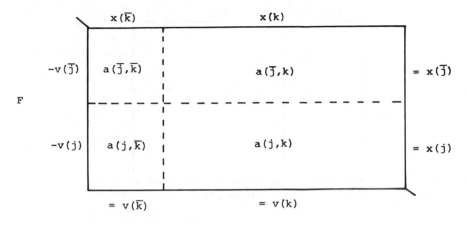

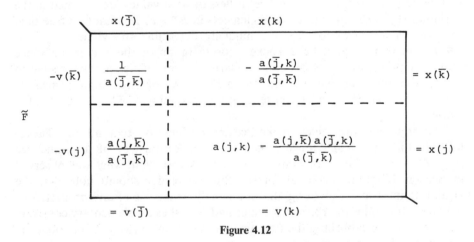

Figure 4.12

Starting with the tableau for F, pivot on any pair $(\bar{j}, \bar{k})$ such that $\bar{j} \in F \setminus F^*$ and $\bar{k} \in F^* \setminus F$, and keep repeating this step until no longer possible (i.e., transform column symbols corresponding to elements of F^* into row symbols, trading them for row symbols in F exclusively, until this is no longer possible). ***Claim***:   *At this stage the tableau must be the one for F^*, if there is one for F^* at all* (i.e., if F^* is really a maximal forest).

For if this is not the tableau for F^*, it must be of the form in Figure 4.13 with at least one column in $F^* \setminus F$ or row in $F \setminus F^*$. (No elements in F can correspond to columns except those that were moved from rows to columns by pivoting, and none of these was in F^*; similarly, no elements not in F can

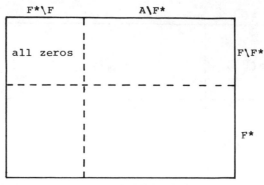

Figure 4.13

correspond to rows except those that were moved from columns to rows by pivoting, and all of these were in F^*.) If some column in Figure 4.12 belongs to an element of $F^* \setminus F$, the row system in this tableau shows that the variables $x(k)$ for columns not in F^* can all be set equal to zero, and then one will have $x(k) = 0$ for *all $k \notin F^*$* regardless of what values are assigned to the variables $x(j)$ belonging to column elements in $F^* \setminus F$. It cannot be true then that the values $x(k)$ for $k \notin F^*$ uniquely determine the values $x(j)$ for $j \in F^*$, so there cannot be a Tucker representation of the circulation space that corresponds to F^*. Reasoning similarly in terms of the column system in the tableau, one sees that if some row in Figure 4.1 belongs to an element of $F \setminus F^*$, there cannot be a Tucker tableau corresponding to F^*. This proves the claim.

The argument just given makes clear, incidentally, that all the Tucker tableaus for a particular linear system of variables must be of the same size (i.e., $|F| = |F^*|$). This is obvious in the network case from what was observed in Section 4F about maximal forests. But the reader should note that the argument, in terms of pivoting, did not actually make use of any properties of networks. It applies to arbitrary systems and furnishes an elementary constructive method of establishing the fundamental results in linear algebra about the dimension of a vector space.

4J.   EXTREME FLOWS

A feasible distribution problem as in Section 3H often has many solutions. But they can all be represented as weighted combinations of certain "extreme" solutions (plus elementary flows around special circuits of "unlimited capacity"), as will be seen in this section.

Geometrically, the representation in question could be developed in terms of the fact that the set of all solutions forms a certain convex polyhedron in R^A. The "extreme" solutions to be introduced could be shown to correspond to the extreme points (vertexes) of the polyhedron (see Exercise 4.38), and general theorems familiar in linear programming could be invoked. However, the discussion will instead be entirely algebraic and algorithmic.

Let the network G have a capacity interval $[c^-(j), c^+(j)]$ for each arc $j \in A$ and a supply $b(i)$ for each node $i \in N$. An *extreme flow*, relative to the corresponding feasible distribution problem, is defined to be a feasible flow x such that the arc set

$$F_x = \{ j \in A \mid c^-(j) < x(j) < c^+(j) \} \tag{1}$$

forms a forest. This condition is equivalent to the existence of a set $F \subset A$ forming a maximal forest such that, for every arc $j \notin F$, either $x(j) = c^+(j)$ or $x(j) = c^-(j)$. In the case of $b = 0$ extreme flows are called *extreme circulations* relative to the given capacity intervals.

A method of calculating an extreme flow, if one exists, by a refinement of the feasible distribution algorithm is described in Exercise 4.42.

In the statement of the main result about extreme flows, an elementary circuit P is said to be *of unlimited capacity* if $c^+(j) = +\infty$ for all $j \in P^+$ and $c^-(j) = -\infty$ for all $j \in P^-$. It is said to be *of doubly unlimited capacity* if P and its reverse are both of unlimited capacity, or in other words, if every arc j in P has capacity interval $(-\infty, +\infty)$.

An elementary circuit of doubly unlimited capacity exists if and only if the arc set

$$\{ j \in A \mid c^+(j) = +\infty \text{ and } c^-(j) = -\infty \}$$

fails to be a forest. (This can be tested algorithmically; see Exercise 4.24.) Since this set is included in F_x for every x, it is clear that the existence of such a circuit precludes the existence of any extreme flows at all. However, circuits of (singly) unlimited capacity play a definite role along with extreme flows in the representation now to be described.

Extremal Representation Theorem for Flows. *Assume the feasible distribution problem has at least one solution, but there are no elementary circuits of doubly unlimited capacity. Then*:

1. *There are only finitely many extreme flows (at least one). They are all integral, if the capacities $c^+(j), c^-(j)$ and supplies $b(i)$ are all integral.*
2. *A flow x is a solution to the feasible distribution problem if and only if it can be represented as*

$$x = \sum_{k=1}^{r} \lambda_k x_k + \sum_{l=1}^{q} \mu_l e_{P_l}, \tag{2}$$

where each x_k is an extreme flow $\lambda_k \geq 0, \sum_{k=1}^{r} \lambda_k = 1$, each P_l is an elementary circuit of unlimited capacity, and $\mu_l \geq 0$.

(*Note*: *The second sum in* (2) *may be vacuous*.)
The proof of the theorem will be given in Section 4K.

EXAMPLE 6.   (*Doubly Stochastic Matrices*)

A doubly stochastic matrix of order n is an array of numbers $x(k, l)$ for $k = 1, \ldots, n$ and $l = 1, \ldots, n$ such that

$$x(k, l) \geq 0 \quad \text{for all } k, l,$$

$$\sum_{l=1}^{n} x(k, l) = 1 \quad \text{for all } k,$$

$$\sum_{k=1}^{n} x(k, l) = 1 \quad \text{for all } l,$$

or in other words, such that each row and column is a probability vector in R^n. The set of all such matrices can be identified with the set of all flows that are solutions to the feasible distribution problem displayed in Figure 4.14.

According to the theorem, any such flow can be represented as a *convex combination* (weighted average) of extreme flows, each of which is integral. (There are no circuits of unlimited capacity in this network.) From the nature of the constraints, it is apparent that the integral solutions to the problem may be identified with matrices having exactly one $+1$ in each row and in each column, all other entries being 0, that is, the permutation matrices of order n. It is also evident that each of these flows is extreme.

Thus one obtains the following fact (a theorem of G. Birkhoff): *Every doubly stochastic matrix can be expressed as a convex combination of permutation matrices*,

$$x = \lambda_1 x_1 + \cdots + \lambda_r x_r \quad \text{with} \quad \lambda_p \geq 0, \Sigma \lambda_p = 1.$$

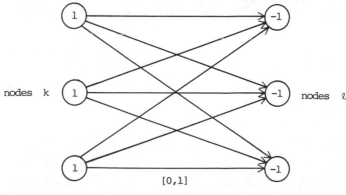

Figure 4.14

Doubly stochastic matrices often arise in situations where a system has n possible states and passes randomly between them with known probabilities. The probability that the system, if in state k at time t, will be in state l at time $t + 1$, is $x(k, l)$. The extreme case where the successor state l is uniquely determined by k corresponds to a permutation matrix. In this context Birkhoff's theorem says that the general case can be interpreted as the net effect of several deterministic mechanisms operating according to certain probabilities λ_p.

EXAMPLE 7.

The decomposition in the extremal representation theorem is not always unique. This may be seen, for instance, in the special case of Example 6 where one has the two representations:

$$
\begin{bmatrix} \frac{1}{3} & \frac{1}{3} & \frac{1}{3} \\ \frac{1}{3} & \frac{1}{3} & \frac{1}{3} \\ \frac{1}{3} & \frac{1}{3} & \frac{1}{3} \end{bmatrix} = \frac{1}{3}\begin{bmatrix} 0 & 0 & 1 \\ 0 & 1 & 0 \\ 1 & 0 & 0 \end{bmatrix} + \frac{1}{3}\begin{bmatrix} 0 & 1 & 0 \\ 1 & 0 & 0 \\ 0 & 0 & 1 \end{bmatrix} + \frac{1}{3}\begin{bmatrix} 1 & 0 & 0 \\ 0 & 0 & 1 \\ 0 & 1 & 0 \end{bmatrix}
$$

$$
= \frac{1}{3}\begin{bmatrix} 1 & 0 & 0 \\ 0 & 1 & 0 \\ 0 & 0 & 1 \end{bmatrix} + \frac{1}{3}\begin{bmatrix} 0 & 1 & 0 \\ 0 & 0 & 1 \\ 1 & 0 & 0 \end{bmatrix} + \frac{1}{3}\begin{bmatrix} 0 & 0 & 1 \\ 1 & 0 & 0 \\ 0 & 1 & 0 \end{bmatrix}
$$

4K.*   EXTREMAL REPRESENTATION ALGORITHM

The theorem in Section 4J will now be proved. The last part of the argument will be presented as an algorithm for expressing a given flow in the manner described.

First consider $b = 0$ (extreme circulations). Then, except for the existence assertion, Part 1 can be derived by considering the various (finitely many) Tucker representations of the circulation space that correspond to maximal forests. The forest F_x corresponding to any extreme solution x is contained in some maximal forest F, and therefore

$$
x(j) = \sum_{k \notin F} a(j, k)x(k) \quad \text{for all } j \in F
$$

(where every $a(j, k)$ is $+1$, -1, or 0). For each $k \notin F$, one has $k \notin F_x$, and hence $x(k)$ is either $c^+(k)$ or $c^-(k)$. It follows that there are only finitely many extreme solutions x with $F_x \subset F$, and these are all integral if the capacity bounds are all integral.

These facts may be applied to the general case where $b \neq 0$ by regarding an extreme flow in G as an extreme circulation in the augmented network $\overline{G}$. Each distribution arc j_i is assigned $[b(i), b(i)]$ as its capacity interval. (Any forest in G remains a forest in $\overline{G}$.) This establishes Part 1, apart from the existence of at least one extreme flow, which of course will follow from establishing the necessity of the representation in Part 2.

The sufficiency of the representation in Part 2 is easy to verify. Since

$$c^-(j) \le x_k(j) \le c^+(j) \quad \text{for all } j, k,$$

one also has

$$c^-(j) \le \sum_{k=1}^{r} \lambda_k x_k(j) < c^+(j) \quad \text{for all } j,$$

since $\lambda_k \ge 0, \Sigma \lambda_k = 1$. Furthermore if for a particular arc j one of the circuits P_l has $e_{P_l}(j) = 1$, then $c^+(j) = +\infty$ by the definition of P_l being of unlimited capacity, whereas if $e_{P_l}(j) = -1$, then $c^-(j) = -\infty$. Therefore

$$\sum_{l=1}^{q} \mu_l e_{P_l}(j) > 0 \quad \text{implies} \quad c^+(j) = +\infty,$$

$$\sum_{l=1}^{q} \mu_l e_{P_l}(j) < 0 \quad \text{implies} \quad c^-(j) = -\infty.$$

It follows that $c^-(j) \le x(j) \le c^+(j)$ for all j. At the same time, since each P_l is a circuit, the divergence of e_{P_l} vanishes, and consequently

$$(\text{div } x)(i) = \sum_{k=1}^{r} \lambda_k (\text{div } x_k)(i) = \sum_{k=1}^{r} \lambda_k b(i) = b(i).$$

Thus x is a solution to the problem.

It remains only to be demonstrated that an arbitrary solution x to the feasible distribution problem can be represented as claimed. This can be carried out in constructive fashion.

Algorithm

Starting with any solution x to the feasible distribution problem, form the arc set F_x defined in 4J(1) and test whether or not it constitutes a forest (see Section 4F). If so, x is an extreme flow, and the desired representation in the theorem in Section 4J is at hand. If not, one obtains an elementary circuit P such that

$$c^-(j) < x(j) < c^+(j) \quad \text{for all arcs } j \text{ in } P.$$

Since by hypothesis P cannot be of doubly unlimited capacity, it can be supposed, passing to the reverse circuit if necessary, that the number

$$\mu' = \max\{\mu \in R \mid c^-(j) \le x(j) - \mu e_P(j) \le c^+(j), \forall j \in A\} \in (0, \infty)$$

exists. Let $x' = x - \mu' e_P$. Then x' is another solution to the feasible distribution problem with $F_{x'}$ strictly contained in F_x, and of course

$$x = x' + \mu' e_P. \tag{1}$$

If P is of (singly) unlimited capacity, the procedure is next repeated for x'. If not, then the number

$$\mu'' = \max\{\mu \in R \mid c^-(j) \le x(j) + \mu e_P(j) \le c^+(j), \forall j \in A\} \in (0, \infty)$$

also exists, and for $x'' = x + \mu'' e_P$ one has

$$x = \lambda' x' + \lambda'' x'' \quad \text{with } \lambda' > 0, \, \lambda'' > 0, \, \lambda' + \lambda'' = 1, \tag{2}$$

where $\lambda' = \mu''/(\mu' + \mu'')$ and $\lambda'' = \mu'/(\mu' + \mu'')$. Moreover x'' too is a solution to the problem, with $F_{x''}$ strictly contained in F_x. In this case the procedure is applied to x'' as well as x'.

Due to the strict decrease in the arc set under consideration in each iteration, every branch followed under the procedure must eventually terminate. In other words, after a finite sequence of decompositions of the form (1) or (2), all the remaining flows must be extreme. It is readily seen that, when these are substituted back to get an expression for the original flow, a representation of the desired sort is obtained.

Determining an Extreme Flow

If the aim is merely to find a solution to the feasible distribution problem that happens to be an extreme flow, rather than to represent a given solution, this can be accomplished in simpler fashion. The procedure is the same, except that there is no need to branch to a pair of flows x' and x'' at each iteration. Forget about x'', and just repeat the reduction step immediately with x'.

4L.*   EXERCISES

4.1. (*Conformal Realization*). Give a conformal realization of the flow in Figure 4.15. (This can be done by inspection without going through the formal details of the algorithm.)

4.2. (*Conformal Realization*). State the special case of the conformal realization theorem corresponding to *circulations* x, and prove by way of the

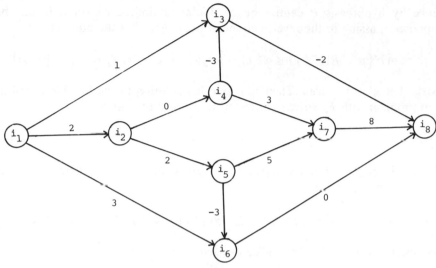

Figure 4.15

augmented network that it is actually equivalent to the general case (except for the part about anticirculations).

4.3. (*Conformal Realization*). Show that for a flow $x \neq 0$ to be an integral circulation satisfying $|x(j)| \leq 1$ for all $j \in A$, it is necessary and sufficient that it be of the form $x = e_{P_1} + \cdots + e_{P_r}$ for certain elementary circuits $P_1, \ldots, P_r$ which are disjoint with respect to arcs.

4.4. (*Conformal Realization*). Let x be an integral anticirculation that is conserved at all nodes except for a source s and sink s' and whose flux from s to s' is $+1$. Prove that $x = e_P$ for some elementary path P: $s \to s'$.

4.5. (*Decomposition into Disjoint Circuits*). Show that the arc set of a network can be expressed as the union of a collection of elementary circuits that are disjoint with respect to arcs, if and only if there is a circulation x satisfying $|x(j)| = 1$ for all arcs j.

4.6. (*Elementary Circulations*). Prove that a circulation x is an elementary flow if and only if its support is nonempty and minimal, in the sense that there is no circulation x' whose support is nonempty and strictly included in the support of x.

4.7. (*Elementary Circulations*). Let $x \neq 0$ be an elementary circulation, and let $x' \neq 0$ be a circulation whose support is included in that of x. Prove that $x' = \lambda x$ for some $\lambda \neq 0$ (see the preceding exercise).

4.8. (*Anticirculations*). Show that if a flow $x \neq 0$ is an anticirculation, then it can be represented as $\sum_{k=1}^{r} \alpha_k e_{P_k}$, where $\alpha_k > 0$ and each path P_k conforms to x and goes from a source to a sink. But the converse is false.

4.9. (*Max Flow*). When a solution to a max flow problem is calculated by means of the max flow algorithm, must it necessarily turn out to be an anticirculation? Show by counterexample that this is false even in the case where $N^+ = \{s\}$, $N^- = \{s'\}$, all capacity intervals are of the form $[0, c^+(j)]$, and the algorithm is initiated with the zero flow.

4.10. (*Max Flow*). Consider a max flow problem with $c^-(j) \le 0 \le c^+(j)$ for all arcs j. Show that, despite the existence of the sort of counterexample in the preceding exercise, if there exists a solution at all, then there exists one that is an anticirculation.

4.11. (*Max Flow*). Consider a max flow problem with all capacity intervals of the form $[0, c^+(j)]$. Prove that any solution calculated by the max flow algorithm will be an anticirculation if the algorithm is initiated with the zero flow and the following refinement (which is a complementary version of arc discrimination) is adhered to. When the painted network algorithm is executed as a subroutine, use white arcs only as a last resort (i.e., prefer a green or black arc to a white one, whenever the choice presents itself).

4.12. (*Feasible Distributions*). Show that the facts in the three preceding exercises carry over to the feasible distribution problem and algorithm.

4.13. (*Acyclic Networks*). Prove that a network is acyclic if and only if it has an anticirculation x such that $x(j) > 0$ for every arc j.

   (*Note.* This is dual to the characterization of strongly connected networks in Exercise 3.19.)

4.14. (*Realization Algorithm*). Formally calculate a conformal realization of the flow in Figure 4.16 using the algorithm.

4.15. (*Realization Algorithm*). Prove that the elementary flows $\alpha_k e_{P_k}$ calculated by the realization algorithm in representing a flow x are linearly independent as elements of the vector space R^A.

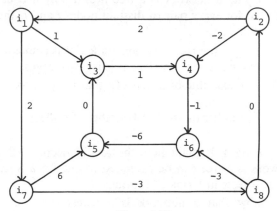

Figure 4.16

4.16. (*Conformal Realization of Differentials*). A cut Q is said to conform to a differential v if $v(j) > 0$ for every $j \in Q^+$ and $v(j) < 0$ for every $j \in Q^-$. Develop a realization algorithm for differentials that is analogous to the one for circulations and leads to the result that every differential $v \neq 0$ can be expressed in the form

$$v = \alpha_1 e_{Q_1} + \cdots + \alpha_r e_{Q_r},$$

where $\alpha_k > 0$ and each Q_k is an elementary cut that conforms to v. Show as part of this that the coefficients α_k can be taken to be integral if v is integral (i.e., if $v(j)$ is an integer for all $j \in A$).

4.17. (*Differentials*). Establish the conformal realization theorem for differentials in the preceding exercise nonalgorithmically, working instead with the fact that every differential is of the form $v = \Delta u$ for some potential u.

4.18. (*Differentials*). Show that for $v \neq 0$ to be an integral differential satisfying $|v(j)| \leq 1$ for all $j \in A$, it is necessary and sufficient that $v = e_{Q_1} + \cdots + e_{Q_r}$, where $Q_1, \dots, Q_r$ are disjoint elementary cuts.

(*Hint.*   Use the result in Exercise 4.16.)

4.19. (*Elementary Differentials*). A differential v is said to be *elementary* if it is of the form αe_Q, where Q is an elementary cut and $\alpha > 0$. Show that this is true if and only if the *support* of v (defined in the same way as in Section 4C for flows) is nonempty and *minimal*, in the sense that there is no differential v' whose support is nonempty and strictly included in the support of v. (Use the result of Exercise 4.16.)

4.20. (*Elementary Differentials*). Let $v \neq 0$ be an elementary differential, and let v' be a differential whose support is included in that of v. Prove that $v' = \lambda v$ for some λ (see the preceding exercise).

4.21. (*Paths in a Tree*). Prove directly from the original definition of a "tree" in Section 4E that a network is a tree if and only if it has at least two nodes and that for every pair of distinct nodes i and i' there is a *unique* elementary path $P: i \to i'$.

4.22. (*Leaves*). A node i in a tree T is called a *leaf* if it is incident to only one arc of T. Prove that every tree has at least two leaves.

4.23. (*Trees*). Give an example of a network that is not a tree and yet has

$$[\text{number of arcs}] = [\text{number of nodes}] - 1.$$

4.24. (*Test for a Forest*). Explain how the test in Section 4E for whether a given network is a tree can be extended to one for a forest. Use this to characterize forests in terms of routings.

4.25. (*Forests*). Show that a network is a forest (i.e., has only trees as components) if and only if it has no isolated nodes and satisfies the relation:

$$[\text{number of arcs}] = [\text{number of nodes}] - [\text{number of components}].$$

4.26. (*Forests*). Prove that an arc set F forms a forest in the network G if and only if there is no circulation $x \neq 0$ whose support is included in F.

4.27. (*Forests*). Give an example of an anticirculation whose support is not included in any forest.

4.28. (*Coforests*). An arc set F' in a network G is said to form a *coforest* if the deletion of all the arcs in F' would not increase the number of components of G. Show that this is true if and only if no elementary cut of G is included in F', or equivalently, there is no differential $v \neq 0$ whose support is included in F' (see Exercise 4.16).

4.29. (*Maximal Coforests*). A coforest in a network G (see the preceding exercise) is *maximal* if it is not strictly included in any other coforest in G. Prove that F' is a maximal coforest if and only if its complement $F = A \setminus F'$ is a maximal forest.

(*Note.* This says that the sets F' appearing in the Tucker representations of the circulation and differential spaces for G are precisely the maximal coforests for G.)

4.30. (*Rows of the Incidence Matrix*). For each node $i \in N$, let e_i be the *row* of the incidence matrix E corresponding to i: $e_i(j) = e(i, j)$ for all $j \in A$. Develop a necessary and sufficient condition on a set $S \subset N$ in order that the rows e_i, $i \in S$, be linearly independent as elements of R^A. When will they form a basis for the differential space $\mathcal{D}$?

4.31. (*Cyclomatic Number*). Show that for a connected network G the dimensions of the circulation space $\mathcal{C}$ and differential space $\mathcal{D}$ are given by

$$\dim \mathcal{C} = |A| - |N| + 1, \qquad \dim \mathcal{D} = |N| - 1.$$

(The first of these is called the cyclomatic number of G.) How does this generalize to a network with p components?

4.32. (*Basis Theorem*). Prove Part 2 of the basis theorem in Section 4H.

4.33. (*Basis Theorem*). For the network shown in Figure 4.17 calculate as in Example 5 in Section 4H the Tucker representations of $\mathcal{C}$ and $\mathcal{D}$ corresponding to the spanning tree F formed by the darkened arcs.

4.34. (*Pivoting*). For the network in Figure 4.11 obtain, by pivoting in the Tucker tableau at the end of Section 4H, the Tucker tableau that corresponds to the spanning tree

$$F = \{ j_1, j_2, j_4 \cdot j_7, j_8, j_{10} \}.$$

4.35. (*Pivoting*). The pivoting transformation in a connected network takes us from one spanning tree F and its associated basis of fundamental

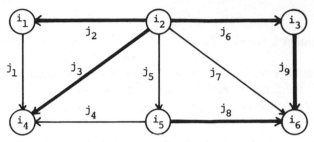

Figure 4.17

circuits P_k (as in Part 1 of the basis theorem in Section 4H) to an adjacent spanning tree $\tilde{F}$ and its circuits $\tilde{P}_k$. The new circuits can be read from the columns of coefficients in the transformed tableau (according to the basis theorem), but explain in more direct terms what happens. How are the new fundamental circuits constructed from the old ones?

4.36. (*Pivoting*). Let F and $\tilde{F}$ be adjacent maximal forests with $\tilde{F} = [F \setminus \bar{j}] \cup \bar{k}$. Let $\bar{j} \in F$ and $\bar{k} \in F$ be such that both $\bar{j}$ and j belong to both of the circuits P_k and $P_{\bar{k}}$ in the basis associated with F (where $\bar{j} \neq j$, $\bar{k} \neq \bar{k}$). Prove that j *cannot* belong to the circuit $P_{\bar{k}}$ in the basis associated with $\tilde{F}$.

(*Hint.* Translate this into an assertion about coefficients in the Tucker tableaus corresponding to F and $\tilde{F}$, and use the fact that these coefficients can have no values other than $+1$, -1 or 0.)

4.37. (*Extreme Flows*). Let x be a solution to a given feasible distribution problem. Prove that x is an extreme flow if and only if (there are no circuits of doubly infinite capacity and) the arc set F_x (see 4J(1)) is *minimal*, in the sense that there is no solution x' different from x and having $F_{x'} \subset F_x$.

4.38. (*Extreme Flows as Extreme Points*). Let x be a solution to a given feasible distribution problem. Prove that x is an extreme flow if and only if it cannot be expressed as $(1 - \lambda)x' + \lambda x''$, where x' and x'' are also solutions to the problem (not both identical to x) and $0 < \lambda < 1$.

(*Note.* This says that x is an extreme point of the convex polyhedron formed by all the solutions to the problem.)

4.39. (*Extreme Flows as Non-Negative Basic Solutions*). Consider a feasible distribution problem in which all the capacity intervals are $[0, +\infty)$. Show that x is an extreme flow if and only if x is a solution whose support forms a forest. Moreover the latter is equivalent to x being a non-negative basic solution to the system

$$\sum_{j \in A} x(j)e^j = b$$

where e^j is the "column" of the incidence matrix corresponding to the arc j. (A *basic* solution to such a system is one obtained as follows. Select a set $F \subset A$ such that $\{e^j | j \in F\}$ is a basis for the column space of the incidence matrix. For $j \in F$ the values $x(j)$ are the unique coefficients such that

$$\sum_{j \in F} x(j)e^j = b,$$

whereas for $j \notin F$ they are all 0. This is a fundamental concept in the theory of the simplex method in linear programming.)

4.40.   (*Circuits of Doubly Infinite Capacity*). Let x be a solution to a given feasible distribution problem. Let L be the set of all flows x' such that, for all $\lambda \in R$, $x + \lambda x'$ is also a solution to the problem. Prove that $x' \in L$ if and only if x' is a circulation whose support consists entirely of arcs with capacity interval $(-\infty, \infty)$. Moreover $x' \in L$ if and only if

$$x' = \alpha_1 e_{P_1} + \cdots + \alpha_r e_{P_r},$$

for certain coefficients $\alpha_k > 0$ and elementary circuits P_k of doubly infinite capacity.

4.41.   (*Extremal Representation*). Calculate an extremal representation for the flow and distribution problem depicted in Figure 4.18.

4.42.   (*Feasible Distribution Algorithm*). Suppose the feasible distribution algorithm is initiated with a flow x such that the set F_x in 4J(1) is a forest. Prove that if arc discrimination is used in the painted network subroutine, then all the flows in the sequence generated by the algo-

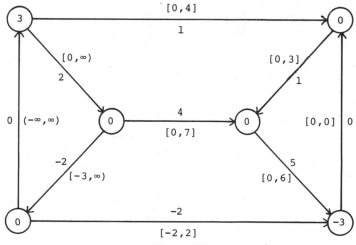

Figure 4.18

rithm will have this same property, and hence that when a solution is attained, it will be an extreme flow. How might an initial flow with the desired property be determined?

4.43. (*Generalized Extreme Flows*). Extend the concept of an "extreme flow" from the feasible distribution problem to the superficially more general "feasible flow problem" in Section 3H, where instead of a fixed value $b(i)$ at each node there is an interval $[c^-(i), c^+(i)]$. State the corresponding generalization of the extremal representation theorem.

4.44. (*0–1 Matrices*). Let m be the set of all $m \times n$ matrices X whose entries $x(k, l)$ are either 0 or 1 and whose rows and columns have certain prescribed sums:

$$\sum_{l=1}^{n} x(k, l) = r(k) \quad \text{for } k = 1, \ldots, m,$$

$$\sum_{k=1}^{m} x(k, l) = s(l) \quad \text{for } l = 1, \ldots, n,$$

where $r(k)$ and $s(l)$ are non-negative integers. Show that M can be identified with the set of all extreme flows in a certain feasible distribution problem. What does the extremal representation theorem say in this case?

4.45. (*0–1 Matrices*). Derive a necessary and sufficient condition for the nonemptiness of the set M in Exercise 4.44 by applying the feasible distribution theorem to the problem in question.

4.46. (*0–1 Matrices*). Extend Exercise 4.44 to the case of constraints of the form

$$r^-(k) \le \sum_{l=1}^{n} x(k, l) \le r^+(k) \quad \text{for } k = 1, \ldots, m,$$

$$s^-(l) \le \sum_{k=1}^{m} x(k, l) \le s^+(l) \quad \text{for } l = 1, \ldots, n,$$

where the bounds are all non-negative integers (or $+\infty$).

4.47. (*Optimal Spanning Trees; Kruskal's Algorithm*). Let G be a connected network with at least one arc. Suppose a real number $r(j)$ is given for each arc j of G, and associate with each spanning tree F the number

$$r(F) = \sum_{j \in F} r(j).$$

Prove that the following algorithm determines a spanning tree F for which $r(F)$ is maximal. Index the arcs of G so that

$$r(j_1) \geq r(j_2) \geq \cdots \geq r(j_n), \quad \text{where } n = |A|.$$

Initially, set $F_1 = \{j_1\}$. In iteration k for $k > 1$, a forest F_{k-1} is on hand; let $F_k = F_{k-1} \cup \{j_k\}$ if F_k would then still be a forest, but otherwise just set $F_k = F_{k-1}$. The desired spanning tree is $F = F_n$.

(*Remark.* A commonly cited example of the optimal spanning tree problem is the following. The nodes of the network represent localities, and the arcs are roads joining them. Transmission cables must be laid along various roads in such a way that every locality can communicate with every other locality. The problem then is to find an optimal spanning tree under the interpretation that $-r(j)$ is the cost of laying the cable along road j.)

4M.* COMMENTS AND REFERENCES

The conformal realization theorem embodies facts that have been recognized by many authors for their usefulness but have been expressed with varying degrees of completeness and constructivity. For instance, Ford and Fulkerson [1962, pp. 6–8] treat only the case of a flow having a single source and sink and are content to throw away the circulation component; however, they give an algorithm for determining the anticirculation component. Busacker and Saaty [1965, p. 243] state of version of the theorem that corresponds almost exactly to the one in Section 4B (except that the concept of an anticirculation is missing), but their proof is rather indirect and not so readily translated into an algorithm for obtaining the desired realization. The constructive approach in Section 4C in terms of the painted network theorem (and Minty's lemma) is very simple in comparison.

Trees and their associated fundamental circuits have long been of importance in electrical circuit analysis. It was G. Kirchhoff [1847] who introduced them and proved the basis theorem. For a computer-oriented discussion of trees and algorithms involving trees, see Christophides [1975, Chap. 7].

Tucker representations are a modern invention spawned by linear programming (see Tucker [1963] for the theory). Two matrices are said to be *combinatorially equivalent* if they can be obtained, one from the other, by a sequence of pivoting transformations and permutations of rows and columns. The corollary to the basis theorem in Section 4H asserts then that a matrix of the kind appearing in a Tucker tableau for the circulation and differential spaces of a network G has the special property that all the matrices in its combinatorial equivalence class have as entries only 0, $+1$, or -1. It can be shown that this holds if and only if every square submatrix of the given matrix has determinant 0, $+1$, or -1. The latter property is called *total unimodularity*. Note that the corollary gives not only the total unimodularity of the matrices in the Tucker tableaus for G but also the total unimodularity of the incidence matrix E, since E corresponds to a Tucker tableau for the augmented network $\bar{G}$, as explained at the end of Section 4G.

This business of total unimodularity is very closely tied in with the integrality results in Sections 4A, 4B, and 4J. The theorem of G. Birkhoff [1946] in Section 4J (see also Frobenius [1912]) is perhaps the earliest result of the integrality of extreme points of a convex polyhedron associated with a flow problem, although it was not originally conceived in terms of flows. This theorem is equivalent to an observation of Dantzig [1951] that did explicitly concern flows, namely, the integrality of basic solutions to the classical transportation problem (Hitchcock problem) in the case of integral supplies and demands. Hoffman and Kruskal [1956] showed that for the extreme points of the convex polyhedron $\{ x | c^- \leqq x \leqq c^+, b^- \leqq Ex \leqq b^+ \}$ all to be integral for every choice of integral c^-, c^+, b^-, b^+, it is necessary and sufficient that E be totally unimodular. This applies to the case where E is the incidence matrix of a network and $b^+ = b^-$, and one then gets the external representation theorem in Section 4J. Consult Lawler [1976, pp. 160–165] for more on unimodularity and Rockafellar [1970, Sec. 18] for extreme points of general convex polyhedra.

The fact that in the network context a representation in terms of extreme flows can be obtained so quickly and constructively by means of Minty's lemma, as in Section 4J, does not seem to have been noticed before.

In the linear programming approach to optimization problems in networks, Tucker representations of the spaces $\mathscr{C}$ and $\mathscr{D}$ are fundamental. The role of such representations in finding solutions by means of the simplex method and its extensions will be discussed in Chapter 7. The fact that pivoting can be carried out combinatorially in terms of operations on trees, as explained in Section 4I, is extremely important to this approach. Clever ways of dealing with trees in a computer lead to tremendous improvement in algorithmic efficiency. See Ali, Helgason, Kennington, and Hall [1977] for a description of such techniques.

The optimal spanning tree algorithm of Kruskal [1956] (see Exercise 4.47) has been of great interest in combinatorial optimization. This subject is treated at length by Lawler [1976, Chap. 7].

The round-off result of Example 1 in Section 4A comes from an article of Baranyai [1973]. He credits it to an observation by Lovász (unpublished).

5

MATCHING THEORY
AND ASSIGNMENT
PROBLEMS

The remarkable facts about integral solutions to the max flow and feasible distribution problems lead to a number of applications of network optimization theory to combinatorial problems. One of the characteristic features of such problems is the powerful and constructive role of duality. This role has already been observed in the min cut problem—a combinatorial optimization problem with a surprisingly efficient method of solution. It will also be seen in the corresponding min path problem, to be studied in Chapter 6.

The basic notion of a "match" relative to a compatibility relation is developed. Several applications are described that involve the existence or optimality of special matches called "assignments."

The results should be viewed more as a matter of reaping a harvest than of sowing a crop. Although they are not needed for subsequent chapters, they are intriguing and useful in themselves and provide added motivation for some of the earlier and later theory. Our real goal is still the study of optimal flows and potentials along with the insights they provide for monotropic programming. In keeping with this goal, we do not attempt here to treat matching and assignment problems in full generality but concentrate on those aspects that best illustrate the usefulness of network flow theory.

The sections deserving the most attention are 5A, 5C, 5E, and 5F. The main requirement for understanding the material is familiarity with the max flow problem and algorithm and the integrality theorem in Section 4A.

5A. MATCHING PROBLEM

Imagine that a government wants to select from a certain list of diplomats its ambassadors for a certain group of countries. For various reasons not every diplomat is eligible or willing to serve in every country. How many of the posts can be filled under such a restriction?

In the abstract version of the problem, there are two nonempty finite sets K and L and a subset $H \subset K \times L$. The pairs $(k, l) \in H$ are said to be *compatible*, whereas those not in H are *incompatible*. A *match* M is a collection of compatible pairs $(k_1, l_1), \ldots, (k_p, l_p)$ such that the elements $k_1, \ldots, k_p$ are all different, as well as the elements $l_1, \ldots, l_p$. In other words, a match is a subset of H that defines a one-to-one correspondence between a subset of K and a subset of L.

Matching Problem.   *Determine a match M that maximizes* $|M|$.

Of course, $|M|$ denotes as usual the *cardinality* of the set M, that is, the number of pairs in the match. Note that maximizing $|M|$ is harder than just finding a match M that is maximal in the sense of not being properly included in any other match. Initially, it may be easy to pair off a number of compatible elements, but later an impass will be reached when no more compatible pairs are available among the elements still unused. At this stage the match M is "maximal," but it may not have maximum cardinality $|M|$ due to unfortunate choices at the beginning of the process that tied up especially "versatile" elements too soon. A substantial reorganization of the early pairings might open the way to a match with greater cardinality.

Some generalizations of the problem where certain pairings are more valuable than others or where the one-to-oneness of the match is relaxed will be discussed in Section 5G.

The matching problem can be solved efficiently using the max flow algorithm. The trick is to consider the network in Figure 5.1, whose node set is

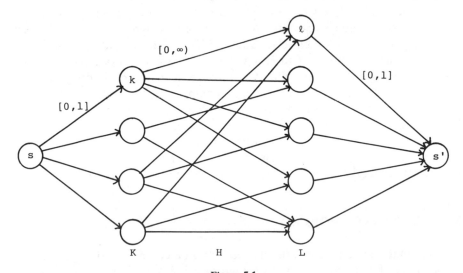

Figure 5.1

formed by K, L, and two extra elements s and s'. Arcs with capacity interval $[0, 1]$ correspond to all the pairs (s, k) and (l, s') with $k \in K$ and $l \in L$, whereas arcs with capacity interval $[0, \infty)$ correspond to all the compatible pairs $(k, l) \in H$. (It would be possible to use $[0, 1]$ for the latter arcs too, but the choice of $[0, \infty)$ simplifies the proof of a theorem to be stated here.)

The corresponding max flow problem with $N^+ = \{s\}$ and $N^- = \{s'\}$ has a solution, according to the max flow min cut theorem. Since the capacity bounds are all integral, it even has an integral solution (by the integrality theorem for flows in Section 4A). Observe now that an integral flow x that satisfies the constraints of the max flow problem corresponds exactly to a match M such that the flux of x from s to s' is $|M|$. Indeed, arcs of the type (s, k) and (l, s') can only have flux 0 or 1. If an arc of the type $(k, l) \in H$ has flux different from 0, then the non-negativity and integrality of x and the conservation conditions at the nodes k and l imply that it is the only such arc incident to k or l, and that the arcs (s, k), (k, l), and (l, s') all have flux 1. Thus the arcs of type (k, l) having nonzero flux constitute the pairs of a match M, and the number of pairs equals the flux of x from s to s'. Conversely, of course, every match M corresponds to a flow x. Figure 5.2 shows the flow corresponding to a match $M = \{(k_1, l_2), (k_3, l_5), (k_4, l_1)\}$. (Darkened arcs have flux 1, all others flux 0. The letters g, w, b, correspond to a painting that will play a role in Section 5B.)

In this way the solutions to the matching problem are identified with the integral solutions to the max flow problem. The max flow algorithm, initiated with the zero flow, will terminate in finitely many iterations with an integral flow (see Sections 4A and 3G), thereby yielding a solution to the matching problem. As a matter of fact, if the algorithm is initiated with *any* integral flow satisfying the constraints, it will generate a sequence of such flows ending in an integral solution. With each iteration the flux from s to s' is increased by one unit. In view of the correspondence just outlined, this means that the algorithm can be initiated with any match and will generate a sequence of matches of increasing cardinality until one of maximum cardinality is attained. This suggests the possibility of translating the procedure into some sort of direct manipulation of matches in which flows are not mentioned explicitly. This will be explored in Section 5B.

The max flow problem corresponding to the matching problem is dual to a certain min cut problem, and this in turn can be identified with a combinatorial problem in the matching context. A duality theorem for the matching problem is thereby obtained. To formulate it, call a subset B of $K \cup L$ a *block* relative to the compatibility relation H, if for every $(k, l) \in H$ either $k \in B$ or $l \in B$ (or both).

Blocking Problem.   *Determine a block B that minimizes $|B|$.*

Incidentally, it must be understood here that K and L are separate sets with no elements in common, as is already implicit in the network set up of Figure

5.1. In the circumstances where a model with $K = L$ is tempting, L should be thought of instead as a "second copy" of K.

König-Egerváry Theorem

$$\left[\begin{array}{c} \text{max } M \text{ in the} \\ \text{matching problem} \end{array} \right] = \left[\begin{array}{c} \text{min } B \text{ in the} \\ \text{blocking problem} \end{array} \right].$$

Proof. It suffices to establish an equivalence between the blocking problem and the min cut problem for the network of Figure 5.1, showing that both have the same minimum. Attention can be restricted in the min cut problem to cuts $Q: s \downarrow s'$ such that $c^+(Q) < +\infty$ (since there is at least one such cut). However, the cuts with this property are those of the form $Q = [S, N \setminus S]$, where $s \in S$, $s' \notin S$, and there is no pair $(k, l) \in H$ with $k \in S$ and $l \notin S$, in which event

$$c^+(Q) = |K \setminus S| + |L \cap S|.$$

Thus they correspond one to one with blocks B under the rule $B = (K \setminus S) \cup (L \cap S)$, $c^+(Q) = |B|$, and the equivalence of the two problems is evident.

Matrix Representation

The matching and blocking problems can be visualized in terms of the *compatibility matrix* for the relation H, whose entries $h(k, l)$ are 1 or 0, depending on whether (k, l) is compatible or not. In this setting a match M corresponds to a set of 1's in the matrix, no two of which are in any single line (a *line* being either a row or a column). A block B, on the other hand, corresponds to a set of lines that covers all the 1's in the matrix. The König-Egerváry theorem says that the maximum number of "noncollinear" 1's equals the minimum number of lines capable of covering all the 1's.

5B.*   MATCHING ALGORITHM

When the max flow algorithm is applied to the matching problem as formulated in terms of the network in Figure 5.1, the pattern of execution has a number of special features. These lead to shortcuts in the calculations. In this way the algorithm can be refined into a procedure that solves the matching problem directly, that is, without any mention of networks or flows. We now work through the main features of such a refinement in order to illustrate how the max flow algorithm can serve as the generator of special combinatorial algorithms that might seem totally unrelated to it.

The first thing to observe is the nature of the flow-augmenting paths in this context. The painting corresponding to a match M is illustrated in Figure 5.2. The pairs in M are represented by green "intermediate" arcs that are con-

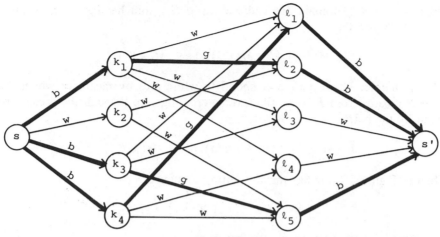

Figure 5.2

nected to s and to s' by black arcs; all other arcs are white. The simplest kind of flow-augmenting path would be of the form $P: s \to k \to l \to s'$, where (k, l) is a compatible pair having neither k nor l already used by M. The general case is represented by a path P from s to s' that alternates between K and L, crossing each time from L back to K through some green arc, but otherwise using only white arcs.

In describing such paths, it is convenient to adopt function notation for the match M: for each pair $(k, l) \in M$, the element l is denoted by $M(k)$. Furthermore

$$\text{domain } M = \{ k \in K \mid \exists l \in L \text{ with } (k, l) \in M \},$$

$$\text{range } M = \{ l \in L \mid \exists k \in K \text{ with } (k, l) \in M \}.$$

For example, in Figure 5.2 one has

$$M(k_1) = l_2, \; M(k_3) = l_5, \; M(k_4) = l_1,$$

$$\text{domain } M = \{ k_1, k_3, k_4 \},$$

$$\text{range } M = \{ l_1, l_2, l_5 \}.$$

Let K_M be the set of all $k \in K$ such that there exists $l \in L \setminus (\text{range } M)$ with (k, l) compatible. The simplest kind of flow-augmenting path is obtained as soon as one has located an element $k \in K \setminus (\text{domain } M)$ that belongs to K_M. To construct a general flow-augmenting path, however, what one needs is a finite alternating sequence

$$k, M(k'), k', M(k''), k'', M(k'''), k''', \ldots \tag{1}$$

beginning in $K \setminus$ (domain M) and ending in K_M, and having its consecutive pairs of the form

$$(k, M(k')), (k', M(k'')), (k'', M(k''')), \ldots$$

all compatible. These pairs correspond to white arcs by means of which the path P crosses from K to L; the reverse crossings are through the green arcs given by the pairs

$$(k', M(k')), (k'', M(k'')), (k''', M(k''')), \ldots$$

Thus in Figure 5.2 one has the alternating sequence

$$k_2, M(k_3), k_3, M(k_4), k_4,$$

and this yields the flow-augmenting path

$$P: s \to k_2 \to l_5 \leftarrow k_3 \to l_1 \leftarrow k_4 \to l_4 \to s'.$$

(Here l_3 could be replaced by l_4. Note also that the path is unnecessarily long: in fact $k_3 \in K_M$, and one could have gone immediately from k_3 to s' by way of l_3.)

Because the flow-augmenting paths have this special structure, the painted network algorithm for determining them can be streamlined in certain respects. For instance, the routing can be represented entirely in terms of elements of K. Indeed, to reconstruct the sequence (1) we simply need to know that k' was reached from k, k'' from k', and so forth, and a labeling function of the type $\phi(k') = k$, $\phi(k'') = k'$, therefore suffices.

At a general stage of the painted network algorithm where S is the set of nodes that has been reached, we can select any $k \in S \cap K$ and check whether $k \in K_M$. If so, a flow-augmenting path is obtained. If not, we look for every element $k' \in$ (domain M) $\setminus S$ such that the pair $(k, M(k'))$ is compatible; then k' and $M(k')$ can be added to S with k' labeled by $\phi(k') = k$. Having inspected k in this fashion, we may regard it as "thoroughly scanned" and never look at it again.

As for the node set S, we can replace it in our description of this procedure by the associated set $B = (K \setminus S) \cup (L \cap S)$. This is more appropriate because it will give the solution to the blocking problem when $Q = [S, N \setminus S]$ is the solution to the min cut problem (see the proof of the König-Egerváry theorem in Section 5A).

With these observations we arrive at a method adapted completely to the framework of the matching and blocking problems. This takes a match M and either verifies that it solves the matching problem (in which event a solution B to the blocking problem is produced too) or constructs a match M' with $|M'| = |M| + 1$. The computations involve a variable set $B \subset K \cup L$ with $B \cap K \subset$ domain M and a labeling function ϕ from (domain M) $\setminus (B \cap K)$ to

K. There is also a variable set W (corresponding in the painted network subroutine of the max flow algorithm to the nodes of K that have been reached but not yet "thoroughly scanned" in the sense of the implementation in Section 2G).

Multipath implementation along the lines of the Dinic-Karzanov approach to the max flow algorithm in Section 3J could also be incorporated into the procedure that follows. This would greatly increase the efficiency; see Exercise 5.49. We forgo giving the details here, since the ideas already covered serve adequately for our purpose of illustration.

STEP 0.   A match M is at hand. Let $B =$ domain M and $W = K \setminus$ (domain M). (The function ϕ is vacuous.) Go to Step 1.

STEP 1.   Is $W = \varnothing$? If yes, then M solves the matching problem, and B solves the blocking problem. If no, go to Step 2.

STEP 2.   Select any $k \in W$. Does there exist an $l \in L \setminus$ (range M) with $(k, l) \in H$? If yes, go to Step 4. If no, go to Step 3.

STEP 3.   For each $k' \in B \cap K$ with $(k, M(k')) \in H$, add $M(k')$ to B and transfer k' from B to W, defining $\phi(k') = k$. Delete k from W, and return to Step 1.

STEP 4.   Generate the sequence $k = k^0, k^1, \ldots, k^r$, where $k^i = \phi(k^{i-1})$ for $i = 1, \ldots, r$ and k^r is not in the domain of ϕ. (If k itself is not in the domain of ϕ, the sequence consists of just $k = k^0$.) Delete the pairs $(k^0, M(k^0)), \ldots,$ $(k^{r-1}, M(k^{r-1}))$ from M, and replace them by the pairs $(k^0, l), (k^1, M(k^0)), \ldots, (k^r, M(k^{r-1}))$. The altered set M is then another match with cardinality greater by 1. Return to Step 0.

The justification of the claims in Step 4 of the algorithm is left for the reader as Exercise 5.10.

For purposes of illustration the calculations can be carried out in terms of the compatibility matrix representation described at the end of Section 5A.

EXAMPLE 1

The first tableau in Figure 5.3 displays a compatibility matrix and an initial match (the entries with solid circles). Application of the algorithm leads immediately (by the "yes" branch in Step 2) to the addition of another element (entry with dashed circle) to the match.

On the next iteration, still in the first tableau, an improvement is not immediate, and the construction of B and ϕ proceeds through several stages, indicated along the right margin. Thus $\phi(k_5) = k_1$, and so on. The rows labeled with * are the initial elements of W, whereas those with two labels are the ones that have been selected and processed in Step 2; the number label

	ℓ_1	ℓ_2	ℓ_3	ℓ_4	ℓ_5	ℓ_6	ℓ_7		
k_1	0	1	0	0	1	0	0	*	1
k_2	1	(1)	0	1	0	0	1	k_1	4
k_3	0	0	0	0	1	0	0	*	2
k_4	1	0	(1)	1	0	1	0		
k_5	1	0	1	0	(1)	0	0	k_1	
k_6	0	1	0	0	1	0	0	*	3

	ℓ_1	ℓ_2	ℓ_3	ℓ_4	ℓ_5	ℓ_6	ℓ_7		
k_1	0	(1)	0	0	1	0	0	k_6	3
k_2	1	1	0	(1)	0	0	1		
k_3	0	0	0	0	1	0	0	*	1
k_4	1	0	(1)	1	0	1	0		
k_5	1	0	1	0	(1)	0	1	k_3	4
k_6	0	1	0	0	1	0	0	*	2

	ℓ_1	ℓ_2	ℓ_3	ℓ_4	ℓ_5	ℓ_6	ℓ_7		
k_1	0	(1)	0	0	1	0	0	k_6	2
k_2	1	1	0	(1)	0	0	1		
k_3	0	0	0	0	(1)	0	0	k_6	3
k_4	1	0	(1)	1	0	1	0		
k_5	(1)	0	1	0	1	0	1		
k_6	0	1	0	0	1	0	0	*	1

Figure 5.3

records the order of selection. (Thus at every stage W consists of the labeled rows having one label, whereas B consists of the unlabeled rows and the columns having a circled entry in a row with two labels.) When Step 2 is performed on k_2, the "yes" branch occurs with l_4. Step 3 then produces the larger match in the second tableau.

The calculations are similar in the next iteration and lead to the still larger match in the third tableau. The subsequent iteration winds up with the "yes" branch of Step 1. Thus the match M represented in the third tableau solves the matching problem, whereas the blocking problem is solved by

$$B = \{k_2, k_4, k_5, l_2, l_5\},$$

with max = min = 5.

5C. ASSIGNMENTS

As in the matching problem, consider two nonempty finite sets K and L and a subset $H \subset K \times L$ giving a "compatibility relation." An *assignment of K to L*, compatible with H, is a match M with $|M| = |K|$. Thus it is a one-to-one function, from *all* of K into L, constituted entirely from pairs in H.

In Section 5E the question of finding an "optimal" assignment will be discussed. For the moment the issue is the existence of any assignment at all. This is easily settled in any specific case by solving the matching problem for K, L, and H. If the maximum cardinality is $|K|$, the optimal match M determined by the algorithm is an assignment. If the maximum is less than $|K|$, then no assignment exists.

The König-Egerváry theorem in Section 5A yields in this way a criterion for whether an assignment exists.

Hall's Existence Theorem. *For there to be an assignment of K into L, relative to a relation $H \subset K \times L$, it is necessary and sufficient that for every subset $T \subset K$ the set*

$$H(T) = \{l \in L | \exists k \in T \text{ with } (k, l) \in H\}$$

satisfies $|H(T)| \geq |T|$.

Proof. The necessity is elementary, for if M is an assignment, one has $|T| = |M(T)|$ and $M(T) \subset H(T)$, implying $|T| \leq |H(T)|$. To prove the sufficiency, consider an arbitrary block $B \subset K \cup L$. By definition, there are no pairs $(k, l) \in H$ such that $k \in K \setminus B$ and $l \in L \setminus B$. In other words, the set $T = K \setminus B$ has $H(T) \subset L \cap B$. Invoking the condition $|H(T)| \geq |T|$, one calculates

$$|B| = |K \cap B| + |L \cap B| \geq |K \cap B| + |H(T)|$$

$$\geq |K \cap B| + |T| = |K \cap B| + |K \setminus B| = |K|.$$

Since this is true for an arbitrary block B, the minimum in the blocking problem is no less than $|K|$. Therefore, by the König-Egerváry theorem, the maximum in the matching problem is at least $|K|$ (hence equal to $|K|$), and an assignment exists.

Remark.   For later computational purposes it should be observed that the matching algorithm produces, in the event that no assignment exists, a set $T \subset K$ with $|H(T)| < |T|$, namely, $T = K \setminus B$ for the solution B to the blocking problem. This is clear from the proof, which is therefore constructive.

Systems of Distinct Representatives

An important case is that where K is a collection of subsets of a finite set L, and the pairs (k, l) which are deemed compatible are those such that $l \in k$. An assignment is then called a *system of distinct representatives*: each subset k is represented by one of its elements l in such a way that the representatives are all different. In this context $|H(T)|$ is the total number of elements belonging to the sets k in the subcollection $T \subset K$.

Corollary.   *A system of distinct representatives exists for a collection K of subsets of a finite set L if and only if, for every subcollection $T \subset K$, one has*

$$\begin{bmatrix} number\ of\ elements\ in\ the \\ union\ of\ the\ sets\ in\ T \end{bmatrix} \geq \begin{bmatrix} number\ of \\ sets\ in\ T \end{bmatrix}$$

EXAMPLE 2

A new committee is to be formed in the United Nations. It will consist of a representative from each of the existing committees, many of which have overlapping membership. Is this possible without any one individual having to represent two or more of the committees to which he or she belongs? According to the corollary, it is impossible if and only if there is some set of n committees containing altogether less than n members.

5D.* APPLICATION TO PARTIALLY ORDERED SETS

An interesting combinatorial optimization problem to which matching theory can be applied is the chain decomposition of partially ordered sets.

Let Z be a finite set with a strict partial ordering (i.e., a transitive and antisymmetric relation $<$, as in Section 2J). Two elements z_1 and z_2 of Z are said to be *comparable* if either $z_1 < z_2$ or $z_2 < z_1$; otherwise they are *noncomparable*.

A *chain* is a subset of Z whose elements are mutually comparable. In other words, a chain is a set of elements that can be arranged in an ascending sequence: $z_1 < z_2 < \cdots < z_r$. (A chain in this sense corresponds to a positive path in the acyclic network associated with the relation $<$ in Section 2J; here $<$ coincides with its transitive closure.) A *chain decomposition* of Z is a collection of disjoint chains whose union is all of Z.

Chain Decomposition Problem. *Determine a chain decomposition D that minimizes $|D|$ (the number of chains).*

EXAMPLE 3

A strictly partially ordered set is schematized in Figure 5.4. Here $z_1 < z_2$ means it is possible to reach z_2 from z_1 by an entirely *upward* route along the connecting lines (z_1 and z_2 do not have to be "consecutive"). A particular chain decomposition D with $|D| = 3$ is

$$D = \{\{j, h, f, d, b\}, \quad \{k, i, c, a\}, \quad \{l, m, g, e\}\}.$$

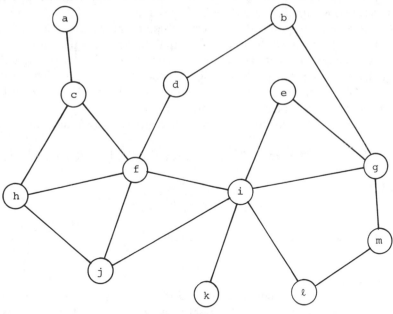

Figure 5.4

EXAMPLE 4

Let Z be a collection of bowls, with the relation $z_1 < z_2$ taken to mean that the bowl z_1 fits inside the bowl z_2. This is a strictly partially ordered set, and a chain is a set of bowls that can form a "nest." The problem is to arrange the collection of bowls into the fewest nests. Instead of bowls, Z could consist of subsets of a given finite set, with the relation $<$ interpreted as strict inclusion.

EXAMPLE 5

Let Z denote a set of meetings scheduled at certain times and of known durations. Define $z_1 < z_2$ to mean that z_2 is scheduled to start at some time after z_1 has finished. A chain is then a set of meetings that do not overlap in time and can therefore be covered by the same newspaper reporter. The chain decomposition problem is that of finding the smallest number of reporters able to cover all the meetings.

Reduction to a Matching Problem

Let K and L be two extra copies of the set Z. The copy in K of an element $z \in Z$ will be denoted by z', whereas the copy of z in L will be z''. The pair (z_1', z_2'') in $K \times L$ (consisting of respective copies of two elements z_1 and z_2 of Z) is deemed compatible if and only if $z_1 < z_2$.

A match M corresponds then to a set of comparable pairs in Z, such that no element z appears more than once as a first element of a pair, nor more than once as a second element. The pairs in such a set can be "strung together" to form disjoint chains, by identifying the common elements. The remaining elements of Z can then be regarded as chains of length 1, and in this way a chain decomposition of Z will be obtained.

For example, if Z consists of eleven elements $z_1, \ldots, z_{11}$ and the match is

$$M = \{(z_1', z_2''), (z_2', z_3''), (z_4', z_5''), (z_6', z_7''), (z_7', z_8''), (z_8', z_9'')\},$$

one has

$$z_1 < z_2 < z_3, \qquad z_4 < z_5, \qquad z_6 < z_7 < z_8 < z_9,$$

and consequently the chain decomposition

$$D = \{\{z_1, z_2, z_3\}, \{z_4, z_5\}, \{z_6, z_7, z_8, z_9\}, \{z_{10}\}, \{z_{11}\}\}.$$

Conversely, every chain decomposition D corresponds in this way to a uniquely determined match M. Observe furthermore that, under this correspondence, each chain in D of length m is associated with $m - 1$ pairs in M. If $C_1, \ldots, C_k$ are the chains in D, one has

$$|M| = (|C_1| - 1) + \cdots + (|C_k| - 1)$$

$$= (|C_1| + \cdots + |C_k|) - k = |Z| - |D|.$$

Thus there is a one-to-one correspondence between chain decompositions D and matches M, with

$$|D| + |M| = |Z|.$$

A solution to the problem therefore provides a solution to the chain decomposition problem. The König-Egerváry theorem in Section 5A thus has a consequence for partially ordered sets. Let us call a subset N of Z a *noncomparable set* if its elements are mutually noncomparable.

Noncomparability Problem. *Determine a noncomparable set N that maximizes $|N|$.*

A particular noncomparable set in Example 3 is $N = \{h, i, m\}$. Note that $|N| = 3 = |D|$ for the chain decomposition already mentioned in this example.

In the case of Example 5, a noncomparable set is a set of meetings, every two of which overlap in time and therefore cannot be assigned to the same reporter. Actually, since each meeting lasts an interval of time, this condition means there is some moment when all the meetings in the set are in progress.

Dilworth's Theorem. *If Z is any finite set with a strict partial ordering, then*

$$\begin{bmatrix} \text{minimum } |D| \text{ in the} \\ \text{decomposition problem} \end{bmatrix} = \begin{bmatrix} \text{maximum } |N| \text{ in the} \\ \text{noncomparability problem} \end{bmatrix}$$

Proof. If D is any chain decomposition and N is any noncomparable set, then certainly $|D| \geq |N|$ because no chain D can contain more than one element of N. It follows that the inequality $\geq$ holds in the theorem.

Now suppose D is a chain decomposition yielding the minimum. To complete the proof, a noncomparable set N will be displayed with $|N| = |D|$.

Let M be the match that corresponds to D as before, with $|M| = |Z| - |D|$. According to the König-Egerváry theorem, there is a block B with $|B| = |Z| - |D|$ (but no block of smaller cardinality). The blocking property means in this context that whenever $z_1 < z_2$ in Z, then B contains either the first copy z_1' of z_1 or the second copy of z_2'' of z_2.

Let N be the set of elements $z \in Z$ with neither copy z' nor z'' in B. It is clear from what has just been said that N is a noncomparable set, and

$$|Z| - |N| = [\text{number of elements } z \in Z \text{ with } z' \text{ or } z'' \text{ in } B].$$

It will be shown now that there are no elements $z \in Z$ with both z' and z'' in B, so that the number on the right is exactly $|B|$.

Suppose $z \in Z$ has both z' and z'' in B. Since B is optimal, none of its elements can be omitted without losing the blocking property. Thus there must exist $z_1 \in Z$ with (z_1', z'') compatible (i.e., $z_1 < z$) but $z_1' \notin B$. Similarly, there must exist $z_2 \in Z$ with (z', z_2'') compatible (i.e., $z < z_2$) but $z_2'' \notin B$. Then

$z_1 < z_2$ by transitivity, so that (z_1', z_2'') is a compatible pair not blocked by B. This is a contradiction leading to the conclusion that there can be no such element z.

Thus $|Z| - |N| = |B| = |Z| - |D|$, and N is therefore a noncomparable set with $|N| = |D|$. This completes the proof.

5E. OPTIMAL ASSIGNMENTS

In the notation and terminology of Section 5C assume there is at least one assignment M from K to I, relative to the relation H. Let q be an integer-valued function on H, and associate with each assignment M the number

$$q(M) = \sum_{(k,l) \in M} q(k,l).$$

Here $q(k,l)$ can be interpreted as the "cost" of incorporating the pair (k,l), and $q(M)$ as the "cost" of M. However, the possibility of negative costs is not excluded. (The assumption that the costs $q(k,l)$ are all integral is a commensurability condition to be used in the algorithm in Section 5F. It could just as well be assumed that the costs are all rational.)

Optimal Assignment Problem. *Minimize $q(M)$ over all assignments M, under the normalizing assumption that $|K| = |L|$.*

The assumption that $|K| = |L|$ is convenient and involves no real loss of generality. The existence of an assignment already implies $|K| \le |L|$. If $|K| < |L|$, it is possible to augment K by dummy elements k (the number of them being $|L| - |K|$), such that k is compatible with every $l \in L$ with corresponding costs $q(k,l) = 0$. This would have no real effect on the problem.

EXAMPLE 6

The elements of K represent certain construction projects to be contracted out by government agencies, whereas the elements of L are construction companies. The pair (k,l) is compatible if company l is capable of doing project k and has expressed its willingness to do so by making a bid $q(k,l)$ (the amount of money it would charge as a fee). Each project must be done, but no company can do more than one project. Thus an assignment M is desired for which $q(M)$, the total amount to be paid for all the projects, is minimal.

The assumption that no company can do more than one project is no real restriction, since if a company can do two projects, for instance, two copies of the same company can be introduced into the set L. As already noted, dummy projects can be invented, if necessary, to ensure that $|K| = |L|$. (A situation without enough willing companies could be modeled similarly with dummy companies whose bids represent the costs of abandoning certain projects.)

The algorithm to be presented in Section 5F for solving the optimal assignment problem depends on duality in the following sense. (The full mechanism for generating such duality will not emerge until Chapters 7 and 8.)

Dual Problem.   *Maximize*

$$r(u) = \sum_{l \in L} u(l) - \sum_{k \in K} u(k)$$

over all functions u: $K \cup L \to R$ satisfying $u(l) - u(k) \le q(k, l)$ for all pairs (k, l) in the relation H. (Here the elements of K and L are supposed to be distinct, and it is assumed as in the optimal assignment problem that $|K| = |L|$.)

This is clearly a linear programming problem. Its constraints can easily be satisfied, for example, by choosing the values $u(l)$ arbitrarily for $l \in L$ and then choosing for each $k \in K$ a value $u(k)$ greater than all the quantities $u(l) - q(k, l)$ corresponding to compatible pairs (k, l). Furthermore if u is any function satisfying the constraints and M is any assignment, one has

$$r(u) = \sum_{(k, l) \in M} [u(l) - u(k)] \le q(M),$$

since every $k \in K$ and $l \in L$ appears exactly once as an element of a pair in M. Thus $q(M)$ furnishes an upper bound to the supremum in the dual problem, whereas $r(u)$ furnishes a lower bound to the minimum in the optimal assignment problem.

The following fact will be established algorithmically in the next section. (The " $\ge$ " part has just been proved.)

Duality Theorem for Optimal Assignments

$$\begin{bmatrix} \min q(M) \text{ in the} \\ \text{optimal assignment problem} \end{bmatrix} = \begin{bmatrix} \max r(u) \text{ in the} \\ \text{dual problem} \end{bmatrix}$$

This relation is based of course on the assumption at the beginning of this section that at least one assignment exists. Otherwise the dual would have sup $= +\infty$ (Exercise 5.27).

Interpretation of the Dual Problem

It is not so easy to get an intuitive feeling for the dual problem. But in the setting of Example 6 it can be viewed as a problem of evaluating the "economic worth" of the projects and companies, relative to the specific context of the bids that were made. This may be phrased in terms of a program of subsidy and taxation conducted by a regulatory agency.

The company l that is selected for project k is to be paid an amount $u(k)$ as a subsidy, in addition to its fee $q(k, l)$. On the other hand, it is to be taxed an amount $u(l)$. The taxes are not permitted to draw on revenue beyond what the companies receive for their projects, nor are the taxes permitted to diminish through this restriction the set of companies that are eligible for any particular project. Thus relative to the subsidies decided upon, the taxes must satisfy

$$u(l) \leq q(k, l) + u(k) \text{ for all compatible pairs } (k, l).$$

Subject to these limitations, the agency wants to maximize the government's net revenues:

$$[\text{total taxes received}] - [\text{total subsidies paid out}] = r(u).$$

The agency's desires are therefore expressed by the dual problem.

Interestingly enough, the algorithm given in the next section can also be interpreted in this framework. It will be seen to lead to a situation where the managers of the various projects can "independently" come up with the optimal assignment of companies.

5F.   HUNGARIAN ASSIGNMENT ALGORITHM

The optimal assignment problem will now be solved by way of a procedure that iteratively searches for assignments relative to certain "subrelations" of the given compatibility relation H. (The existence of an assignment relative to H itself is presupposed.) The procedure leads at the same time to a solution to the dual problem.

In the general step of the algorithm there is an integer-valued function u on $K \cup L$ that satisfies the constraints of the dual problem. The algorithm can be initiated with any such u, for instance,

$$u(k) = 0, \qquad u(l) = \min_{k \in K} q(k, l). \tag{1}$$

Let

$$H_u = \{(k, l) \in H \mid u(l) - u(k) = q(k, l)\},$$

and test whether there exists an assignment M relative to H_u. If yes, it is claimed that M solves the optimal assignment problem, whereas u solves the dual problem. If no, then a set $T \subset K$ is detected such that $|H_u(T)| < |T|$, where of course

$$H_u(T) = \{l \in L \mid \exists k \in T \text{ with } (k, l) \in H_u\}.$$

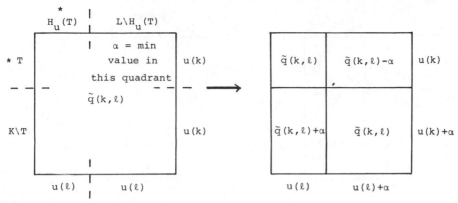

Figure 5.5

In this case let

$$\alpha = \min\{ q(k, l) + u(k) - u(l) \,|\, (k, l) \in H \setminus H_u, k \in T \}$$

and define u' on $K \cup L$ by

$$u'(k) = \begin{cases} u(k) & \text{if } k \in T, \\ u(k) + \alpha & \text{for all other } k \in K, \end{cases}$$

$$u'(l) = \begin{cases} u(l) & \text{if } l \in H_u(T), \\ u(l) + \alpha & \text{for all other } l \in L. \end{cases}$$

Then u' again satisfies the constraints of the dual problem, and $r(u') > r(u)$. Repeat the step with u' in place of u.

Remark. The computation in the algorithm can be carried out in the tableau format of Figure 5.5, where

$$\tilde{q}(k, l) = q(k, l) + u(k) - u(l).$$

The pairs in H_u correspond to the *zeros* in the tableau at each iteration. Pairs not in H are marked by $+\infty$.

EXAMPLE 7

The calculation of an optimal assignment is illustrated in Figure 5.6. The first tableau shows the costs $q(k, l)$ (the relation H consisting of the pairs with finite cost), whereas the other tableaus give the computations in the pattern of the preceding remark. The determination of the set T by way of matching theory (by a direct max flow approach or its refinement in Section 5B, or

Iteration 1
(initial u from (1))

given data

	ℓ_1	ℓ_2	ℓ_3	ℓ_4	ℓ_5	ℓ_6
k_1	4	∞	∞	3	2	7
k_2	∞	1	4	2	∞	∞
k_3	2	6	∞	1	3	5
k_4	3	0	5	2	1	∞
k_5	∞	∞	6	∞	∞	10
k_6	4	∞	3	∞	2	7

Iteration 1 matrix:

*	2	∞	∞	2	1	2	0
*	∞	1	1	1	∞	∞	0
	0	6	∞	0	2	0	0
	1	0	2	1	0	∞	0
*	∞	∞	3	∞	∞	5	0
	2	∞	0	∞	1	2	0
	2	0	3	1	1	5	

$T=\{k_1,k_2,k_5\}$
$H_u(T)=\emptyset$
$\alpha=1$

Iteration 2

1	∞	∞	1	0	1	0
∞	0	0	0	∞	∞	0
0	6	∞	0	2	0	1
1	0	2	1	0	∞	1
* ∞	∞	2	∞	∞	4	0
2	∞	0	∞	1	2	1
3	1	4	2	2	6	

$T=\{k_5\}$
$H_u(T)=\emptyset$
$\alpha=2$

Iteration 3

1	∞	∞	1	0	1	2
∞	0	0	0	∞	∞	2
0	6	∞	0	2	0	3
1	0	2	1	0	∞	3
* ∞	∞	0	∞	∞	2	0
* 2	∞	0	∞	1	2	3
5	3	6	4	4	8	

$T=\{k_5,k_6\}$
$H_u(T)=\{\ell_3\}$
$\alpha=1$

Iteration 4

* 1	∞	∞	1	0	1	3
∞	0	1	0	∞	∞	3
0	6	∞	0	2	0	4
1	0	3	1	0	∞	4
* ∞	∞	0	∞	∞	1	0
* 1	∞	0	∞	0	1	3
6	4	6	5	5	9	

$T=\{k_1,k_5,k_6\}$
$H_u(T)=\{\ell_3,\ell_5\}$
$\alpha=1$

Iteration 5 (terminate)

0	∞	∞	0	(0)	0	3
∞	0	2	(0)	∞	∞	4
(0)	6	∞	0	3	0	5
1	(0)	4	1	1	∞	5
∞	∞	0	∞	∞	(0)	0
0	∞	(0)	∞	0	0	3
7	5	6	6	5	10	

Figure 5.6

154

simply by inspection) is not shown. The circled entries in the last tableau furnish the optimal assignment:

$$M = \{(k_1, l_5), (k_2, l_4), (k_3, l_1), (k_4, l_2), (k_5, l_6), (k_6, l_3)\}$$

with cost $q(M) = 19 = r(u)$.

Justification of the Algorithm

If M is an assignment relative to H_u, then M is of course also an assignment relative to H, and one has

$$q(M) = \sum_{(k,l) \in M} q(k, l) = \sum_{(k,l) \in M} [u(l) - u(k)] = r(u).$$

Since, as already observed in Section 5E, $q(M)$ is a lower bound to the minimum in the optimal assignment problem and $r(u)$ is an upper bound to the supremum in the dual problem, this equation implies M and u are optimal.

If there exists no assignment relative to H_u, there must exist pairs (k, l) in $H \setminus H_u$ with $k \in T$, for otherwise $|H(T)| = |H_u(T)| < |T|$, and one could conclude, contrary to the assumption, that no assignment exists relative to H. For each such (k, l), one has $u(l) - u(k) < q(k, l)$ by definition. Therefore $\alpha > 0$. The formula for u' yields

$$u'(l) - u'(k) = \begin{cases} u(l) - u(k) & \text{if } k \in T, l \in H_u(T), \\ u(l) - u(k) + \alpha & \text{if } k \in T, l \notin H_u(T), \\ u(l) - u(k) - \alpha & \text{if } k \notin T, l \in H_u(T), \\ u(l) - u(k) & \text{if } k \notin T, l \in H_u(T), \end{cases}$$

where only the second alternative offers any threat of u' failing to satisfy the constraints, like u. This threat is eliminated by the choice of α, since the pairs $(k, l) \in H$ with $k \in T$, $l \notin H_u(T)$ belong to $H \setminus H_u$ and consequently have $u(l) - u(k) + \alpha \le q(k, l)$.

Finally, observe from the definition of u' and the relation $|K| = |L|$ that

$$r(u') = \sum_{l \in H_u(T)} u(l) + \sum_{l \notin H_u(T)} [u(l) + \alpha] - \sum_{k \in T} u(k) - \sum_{k \notin T} [u(k) + \alpha]$$

$$= \sum_{l \in L} u(l) + \alpha[|L| - |H_u(T)|] - \sum_{k \in K} u(k) - \alpha[|K| - |T|]$$

$$= r(u) + \alpha[|T| - |H_u(T)|].$$

Since $|H_u(T)| < |T|$, it may be concluded that $r(u') > r(u)$.

Termination

Suppose the algorithm is initiated with a function u whose values are all integral (or even just rational). This is always possible; for instance, one could start with arbitrary integer values for $u(k)$ (e.g., all zeros) and let

$$u(l) = \min\{q(k,l) + u(k) | (k,l) \in H \quad \text{for all } l \in L\}.$$

(The latter minimum is over a nonempty set, because at least one assignment relative to H exists by assumption.) Since the costs $q(k, l)$ are all integral, the numbers α, $u'(k)$, $u'(l)$, and $r(u') - r(u)$ calculated in the course of the iteration will all be integral (or if u is just rational valued, at least in the same commensurability class as the values of u). Since the supremum in the dual problem is finite (due again to the assumed existence of at least one assignment relative to H), it follows that there cannot be an infinite sequence of iterations.

Proof of the Duality Theorem

If the algorithm terminates, it does so with solutions M and u which have been shown to satisfy $q(M) = r(u)$. Then the relation in the theorem in Section 5E must hold, by virtue of the general inequality derived just before the theorem. Since the algorithm can always be initialized so that termination is assured, as just shown, this proves the theorem.

Interpretation of the Algorithm

The algorithm takes on a natural meaning in terms of Example 6 and the interpretation given at the end of Section 5E for the dual problem. In the context of the primal problem, the manager of project k, trying from an individual perspective to do the best for the unit of government he or she represents, would aim at selecting the company that entered the lowest bid for the project in question. Thus the manager would aim at minimizing $q(k, l)$ over all $l \in L$ such that $(k, l) \in H$. But as we know, this could well lead to conflicts with the choices of the managers of other projects, because the overall situation is not taken into account.

The perspective of the managers is different, however, after taxes have been announced by the regulatory agency, especially if the amount $u(l)$ levied on company l is earmarked for the budget of the manager of the project to be undertaken by company l. Then the manager of project k will aim instead at maximizing $u(l) - q(k, l)$ over all $l \in L$ such that $(k, l) \in H$. This maximum value is of course the regulatory agency's choice for the subsidy $u(k)$. The set of all l for which the maximum is attained can thus be identified with $H_u(k)$, the set of all $l \in L$ such that $(k, l) \in H_u$.

The individual preferences of the project managers therefore lead to their specification of the sets $H_u(k)$ for $k \in K$ that together constitute the relation

H_u. The manager of project k will accept any of the companies l such that $l \in H_u(k)$. If the companies can work out an assignment that fits this prescription, it will be an optimal assignment as just proved. If not, the regulatory agency is notified of a conflict (i.e., told of a set T with $|H_u(T)| < |T|$), and then it acts to take advantage of this by adjusting taxes so as to obtain increased revenues over the current amount $r(u)$.

Other Computational Approaches

The optimal assignment problem can also be solved by algorithms that at each step replace an assignment M relative to H by a better assignment M'. Such a procedure will appear in Exercise 7.42 as a special case of a more general procedure for the optimal distribution problem.

5G.* MATCHING WITH MULTIPLICITIES

Up until now matches and assignments have been subjected to the requirement that no element $k \in K$ or $l \in L$ can belong to more than one of the pairs (k, l) selected from the compatibility relation $H \subset K \times L$. This can easily be broadened, without sacrificing the main features of the theory, so that even the pairs themselves can be present in the selection with certain multiplicities $x(k, l)$.

Suppose that positive integers $m(k)$ and $n(l)$ are given for each $k \in K$ and $l \in L$. A *multimatch* is a function x on H whose values $x(k, l)$ are non-negative integers such that for each $k \in K$ the total of $x(k, l)$ over all l compatible with k is at most $m(k)$, whereas for each $l \in L$ the total of $x(k, l)$ over all $k \in K$ compatible with l is at most $n(l)$. The *cardinality* of the multimatch x is the number

$$c(x) = \sum_{(k, l) \in H} x(k, l).$$

A *multiassignment* is a multimatch x such that for each $k \in K$ the sum of $x(k, l)$ over all compatible l actually equals $m(k)$. This is equivalent to the condition that $c(x) = m(K) \le n(L)$, where

$$m(K) = \sum_{k \in K} m(k), \qquad n(L) = \sum_{l \in L} n(l).$$

Multimatching Problem. *Determine a multimatch x of maximum cardinality $c(x)$.*

Like the ordinary matching problem, which corresponds to $m(k) = n(l) = 1$, this can be solved as a max flow problem; see the network of Figure 5.7. If the maximum cardinality is $m(K)$, the solutions are multiassignments, although otherwise no multiassignment exists.

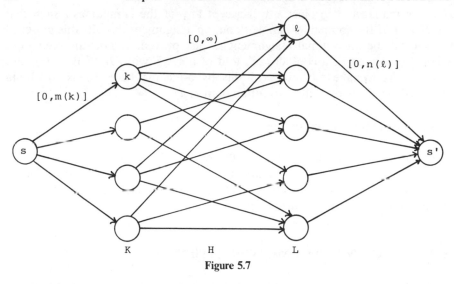

Figure 5.7

EXAMPLE 8

A company is faced with repairing a large fleet of trucks, which can be divided into a number of categories according to the different models and different kinds of repairs that are needed. There are several repair centers that can work on limited numbers of trucks, but they are not all equipped to handle all the categories. How should the trucks be allocated among the centers so as to get as many as possible repaired? (This should be thought of as relative to a "unit time period.")

Let the elements of K be the categories of trucks and the elements of L the repair centers: (k, l) is compatible if center l can handle category k. There are $m(k)$ trucks in category k. The maximum number of trucks that can be handled by center l is $n(l)$. The multimatching problem is thus to be solved, where $x(k, l)$ is the number of trucks of category k routed to center l.

EXAMPLE 9

A mother of several small children has a limited stock of several kinds of candy. As a special treat, she has promised that each child can have four pieces after lunch. Unfortunately only certain kinds are acceptable to each child. Will it be possible for her to keep her promise under this restriction?

Let the elements of K represent the children, whereas those of L represent the kinds of candy. Define the pair (k, l) to be compatible if candy l is acceptable to child k. Let $n(l)$ be the number of pieces of candy l in stock; $m(k) = 4$ for all k. The existence of a multiassignment is in question.

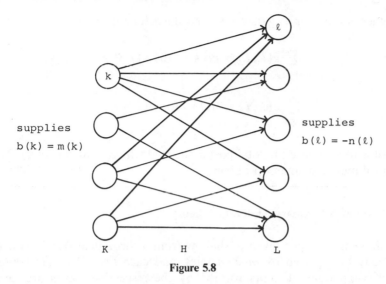

Figure 5.8

Another extension of earlier theory is obtained by supposing each $(k, l) \in H$ to have an integral "cost" $q(k, l)$ associated with it and defining the cost of a multimatch x as

$$q(x) = \sum_{(k, l) \in H} q(k, l) x(k, l).$$

(Again the assumption that the costs $q(k, l)$ are integral is no more stringent really than the assumption that they are all rational.)

Optimal Multiassignment Problem.   *Minimize $q(x)$ over all multiassignments x, under the normalizing assumption that $m(K) = n(L)$.*

The normalizing assumption makes it possible to identify the multiassignments with the integral solutions to the feasible distribution problem in Figure 5.8.

EXAMPLE 10.   (*The Classical "Transportation" Problem*)

Each $k \in K$ is a warehouse with supply $m(k)$, whereas each $l \in L$ is a customer with demand $n(l)$. The total supply $m(K)$ equals the total demand $n(L)$. Each warehouse k is able to ship to any customer l, the cost per unit being $q(k, l)$. The number of units to be sent from k to l is $x(k, l)$. Thus the feasible shipping schedules are the multiassignments x relative to $H = K \times L$, or in other words, non-negative integer-valued functions on $K \times L$ such that

$$\sum_{l \in L} x(k, l) = m(k) \quad \text{for all } k \in K,$$

$$\sum_{k \in K} x(k, l) = n(l) \quad \text{for all } l \in L.$$

The goal is to minimize the total cost associated with x, which is $q(x)$. This is an optimal multiassignment problem.

Equivalence with Ordinary Matching Theory

The extra generality just given is shallow from a theoretical standpoint. Each $k \in K$ may be replaced by $m(k)$ copies, and each $l \in L$ by $n(l)$ copies; if (k, l) is compatible, all pairs formed by the respective copies are deemed compatible. Generalized matches may then be identified with ordinary matches in the new framework.

This would not be practical in most cases because the size of the model would be increased enormously. Nevertheless, it does suggest how results of ordinary matching theory can be carried over to generalized matches with relatively few changes (see Exercises 5.33 through 5.37).

Further Generalization

Integral upper bounds $p(k, l)$ could also be placed on the multiplicities $x(k, l)$. This would amount to replacing $[0, \infty]$ by $[0, p(k, l)]$ as capacity interval for the various middle arcs in Figure 5.7.

In Example 8, $p(k, l)$ would be the maximum number of trucks of category k that can be handled at repair center l. If the children in Example 9 insist that all four pieces of candy they get must be different, then $p(k, l) = 1$ for all compatible pairs (k, l).

The generalizing does not have to stop here, but there comes a stage when it makes more sense simply to pass to the fundamental theory already established for max flow problems and feasible distribution problems, as well as the general linear optimization theory to be developed in Chapter 7.

5H.*   BOTTLENECK OPTIMIZATION

Another kind of optimization problem corresponds to the philosophy that "a chain is only as strong as its weakest link." Suppose that each of the pairs (k, l) in the compatibility relation $H \subset K \times L$ has a number $r(k, l)$ associated with it, called its *rating*. (This could be a real number or $\pm \infty$.) The rating of an assignment M (relative to H) is defined as

$$f(M) = \min_{(k,l) \in M} r(k,l).$$

It is assumed in what follows that at least one assignment M exists, and that the numbers $r(k,l)$ are not all the same.

Bottleneck Assignment Problem. *Determine an assignment M of maximum rating $f(M)$.*

The "bottleneck" terminology was suggested originally by the following setting.

EXAMPLE 11

The elements of L are workers, whereas those of K are positions in an assembly line. The pair (k,l) is "compatible" if worker l is willing and competent to function in position k; the rating $r(k,l)$ is the number of units per hour that can be handled by this worker in this position. An assignment M is obtained by putting a "compatible" worker in each position, and its rating $f(M)$ is the number of units per hour that can then be assembled by the line as a whole. This is to be maximized.

Threshold Algorithm

There are only finitely many numbers occurring as ratings $r(k,l)$ for the different pairs $(k,l) \in H$. Let them be $\alpha_1, \ldots, \alpha_s$ in decreasing order. For $p = 1, \ldots, s$, let H^p be the set of all pairs $(k,l) \in H$ such that $r(k,l) \geq \alpha_p$. Obviously, M is an assignment relative to H^p if and only if M is an assignment relative to H having $f(M) \geq \alpha_p$. Therefore one simply has to determine the first index p such that an assignment exists relative to H^p (as can be tested by the matching algorithm), and any corresponding assignment will then solve the problem. The testing can be done systematically in various ways. One can work forward from $p = 1$ or backward from $p = s$. Or one can jump around in a method of bracketing the desired index p.

Duality

The algorithm suggests a form of duality based on the fact that for each index p either there does exist an assignment M relative to H^p, or there is a blocking set B relative to H^p with $|B| < |K|$ (see the König-Egerváry theorem in Section 5A). The latter means that B is a subset of $K \cup L$ with $|B| < |K|$ and $r(k,l) < \alpha_p$ for all $(k,l) \in H$ having $k \in K \setminus B$ and $l \in L \setminus B$.

To put this in an agreeable form, let us define the *rating* of a set $B \subset K \cup L$ with $|B| < |K|$ by

$$g(B) = \max\{r(k, l) | (k, l) \in H, k \in K \setminus B, l \in L \setminus B\}.$$

Bottleneck Blocking Problem.   *Minimize the rating $g(B)$ over all sets $B \subset K \cup L$ with $|B| < |K|$.*

Bottleneck Duality Theorem.

$$\begin{bmatrix} \max f(M) \text{ in the bottleneck} \\ \text{assignment problem} \end{bmatrix} = \begin{bmatrix} \min g(B) \text{ in the bottleneck} \\ \text{blocking problem} \end{bmatrix}$$

Proof.   As in the algorithm, let $\alpha_1, \ldots, \alpha_s$ be the various values of the ratings $r(k, l)$, arranged in descending order. Of course the only possible values for $f(M)$ and $g(B)$ in the two problems are among these. It has been seen previously that for each index p there is either an assignment M relative to H with $f(M) \geq \alpha_p$ or a set $B \subset K \cup L$ with $|B| < |K|$ and $g(B) < \alpha_p$ (but not both). In other words, either $\max f \geq \alpha_p$ or $\min g < \alpha_p$, but not both. Therefore the highest α_p for which the first alternative holds must be the common extremum in the two problems. (Note that a solution to the bottleneck blocking problem is then obtainable constructively by solving the ordinary blocking problem for the relation H^{p-1}.)

5I.*   EXERCISES

5.1.   (*Matching Problem*). Figure 5.9 displays a compatibility relation and a match M (in the notation explained at the end of Section 5A). Verify that M is of maximum cardinality.

(*Hint.*   Determine by inspection a block B with $|B| = |M|$.)

	ℓ_1	ℓ_2	ℓ_3	ℓ_4	ℓ_5	ℓ_6	ℓ_7	ℓ_8
k_1	0	1	0	0	①	0	0	0
k_2	1	0	①	0	0	1	0	1
k_3	0	①	0	0	1	0	0	0
k_4	0	1	0	0	0	0	1	0
k_5	1	0	1	1	0	①	0	0
k_6	0	0	0	0	1	0	①	0
k_7	0	1	0	0	0	0	1	0
k_8	0	1	0	①	0	0	0	1

Figure 5.9

5.2. (*Connection Problem*). Let N^+ and N^- be two nonempty disjoint node sets in a network G. A *connection* from N^+ to N^- is a collection M of elementary paths from N^+ to N^- (with only their initial node in N^+ and only their terminal node in N^-) such that no two of the paths in M have any arc in common. The paths may intersect at various nodes, however. The connection problem consists of determining a connection M of maximum cardinality $|M|$ (number of paths). Show how this can be solved by max flow theory.

(*Hint*. Introduce appropriate capacity intervals, consider conformal realizations as in Section 4B of the corresponding integral feasible flows.)

5.3. (*Connection Problem*). Prove that the maximum of $|M|$ in the preceding exercise equals the minimum of $|Q|$ over all cuts $Q: N^+ \downarrow N^-$.

(*Hint*. Apply the max flow min cut theorem.)

5.4. (*Positive Connections*). Give the appropriate modifications of Exercises 5.2 and 5.3 for the case where only *positive* paths are permitted (i.e., paths $P: N^+ \to N^-$ with $P^+ = P$, $P^- = \varnothing$).

5.5. (*Strong Connections*). In the context of Exercise 5.2 a *strong* connection from N^+ to N^- is a connection M consisting entirely of positive paths, no two of which have any nodes in common (including initial or terminal nodes). The strong connection problem maximizes $|M|$ over all such M. Explain how this could be solved by max flow theory.

(*Hint*. Use node capacities in the sense of Section 3L.)

5.6. (*Menger's Theorem*). Let N^+ and N^- be nonempty disjoint node sets in a network G. A note set B is called a *block* from N^+ to N^- if every positive path from N^+ to N^- contains at least one node of B. Prove that the minimum of $|B|$ over all such blocks B equals the maximum of $|M|$ over all strong connections M from N^+ to N^- (see Exercise 5.5).

5.7. (*Menger's Theorem*). Show that the König-Egerváry theorem can be identified with a special case of the result in the preceding exercise.

5.8. (*Dynamic Paths*). Suppose the network in Figure 1.10 represents a system of unidirectional bicycle trails in a park, the durations being in terms of the minutes it takes a bicycle to traverse the trail. What is the maximum number of bicyclists starting at point a at time $t = 0$ that can arrive at point d by $t = 5$ without having encountered each other en route? This is a model in discrete time periods; two cyclists have "encountered" each other if in some period they enter the same trail or wait over at the same intermediate node b or c.

(*Hint*. Set this up in a manner resembling Exercise 5.5 in terms of the dynamic network G_T for $T = 5$. Apply the max flow algorithm.)

5.9. (*Matching Problem*). Solve the matching and blocking problems relative to the compatibility relation indicated in Figure 5.10 using for the initial match the pairs corresponding to the circled entries. (This can be done

	ℓ_1	ℓ_2	ℓ_3	ℓ_4	ℓ_5	ℓ_6	ℓ_7	ℓ_8
k_1	0	1	1	0	1	1	0	0
k_2	1	0	0	0	0	1	0	0
k_3	0	0	1	1	1	0	1	1
k_4	0	0	0	1	0	0	0	1
k_5	1	0	0	0	0	1	0	1
k_6	0	1	0	0	0	0	1	0
k_7	1	0	0	1	0	0	0	0
k_8	0	0	0	0	0	1	0	1

Figure 5.10

either by the max flow approach as in Section 5A or by its refinement, the matching algorithm explained in Section 5B.)

5.10. (*Matching Algorithm*). Show that Step 4 of the algorithm in Section 5B does produce a match M' with $|M'| = |M| + 1$. (Relate this to the step $x \to x' = x + \alpha e_p$ in the max flow algorithm.)

5.11. (*Matching with Obligatory Elements*). Let two subsets $K_0 \subset K$ and $L_0 \subset L$ be designed as obligatory, in the sense that one is only interested in matches that use *all* the elements $k \in K_0\ l \in L_0$. Prove that if there exists such a match at all, then there is one that at the same time solves the matching problem, as previously stated. Thus this side condition, if satisfiable, imposes no diminution of the maximum cardinality when it is added to the matching problem.

(*Hint.*   Initiate the algorithm with a match satisfying the condition.)

5.12. (*Matching with Obligatory Elements*). Prove that the following condition is necessary and sufficient for the existence of at least one match using all the elements in K_0 and L_0, where $K_0 \subset K$ and $L_0 \subset L$ are given sets. For every block B, both $|K_0 \setminus B| \leq |L \cap B|$ and $|L_0 \setminus B| \leq |K \cap B|$.

(*Hint.*   Apply the feasible circulation theorem in Section 3H to the network which is like the one in Figure 5.1, except that there is a feedback arc $\tilde{j} \sim (s', s)$ with capacity interval $[0, \infty)$, and the interval $[0, 1]$ is replaced by $[1, 1]$ for arcs $j \sim (s, k)$ with $k \in K_0$, or $j \sim (l, s')$ with $l \in L_0$. Pass from a node set $S \subset K \cup L \cup \{s, s'\}$ to a block B much as in the proof of the König-Egerváry theorem in Section 5A.)

5.13. (*Matching with Obligatory Elements*). Derive the following result from the result in Exercise 5.12. If there is a match M_1 using all the elements of a set $K_0 \subset K$ and also a match M_2 using all the elements of $L_0 \subset L$, then there is a match $M \subset M_1 \cup M_2$ using all of both K_0 and L_0.

(*Hint.*   Replace H by $H_0 = M_1 \cup M_2$.)

5.14. (*Strong Connections*). Develop a special procedure for solving the problem in Exercise 5.5 (and its dual in Exercise 5.6), much as the matching algorithm was obtained as a specialization of the max flow algorithm.

5.15. (*Assignments*). Show that Hall's existence theorem may in turn be derived from its corollary in Section 5C. Thus the corollary is an equivalent form of the same result.

(*Hint.* Take L to be the same set in both cases.)

5.16. (*Assignments*). Let A be an $n \times n$ matrix of 0's and 1's such that the number of 1's in each row is exactly r, and the number of 1's in each column is exactly r. Prove that A is the sum of r permutation matrices.

5.17. (*Riding Problem*). A group of n girls is planning a series of r pony rides together. There are n ponies available, and each girl has listed the r different ponies she wants to ride; the order doesn't matter. Show that the rides can be organized in accordance with these wishes, if each pony has been listed by r different girls.

(*Hint.* See the preceding exercise.)

5.18. (*Assignments*). Let X be a finite set partitioned into disjoint subsets $X_1, \ldots, X_n$. Let Y be a collection of subsets of X (not necessarily disjoint). Give a necessary and sufficient condition for the existence of sets $Y_1, \ldots, Y_n$ in Y, all different, such that $X_i \cap Y_i \neq \varnothing$ for $i = 1, \ldots, n$.

5.19. (*Group Theory*; *P. Hall*). Let G be a finite noncommutative group, and let S be a subgroup of G. Prove there is a subset A of G such that each right coset of G with respect to S contains exactly one element of A, and the same for each left coset.

(*Hint.* Apply Hall's existence theorem in Section 5C in a certain way, with K the collection of right cosets and L the collection of left cosets.)

5.20. (*Assignments*). Generalize Hall's existence theorem to the case where K and L are infinite sets, but still there are only finitely many elements $l \in L$ compatible with any one $k \in K$.

5.21. (*Linear Algebra*). Use Hall's existence theorem to show that any two bases of a finite-dimensional vector space have the same cardinality.

	ℓ_1	ℓ_2	ℓ_3	ℓ_4	ℓ_5	ℓ_6
k_1	0	1	0	1	0	0
k_2	1	0	1	1	0	1
k_3	0	0	0	1	0	0
k_4	1	0	1	0	1	0
k_5	0	1	0	0	0	0

Figure 5.11

5.22. (*Assignments*). For the compatibility relation H indicated in Figure 5.11 (in the notation explained at the end of Section 5A), demonstrate that no assignment exists. (Determine by inspection a set T that violates the condition in Hall's existence theorem.)

5.23. (*Hoffman-Kuhn Theorem*). Consider the problem in Section 5C of finding a system of distinct set representatives, but under the side condition that all the elements of a certain set $L_0 \subset L$ must occur among the representatives. Prove that a solution exists if and only if the following condition holds in addition to the one in the corollary to Hall's existence theorem: for every subset $N \subset L_0$, the number of sets in K meeting N is at least $|N|$.

(*Hint.* See Exercises 5.11 and 5.12.)

5.24. (*Chain Decomposition*). What does Dilworth's theorem in Section 5D say in the case of Example 4 and Example 5?

5.25. (*Chain Decomposition*). As an improvement in the model of Example 5 in Section 5D, suppose that for the same reporter to cover meeting z_2 after meeting z_1, a gap of at least $t(z_1, z_2)$ hours is necessary between the end of z_1 and the start of z_2. (Regard $t(z_1, z_2)$ as $+\infty$ if z_2 starts before the end of z_1.) Demonstrate that the theory of chain decompositions is still applicable, if the time requirements satisfy the reasonable rule that:

$$0 \le t(z_1, z_2) \le t(z_1, z_3) + t(z_3, z_2) \quad \text{for all } z_1, z_2, z_3$$

5.26. (*Partially Ordered Sets*). Show that in the context of the application of matching theory made in Section 5D an assignment cannot exist.

5.27. (*Optimal Assignments*). Prove that the dual problem in Section 5F has supremum $+\infty$ in the case where no assignment relative to H exists.

(*Hint.* This is not covered by the duality theorem, as formulated. Use an argument similar to that for the "improvement" step of the Hungarian assignment algorithm in Section 5F.)

5.28. (*Linear Programming*). What is the dual of the dual problem in Section 5E according to linear programming theory, and how is it related to the optimal assignment problem?

(*Note.* The extremal representation theorem of Section 4J sheds further light in this context.)

5.29. (*Hungarian Assignment Algorithm*). Solve the optimal assignment problem corresponding to the cost matrix in Figure 5.12. Here the number in entry (k, l) is $q(k, l)$, and ∞ is used to mark the pairs which are incompatible (see Example 7 in Section 5F).

5.30. (*Fortified Assignment Algorithm*). Fix any $\varepsilon > 0$, and modify the assignment algorithm by instead taking H_u to be the set of all pairs $(k, l) \in H$ such that $u(l) - u(k) \ge q(k, l) - \varepsilon$. Everything else is to be left the

	ℓ_1	ℓ_2	ℓ_3	ℓ_4	ℓ_5	ℓ_6
k_1	3	∞	0	2	6	4
k_2	2	3	2	1	∞	3
k_3	6	6	∞	7	4	∞
k_4	∞	2	0	2	3	5
k_5	5	∞	6	5	4	7
k_6	7	7	1	∞	6	8

Figure 5.12

same. Show that then termination is inevitable, and when it arrives, one has an assignment M and function u (satisfying the constraints of the dual problem) such that

$$q(M) - r(u) < \varepsilon|K|.$$

Show further that the latter implies M comes within $\varepsilon' = \varepsilon|K|$ of attaining the infimum in the optimal assignment problem, whereas u comes within ε' of attaining the supremum in the dual problem.

5.31. (*Optimal Assignments*). Prove that the duality theorem for optimal assignments is valid even if the costs $q(k, l)$ are arbitrary real numbers. (Make use of the result in the preceding exercise, which does not depend on integrality, and invoke linear programming theory to get the attainment of the supremum in the dual problem.)

5.32. (*Matching with Multiplicities*). Verify in detail that the multimatching problem can be reduced to solving the max flow problem in Figure 5.7.

5.33. (*Matching with Multiplicities*). Formulate and prove the appropriate extension of the König-Egerváry theorem in Section 5A to the multimatching problem. (This can be done by a direct application of max flow theory, or easier, by way of the equivalence mentioned at the end of Section 5G).

5.34. (*Existence of Multiassignments*). Prove the following extension of Hall's existence theorem in Section 5C to the context of Section 5G: a multiassignment exists if and only if for every set $T \subset K$ one has $n(H(T)) \geq m(T)$, where

$$m(T) = \sum_{k \in T} m(k), \qquad n(H(T)) = \sum_{l \in H(T)} n(l).$$

5.34. (*Multimatching Algorithm*). Develop a specialized procedure for solving the multimatching problem (based on the max flow algorithm like the ordinary matching algorithm).

5.36. (*Optimal Multiassignments*). Formulate and prove the appropriate extension of the duality theorem in Section 5E to optimal multiassignments. (The proof can be based on the earlier duality theorem. Alternatively, it could be based on an extended algorithm; see the next exercise.)

5.37. (*Optimal Multiassignment Algorithm*). Develop the appropriate extension of Hungarian assignment algorithm to the optimal multiassignment problem.

5.38. (*Transportation Problem*). Interpret the results in the last two exercises in the case of Example 10 in Section 5G.

5.39. (*Matching with Multiplicities*). Demonstrate that the multimatching problem can be identified with the special case of the optimal multiassignment problem where all pairs are compatible, but their costs are all either 0 or -1.

5.40. (*Optimal Multiassignments*). Show that there is no real effect on the optimal multiassignment problem if the costs $q(k, l)$ are all modified to

$$q'(k, l) = q(k, l) + a(k) + b(l)$$

for arbitrary numbers $a(k)$ and $b(l)$. (In particular, then, there would be no real loss of generality in supposing all costs in the problem to be non-negative.)

5.41. (*Set Representatives*). Consider the problem of finding a system of distinct set representatives as in Section 5C, but with L partitioned into disjoint subsets $L_1, \ldots, L_r$ and the side condition that at least n_j of the representatives must belong to L_j. Show that this can be solved as a max flow problem. What if, instead, at most m_j representatives may belong to L_j?

5.42. (*Set Representatives*). Derive a necessary and sufficient condition for the existence of a system of distinct representatives under the two side conditions in the preceding exercise.

5.43. (*Seating Problem*). Each $k \in K$ is a family with $m(k)$ members, whereas each $l \in L$ is a table with $n(l)$ seats. The families are to be seated at the tables in such a way that no two members of the same family are at the same table. Show that this can be set up as a max flow problem.

5.44. (*Ore's Theorem on Subnetworks*). Associate with each node i of a network G a pair of non-negative integers $r(i)$ and $s(i)$. Is it possible, by removing arcs of G, to obtain a "subnetwork" G' with the same node set N, such that each $i \in N$ is the initial node of exactly $r(i)$ arcs in G' and the terminal node of exactly $s(i)$? Prove that this is true if and only if $r(N) = s(N)$ and

$$q(T', T'') \geq r(T') - s(T'') \quad \text{for all } T' \subset N, T'' \subset N,$$

where $q(T', T'')$ is the number of arcs in G with initial node in T' and

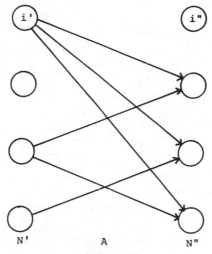

Figure 5.13

terminal node *not* in T'', and

$$r(T') = \sum_{i \in T'} r(i), \qquad s(T'') = \sum_{i \in T''} s(i).$$

(*Hint.* Set this up as an application of the feasible distribution theorem to the bipartite network in Figure 5.13, whose node set is comprised of two copies of the node set of G and whose arcs correspond one to one with those of G.)

5.45. (*Gale-Ryser Theorem on 0–1 Matrices*). For non-negative integers $r_1 \geq r_2 \geq \ldots \geq r_m$ and $s_1 \leq s_2 \leq \ldots \leq s_n$, does there exist an $m \times n$ matrix of 0's and 1's such that the number of 1's in row k is r_k, whereas the number of 1's in column l is s_l? Prove that this is true if and only if

$$\sum_{k=1}^{p} r_k - \sum_{l=1}^{q} s_l \leq p(n-q) \quad \text{for } 0 \leq p \leq m \quad \text{and} \quad 0 \leq q \leq n,$$

where the corresponding sum is regarded as 0 if $p = 0$ or $q = 0$. This condition can be simplified further by showing that

$$\min_{0 \leq q \leq n} \left\{ p(n-q) + \sum_{l=1}^{q} s_l \right\} = \sum_{k=1}^{p} s_k^*,$$

where s_k^* is number of indices l such that $s_l \geq k$.

(*Hint.* Obtain the first form of the condition by an application of the feasible distribution theorem to the network in Figure 5.14.)

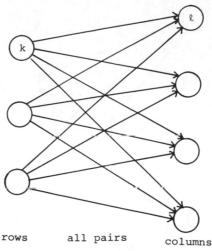

rows   all pairs   columns

Figure 5.14

5.46. (*Bottleneck Path Problem*). Let N^+ and N^- be nonempty disjoint node sets of an arbitrary network G, and let there be associated with each arc j of G two numbers $r^+(j)$ and $r^-(j)$. (These may be real or $\pm\infty$, and no relation between them is required.) For a path $P: N^+ \to N^-$, let $f(P)$ denote the *minimum* of the values $r^+(j)$ for $j \in P^+$ and $r^-(j)$ for $j \in P^-$. For a cut $Q: N^+ \downarrow N^-$, let $g(Q)$ denote the *maximum* of the values $r^+(j)$ for $j \in Q^+$ and $r^-(j)$ for $j \in Q^-$. Prove that

$$\max_{P: N^+ \to N^-} f(P) = \min_{Q: N^+ \downarrow N^-} g(Q).$$

(*Hint*. The painted network algorithm can be used to test whether there exists $P: N^+ \to N^-$ with $f(P) \geq \alpha$.)

(*Note*. The problems could be solved by a threshhold algorithm like the one in Section 5H, but a more efficient approach will appear in Exercise 6.35.)

5.47. (*Max Flow Algorithm*). Suppose that in each iteration of the max flow algorithm one tried to determine the *best* flow-augmenting path in the sense of yielding the greatest increase in flux from N^+ to N^-. Show that this operation would amount to solving a certain bottleneck path problem (see preceding exercise).

5.48. (*Bottleneck Tree Problem*). Let G be a connected network, and let there be associated with each arc j of G a number $r(j)$, real or $\pm\infty$. For each spanning tree T and nonempty cut Q, let

$$f(T) = \min\{r(j)|j \in T\}, \qquad g(Q) = \max\{r(j)|j \in Q\}.$$

Prove that

$$\max\{f(T)|T \text{ is a spanning tree}\} = \min\{g(Q)|Q \text{ is a nonempty cut}\}.$$

5.49. (*Matching Algorithm*). The matching algorithm in Section 5B is an adaptation of the max flow algorithm, so it can be refined in accordance with the multipath implementation approach in Section 3J (see Exercise 3.23). Due to the nature of the network model for the matching problem, a very special procedure results. Work out the details and incorporate them into the statement of the algorithm.

5J.* COMMENTS AND REFERENCES

A *bipartite* network is a digraph whose node set is the union of two nonempty, disjoint sets K and L and whose arc set is a subset H of $K \times L$. Such networks underlie much of this chapter, and it is no wonder that the kind of matching theory that has been presented is often referred to as *bipartite matching*.

There is a more general theory that concerns arbitrary networks and likewise has interesting applications. A *match* is an arbitrary network G is defined to be an arc set M such that each node of G is incident to at most one of the arcs in M. Again the problem is to maximize $|M|$ over all matches M. A duality theorem can be established, and solution techniques involving "alternating" paths such as in the ordinary matching problem can be developed, but there is no simple relationship in the nonbipartite case with problems of flows and capacities. Much of the generalization has been achieved by J. Edmonds; see Lawler [1976, Chap. 6] for an exposition.

The history of the bipartite matching problem goes back much farther. The fundamental theorem of König [1931] and Egerváry [1931] in Section 5A was formulated originally in terms of graphs. The more general and seemingly earlier theorem of Menger [1927] (see Exercises 5.6, and 5.7) was expressed in terms of a sort of "curve theory" and in fact did not receive complete proof until later; see Menger [1932], König [1933], and König [1936]. Hall's existence theorem has been presented in Section 5C as a consequence of the König-Egerváry theorem, but its original statement by Hall [1935] dealt with systems of distinct representatives. Actually the problem of finding a system of distinct representatives is *equivalent* to that of determining an assignment relative to some compatibility relation: to reduce the latter to the former (the reverse has already been done in Section 5C), just identify each $k \in K$ with the set of all $l \in L$ such that $(k, l) \in H$, that is, the set $H(k)$.

The theorem of Hoffman and Kuhn [1956] (see Exercise 5.23) about systems of distinct representatives involving obligatory elements sharpens a result of Mann and Ryser [1953]. The generalization corresponding to Exercises 5.11, 5.12, and 5.13 is due essentially to Mendelssohn and Dulmage [1958].

The matching algorithm in Section 5B, which is based on the max flow representation of the bipartite matching problem, is essentially the same as the one given by Ford and Fulkerson [1962, pp. 55–56] and said by them to be more or less implicit in the proof of König [1931]. However, a closely parallel procedure in terms of set representatives was given, for instance, by M. Hall Jr.

[1958]. A highly polished implementation of the algorithm has been provided by Hopcroft and Karp [1973].

The theorem of Dilworth [1950] in Section 5D was first applied to problems in operations research by Dantzig and Fulkerson [1954].

H. W. Kuhn [1955] developed the algorithm for optimal assignments in Section 5F. He dubbed it the "Hungarian method" because it could be regarded as implicit in the proofs of Egerváry [1931]. The approach eventually led to algorithms for the classical transportation problem in Section 5G (see Exercises 5.36, 5.37, and 5.38) and more general distribution problems with linear costs, as will be discussed in Chapters 7 and 9 (the "out-of-kilter" method). The classical transportation problem, also called the Hitchcock problem, has a long history in economic applications, with the optimal assignment problem as a special case; see the book of Dantzig [1963].

The subgraph theorem of O. Ore [1956] (see Exercise 5.44) and the results of D. Gale [1957] and H. J. Ryser [1957] on 0–1 matrices (see Exercise 5.45) are discussed in detail by Ford and Fulkerson [1962, Chap. 2], along with generalizations.

The bottleneck assignment problem in Section 5H was proposed and solved by O. Gross [1959]. The bottleneck path problem in Exercise 5.46 was analyzed by Fulkerson [1966]. Later, Edmonds and Fulkerson [1970] developed a theory that encompassed not only this but other interesting "bottleneck" problems, such as the one for trees in Exercise 5.48.

See Edmonds and Giles [1977] for related combinatorial results based on the theory of submodular functions.

6

POTENTIALS AND SPANS

In the analysis of flow problems the relationship between capacity constraints and flux across various cuts is of great importance, as has been seen in Chapter 3. For problems involving potentials, the constraints are expressed by "span intervals" and are analyzed similarly in terms of the notion of spread relative to a path. In analogy with max flow min cut theory there is a max tension problem whose dual, the min path problem, seeks to determine a path of least "span" from one node set to another. The word "span" covers numerous applications in operations research, where the quantity minimized can represent cost, time, length, the negative of length, and other things.

The basic method for solving the max tension and min path problems can be realized very efficiently as a sort of enriched version of the painted network algorithm. It can also be used iteratively to construct potentials that satisfy various constraints and even routings that are optimal with respect to a given system of "spans." Computations of this kind must be carried out as subroutines in some of the algorithms in Chapters 7 and 9.

There are integrality and extremal representation theorems for potentials corresponding to the ones for flows in Chapter 4.

6A. SPREAD RELATIVE TO A PATH

Recall from Section 1I that any element v of R^A, where A is the arc set of the network G, can be regarded as a "tension," whether or not it is the differential Δu associated with some potential $u \in R^N$. For an arbitrary tension v and path P, the *spread of v relative to P* is defined as the sum of the values $\pm v(j)$ along the arcs of P, with the negative sign taken for arcs that are traversed negatively. In other words, if P is a path without multiplicities,

$$[\text{spread of } v \text{ relative to } P] = \sum_{j \in P^+} v(j) - \sum_{j \in P^-} v(j) = v \cdot e_P.$$

173

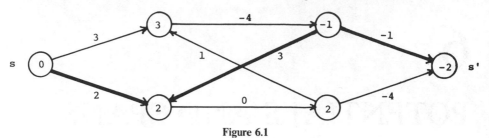

Figure 6.1

If P has multiplicities, it can be decomposed into several paths without multiplicities, each of which contributes an expression of this form.

For example, a tension v is indicated in Figure 6.1 together with a path P: $s \rightarrow s'$. the spread relative to P is $2 - 3 - 1 = -2$.

This concept is the discrete version of a line integral. It comes as no surprise, then, that

$$[\text{spread of } v \text{ relative to } P] = u(i') - u(i)$$

if

$$v = \Delta u \quad \text{and} \quad P: i \rightarrow i'.$$

This formula, called the *integration rule*, can be established merely by the observation that the term $\pm v(j_k)$ corresponding in the sum to the kth arc of P, which joins the kth node i_k with its predecessor i_{k-1}, equals $u(i_k) - u(i_{k-1})$ whichever the orientation of j_k (because of the sign convention). Figure 6.1 provides an illustration; the values at the nodes belong to a potential giving rise to the tension in question.

One consequence of the integration rule is that *when v is a differential, the spread of v relative to P depends only on the endpoints of P* and not on what comes between. An equivalent statement of this property is that for every circuit P, the spread of v relative to P is zero. In fact v satisfies this if and *only* if it is a differential (see Exercise 2.12).

The integration rule can be used to construct for a given tension v a potential u such that $\Delta u = v$, if one exists. Suppose first that G is connected, and fix any $s \in N$. If there is indeed a potential u with v as its differential, its values at all nodes other than s are uniquely determined by

$$u(i) = u(s) + [\text{spread of } v \text{ relative to } P],$$

where P may be any path from s to i. Of course $u(s)$ can itself take on any value, since a constant can be added to a potential function without affecting its differential. This observation leads to the following procedure for the general case, where it is not necessarily known in advance whether v is a differential. Take any routing θ of all of N with base s and any $\alpha \in R$; define

$$u(i) = \begin{cases} \alpha & \text{for } i = s \\ \alpha + [\text{spread of } v \text{ relative to } P_i] & \text{for } i \neq s, \end{cases}$$

where P_i is the θ-path from s to i. If v is a differential, u must be the unique potential with $\Delta u = v$ and $u(s) = \alpha$. If v is not a differential, this will be evidenced by having $(\Delta u)(j) \neq v(j)$ for some arc j (in fact for some arc j not used by the routing θ).

For networks G that are not connected, the construction can be repeated for each component. Then there is not just one α to choose but a separate "constant of integration" for each component.

In line with the aim of developing results about tensions and paths that are parallel to those about flows and cuts, we now suppose each arc j has associated with it a nonempty real interval $D(j)$, called its *span interval*. Nonclosed intervals will be treated in a number of exercises and will enter into some results in Chapter 8, but for the time being $D(j)$ is assumed to be closed. Thus we write $D(j) = [d^-(j), d^+(j)]$, with the same notational conventions as with capacity intervals. The number $d^+(j)$ is the *upper span* for the arc j, whereas $d^-(j)$ is the *lower span*.

Since $D(j)$ can be any nonempty closed real interval, the only requirements on $d^+(j)$ and $d^-(j)$ are that

$$-\infty \leq d^-(j) \leq d^+(j) \leq +\infty, \qquad d^-(j) < +\infty, \qquad d^+(j) > -\infty. \quad (1)$$

It must be remembered, however, that $+\infty$ and $-\infty$ themselves are never elements of $D(j)$, not being real numbers. The generic notation $D(j) = [d^-(j), d^+(j)]$ therefore is not strictly appropriate in the case where $d^+(j)$ or $d^-(j)$ might be infinite, but it will be used nonetheless.

A tension v will be called *feasible with respect to spans* if

$$v(j) \in D(j) \quad \text{for all arcs } j \in A.$$

Mostly we shall be concerned with the feasibility of differentials, and an existence theorem for feasible differentials will be studied in Section 6F.

Some insights into the nature of the feasibility constraints can be gained by thinking of a potential u as a mapping that projects the network onto the real line, with $u(i)$ the coordinate of the point assigned to node i. For $v = \Delta u$, the condition $v(j) \in D(j)$, $j \sim (i, i')$, means that the image of i must be placed relative to the image of i' in such a way that

$$u(i) + d^-(j) \leq u(i') \leq u(i) + d^+(j).$$

An arc $j \sim (i, i')$ with its span interval $D(j)$ is thus seen as representing a constraint on the positions relative to each other that can be assigned to i and i' on some linear scale.

Feasibility with respect to span intervals implies, by the integration rule, certain constraints with respect to paths, as will now be explained.

The *upper span* $d^+(P)$ of a path P is defined as the sum of the values $d^+(j)$ or $-d^-(j)$ along P, the first taken if the arc j is traversed positively and the second if it is traversed negatively. The *lower span* $d^-(P)$ is defined in the same way, but with $d^-(j)$ and $-d^+(j)$ in the two cases. Thus if P is without multiplicities, one has

$$d^+(P) = \sum_{j \in P^+} d^+(j) - \sum_{j \in P^-} d^-(j),$$

$$d^-(P) = \sum_{j \in P^+} d^-(j) - \sum_{j \in P} d^+(j).$$

Various interpretations of these quantities will be discussed in the next section, but for now only the close connection with integration need be observed. If v is a feasible tension, so that $d^-(j) \le v(j) \le d^+(j)$ for all j, then

$$d^-(P) \le [\text{spread of } v \text{ relative to } P] \le d^+(P)$$

for all paths P. If $v = \Delta u$, this condition says that

$$u(i) + d^-(P) \le u(i') \le u(i) + d^+(P)$$

for all paths P: $i \to i'$.

6B. OPTIMAL PATHS

There are many situations where one can naturally associate with each arc j a pair of numbers $d^+(j)$ and $d^-(j)$ satisfying 6A(1) and then form the expressions $d^+(P)$ and $d^-(P)$ for various paths P, without having in mind anything about feasibility of tensions. For instance, $d^+(j)$ could be the cost of traversing j in the positive direction, whereas $-d^-(j)$ is the cost in the negative direction. Then $d^+(P)$ would be the total cost of traversing the path P, and $-d^-(P)$ the cost of traversing the reverse of P. Of course such "costs" might be negative in some contexts, in effect representing "rewards."

Let N^+ and N^- be nonempty disjoint node sets of the network G. The following problem is of evident interest under the interpretation just given.

Min Path Problem.   *Minimize $d^+(P)$ over all paths P: $N^+ \to N^-$.*

If there is no path P: $N^+ \to N^-$, the convention is used that the problem has min $= +\infty$. This conforms to the fact that extra arcs j with $d^+(j) = \infty$, $d^-(j) = -\infty$, can always be added to make the network connected, without really changing anything else.

Since the min path problem could be formulated entirely in terms of forward and backward "costs" for each arc, say, it may seem odd to use the notation $d^+(j)$ and $-d^-(j)$ for such "costs" and to insist on the setting of a network with span *intervals* $D(j) = [d^-(j), d^+(j)]$. But the connection between paths and potentials is very strong, as evidenced by the inequalities at the end of Section 6A, and these intervals will have many roles to play in solving problems that might at first seem purely combinatorial in nature.

Observe that $d^+(P) < +\infty$ if and only if P is compatible with a certain painting of the network, called the *painting associated with the spans*:

$$\begin{array}{ll} \text{green} & \text{if } d^+(j) < +\infty \text{ and } d^-(j) > -\infty \\ \text{white} & \text{if } d^+(j) < +\infty \text{ and } d^-(j) = -\infty \\ \text{black} & \text{if } d^+(j) = +\infty \text{ and } d^-(j) > -\infty \\ \text{red} & \text{if } d^+(j) = +\infty \text{ and } d^-(j) = -\infty \end{array}$$

Therefore the min path problem is equivalent to minimizing $d^+(P)$ over all solutions to the painted path problem (as in Section 2D) relative to this painting, if at least one exists. It can be interpreted, in other words, in terms of *finding a "cheapest" solution to a certain painted path problem*. This recalls the interpretation given in Section 3E for the min cut problem.

The ordinary painted path problem, without "costs," can always be reduced to the case of elementary paths $P: N^+ \rightarrow N^-$ having only the first node in N^+ and only the last in N^-. Such a reduction is not automatically possible for the min path problem, since there can be negative "costs," and these might make it profitable to include redundant segments and side circuits as shown in Figure 6.2. The natural conditions to impose on the spans, in order for the reduction to be possible, turn out to be the conditions for the existence of a potential satisfying the constraints of the "max tension problem" that will be introduced in the next section. This topic will be treated further in Section 6F.

In view of these complications it is well to observe that the general case of the min path problem can always be reformulated, if necessary, as a problem with $N^+ = \{s\}$, $N^- = \{s'\}$. The trick is displayed in Figure 6.3. In this context

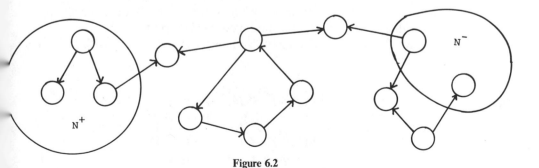

Figure 6.2

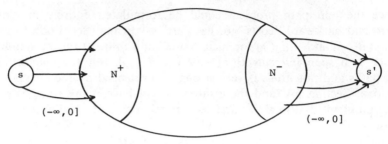

$(-\infty, 0]$

$(-\infty, 0]$

Figure 6.3

the only thing to worry about is the possible presence of circuits P with $d^+(P) < 0$ (see Section 6F).

Although the cost idea seems to provide the best insights into the general min path problem, there are special cases having a greater geometric appeal. There are also applications where the relevance of the min path model is not immediately apparent. (One of these, the technique known as PERT, will not be discussed until Section 6C.) The examples that follow make clear, at any rate, that positive, negative, and infinite values for both $d^+(j)$ and $d^-(j)$ are all of solid importance to the theory.

EXAMPLE 1.   (*Shortest Paths*)

Suppose the span intervals are of the form $D(j) = [-l(j), l(j)]$, where $l(j)$ is a non-negative real number called the *length* of the arc j. Then $d^+(P) = l(P)$ where (assuming P is without multiplicities)

$$l(P) = \sum_{j \in P} l(j)$$

is the *length* of the path P. The min path problem consists of finding the shortest path from N^+ to N^-. If $D(j) = (-\infty, l(j)]$ instead, it is the problem of finding the shortest *positive* path from N^+ to N^-. Thinking of length in terms of the time it takes to traverse an arc, rather than the distance along it, one has in either case a problem of finding a *quickest* path.

More generally, if some painting of the network is given along with numbers $l(j) \geq 0$, one can define

$$D(j) = \begin{cases} [-l(j), l(j)] & \text{if } j \text{ is green,} \\ (-\infty, l(j)] & \text{if } j \text{ is white,} \\ [-l(j), \infty) & \text{if } j \text{ is black,} \\ (-\infty, \infty) & \text{if } j \text{ is red.} \end{cases}$$

The min path problem then consists of minimizing $l(P)$ over all paths P: $N^+ \to N^-$ that are compatible with the painting, that is, of finding the shortest solution to a given painted path problem.

If $l(j) = 1$ for all j, the length $l(P)$ is just the number of arcs in the path P.

EXAMPLE 2.   (*Longest Paths*)

Suppose $D(j) = (-\infty, -l(j)]$, with $l(j)$ the non-negative length of j as in Example 1. Then $d^+(P) = -l(P)$ for paths with $P^+ = P$, $P^- = \varnothing$, whereas $d^+(P) = +\infty$ for all other paths. The min path problem is therefore equivalent to minimizing $-l(P)$ over all "positive" paths P: $N^+ \to N^-$. In other words, it consists of finding a *longest positive* path from N^+ to N^-.

EXAMPLE 3.   (*Knapsack Problem*)

This is an example of an important combinatorial problem that can be reformulated in terms of finding a longest path in a certain network (and thus becomes a special case of the min path problem; see Example 2). A finite set $\{z_1, \ldots, z_m\}$ is given, along with the *value* r_k and *weight* w_k of each element z_k. The values are non-negative real numbers, but the weights are positive integers. The problem is to choose a subset $Z \subset \{z_1, \ldots, z_m\}$ so as to maximize the total value $r(Z)$ subject to a weight restriction $w(Z) \leq n$, where

$$r(Z) = \sum_{z_i \in Z} r_i, \qquad w(Z) = \sum_{z_i \in Z} w_i,$$

and n is a given positive integer. (The restriction could be thought of in terms of "volume," instead of "weight." Applications include cargo loading of ships and planes.)

At a first approach, picture this problem in terms of a decision process in discrete time. At time k, the element z_k is designated as an element of the subset Z or its complement. Use the pairs (k, q) with $k = 1, \ldots, m + 1$ and $q = 0, \ldots, n$ to represent states in this process, where k is the time and q is an index recording the total weight that has been accumulated in the selection of elements for Z up to time k. From the state (k, q), one can pass either to $(k + 1, q + w_k)$ if z_k is designated to be in Z, or to $(k + 1, q)$ if not. In the first case, which is permitted only if $q + w_k \leq n$, the cumulative value of the selected elements is increased by r_k, whereas in the second case it is left the same.

It is easy now to place the problem in a network setting; see Figure 6.4. The states (k, q) are taken as the nodes. Each permissible transition $(k, q) \to (k + 1, q + w_k)$ is represented by an arc j whose length $l(j)$ is taken to be r_k. Transitions $(k, q) \to (k + 1, q)$ correspond to arcs of length $l(j) = 0$. The

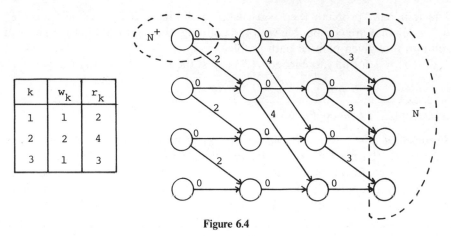

k	w_k	r_k
1	1	2
2	2	4
3	1	3

Figure 6.4

problem is thereby reduced to that of finding a longest positive path from $N^+ = \{(1,0)\}$ to $N^- = \{(m+1,0),\ldots,(m+1,n)\}$.

EXAMPLE 4. (*Reliability*)

Imagine the arcs of the network G as subject to random failure (blockage), like the channels of a communication network. Associated with each arc j is a number $p(j) \in [0,1]$, the probability that j will *not* be blocked. For a path P with arcs $j_1,\ldots,j_r$, the probability that P will *not* be blocked is

$$p(P) = p(j_1)p(j_2)\cdots p(j_r)$$

(if arc failures are assumed to be independent of each other). The problem is to maximize $p(P)$ over all paths $P\colon N^+ \to N^-$.

This can be converted to a shortest path problem (see Example 1) by defining $l(j) = -\log p(j) \in [0, +\infty]$, so that $l(P) = -\log p(P)$.

Very similar is the case where each arc j has an associated *gain* $\gamma(j) \in (0,\infty)$ representing amplification or, if $\gamma(j) < 1$, attenuation. The gain of a path is the product of the gains of the arcs in the path. Again one can use logarithms to transform products into sums that fit the framework of the min path problem.

6C. MAX TENSION MIN PATH

There is a problem for potentials that is analogous to the max flow problem and likewise concerns two nonempty disjoint node sets N^+ and N^- of a network G.

Let u be a potential that is constant on N^+ and constant on N^-. The value $u(i') - u(i)$ is then the same for any $i \in N^+$ and $i' \in N^-$, and it is called the *spread of u from N^+ to N^-*. This value equals the spread of the differential $v = \Delta u$ relative to any path $P: N^+ \to N^-$, according to the integration rule in Section 6A.

Max Tension Problem.   *Maximize the spread of u from N^+ to N^- over all potentials u that are constant on N^+, constant on N^-, and such that $v = \Delta u$ is feasible with respect to spans.*

In order for there to exist a potential satisfying the constraints, the spans must satisfy a certain regularity condition that corresponds to the "reducibility" of the min path problem; see Section 6F.

The main result uses the concept of a *cut of unlimited span* from N^+ to N^-. By this is meant a cut $Q: N^+ \downarrow N^-$ such that $d^+(j) = +\infty$ for all $j \in Q^+$ and $d^-(j) = -\infty$ for all $j \in Q^-$ (or one such that $Q = \varnothing$).

Max Tension Min Path Theorem (Minty).   *Suppose there is at least one potential satisfying the constraints of the max tension problem. Then*

$$[\text{sup } in\ max\ tension\ problem] = [\text{min } in\ min\ path\ problem].$$

This common value is $+\infty$ if there is a cut $Q: N^+ \downarrow N^-$ of unlimited span, whereas if there is no such cut, it is finite, and the max tension problem has a solution.

The inequality $\leq$ in the theorem follows at once from the fact that if u is any potential satisfying the constraints of the max tension problem and P is any path from N^+ to N^-, one has

$$[\text{spread of } u \text{ from } N^+ \text{ to } N^-] = [\text{spread of } \Delta u \text{ relative to } P] \leq d^+(P).$$

The rest of the proof will be given algorithmically in Section 6D. Note that the validity of the theorem is unaffected by the addition of extra arcs with span interval $(-\infty, \infty)$. Thus everything can be reduced to the case of a connected network.

EXAMPLE 5.   (*PERT = Project Evaluation and Review Technique*)

Consider the task of putting together an efficient timetable for a large-scale project such as the development of a new kind of space vehicle. Certain "events" or "milestones" in the course of the project are selected as reference points in the planning. These typically correspond to the start or finish of some activity, such as the start of the survey of the construction site or the finish of engine assembly. Various relationships naturally pose constraints on the timing

of the events. For instance, the vehicle cannot be launched before the finish of
the training program for the launch personnel, and that in turn cannot take
place until a certain number of days after the start of the training program.
Perhaps also the training program cannot be finished too early before the
launch, or the lessons might be forgotten.

A network G is formed whose node set N consists of the "events." A
schedule is a function $u: N \to R$, with $u(i)$ the time when event i is to take
place. For each pair of events i and i' whose timing involves a direct
constraint, an arc j is introduced with a span interval $[d^-(j), d^+(j)]$, so that
the constraint can in general be written as

$$u(i) + d^-(j) \leq u(i') \leq u(i) + d^+(j).$$

The schedules satisfying these constraints can then be regarded as potentials
feasible with respect to certain spans. (Of course only Δu is really involved; in
other words, the choice of the origin on the time scale is not important.)

The problem is to determine, from among the schedules meeting the
constraints, one for which the overall time taken by the project is a minimum.
This can be converted into a max tension problem by letting $N^+ = \{s\}$ and
$N^- = \{s'\}$, where s is the event signifying the *finish* of the whole project and s'
is the *start*. Then $u(s') - u(s)$ is the *negative* of the overall time.

A solution to the corresponding min path problem is called a *critical path*
for the project. Knowledge of such paths helps to identify which activities must
receive especially close attention from managers, if delays are to be avoided.

The next two examples explore the meaning of max tension min path
duality for cases of the min path problem that have already been discussed.

EXAMPLE 6.   (*Costs*)

Suppose $d^+(j)$ is the cost of traversing the arc j positively, whereas $-d^-(j)$ is
the cost in the opposite direction. Viewing G as a transportation network, think
of these as costs per unit for transporting a certain good whose value or price
at node i is $u(i)$. A fundamental economic requirement is that the price of one
unit of the good at a location i' cannot exceed the price of one unit at an
adjacent location i, plus the cost of transporting one unit from i to i'. Thus for
each arc $j \sim (i, i')$ one must have $u(i') \leq u(i) + d^+(j)$, and likewise $u(i) \geq
u(i') - d^-(j)$. The price systems u meeting this requirement are the potentials
on G that are feasible with respect to the intervals $[d^-(j), d^+(j)]$ as span
intervals. In the max tension problem one seeks to determine the maximum
price difference that can exist between two "regions" N^+ and N^- in which
prices are constant.

EXAMPLE 7.   (*Lengths*)

Suppose the span intervals are of the form $[-l(j), l(j)]$, where $l(j) \in [0, \infty)$
is the length of the arc j. The max tension problem can then be seen as an

analogue approach to solving the shortest path problem. For simplicity, suppose that $N^+ = \{s\}$ and $N^- = \{s'\}$, and imagine that a model of the network has been constructed with the nodes represented by small, heavy beads and the arcs by inelastic threads of lengths $l(j)$ joining the beads. One way to find the shortest path, in principle, is to take hold of the bead representing s and let the whole model hang. If nothing gets tangled (!), the difference in height between the s' bead and the s bead should be the distance from s to s' by the shortest path, and the path itself should be visible among the threads that are taut. Incidentally, this method would in fact solve the problem simultaneously for *all* nodes $s' \neq s$, a phenomenon that will be observed also in the algorithms of this chapter.

The mathematical version of the bead-and-thread method, of course, is to assign to each node i a number $u(i)$, giving its coordinate on a linear scale (in the preceding model, a vertical scale directed downward). This is to be done in such a way that the numbers $u(i) = u(s)$ for $i \neq s$ are all to be as large as possible, consistent with the constraint for each arc $j \sim (i, i')$ that $|u(i') - u(i)| \leq l(j)$. In particular, $u(s') - u(s)$ is then the maximum in the max tension problem as well as the min in the min path problem from s to s'.

6D. MIN PATH ALGORITHM

If a potential u_0 is known that satisfies the constraints of the max tension problem, it can be employed in the procedure described in this section to calculate solutions to the max tension problem and min path problem simultaneously. One can of course take $u_0 \equiv 0$ if $d^-(j) \leq 0 \leq d^+(j)$ for all j. Another case where a suitable potential u_0 is readily available is that of an acyclic network with $d^-(j) = -\infty$ for all j (see Examples 2, 3, and 5) and having single nodes for N^+ and N^-, which may involve no loss of generality; see Figure 6.3. (This is covered in Exercise 6.19; for a combination of the two cases, see Exercise 6.23.) In general, u_0 could always be constructed by one of the techniques developed later in this chapter, starting in Section 6F, although this would involve computations of a higher order of complexity.

An interesting fact theoretically is that although the procedure appears as a more elaborate form of the painted network algorithm, it can also be viewed as an implementation of a certain "max tension algorithm" that is very similar in form to the max flow algorithm but turns out not to suffer from the same troubles over termination (see Exercises 6.17 and 6.18).

Statement of the Algorithm

Given any potential u_0 that is constant on N^+, constant on N^-, and whose differential v_0 is feasible with respect to spans, let

$$d_0^+(j) = d^+(j) - v_0(j), \qquad d_0^-(j) = d^-(j) - v_0(j). \tag{1}$$

(Then $d_0^-(j) \leq 0 \leq d_0^+(j)$ for all $j \in A$.)

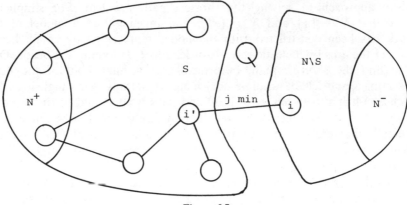

Figure 6.5

In the general step (schematized in Figure 6.5) there is a node set S satisfying $N^+ \subset S \subset N \setminus N^-$ and a routing θ of S with base N^+. There is also a function $w: S \to R$. Initially, $w \equiv 0$ on $N^+ = S$.

For the cut $Q = [S, N \setminus S]$, calculate

$$\beta = \min \begin{cases} w(i') + d_0^+(j) & \text{for } (i', i) \sim j \in Q^+, \\ w(i') - d_0^-(j) & \text{for } (i, i') \sim j \in Q^-. \end{cases} \tag{2}$$

If $\beta = \infty$ (as is true in particular by convention if $Q = \varnothing$), terminate; Q is a cut of unlimited span, and $[\sup] = [\min] = \infty$ in the two problems. If $\beta < \infty$, take any arc j achieving the minimum in (2) and add to S the corresponding node i, defining $\theta(i) = j$ and $w(i) = \beta$. (In case of a tie for the minimum, several nodes can be added in this manner at once.) If $i \in N^-$, terminate; the potential $u = u_0 + w$, with w taken to have the value β everywhere outside S, solves the max tension problem, and the θ-path $P: N^+ \to i$ solves the min path problem with

$$d^+(P) = [\text{spread of } u_0 \text{ from } N^+ \text{ to } N^-] + \beta$$

$$= [\text{spread } u \text{ from } N^+ \text{ to } N^-]. \tag{3}$$

Otherwise repeat the procedure with the extended S, θ, and w.

EXAMPLE 8

The algorithm can be applied to the network in Figure 6.6 with $N^+ = \{s\}$, $N^- = \{s'\}$, and $u_0 \equiv 0$ (since $d^-(j) \leq 0 \leq d^+(j)$ for all $j \in A$). The calculations are summarized in the accompanying table. Figure 6.7 displays the final routing θ and values $w(i)$. These values also give the solution u obtained for

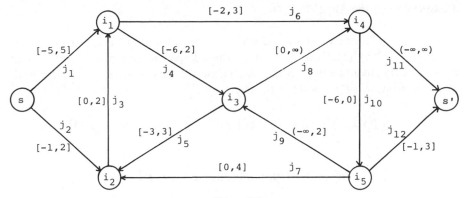

Figure 6.6

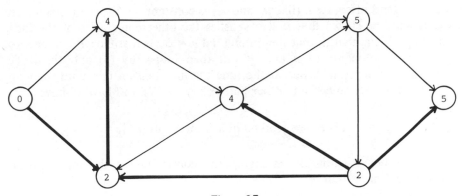

Figure 6.7

the max tension problem (since $u = u_0 + w$ and $u_0 \equiv 0$). The solution to the min path problem is $P: s \rightarrow i_2 \leftarrow i_5 \rightarrow s'$. One has [max] = [min] = 5.

Iteration	Numbers Compared in Determining β	Added Node i	$\theta(i)$	$w(i): = \beta$
0		s		0
1	$0 + 5, 0 + 2$	i_2	j_2	2
2	$0 + 5, 2 + 2, 2 + 3,$ $2 + 0$	i_5	j_7	2
3	$0 + 5, 2 + 2, 2 + 3,$ $2 + 2, 2 + 6, 2 + 3$	i_1 i_3	j_3 j_4	4 4
4	$4 + 3, 4 + \infty, 2 + 6,$ $2 + 3$	s'	j_{12}	5

Justification of the Algorithm

If $\beta = \infty$ in (2), one must have $d^+(j) = \infty$ for all $j \in Q^+$ and $d^-(j) = -\infty$ for all $j \in Q^-$ by (1). Thus Q is of unlimited span in this case. Another fact easy to verify (by induction) is that each time a node i is added to S, the θ-path $P: N^+ \to i$ has $d_0^+(P) = w(i) = \beta$, where by (1)

$$d_0^+(P) = \sum_{j \subset P^+} \left[d^+(j) - v_0(j)\right] - \sum_{j \in P^-} \left[d^-(j) - v_0(j)\right]$$

$$= d^+(P) - \left[\text{spread of } v_0 \text{ relative to } P\right].$$

Therefore when the algorithm terminates with $i \in N^-$ (and $w \equiv \beta$ on N^-), (3) does hold, and of course u (like u_0 and w) is constant on N^+ and constant on N^-. It must be shown that u also satisfies the other constraints of the max tension problem, that is, that the differential $v = \Delta u$ is feasible with respect to spans. The optimality of u and P will then follow by (3) and the basic inequality already established at the time the max tension min path theorem was stated in Section 6C: any other path $P': N^+ \to N^-$ will have to have

$$d^+(P') \geq \left[\text{spread of } u \text{ from } N^+ \text{ to } N^-\right],$$

and any other potential u' satisfying the constraints of the max tension problem will have to have

$$\left[\text{spread of } u' \text{ from } N^+ \text{ to } N^-\right] \leq d^+(P).$$

To see the feasibility of the terminal differential v and obtain other insights as well, it is useful to look at the procedure in a slightly different way. Although no extension of w beyond S is needed except at the very end, there is no harm in imagining that in each iteration w is given the value β everywhere on $N \setminus S$ just before the new nodes are added to S, and in letting u denote the corresponding potential $u_0 + w$. (Initially, $w \equiv 0$ on all of N.)

In this formulation u starts out equal to u_0 but is modified in each iteration by adding a certain constant α to its values at all nodes not in S. (The same α is added to the constant value of w on $N \setminus S$.) This α is the difference between the current β value given by (2) (if not $+\infty$) and the β in the previous iteration (taken to be 0 if there was no previous iteration), and since the latter is the value of $w(i)$ for all $i \in N \setminus S$, one has

$$\alpha = \min \begin{cases} w(i') + d_0^+(j) - w(i) & \text{for } (i', i) \sim j \in Q^+, \\ w(i') - d_0^-(j) - w(i) & \text{for } (i, i') \sim j \in Q^-. \end{cases}$$

In terms of $u = u_0 + w$ and (1), the formula is

$$\alpha = \min \begin{cases} d^+(j) - v(j) & \text{for } j \in Q^+, \\ v(j) - d^-(j) & \text{for } j \in Q^-. \end{cases} \tag{4}$$

The general step can therefore be recast in the following form. Calculate α by (4); if $\alpha = \infty$, terminate because Q is of unlimited span. Otherwise replace u by $u + \alpha e_{N \setminus S}$ (and v by $v + \alpha e_Q$), and for any arc j giving the minimum in (4) add the corresponding node $i \in N \setminus S$ to S with $\theta(i) = j$. Terminate when a node $i \in N^-$ is added to this manner.

If v is feasible with respect to spans, one has $\alpha \geq 0$ in (4), and the new differential $v + \alpha e_Q$ is likewise feasible. Since the initial differential $v = v_0$ is feasible with respect to spans, feasibility is preserved in each iteration, and the final potential does satisfy the constraints of the max tension problem as claimed.

Monotonicity Property

The sequence of β values generated by the algorithm is always nondecreasing. This important fact is obvious from the argument just given: α is the increment in β, and since feasibility of the differential v is maintained, one always has $\alpha \geq 0$ in (4).

Proof of the Max Tension Min Path Theorem

The algorithm must terminate eventually, because the node set S is enlarged in every iteration. If it terminates by reaching N^-, we have $[\text{max}] - [\text{min}] < \infty$, as seen in the justification. Otherwise it terminates because $\alpha = \infty$ in Equation (4), where u is a certain potential satisfying the constraints of the max tension problem. Then for arbitrary $\alpha' \in (0, \infty)$ the potential $u' = u + \alpha' e_{N \setminus S}$ would also satisfy the constraints and have

$$[\text{spread of } u' \text{ from } N^+ \text{ to } N^-] = [\text{spread of } u \text{ from } N^+ \text{ to } N^-] + \alpha'.$$

This implies $[\text{sup}] = +\infty$ in the max tension problem, and in view of the basic inequality in Section 6C, also $[\text{min}] = +\infty$ in the min path problem.

6E.*   NODE-SCANNING IMPLEMENTATION

In the formula for β in the min path algorithm, certain numbers $w(i') + d_0^+(j)$ or $w(i') - d_0^-(j)$ may appear in several iterations because the arc j may belong to all the corresponding cuts Q. It would be wasteful to compute them over and over again. Fortunately once is enough. It is possible to keep track of the data gathered up to any moment in terms of "tentative" extensions of θ and w from S to a larger node set T (not to be confused with the extension of w to all of N that was used in the justification of the min path algorithm in the preceding section).

Dijkstra's Method

The setting and notation are the same as in the form of the min path algorithm already given, but in the general step there are two sets S and T satisfying $N^+ \subset S \subset T \subset N \setminus N^-$; θ is a routing of T with base N^+ whose restriction to S is a routing of S with base N^+, and w is defined on all of T. Initially, $T = S = N^+$, θ is vacuous, and $w \equiv 0$. To begin with, all nodes are "unscanned." Nodes will be processed one by one, and after this has been done, a node will be said to have been "scanned."

STEP 1.   If there are no "unscanned" nodes in S (which is false initially), then go to Step 3. Otherwise select any "unscanned" node $i' \in S$ and go to Step 2.

STEP 2.   If there is an arc j incident to i' and belonging to $Q = [S, N \setminus S]$, calculate

$$\gamma = \begin{cases} w(i') + d_0^+(j) & \text{if } j \sim (i', i), \\ w(i') - d_0^-(j) & \text{if } j \sim (i, i'), \end{cases}$$

where i denotes the other node of j. If $i \in T$ and $\gamma < w(i)$, redefine $\theta(i) = j$ and $w(i) = \gamma$. If $i \notin T$ and $\gamma < \infty$, add i to T with $\theta(i) = j$ and $w(i) = \gamma$. If $\gamma = \infty$ or $i \in T$ with $\gamma \geq w(i)$, do not change T, θ, or w. Repeat this step for each arc j incident to i' and belonging to Q, and then return to Step 1 (with i' henceforth regarded as "scanned").

STEP 3.   Calculate

$$\beta = \min\{w(i) | i \in T \setminus S\}.$$

If $T \setminus S = \varnothing$, regard β as $+\infty$ and terminate; Q is a cut of unlimited span (so [sup] = [min] = $+\infty$ in the two problems). Otherwise add to S the nodes of $T \setminus S$ for which the minimum defining β is achieved. If a node $i \in N^-$ is among these, terminate; the θ-path $P: N^+ \to i$ solves the min path problem, and the potential $u = u_0 + w$ on S, $u = u_0 + \beta$ on $N \setminus S$, solves the max tension problem. Otherwise return to Step 1.

Justification

At every stage S and Q correspond to the sets of that designation in the earlier version of the min path algorithm. For a node $i \in T \setminus S$, $w(i)$ always gives the lowest of the values $w(i') + d_0^+(j)$ for $(i, i') \sim j \in Q^+$ or $w(i') - d_0^-(j)$ for

$(i', i) \sim j \in Q^-$ that has yet been discovered through inspection of arcs, and $\theta(i)$ is an arc j yielding this value. The inspection is carried out in the scanning routine in Steps 1 and 2, and when Step 3 is reached; that is, when there are no more nodes to scan in S, there are no arcs in Q that have not been inspected. At that time $T \setminus S$ consists therefore of all the nodes $i \in N \setminus S$ that can be reached from S by arcs $j \in Q^+$ with $d_0^+(j) < \infty$, or $j \in Q^-$ with $d_0^-(j) > -\infty$, whereas $w(i)$ and $\theta(i)$ record the information about how this can be done optimally. Step 3 merely executes the general step in the earlier form of the algorithm in terms of this information.

EXAMPLE 9

When Dijkstra's method is applied to the network in Figure 6.6 with $N^+ = \{s\}$, $N^- = \{s'\}$ and $u_0 \equiv 0$, the result is of course the same as in Example 8 of Section 6D with the original form of the min path algorithm. Labor is saved, however, because no arc is inspected more than once. The Table 6.1 records the details of the calculations.

Table 6.1

$T \setminus S$	Added to S	Node i' Scanned	Arc j Inspected	Node i' Affected	γ	$w(i)$ Previous	$w(i)$ Updated	$\theta(i)$ Updated
		s	j_1	i_1	5		5	j_1
			j_2	i_2	2		2	j_2
i_1, i_2 $\beta = 2$	i_2	i_2	j_3	i_1	4	5	4	j_3
			j_5	i_3	5		5	j_5
			j_7	i_5	2		2	j_7
i_1, i_3, i_5 $\beta = 2$	i_5	i_5	j_9	i_3	4	5	4	j_9
			j_{10}	i_4	8		8	j_{10}
			j_{12}	s'	5		5	j_{12}
i_1, i_3, s' $\beta = 4$	i_1, i_3	i_1	j_6	i_4	7	8	7	j_6
		i_3	j_8	i_4	∞			
i_4, s' $\beta = 5$	s'							

Computational Complexity

At most once for any arc j is a value $w(i') + d_0^+(j)$ or $w(i') - d_0^-(j)$ computed, and j is never encountered in any other way (except perhaps when its other node i is scanned and j is seen not to be in Q). The phase of Dijkstra's method concerned with arc inspection therefore has complexity of order $0(|A|)$.

The minimization step that determines what new nodes should be added to S requires a look at each node of $T \setminus S$. The number of such nodes is bounded by $|N| - |S|$. Here $|S|$ begins as $|N^+|$ and increases by at least 1 in each iteration, so the number of iterations is bounded by $|N| - |N^+| - |N^-| + 1$. At all events the minimization step is no worse than $0(|N|^2)$. In the digraph case, where $|A| \leq |N|^2$, one therefore has $0(|N|^2)$ as a rough estimate for the entire algorithm.

Shortest Solutions and Multiple Criteria

In Dijkstra's method as stated, the choice of which available node to scan is arbitrary, but priorities could be introduced as in the case of the painted network algorithm in Section 2G. For example, one could follow the rule of breadth-first search, where new nodes added to S are always assigned a lower priority than nodes already in S. This would ensure that when a solution path P is obtained, it will also involve *as few arcs as possible*. (The argument is the same as for the painted network algorithm; see Section 2G and Exercise 2.20.) In effect we are able to solve by this simple device a special type of problem where there are *two* criteria for optimality: among the paths $P: N^+ \to N^-$, our primary aim is to minimize $d^+(P)$, but we wish to resolve ties in terms of minimizing a second quantity, the number of arcs used by P.

More general cases of multiple criteria can be handled by Dijkstra's method (or for that matter by the basic version of the min path algorithm) with hardly any more difficulty. Suppose each arc j has associated with it not only $d^+(j)$ and $d^-(j)$, but additional numbers $d_k^+(j) > -\infty$ and $d_k^-(j) < +\infty$ for $k = 1, \dots, m$. The problem is to minimize $d^+(P)$ over all paths $P: N^+ \to N^-$; among the solutions we want the ones yielding the lowest value of $d_1^+(P)$ and among those the ones with $d_2^+(P)$ lowest, and so forth.

The multiple criterion problem can be formulated instead in terms of the vectors

$$\bar{d}^+(j) = [d^+(j), d_1^+(j), \dots, d_m^+(j)],$$

$$\bar{d}^-(j) = [d^-(j), d_1^-(j), \dots, d_m^-(j)],$$

and corresponding expressions

$$\bar{d}^+(P) = \sum_{j \in P^+} \bar{d}^+(j) - \sum_{j \in P^-} \bar{d}^-(j)$$

$$= [d^+(P), d_1^+(P), \dots, d_m^+(P)].$$

These are well-defined elements of the space $[-\infty, \infty]^{m+1}$, which can be endowed with the *lexicographic ordering*:

$$[\lambda, \lambda_1, \dots, \lambda_m] \leq [\mu, \mu_1, \dots, \mu_m]$$

if and only if $\lambda < \mu$ or ($\lambda = \mu$ and $\lambda_1 < \mu_1$) or ($\lambda = \mu, \lambda_1 = \mu_1$ and $\lambda_2 < \mu_2$)...
or ($\lambda = \mu, \lambda_1 = \mu_1, \lambda_2 = \mu_2, \ldots, \lambda_m = \mu_m$). This is a total ordering that obeys
rules similar enough to those for ordinary numbers that it can be used in all
the preceding computations.

We are interested in minimizing $\bar{d}^+(P)$ in the lexicographic ordering, and
we can do this simply by executing Dijkstra's method with functions w and
quantities γ and β that are vector valued. Comparisons are made lexicographically. Everything else stays the same, including the justification. The procedure
can be initiated with any vector-valued potential $\bar{u}_0$ whose differential $\bar{v}_0$
satisfies $\bar{d}^-(j) \leq \bar{v}_0(j) \leq \bar{d}^+(j)$ lexicographically for all $j \in A$.

The example we started from, where the secondary concern was the number
of arcs in P, corresponds to $m = 1$, $d_1^+(j) = 1$ and $d_1^-(j) = -1$, with $\bar{u}_0 = (u_0, 0)$ initially. Note, however, that we do not need $d_k^-(j) \leq d_k^+(j)$ to make
things work in the general case, merely $\bar{d}^-(j) \leq \bar{d}^+(j)$.

Multipath Version

In case of ties in the construction of the routing, where for a node i there are
several arcs that could be taken as $\theta(i)$ (whether in the basic form of the min
path algorithm or in Dijkstra's method), the competing arcs can all be retained
as a set $\Theta(i)$. If this is done at every step, one obtains in place of a routing θ a
multirouting Θ that on termination embodies *all* solutions to the min path
problem: P solves the problem if and only if P is a Θ-path from N^+ to N^-.
(See Section 2G for "multiroutings.")

The same idea carries over to problems with multiple criteria using lexicographic ordering, as just explained. Thus, for example, one can construct by
this method a multirouting that represents all solutions employing the fewest
arcs. (Breadth-first search would not suffice in the multirouting case.)

6F.   FEASIBLE POTENTIALS

The min path algorithm makes use of an initial potential satisfying the
constraints of the max tension problem, and this feature leads to the question
of how such a potential might be constructed in general, when none may be
evident. The constraints involving N^+ and N^- cause no special difficulty,
because these sets could always be replaced by single nodes s and s' added to
the network as in Figure 6.8. The following problem is therefore fundamental.
The algorithms developed for it will later be found to have applications not
just to the max tension and min path problems but also to many other
questions in network optimization.

Feasible Differential Problem. *In a network with span intervals* $D(j) = [d^-(j), d^+(j)]$, *determine a potential u whose differential v satisfies $v(j) \in D(j)$
for all $j \in A$.*

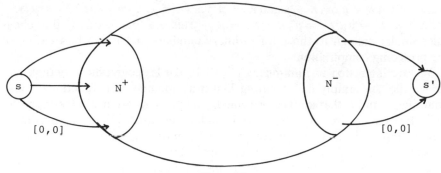

Figure 6.8

A more general model, called the *feasible potential problem*, can also be considered. In this, one asks for a potential u that is not only feasible with respect to the spans associated with the arcs but also satisfies $u(i) \in D(i)$ for all $i \in N$, where $D(i)$ is a nonempty closed real interval associated with the node i. The feasible differential problem corresponds to $D(i) = (-\infty, \infty)$ for all $i \in N$.

However, the feasible potential problem can easily be reduced to a feasible differential problem in the augmented network $\overline{G}$. Recall from Section 1I that the values $u(i)$ and $v(j)$ can be regarded as tensions with respect to a potential $\overline{u}$ in $\overline{G}$. If $D(i)$ is taken as the span interval for the distribution arc j_i, the conditions on $u(i)$ and $v(j)$ are turned into conditions on $\overline{v} = \Delta\overline{u}$, and one simply has a feasible differential problem for $\overline{G}$. See Figure 6.9.

If the feasibility conditions of interest can be expressed entirely in terms of differentials, what is the point of formulating the problem in terms of potentials at all? Why not just state it in terms of elements v of the differential space

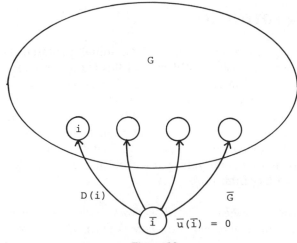

Figure 6.9

$\mathscr{D}$? The main reason is that potentials offer the most convenient means of representing differentials, despite nonuniqueness. In algorithmic applications, for example, it is usually potentials that are in the forefront, because they are easier to manipulate. Also they are often easier to interpret.

A necessary and sufficient condition for the existence of a solution to a feasible differential problem will now be formulated. Observe that the problem is not at all affected by adjoining to the network extra arcs having span interval $(-\infty, \infty)$, just as the feasible distribution problem is not affected by extra arcs with capacity interval $[0, 0]$. In particular, there is no loss of generality in assuming, when convenient, that the network is connected.

To motivate the condition, recall from Section 6A that for any path P and any tension v that is feasible with respect to spans, one has

$$d^-(P) \leq [\text{spread of } v \text{ relative to } P] \leq d^+(P).$$

If P is a circuit and v a differential, the quantity in the middle vanishes, and this says something about the nature of $d^+(P)$ and $d^-(P)$.

The following result may be compared with the feasible distribution theorem in Section 3H in the case of circulations ($b = 0$).

Feasible Differential Theorem.   *The feasible differential problem has a solution if and only if $d^+(P) \geq 0$ for all [elementary] circuits P.*

The sufficiency of the condition $d^+(P) \geq 0$, complementing the necessity just observed, will be established in Section 6G. Note that it is not necessary to include $d^-(P) \leq 0$ too, since $d^-(P) = -d^+(P')$ for the reverse path P'. Also if $d^+(P) \geq 0$ for all elementary circuits P, then the same holds for *all* circuits.

A related condition plays a role in both the min path problem and the max tension problem. The spans $d^+(j)$, $d^-(j)$, will be said to be *regular with respect to N^+ and N^-*, where N^+ and N^- are nonempty disjoint node sets, if $d^+(P) \geq 0$ not only for all circuits P but also for all paths $P: N^+ \to N^+$ and $P: N^- \to N^-$, and if furthermore for any two paths $P': N^+ \to N^-$ and $P'': N^- \to N^+$ one has $d^+(P') + d^+(P'') \geq 0$. This is trivially true if $d^-(j) \leq 0 \leq d^+(j)$ for all arcs j. (If $N^+ = \{s\}$, and $N^- = \{s'\}$, regularity reduces to the condition that $d^+(P) \geq 0$ for all circuits.)

Proposition 1.   *The spans are regular with respect to N^+ and N^- if and only if there is a potential u that solves the feasible differential problem and is constant on N^+ and constant on N^-, that is, satisfies the constraints of the max tension problem.*

Proof.   Apply the feasible differential theorem to the extended network shown in Figure 6.8.

As was observed in the discussion of the min path problem in Section 6B, the minimum might be $-\infty$, or one might have to resort to paths like those in

Figure 6.2 in order to have optimality, unless some restriction is made in the nature of the spans. The right restriction turns out to be "regularity."

Proposition 2.   *If the spans are regular with respect to N^+ and N^- (and there is at least one path from N^+ to N^-), then the min path problem has a solution, which moreover can be taken to be an elementary path with no intermediate nodes in N^+ or N^-.*

Proof 1.   Any path $P: N^+ \to N^-$ can be reduced to a path with these extra properties by deleting from it certain subpaths that are circuits, paths from N^+ to N^+, or paths from N^- to N^-. The regularity condition implies that such subpaths contribute a non-negative term to $d^+(P)$, so their deletion poses no disadvantage as far as minimization is concerned. (If P is formed by certain paths $P_1, \ldots, P_q$ in succession, then $d^+(P) = d^+(P_1) + \cdots + d^+(P_q)$.) In particular, there is no loss of generality in restricting the min path problem to elementary paths. There are only finitely many of these, so the minimum must be attained.

Proof 2.   The preceding argument was direct, but the result also follows constructively from Proposition 1. The existence of a potential satisfying the constraints in question means that the min path algorithm (in one form or another) can be used to obtain a solution to the min path problem. This solution, by the nature of the procedure, is an elementary path having only its initial node in N^+ and only its terminal node in N^-.

Irregular Problems

Of course there may well be circumstances where the min path problem is of interest even without spans being regular with respect to N^+ and N^-, so that paths might have more than one node in N^+ or in N^-. The nonexistence of a circuit with negative upper span would still be a prerequisite for the problem to be reasonable, but as long as that holds, the problem can be tackled by reformulating it as in Figure 6.3 and thereby transforming it to the "regular" case.

6G.   FEASIBLE DIFFERENTIAL ALGORITHM

A procedure will now be given that produces either a solution u to the feasible differential problem or an elementary circuit P with $d^+(P) < 0$. It always accomplishes this in a finite number of iterations and therefore furnishes a constructive proof of the feasible differential theorem. Each iteration involves a modified min path subroutine that will be seen, in the justification of the procedure, to be the ordinary min path algorithm applied to an auxiliary network.

A clear virtue of the method is that the amount of effort involved is roughly proportional to the extent to which the initial potential fails to be feasible. This could be valuable in applications like some of those in Chapters 7 and 9, where it is necessary to deal with a sequence of feasible differential problems differing only slightly from each other. The last potential examined in one problem can be used as the initial potential in the next.

Statement of the Algorithm

Starting with an arbitrary potential u, set

$$d_0^+(j) = d^+(j) - v(j), \qquad d_0^-(j) = d^-(j) - v(j), \qquad (1)$$

and check whether $\rho(i) \geq 0$ for all $i \in N$, where

$$\rho(i) = \min \begin{cases} d_0^+(j) & \text{for } j \in [i, N \setminus i]^+, \\ -d_0^-(j) & \text{for } j \in [i, N \setminus i]^-, \\ 0. \end{cases} \qquad (2)$$

If so, terminate; u solves the feasible differential problem. Otherwise select any node i such that $\rho(i) < 0$, and apply the min path algorithm as follows:

1. Take u as the initial potential u_0.
2. Take both $N^+ = \{\bar{i}\}$ and $N^- = \{\bar{i}\}$. In other words, treat $\bar{i}$ throughout the computations both as an element of S and as an element of $N \setminus S$ (as if it appeared in two copies), so that $\bar{i}$ might be reached a second time by some other node added to S, thereby generating a circuit.
3. Except in traversing an arc j from $\bar{i}$ outward, treat any *negative* terms $d_0^+(j)$ or $-d_0^-(j)$ that arise as if they were 0.

If $\bar{i}$ is reached (a second time) with $\beta < 0$, terminate; the θ-path $P: \bar{i} \to \bar{i}$ is a circuit with $d^+(P) < 0$, and the feasible differential problem has no solution.

Otherwise stop the min path algorithm as soon as a value $\beta \geq 0$ has been produced (which must happen sooner or later), even if $\bar{i}$ has not yet been reached (a second time). In this event do not proceed to add corresponding new nodes to S, but merely set $w(i) = 0$ for all $i \notin S$. The potential $u' = u + w$ then improves u, in the sense that the new ρ values are no lower than they were for u, and now $\rho(\bar{i}) = 0$. Start over with u'.

Remark 1. As will be explained in Section 6H, rules 2 and 3 are easily taken care of when Dijkstra's method Section 6E, is used to implement the min path algorithm: after an initial stage when all the arcs from $\bar{i}$ to adjacent nodes are inspected, simply regard $\bar{i}$ as deleted from S, and henceforth replace $d_0^+(j)$ by $\max\{d_0^+(j), 0\}$ and $d_0^-(j)$ by $\min\{d_0^-(j), 0\}$.

Remark 2. In updating the values of $d_0^+(j)$, $d_0^-(j)$ and $\rho(i)$ after each iteration, i.e., each application of the min path algorithm to a node $\bar{i}$, it is not necessary to refer back to the original spans $d^+(j), d^-(j)$, and the new potential values $u'(i)$. The new values for $d_0^+(j)$ and $d_0^-(j)$ can be obtained by subtracting $\Delta w(j)$ from the old ones, since $\Delta u' = \Delta u + \Delta w$, and the new values for $\rho(i)$ can then be read off from Equation (2). Of course, none of these values actually has to be computed until needed. When the algorithm terminates with a solution to the feasible differential problem, the potential in question is $u + w_1 + \cdots + w_m$, where u is the original potential and w_k is the function constructed in iteration k.

Other possible shortcuts are explained in Section 6H.

EXAMPLE 10

Figure 6.10 gives the data for a feasible differential problem. Taking $u \equiv 0$ initially, one obtains $[d_0^-(j), d_0^+(j)] = [d^-(j), d^+(j)]$. These intervals are repeated in Figure 6.11 along with the corresponding ρ values, calculated from

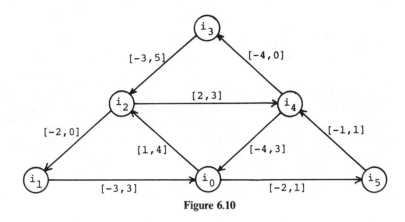

Figure 6.10

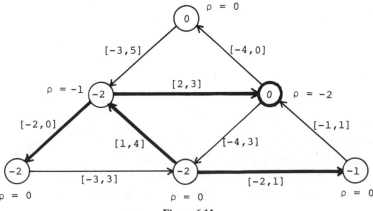

Figure 6.11

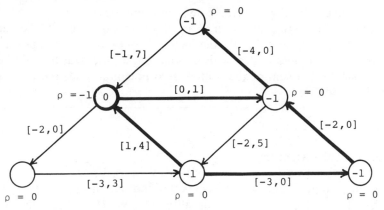

Figure 6.12

Equation (2). The nodes with ρ negative are i_2 and i_4, and i_4 is arbitrarily selected as $\bar{i}$.

The (modified) min path subroutine then produces the routing indicated by the heavy arcs. The numbers at the nodes in Figure 6.11 are the corresponding values $w(i)$. The first iteration ends without $\bar{i}$ being added to S (a second time), despite the fact that the last β value calculated is 0 and would, without the special stopping criterion for the min path subroutine, correspond to reaching $\bar{i} = i_4$ from i_5.

For the next iteration the intervals $[d_0^-(j), d_0^+(j)]$ and quantities $\rho(i)$ are the ones displayed in Figure 6.12. (The intervals are obtained by subtracting Δw from those in Figure 6.11, as noted in the foregoing Remarks.) This time i_2 is the only candidate for $\bar{i}$, and on application of the min path subroutine it turns out to be reached with $\beta = -1$. The corresponding θ-path is the circuit $P: i_2 \leftarrow i_0 \rightarrow i_5 \rightarrow i_4 \leftarrow i_2$ with $d^+(P) = -1$. The feasible differential problem has no solution.

If the original span interval for $j \sim (i_5, i_4)$ had been $[-1, 3]$ say, instead of $[-1, 1]$, the first iteration would not have been affected, and Figure 6.12 would

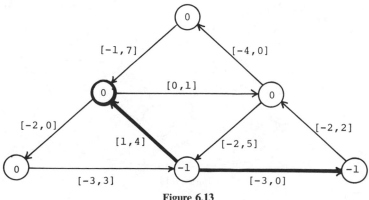

Figure 6.13

have been replaced by Figure 6.13. When the new intervals $[d_0^-(j), d_0^+(j)]$ are calculated (by subtracting Δw from w as shown in Figure 6.13), they are all found to contain 0, that is, all the new ρ values are non-negative. The algorithm therefore terminates with a solution to the feasible differential problem, namely, the sum of the (initial, zero potential, and) the w values in Figures 6.11 and 6.13.

Justification of the Algorithm

From the definition it is clear that $\rho(i) = 0$ if and only if $v(j) \le d^+(j)$ for all $j \in [i, N \setminus i]^+$, and $v(j) \ge d^-(j)$ for all $j \in [i, N \setminus i]$. This holds for all $i \in N$ if and only if $d^-(j) \le v(j) \le d^+(j)$ for all $j \in A$.

Supposing now that i is a node with $\rho(i) < 0$, consider the auxiliary network G^* in Figure 6.14. This is constructed by adding to G a second copy i^* of i attached to second copies j^* of all the arcs j incident to i, as well as a special arc $\bar{j} \sim (i^*, i)$ with span interval $(-\infty, 0]$. For $j \in [i, N \setminus i]^+$, j^* is assigned $(-\infty, d^+(j)]$, whereas for $j \in [i, N \setminus i]^-$, it is assigned $[d^-(j), \infty)$. Every arc j of G is given the *enlarged* span interval $[\hat{d}^-(j), \hat{d}^+(j)]$, where

$$\hat{d}^+(j) = \max\{d^+(j), v(j)\}, \qquad \hat{d}^-(j) = \min\{d^-(j), v(j)\}. \qquad (3)$$

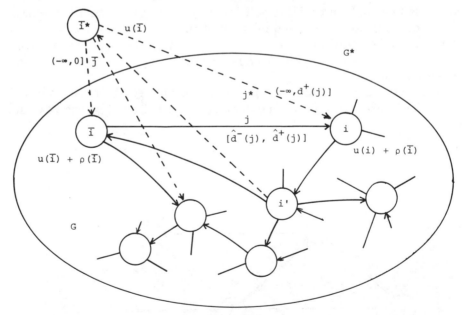

Figure 6.14

For the potential

$$u_0(i) = \begin{cases} u(i) & \text{for } i = i^*, \\ u(i) + \rho(i) & \text{for all other } i, \end{cases}$$

on G^*, the differential v_0 agrees with v on G and hence is feasible with respect to the enlarged spans for arcs of G. It is also feasible with respect to the span intervals for the new arcs j by the definition of $\rho(i)$. Thus v_0 is feasible with respect to all the spans in G^*.

It will be demonstrated later that the general step in the feasible differential algorithm amounts to applying the min path algorithm to solve the max tension problem in G^* for $N^+ = \{i^*\}$ and $N^- = \{\bar{i}\}$, with u_0 as the initial potential. Because of the arc $\bar{j}$ one certainly has [max] ≤ 0 in this problem. If [max] < 0, the min path algorithm furnishes as a solution to the dual problem a path from i^* to $\bar{i}$ with negative upper span, and this can be identified with an elementary circuit P in G with $d^+(P) < 0$ (since $\hat{d}^+(j) \geq d^+(j), -\hat{d}^-(j) \geq -d^-(j)$). If [max] $= 0$, the solution u' that is obtained for the max tension problem satisfies $u'(\bar{i}) - u'(i^*) = 0$, and its differential v' consequently has $v'(j) = v'(j^*)$ for each of the arcs j incident to $\bar{i}$ in G. Since v' is feasible with respect to the spans in G^*, it follows that the restriction of v' to G (which is the differential of the restriction of u' to G), is feasible with respect to the enlarged spans and has $v'(j) \leq d^+(j)$ for all $j \in [\bar{i}, N \setminus \bar{i}]^+, v'(j) \geq d^-(j)$ for all $[\bar{i}, N \setminus \bar{i}]^-$. Thus u' yields values of ρ that in no case are worse than the ones for u, and for u' one has $\rho(\bar{i}) = 0$.

The min path algorithm in G^* works in terms of shifted spans obtained by subtracting off the values of v_0. These can be interpreted as "relative costs" for traversing the various arcs. For the arcs of G^* belonging to G, the forward and backward "relative costs" are

$$\hat{d}_0^+(j) = \max\{d_0^+(j), 0\}, \qquad -\hat{d}_0^-(j) = \max\{-d_0^-(j), 0\} \qquad (4)$$

by (3). For the arcs j^* incident to i^*, only the "relative costs" for traversal away from i^* are of interest, and these are

$$d_0^+(j) - \rho(\bar{i}) \quad \text{if } j^* \leftrightarrow j \in [\bar{i}, N \setminus \bar{i}]^+,$$

$$-d_0^-(j) - \rho(\bar{i}) \quad \text{if } j^* \leftrightarrow j \in [\bar{i}, N \setminus \bar{i}]^-. \qquad (5)$$

For the special arc $\bar{j}$ the "relative cost" from i^* to $\bar{i}$ is $-\rho(\bar{i}) = |\rho(\bar{i})|$.

Recall now that the min path algorithm constructs the function w from a *nondecreasing* sequence of values $\beta \geq 0$, representing higher and higher levels of "relative cost" as nodes are added to S. Let $\bar{\beta}$ denote the value when $\bar{i}$ is reached; then [max] = [min] = $\rho(\bar{i}) + \bar{\beta}$, because [spread of u_0 from i^* to

$\bar{i}] = \rho(\bar{i})$ (see 6D(3)). Thus if $\bar{\beta} < |\rho(\bar{i})|$, a circuit of negative upper span in G has been detected; otherwise $\bar{\beta} = |\rho(\bar{i})|$, and then $\bar{i}$ can be regarded as having been reached by way of $\bar{j}$. In this event w is assigned the value $|\rho(\bar{i})|$ at $\bar{i}$ and all nodes not yet reached, and the potential $u' = u_0 + w$ is formed.

Only the restriction of u' to G is of interest, and this agrees with $u + w + \rho(\bar{i})$, since $u_0 = u + \rho(\bar{i})$ on G. All that remains is the observation that the constant $\rho(\bar{i})$ can be handled more efficiently by absorbing it into w. Specifically, the $\rho(\bar{i})$ terms can be dropped from the "relative costs" in (5), and since exactly one of the arcs covered by (5) must be traversed by each θ-path, this will have the effect of subtracting $|\rho(\bar{i})|$ from every value of w (except at $\bar{i}*$) and from every β. The threshhold level for $\bar{\beta}$ will then be 0 instead of $|\rho(\bar{i})|$. When this level is reached, one can simply set $u' = u + w$ on G. (Note that $w(\bar{i}) = 0 = w(\bar{i}*)$ at this point.)

In this form the procedure can be carried out perfectly well in terms of G itself, which is the way it has been stated as the general step of the feasible differential algorithm.

6H.*   REFINEMENTS

The min path subroutine in the feasible differential algorithm can be implemented as Dijkstra's method (Section 6E), with the spans modified in accordance with Rule 3 in Section 6G. Besides increasing efficiency in the obvious way, this has several other advantages.

Dijkstra's method begins by scanning $\bar{i}$: all the arcs incident to $\bar{i}$ are examined thoroughly and never again considered in the "outbound" direction. Therefore $\bar{i}$ *can simply be deleted from* S *after this initial step*, and no special effort has to be made over its dual role in the sense of Rule 2.

Another feature is that the explicit test for feasibility at the start of each iteration of the feasible differential algorithm can be streamlined. No matter which node is selected as $\bar{i}$, the first β value calculated in Dijkstra's method will yield $\rho(\bar{i}) = \min\{\beta, 0\}$ by 6G(2). (The rule for modifying spans does not come into play on the initial round.) Therefore the nodes of G can simply be tested by placing them one by one in the role of $\bar{i}$. If the first β value satisfies $\beta \geq 0$, conclude $\rho(\bar{i}) = 0$ and pass to new choice of $\bar{i}$, whereas if $\beta < 0$, continue the procedure to its end (henceforth replacing negative terms $d_0^+(j)$ or $-d_0^-(j)$ by 0) before passing to a new node. If a circuit of negative upper span has not been detected in the meantime, termination will come when all nodes have been tested, and the potential on hand at that time will be a solution to the feasible differential problem.

Computational Complexity

The application of the first step of Dijkstra's method to each node is at worst an $0(|N|)$ computation performed $|N|$ times. Each phase where a node $\bar{i}$ is brought into balance by following Dijkstra's method through several steps (until a positive ρ value has been reduced to 0) can be estimated as $0(|N|^2)$ in the digraph case; see Section 6E. Of course this does not really do justice to the idea, since one expects rarely to have to go through the whole network in this procedure. At any rate such reasoning yields $0(|N_0||N|^2)$ for the whole feasible differential algorithm in the digraph case, where N_0 is the set of nodes i having $\rho(i) < 0$ initially. Thus even if $\rho(i) < 0$ for every node initially and the whole network had to be examined to bring each node in line, the order of complexity could be no worse than $0(|N|^3)$ in the digraph case.

In certain important cases involving acyclicity, the construction of a solution to the feasible differential problem can be accomplished differently by a procedure of order $0(|N|^2)$; see Exercises 6.10 and 6.23.

Restart Provision

We have mentioned that in some applications a sequence of feasible differential problems, differing only slightly from each other, must be tackled. The potential left over from the attempt at solving one problem can be used as the initial potential for the next. However, it is important to gain as much advantage as possible from the information generated in such a process. This means that even when the algorithm terminates with $\beta < 0$ and a circuit of negative upper span, we should not leave u unchanged but make use of the w values at that stage.

The rule is simple. *When $\bar{i}$ is reached* (a second time) *with $\beta < 0$, add $w(i) - \beta$ to $u(i)$ for all $i \in S$ but leave $u(i)$ unaltered for all $i \in N \setminus S$.* (It makes no difference whether the nodes just reached are added to S or not, because either way u will not be changed at such nodes; in particular, $u(\bar{i})$ will stay the same.)

The justification of the algorithm in Section 6G shows that the potential obtained in this way corresponds to a solution to the max tension problem for the network G^* in Figure 6.14. Namely, the restriction to G of the potential constructed by the min path algorithm in G^* for $N^+ = \{\bar{i}^*\}$ and $N^- = \{\bar{i}\}$ (and the indicated starting potential) is $u + w$, where on termination w is given the value β at $\bar{i}$ and every node $i \notin S$. Subtracting the constant β from this potential does not affect its optimality in the corresponding max tension problem but restores the values at $\bar{i}$ and the nodes $i \notin S$ to those of u. (Recall that the $w(i)$ and β values spoken of here have been shifted downward by $|\rho(\bar{i})|$ from the ones appearing in the min path algorithm itself; see the end of Section 6G.)

It follows, just as in the case of termination with $\beta = 0$, that the terminal potential under this rule has $\rho(i)$ no lower for any node i than was true originally. Although $\rho(\bar{i})$ has not been brought to 0, it has at least been raised by the amount $|\rho(\bar{i})| + \beta \geq 0$ to β.

For future purposes (in Section 7C) it should be noted that the following conditions relate the terminal potential u and the circuit $P: \bar{i} \rightarrow \bar{i}$, which of course has $d^+(P) = \beta$. Let j_1 denote the first arc of P after $\bar{i}$. Then

$$
v(j) = \begin{cases}
d^+(j) & \text{for } j = j_1 \text{ if } j_1 \in P^+, \\
d^-(j) & \text{for } j = j_1 \text{ if } j_1 \in P^-, \\
\max\{d^+(j), 0\} & \text{for all } j \neq j_1 \text{ in } P^+, \\
\min\{d^-(j), 0\} & \text{for all } j \neq j_1 \text{ in } P^-.
\end{cases}
$$

This is immediate from the min path characterization in Exercise 6.9, as applied to the special span intervals in G^*. Thus in particular,

$$
v(j) \geq d^+(j) \quad \text{for all } j \in P^+,
$$

$$
v(j) \leq d^-(j) \quad \text{for all } j \in P^-. \tag{1}
$$

Double Version of the Algorithm

The feasible differential algorithm has been developed for the quantities $\rho(i)$ as measures of infeasibility, but a parallel development would be possible in terms of

$$
\sigma(i) = \min \begin{cases}
-d_0^-(j) & \text{for } j \in [i, N \setminus i]^+, \\
d_0^+(j) & \text{for } j \in [i, N \setminus i]^-, \\
0
\end{cases} \tag{2}
$$

(where $d_0^+(j) = d^+(j) - v(j)$ and $d_0^-(j) = d^-(j) - v(j)$). Obviously, v is feasible with respect to spans if and only if $\sigma(i) = 0$ for all $i \in N$. A more interesting observation for the moment, though, is that the double condition $\rho(i) = 0 = \sigma(i)$ is equivalent to $d^-(j) \leq v(j) \leq d^+(j)$ for all arcs j incident to i. When the feasible differential algorithm acts on a node $\bar{i}$, there is no danger of it lowering $\sigma(\bar{i})$ (since, as we know, infeasibility is never worsened in any arc; the new differential is always feasible with respect to the enlarged span intervals corresponding to the old differential). Hence if $\sigma(\bar{i}) = 0$ at the beginning of an iteration, the same is true when the iteration is finished.

We can take advantage of this as follows. Each time a node $\bar{i}$ is up for inspection, *first calculate* $\sigma(\bar{i})$, *and add it to* $u(\bar{i})$. The altered potential will have $\sigma(\bar{i}) = 0$. Infeasibility as measured by $\rho(\bar{i})$ may have been worsened (in particular, if $\rho(\bar{i})$ was negative it has been replaced by $\rho(\bar{i}) + \sigma(\bar{i})$), but arcs not incident to $\bar{i}$ will not have been affected. Next apply the procedure in the feasible differential algorithm to $\bar{i}$ as previously. When this is done, $\rho(\bar{i})$ will have been lifted to 0, and $\sigma(\bar{i})$ will still be 0, so that the differential at that stage will be *feasible with respect to spans in all the arcs incident to* $\bar{i}$.

The calculation of $\sigma(i)$ for every node i is an operation of complexity $0(|N|^2)$, so the feasible differential algorithm with this modification still has the same order of complexity, despite the extra work. In some cases one might be able to compensate for the extra work by treating only "every other node": it suffices to treat merely the nodes in a subset $\overline{N} \subset N$ having the property that every arc $j \in A$ is incident to at least one of the nodes in $\overline{N}$. One can imagine that this typically would cut the number of iterations roughly in half, since every arc is incident to two nodes, but in practice a suitable set $\overline{N}$ may not be so easy to determine; see Exercise 6.35.

Unfortunately too, the algorithm in this "double" form does not lend itself as well to certain other applications, in particular, to the construction of optimal routings (Section 6K), as does the original form.

6I.* TENSION RECTIFICATION ALGORITHM

Another method of solving the feasible differential problem, or producing a circuit of negative upper span when no solution exists, will now be described. Although its efficiency can be brought on a par with that of the feasible differential algorithm, which it resembles almost as a mirror image once a number of refinements are made, it lacks some of the versatility of the latter. It cannot, for example, be used with such natural ease as the feasible differential algorithm in constructing optimal routings (see Section 6K). However, it has definite theoretical interest because it is closely parallel to the flow rectification algorithm in Section 3K and appears as a specialization of the general "out-of-kilter algorithm" that will be studied in Chapter 9.

We shall first examine the idea in a fundamental form and then work our way toward a more polished implementation.

Statement of the Algorithm

Given an arbitrary potential u and its differential v, let

$$A^+ = \{ j \in A \mid v(j) < d^-(j) \}, \qquad A^- = \{ j \in A \mid v(j) > d^+(j) \}.$$

If $A^+ = \varnothing = A^-$, u is a solution to the feasible differential problem, and nothing needs to be done. Otherwise select any arc $\bar{j}$ in $A^+ \cup A^-$ and apply Minty's lemma (and the associated algorithm; see Section 2H) to $\bar{j}$ and the following painting of the arcs j of the network:

red	if $d^-(j) < v(j) < d^+(j)$
black	if $v(j) \leq d^-(j)$, $v(j) < d^+(j)$
white	if $v(j) \geq d^+(j)$, $v(j) > d^-(j)$
green	if $d^-(j) = v(j) = d^+(j)$

(Note that $\bar{j}$ will be black or white.)

If the outcome is a compatible circuit P containing $\bar{j}$, terminate; this circuit has $d^+(P) < 0$.

If the outcome is a compatible cut Q containing $\bar{j}$, calculate

$$\alpha = \min \begin{cases} d^+(j) - v(j) & \text{for } j \in Q^+, \\ v(j) - d^-(j) & \text{for } j \in Q^-, \\ \bar{\alpha}, \end{cases} \tag{1}$$

where

$$\bar{\alpha} = \text{dist}\left(v(\bar{j}), [d^-(\bar{j}), d^{\,\text{l}}(\bar{j})]\right) = \begin{cases} d^-(\bar{j}) - v(\bar{j}) & \text{if } \bar{j} \in A^+, \\ v(\bar{j}) - d^+(\bar{j}) & \text{if } \bar{j} \in A^-. \end{cases} \tag{2}$$

Then $0 < \alpha < \infty$. Let $u' = u + \alpha e_{N \setminus S}$, where $[S, N \setminus S] = Q$. Then u' is better than u in the sense that the distance of $v'(j)$ from the interval $[d^-(j), d^+(j)]$ is no greater than that of $v(j)$ for any arc j, and in the case of $\bar{j}$ it has been diminished by α. (In particular, the new sets A^+ and A^- are included in the old ones.) Repeat the procedure with u'.

Comments

The direct verification of the claims in the statement of the algorithm is Exercise 6.41.

Termination in a finite number of iterations could be established under a commensurability condition, but something much stronger is true: if the same arc $\bar{j}$ is selected as long as it is still in $A^+ \cup A^-$, there can only be a finitely many iterations, fewer than $|N|$ in fact, before $\bar{j}$ is finally removed from $A^+ \cup A^-$. Under this rule therefore the total number of iterations is less than $|N|(|A^+| + |A^-|)$.

The argument establishing this is based on the realization that, as long as $\bar{j}$ is fixed, the procedure is equivalent to applying the max tension algorithm (formulated in Exercise 6.17) in a certain way. With u as the initial potential, we are trying to solve the max tension problem with respect to the modified spans

$$\hat{d}^+(j) = \max\{d^+(j), v(j)\}, \qquad \hat{d}^-(j) = \min\{d^-(j), v(j)\}, \tag{3}$$

and $N^+ = \{s\}, N^- = \{s'\}$, where

$$(s, s') \sim \bar{j} \quad \text{if } \bar{j} \in A^+, \qquad (s', s) \sim \bar{j} \quad \text{if } \bar{j} \in A^-. \tag{4}$$

The inclusion of $\bar{\alpha}$ in (1) (to avoid "overshooting") corresponds to replacing the interval for $\bar{j}$ by $(-\infty, \bar{\alpha}]$ if $\bar{j} \in A^+$, or $[-\bar{\alpha}, \infty)$ if $\bar{j} \in A^-$. The max tension algorithm is just a rudimentary form of the min path algorithm, and it always takes less than $|N|$ iterations (see Exercise 6.18).

This insight suggests the following improvement.

Min Path Implementation

Given any potential u, form the sets A^+ and A^- as before. If $A^+ = \emptyset = A^-$, terminate; u solves the feasible differential problem. Otherwise select any $\bar{j}$ in $A^+ \cup A^-$, choose s and s' as in (4), and apply the min path algorithm to $N^+ = \{s\}$, $N^- = \{s'\}$ and the modified spans in (3) with u as the initial potential u_0.

If s' is reached with a value $\beta < \bar{\alpha}$ (as defined in (2)), terminate; the path $P = P_0, \bar{j}, s$, where P_0 is the θ-path from s to s', is an elementary circuit with $d^+(P) < 0$, and the feasible differential problem has no solution.

Otherwise stop the min path algorithm as soon as a value $\beta \geq \bar{\alpha}$ has been generated, even if s' has not yet been reached. In this case do not proceed to add new nodes to S; just set $w(i) = \bar{\alpha}$ for all $i \notin S$ (including s'), and define $u' = u + w$. Then u' is better than u in the sense that the new sets A^+ and A^- are included in the old ones and no longer contain $\bar{j}$.

Refinements

Dijkstra's method can be used in the min path subroutine, and the algorithm then appears to have order of complexity $0(|A|(|A^+| + |A^-|)) \leq 0(|A|^2)$. This might indicate a disadvantage for the tension rectification algorithm in comparison with the feasible differential algorithm, in which the number of applications of Dijkstra's method (each of order $0(|A|)$) is bounded by the number of nodes initially with $\rho(i) < 0$, rather than the number of arcs initially in $A^+ \cup A^-$.

It turns out, however, that several arcs can sometimes be treated at the same time, and when this feature is incorporated the two algorithms have the same order of complexity.

Specifically, suppose $\bar{j}_1, \bar{j}_2, \ldots, \bar{j}_r$ are arcs in $A^+ \cup A^-$ that correspond to the same s but involve (possibly) different nodes $s'_1, s'_2, \ldots, s'_r$ and $\bar{\alpha}_1 \leq \bar{\alpha}_2 \leq \cdots \leq \bar{\alpha}_r$. The min path subroutine can be executed by focusing first on $\bar{j}_1$, then on $\bar{j}_2$, and so forth, without having to start all over again each time. What this works out to is that the same can be done as if $\bar{j} = \bar{j}_r$, $s' = s'_r$, and $\bar{\beta} = \bar{\beta}_r$, but with the provision that if *any* of the nodes s'_k is reached with $\beta < \bar{\alpha}_k$, the algorithm terminates. In that case the θ-path $P_0 : s \to s'_k$, followed by $\bar{j}_k, s$, is a circuit P with $d^+(P) < 0$. Otherwise one has on termination a potential for which none of the arcs $\bar{j}_1, \ldots, \bar{j}_r$ appears any longer in $A^+ \cup A^-$. (This conclusion is obtained from the relations $u'(s'_k) = w(s'_k) + u(s'_k) \geq \bar{\alpha}_k + u(s'_k)$, $u'(s) = u(s)$, and the definition of $\bar{\alpha}_k$.)

In this form the tension rectification algorithm looks even more like the feasible differential algorithm. The nodes that can serve as s are those for which the quantity

$$\sigma(i) = \min \begin{cases} v(j) - d^-(j) & \text{for } j \in [i, N \setminus i]^+, \\ d^+(j) - v(j) & \text{for } j \in [i, N \setminus i]^-, \\ 0, \end{cases} \tag{5}$$

is negative. (This is the same $\sigma(i)$ as in the double version of the feasible differential algorithm in 6H(2).) When such a node i is selected as s, the arcs $\bar{j}_1, \dots, \bar{j}_r$ that can be treated are those that make negative contributions in (5), and we have $\bar{\alpha}_r = |\sigma(i)|$. It can come as no surprise that with a little extra effort, the difference between the algorithms practically evaporates, except that they "operate in opposite directions" (Exercise 6.42).

6J.* OPTIMAL ROUTINGS

In a network with spans, a routing θ of S with base N^+ (where $N^+ \subset S$) is called *optimal* if for each node $i \in S \setminus N^+$, the θ-path $P: N^+ \to i$ solves the min path problem for $N^- = \{i\}$. Of particular interest are *complete* optimal routings, those for $S = N$. The construction of such routings by extended versions of the min path algorithm and the feasible differential algorithm is the subject of this section and the next.

The min path problem for general N^+ and N^- has a solution if the network is connected and if the feasible differential problem has a solution, or equivalently if spans are regular with respect to N^+ and N^- (see Proposition 2 in Section 6F). In the case of $N^+ = \{i'\}$, $N^- = \{i\}$, the condition is just that $d^+(P) \geq 0$ for all circuits P. For each pair of distinct nodes i and i', denote the minimum in this problem by $d(i', i)$, and define $d(i', i) = 0$ when $i = i'$. Then d is a function on $N \times N$ with values in $(-\infty, +\infty]$. It will be called the *span function* for G.

It is easy to see that the *triangle inequality* is valid, namely

$$d(i_1, i_2) \leq d(i_1, i_3) + d(i_3, i_2) \quad \text{for all nodes } i_1, i_2, i_3. \tag{1}$$

(Exercise 6.43). But d is generally not a "metric," since its values may be negative or $+\infty$, and perhaps $d(i', i) \neq d(i, i')$.

In some applications the entire span function for G must be computed, and one method of doing this is described in Exercise 6.49. Here we are interested mainly in the values

$$d(N^+, i) = \min_{i' \in N^+} d(i', i) \quad \text{for all } i \in N,$$

where N^+ is a given set, possibly consisting of a single node s. For $i \notin N^+$, these correspond to the min path problem for $N^- = \{i\}$. For N^- of such type, regularity of spans with respect to N^+ and N^- means that $d^+(P) \geq 0$ for all circuits and for all paths $P: N^+ \rightarrow N^+$. Since this condition is independent of the choice of i, we shall refer to it simply as *regularity of spans with respect to* N^+. In particular, it implies $d(i', i) \geq 0$ for all $i' \in N^+$ and $i \in N^+$, and since $d(i', i) = 0$ when $i' = i$, one can then conclude

$$d(N^+, i) = 0 \quad \text{for all } i \in N^+.$$

A routing θ is said to be *compatible* with respect to the spans in G if every θ-path P has $d^+(P) < \infty$ (or equivalently, is compatible with respect to the painting associated with the spans in Section 6B). If θ is a compatible routing with base N^+ that is *complete* (i.e., a routing of all of N), it gives for each node $i \in N \setminus N^+$ a θ-path $P_i: N^+ \rightarrow i$ such that $d^+(P_i) < 0$. Defining

$$u(i) = \begin{cases} d^+(P_i) & \text{for } i \in N \setminus N^+, \\ 0 & \text{for } i \in N^+, \end{cases}$$

one obtains a finite function on N called the *potential* associated with θ. The role of this potential is explained by the following result.

Optimal Routing Theorem. *Suppose the spans in G are regular with respect to N^+ and that θ is a complete compatible routing with base N^+. Let u be the associated potential. Then the following are equivalent*:

1. *θ is optimal.*
2. *$u(i) = d(N^+, i)$ for all $i \in N$.*
3. *u is feasible with respect to spans, that is u solves the feasible distribution problem.*

When these hold, u also solves, for every $i \in N \setminus N^+$, the max tension problem for N^+ and $N^- = \{i\}$.

Proof. The equivalence between Assertions 1 and 2 is immediate from the definitions and the earlier observation that $d(N^+, i) = 0$ for all $i \in N^+$ when spans are regular with respect to N^+.

To see that Assertion 2 implies Assertion 3, let $v = \Delta u$, and consider any arc $j \sim (i, i')$. One has $v(j) = d(N^+, i') - d(N^+, i)$ and the triangle inequality yields

$$d(N^+, i') \leq d(N^+, i) + d(i, i') \leq d(N^+, i) + d^+(j),$$

$$d(N^+, i) \leq d(N^+, i') + d(i', i) \leq d(N^+, i') - d^-(j).$$

Therefore $d^-(j) \leq v(j) \leq d^+(j)$.

For the converse, observe that if u is feasible, then for any node $i \in N \setminus N^+$ and any path $P: N^+ \to i$ one has

$$d^+(P) \geq [\text{spread of } \Delta u \text{ relative to } P] = u(i) - 0 = d^+(P_i).$$

Hence the θ-path solves the min path problem for N^+ and $N^- = \{i\}$.

As for the final remark in the theorem we know that u satisfies the constraints of the max tension problem for $N^- = \{i\}$ and has

$$[\text{spread of } u \text{ from } N^+ \text{ to } i] = u(i) = d(N^+, i).$$

Then u solves the problem in question, according to the max tension min path theorem.

Construction of an Optimal Routing

If a potential u_0 that is feasible with respect to spans and identically zero on N^+ is available, the construction of an optimal routing with base N^+ is especially easy. Just apply the min path algorithm with u_0 the initial potential as in Section 6E, but with $N^- = \varnothing$. In other words, continue iterating until $\beta = \infty$, which certainly will happen when $S = N$, if not sooner. Then $S \setminus N^+$ consists of all the nodes of $N \setminus N^+$ that can be reached from N^+ by paths P with $d^+(P) < 0$, and θ is an optimal routing of S with base N^+.

This is seen from the fact that whichever $i \notin N^+$ we might be interested in, if $i \in S$, the θ-path from N^+ to i is the one we would have terminated with if we had had $N^- = \{i\}$, whereas if $i \notin S$, the cut $Q = [S, N \setminus S]$ separates N^+ from i and is of unlimited span.

Note that if $S = N$, the potential $u = u_0 + w$ furnished by the algorithm is the one associated with θ as in the preceding theorem: $u(i) = d(N^+, i)$ for all $i \in N$.

To illustrate, a complete optimal routing with base $N^+ = \{s\}$ for the network in Figure 6.6 is obtained by applying the min path algorithm with $u_0 \equiv 0$ and not just terminating when s' was reached, as was the case in Example 8 in Section 6D, but continuing until there are no new nodes left to reach at all. The resulting routing is shown in Figure 6.15, and the numbers at the nodes are the values $d(s, i)$.

An initial potential u_0 meeting the requirements is readily available if $d^-(j) \leq 0 \leq d^+(j)$ for all $j \in A$ (take $u_0 \equiv 0$) or if the network is acyclic and all span intervals are of the form $(-\infty, d^+(j)]$ (see Exercise 6.19, and for a generalization, Exercise 6.23). If all else fails, u_0 could first be constructed by the feasible differential algorithm. The constancy of u_0 on N^+ can be guaranteed by joining all nodes of N^+ to a special node s by arcs with span interval $[0, 0]$.

The next section describes another approach to the general case. It avoids such a two-stage process and obtains feasibility and optimality at the same time just by executing the feasible differential algorithm with extra care.

6K.*   ROUTING OPTIMIZATION PROCEDURE

In a network with spans, let θ be any complete compatible routing with base N^+. Unless a path violating the condition (in Section 6J) of regularity of spans with respect to N^+ is detected in the meantime, the method about to be explained will transform θ into a complete *optimal* routing with base N^+ and at the same time calculate all values $d(N^+, i)$.

Apply the feasible differential algorithm with the following supplementary provisions, taking as initial potential u the potential associated with the given θ. In each iteration where a node i with $\rho(i) < 0$ is treated and the modified min path subroutine is used to generate a (partial) routing $\bar{\theta}$ with base $\bar{i}$:

1. Replace the current routing θ by $\bar{\theta}$ on the set where $\bar{\theta}$ is defined and as it is generated.
2. Terminate the algorithm not only if $\bar{i}$ is reached (a second time) by $\bar{\theta}$ with $\beta < 0$ but also if any other node in the current θ-path from N^+ to $\bar{i}$, or any node at all in N^+, is reached by $\bar{\theta}$ with $\beta < 0$.

If the broadened termination condition given by Rule 2 does not intervene (so that each iteration ends when a value $\beta \geq 0$ is produced and without modifying the current θ at any of the nodes of the kinds mentioned in Rule 2), the final θ is a complete, optimal routing with base N^+, and the final potential u is the function

$$u(i) = d(N^+, i) \quad \text{for all } i \in N. \tag{1}$$

If termination comes under Rule 2, spans are not regular with respect to N^+. There are two cases. If a node i in the current θ-path from N^+ to $\bar{i}$ has been reached, the redefinition of θ at i has disrupted this path, and θ now gives an elementary circuit P through i and $\bar{i}$ with $d^+(P) < 0$. If a node $i \in N^+$ has

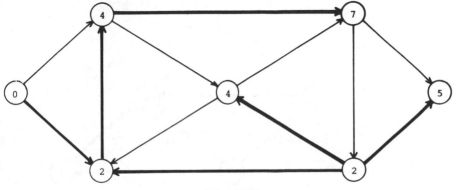

Figure 6.15

been reached, the current θ-path from N^+ to $\bar{i}$ and then to i is a path P: $N^+ \to N^+$ with $d^+(P) < 0$.

EXAMPLE 11

Take G again to be the network in Figure 6.11. Let $N^+ = \{i_0\}$, and take the initial routing θ to be the one in Figure 6.16. The numbers at the nodes in Figure 6.16 give the potential u associated with θ. The nodes i_0 and i_2 are unbalanced. Arbitrarily selecting i_2 as $\bar{i}$, we obtain in the first iteration of the feasible differential algorithm the routing $\bar{\theta}$ and function w displayed in Figure 6.17; the intervals here are $[d_0^-(j), d_0^+(j)]$. At the end of the iteration therefore the routing θ and potential u have been replaced by the ones in Figure 6.18. Now only i_0 is unbalanced. Figure 6.19 shows the next iteration of the feasible differential algorithm with the new intervals $[d_0^-(j), d_0^+(j)]$ and $\bar{i} = i_0$. It turns out that $\bar{i}$ is reached a second time with $\beta = -1$. The circuit $P: i_0 \to i_5 \to i_4 \leftarrow i_2 \leftarrow i_0$ has $d^+(P) = -1$. No optimal routing with base i_0 exists.

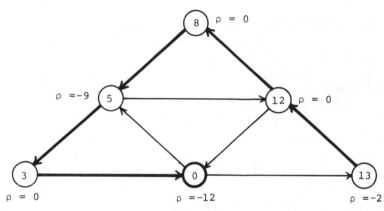

Figure 6.16

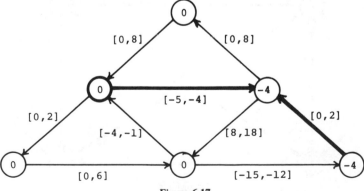

Figure 6.17

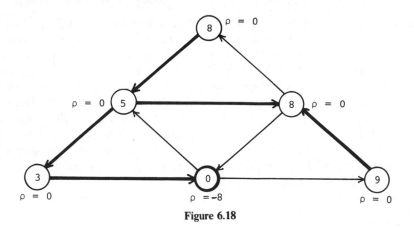

Figure 6.18

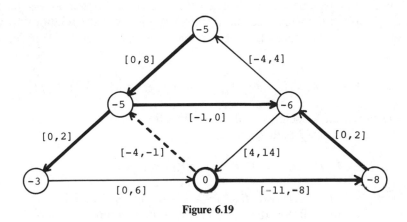

Figure 6.19

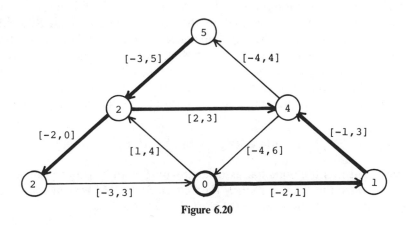

Figure 6.20

If the original span interval for $j \sim (i_5, i_4)$ had been $[-1,3]$ instead of $[-1,1]$, Figures 6.16, 6.17, and 6.18' would have been the same, except for having $[0,4]$ in place of $[0,2]$ in Figure 6.17 for the arc in question and $\rho(i_5) = -4$ in Figure 6.18. Then the w values in Figure 6.19 at nodes other than i_0 and i_5 would all have been raised by 2, and the procedure would have terminated with the routing θ and potential u shown in Figure 6.20. (Here the previous θ was entirely replaced by $\bar{\theta}$ in the final iteration, which is unusual.) The initial span intervals $[d^-(j), d^+(j)]$ are included in Figure 6.20 for comparison. As a check on the optimality of θ in terms of the optimal routing theorem in Section 6J, it can be verified that u is feasible (and is indeed the potential associated with θ).

Justification of the Procedure

It will not necessarily be true at all stages that the current potential u is the one associated with the current routing θ, but a slightly weaker relationship will turn out to be maintained. Let us say that u is *admissible* for θ if for the enlarged spans

$$\hat{d}^+(j) = \max\{d^+(j), v(j)\}, \qquad \hat{d}^-(j) = \min\{d^-(j), v(j)\}, \qquad (2)$$

it is the potential associated with θ, or in other words, if $u(i) = 0$ for all $i \in N^+$, and for each θ-path P one has

$$v(j) \geq d^+(j) \quad \text{for all } j \in P^+, \qquad v(j) \leq d^-(j) \quad \text{for all } j \in P^- \qquad (3)$$

(which implies of course that $d^+(P) < \infty$, and hence that θ is compatible). Certainly this holds if for the original spans θ is compatible and u is the associated potential, and hence it holds initially.

Suppose that at the start of an iteration where an unbalanced node i is treated, θ is a complete routing with base N^+ and u is admissible for θ. It will be demonstrated that the same holds at the end of the iteration, if Rule 2 has not intervened. Therefore when all unbalanced nodes have been treated and the feasible differential algorithm has finished with u feasible with respect to the original spans, θ will still be a complete routing with base N^+ for which u is admissible. But with u now feasible with respect to the original spans, admissibility reduces to the condition that $u(i) = 0$ for all $i \in N^+$, and for each θ-path P,

$$v(j) = d^+(j) \quad \text{for all } j \in P^+, \qquad v(j) = d^-(j) \quad \text{for all } j \in P^-.$$

This says that u is the potential associated with θ with respect to the *original* spans, and from the optimal routing theorem in Section 6J we can conclude in this event that 0 is optimal and $u(i) = d(N^+, i)$ for all $i \in N$.

In the iteration that treats node $\bar{i}$, the routing $\bar{\theta}$ is defined only on a set $S \setminus \{\bar{i}\}$ that (if Rule 2 does not intervene) never meets N^+ or the current θ-path from N^+ to $\bar{i}$. When $\bar{\theta}$ is substituted for θ on this set, the effect is simply that in tracing the routing backward from an arbitrary node in $N \setminus N^+$, if a node of $S \setminus \{\bar{i}\}$ is encountered, one proceeds through $\bar{\theta}$ from there to $\bar{i}$ and then through the unaltered labels from $\bar{i}$ to N^+. Thus the new θ still gives a path from N^+ to each node in $N \setminus N^+$; it is still a complete routing with base N^+.

The modified min path subroutine that generates $\bar{\theta}$ is really just the ordinary min path algorithm applied to a different choice of span intervals, as explained in Section 6H. The remarks in Section 6J about the construction of an optimal routing by the min path algorithm are therefore applicable: $\bar{\theta}$ is an optimal routing of S with base $\bar{i}$ relative to the intervals

$$
\begin{array}{ll}
(-\infty, d^+(j)] & \text{for } j \in [\bar{i}, N \setminus \bar{i}]^+, \\
\,[d^-(j), \infty) & \text{for } j \in [\bar{i}, N \setminus \bar{i}]^-, \qquad (4) \\
\,[\hat{d}^-(j), \hat{d}^+(j)] & \text{for all other arcs } j,
\end{array}
$$

and the potential $\bar{u}$ obtained on S is the one associated with $\bar{\theta}$ with respect to these intervals. Since at the end of the iteration the restrictions of θ and u to S are $\bar{\theta}$ and $\bar{u}$, it follows that the admissibility condition (3) holds for all portions of θ-paths where the nodes are all in S. Of course too, since θ and u are unchanged outside of $S \setminus \{\bar{i}\}$, the same condition still holds for portions of θ-paths that do not meet $S \setminus \{\bar{i}\}$, and u is still identically zero on N^+. It will be argued next that there are in fact no portions of θ-paths not covered by these observations, and hence the new u is admissible for the new θ as claimed.

The question is whether one of the new θ-paths can contain an arc j that is not incident to $\bar{i}$ but belongs to the cut $Q = [S, N \setminus S]$. Suppose there were such an arc j, and let i be its node in $N \setminus S$. Then either $j = \theta(i)$ or $j = \theta(i')$, but the first is impossible because $\theta(i) = \bar{\theta}(i)$ (and in tracing θ-paths backward one stays in S until $\bar{i}$ is reached). Since θ has not been changed on $N \setminus S$, it follows that the old θ-path from N^+ to i' must have traversed j from i to i'. The old u was admissible for the old θ, so there are only two possibilities: $j \sim (i, i')$ and $d_0^+(j) = d^+(j) - v(j) \leq 0$, or $j \sim (i', i)$ and $-d_0^-(j) = v(j) - d^-(j) \leq 0$. Either way, after i was reached by the modified min path subroutine with $\beta < 0$, i' could have been reached on the next step without any increase in β. Thus the subroutine could not have ended with i' still in $N \setminus S$. (This fact leads to a refinement of the procedure, which follows.)

The claims about termination under Rule 2, namely, that $d^+(P) < 0$ in both cases, still need to be verified. Again we rely on the way $\bar{\theta}$ is constructed by the subroutine, namely, in terms of the "relative costs" $d_0^+(j)$ and $-d_0^-(j)$ that are replaced by 0 when they would otherwise turn out negative in an arc j not incident to $\bar{i}$. Each time a node i is added to S with a value $w(i) = \beta$, the corresponding $\bar{\theta}$-path $\tilde{P}: \bar{i} \to i$ has β as its "relative cost," and hence (in terms of the potential at the start of the iteration)

$$\beta \geq \sum_{j \in \tilde{P}^+} [d^+(j) - v(j)] + \sum_{j \in \tilde{P}^-} [v(j) - d^-(j)]$$

$$= d^+(\tilde{P}) - [\text{spread of } v \text{ relative to } P] = d^+(\tilde{P}) - u(i) + u(\bar{i}).$$

With termination under Rule 2 we therefore have

$$0 > d^+(\tilde{P}) - u(i) + u(\bar{i}), \tag{5}$$

where i is either a node belonging to the θ-path from N^+ to $\bar{i}$ (at the start of the iteration), in which case let $\tilde{P}$ denote the portion of this path from i to $\bar{i}$, or i is a node in N^+, in which case let $\overline{P}$ denote the entire θ-path from N^+ to $\bar{i}$. In both cases we are interested in the path P that consists of $\overline{P}$ followed by $\tilde{P}$. Since at the start of the iteration u was admissible for θ, and $\overline{P}$ is (part of) a θ-path, we know that

$$v(j) \geq d^+(j) \quad \text{for all } j \in \overline{P}^+, \qquad v(j) \leq d^-(j) \quad \text{for all } j \in \overline{P}^-.$$

Hence

$$0 \geq \sum_{j \in \overline{P}^+} [d^+(j) - v(j)] + \sum_{j \in \overline{P}^-} [v(j) - d^-(j)]$$

$$= d^+(\overline{P}) - [\text{spread of } v \text{ relative to } \overline{P}] = d^+(\overline{P}) - u(\bar{i}) + u(i).$$

Combining this with (5), we obtain

$$0 > d^+(\overline{P}) + d^+(\tilde{P}) = d^+(P).$$

Termination under Rule 2 therefore does yield a path P that violates the condition of regularity of spans with respect to N^+.

Remarks

This procedure is the same as the feasible differential algorithm, except for a broader termination condition and the retention of the routing constructed in each iteration. It consequently has the same order of complexity $0(|N_0||N|^2)$ in the digraph case, where N_0 is the set of nodes initially having $\rho(i) < 0$.

Like the feasible differential algorithm it could be used to solve the general feasible differential problem or detect a circuit of negative upper span. In this case N^+ could be chosen to consist of a single node s, and a complete, compatible routing with base s would be needed for initialization. A possible advantage, which however could be offset by additional effort in other respects, is that a circuit of negative upper span might be detected more readily.

Termination with such a circuit would occur not only by reaching $\bar{i}$ a second time but by reaching any node on the θ-path from s to $\bar{i}$ while $\beta < 0$. Another possible advantage is the shortcut described next.

Refinements

Besides employing Dijkstra's method (Section 6E) in the min path subroutine to increase efficiency, one can make use of a fact laid bare in the justification of the algorithm. When a node i is added to S with $w(i) = \beta < 0$, any adjacent node i' having i as the predecessor on its θ-path, that is, having i as the other node of the arc $j = \theta(i')$, can be added to S on the next step with the routing unchanged and with $w(i') = \beta$ too. Then certain nodes i'' adjacent to i' can perhaps be added in the same way, and so forth.

There is no reason why this cannot be done all at once. Let us say that a node i' is a θ-*successor* to a node i if i lies *somewhere* on the θ-path from N^+ to i'. The following rule then provides a shortcut in the procedure: each time a node i is added to S with $w(i) = \beta < 0$ (and with a change in the arc designated as $\theta(i)$), add to S at the same time *every* θ-successor i' to i, defining $w(i') = \beta$ too but leaving $\theta(i')$ unchanged.

6L.   INTEGRAL AND EXTREME DIFFERENTIALS

A potential (or differential) is said to be integral if its value at every node (or arc) is an integer. Of course if u is an integral potential, then $v = \Delta u$ is an integral differential. Every integral differential can be expressed as the differential of some integral potential.

Integrality properties of solutions to the problems in this chapter can be derived from properties of the corresponding algorithms. The min path algorithm, for example, as reformulated during its justification in Section 6D, iteratively replaces a potential u by $u' = u + \alpha e_{N \setminus S}$, where

$$\alpha = \min \begin{cases} d^+(j) - v(j) & \text{for } j \in Q^+, \\ v(j) - d^-(j) & \text{for } j \in Q^-, \end{cases}$$

with $Q = [S, N \setminus S]$. If u is integral and all the spans $d^+(j)$ and $d^-(j)$ are integral, it is clear that α is an integer, and hence u' is integral too. This step is therefore integrality preserving. Dijkstra's method simply represents a higher level of detail and efficiency in implementing the min path algorithm, so it follows that in networks with integral spans both of these procedures always terminate with an integral potential when initiated with one.

The feasible differential algorithm consists of a series of applications of the min path algorithm to modified networks in which certain of the values $d^+(j) - v(j)$ or $v(j) - d^-(j)$ may be replaced by 0 when negative. Therefore this too is integrality preserving in each iteration when the spans $d^+(j)$ and $d^-(j)$ are all integral, and if initiated with any integral potential (e.g., $u \equiv 0$), it must terminate with an integral potential.

In particular, the feasible differential algorithm (or its implementation as the routing optimization procedure) can be used to find a potential that satisfies the constraints of the max tension problem if one is not apparent. This, as we have seen in Section 6F, involves solving a feasible differential problem in a certain modified network whose spans are all integral if those in the original network are integral. This method, which can start with an arbitrary integral potential, then produces an integral potential that can be used to initiate the min path algorithm and will subsequently lead to an integral solution to the max tension problem.

Putting these facts together, one has the following result.

Integrality Theorem for Potentials. *If a max tension problem or feasible differential problem has a solution at all, and if all the spans $d^+(j)$ and $d^-(j)$ in the network are integral (where $+\infty$ and $-\infty$ count as integral), then the problem has a solution that is an integral potential.*

The general solution to the feasible differential problem can be represented in terms of certain "extreme" solutions, much as was demonstrated in Section 4J for the corresponding problem for flows. This fact has consequences in the linear optimization theory for potentials.

An arc set F' is said to form a *coforest* in the network G if the deletion of all the arcs in F' would not increase the number of components of G. An equivalent property (see Exercise 4.28) is that there is no elementary cut of G included in F' (in fact no nonempty cut at all; Exercise 2.9). The maximal coforests are the complements of the maximal forests (Exercise 4.29).Thus if G is connected, the coforests are the sets $F' \subset A$ such that $A \setminus F'$ contains a spanning tree for G.

A differential v is *extreme* (with respect to given span intervals) if $d^-(j) \leq v(j) \leq d^+(j)$ for all arcs j, and the set

$$F_v' = \{ j \in A \mid d^-(j) < v(j) < d^+(j)\}$$

is a coforest. This is true if and only if there is a maximal forest F such that for all $j \in F$ either $v(j) = d^+(j)$ or $v(j) = d^-(j)$.

The main theorem about extreme differentials speaks of cuts Q of unlimited span in the sense of Section 6C, that is, with $d^+(j) = +\infty$ for all $j \in Q^+$ and $d^-(j) = -\infty$ for all $j \in Q^-$. If both Q and its reverse are of unlimited span, Q is said to be *of doubly unlimited span*.

Extremal Representation Theorem for Differentials. *Assume the feasible differential problem has at least one solution, but there are no elementary cuts of* *doubly infinite capacity. Then*:

1. *There are only finitely many extreme differentials (at least one). They are all integral, if the spans* $d^+(j)$, $d^-(j)$ *are all integral.*

2. *A differential v is a solution to the feasible differential problem if and only if it can be represented as*

$$v = \sum_{k=1}^{r} \lambda_k v_k + \sum_{l=1}^{q} \mu_l e_{Q_l},$$

where each v_k is an extreme differential, $\lambda_k \geq 0$, $\sum_{k=1}^{r}\lambda_k = 1$, each Q_l is an elementary cut of unlimited span, and $\mu_l \geq 0$.

(*The second sum may be vacuous.*)

Proof. If v is an extreme differential, there is a maximal forest F disjoint from F_v' and a corresponding Tucker representation (Section 4G):

$$v(k) = - \sum_{j \in F} v(j)a(j, k) \quad \text{for all } k \notin F,$$

where the coefficients $a(j, k)$ are all $+1$, -1, or 0. Then either $v(j) = d^+(j)$ or $v(j) = d^-(j)$ for all $j \in F$. There are only finitely many differentials expressible in this manner for a given F, and they are integral when the spans $d^+(j)$, $d^-(j)$, are integral. Furthermore there are only finitely many maximal forests F. This establishes all of Part 1 except the existence assertion, which is covered actually by Part 2.

The sufficiency in Part 2 will be verified next. If v has this form, then $d^-(j) \leq v_k(j) \leq d^+(j)$ for $k = 1,\ldots,r$, and hence

$$d^-(j) \leq \sum_{k=1}^{r} \lambda_k v_k(j) \leq d^+(j).$$

The term $\mu_l e_{Q_l}(j)$ can increase the sum only if $d^+(j) = +\infty$, and it can decrease it only if $d^-(j) = -\infty$. Hence it cannot lead to a violation of the bounds, and one does have $d^-(j) \leq v(j) \leq d^+(j)$ for all $j \in A$.

The fact that every solution to the feasible differential problem can be represented as claimed in Part 2 can be demonstrated by a simple algorithm parallel to the one for flows in Section 4K. This is left to the reader as Exercise 6.54.

Remark

Extreme differentials can be identified with the extreme points (vertexes) of the convex polyhedron consisting of all differentials that are feasible with respect to the given spans (see Exercise 6.56).

6M.*   **EXERCISES**

6.1. (*Integration*). Demonstrate that the integration rule in Section 6A can be regarded as a special case of the conversion formula in Section 1I.

6.2. (*Integration*). Give a potential u whose differential is the tension function displayed in Figure 6.21.

6.3. (*Path Inequality*). Associate with any path P without multiplicities the set of real numbers

$$D(P) = \sum_{j \in P^+} D(j) - \sum_{j \in P^-} D(j)$$

$$- \{e_p \cdot v \,|\, v(j) \in D(j) \text{ for all } j \subset A\}.$$

Show that $D(P)$ is an *interval* whose endpoints are $d^+(P)$ and $d^-(P)$ and that the fundamental condition at the end of Section 6A, which is necessary if a potential u is to be feasible with respect to spans, can be expressed as

$$u(i') - u(i) \in D(P) \quad \text{for any path } P: i \to i'.$$

More important, show that this form of the necessary condition remains valid even if the span intervals $D(j)$ are not necessarily closed, although the interval $D(P)$ may not contain its endpoints in that case.

6.4. (*Min Path*). Show how the min path problem can always be reduced (by "restructuring" the network) to the case where all span intervals are of the form $(-\infty, d^+(j)]$ with $d^+(j)$ finite.

6.5. (*Node Tolls*). For a network whose intervals are all of the form $(-\infty, d^+(j)]$ (see the preceding exercise), show how the min path problem, with $d^+(j)$ interpreted as the cost of traversing j, can be generalized to the case where, in addition to such costs, a cost $t(i)$ is incurred in passing through any node i.

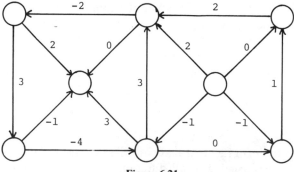

Figure 6.21

(*Hint*. Replace each i by a pair of nodes i^+ and i^- joined by an arc j^i, as in Section 3L.)

6.6. (*Knapsack Problem*). Generalize the approach in Example 3 of Section 6B to the case of a knapsack problem having a "weight" *and* a "volume" constraint.

6.7. (*Max Tension*). As noted in Section 6B, any min path problem for sets N^+ and N^- can be reformulated as a problem for paths $P: s \to s'$ by adding arcs and nodes to the network G as in Figure 6.3. This is in turn dual to the problem of maximizing $u(s') - u(s)$. Show how the latter can be expressed in terms of G itself, without reference to the added arcs and nodes. (This will be a generalized type of max tension problem that does not require u to be constant on N^+ and constant on N^-.)

6.8. (*Generalized Max Tension Min Path Theorem*). Suppose in the max tension problem that there is at least one potential satisfying the constraints. *Then the set J of spread values from N^+ to N^- corresponding to such potentials is an interval, and*

$$ J = \bigcap_{P:\, N^+ \to N^-} D(P) \qquad (P \text{ elementary}). $$

(see Exercise 6.3 for the definition of $D(P)$). *Moreover this is true even if the span intervals $D(j)$ for the various arcs j are not necessarily closed.* Derive this result from the max tension min path theorem as stated in Section 6C (see Exercise 3.4).

6.9. (*Min Path Characterization*). Let u be an arbitrary solution to the max tension problem, $v = \Delta u$. Show that the solutions to the corresponding min path problem are precisely the paths $P: N^+ \to N^-$ such that $v(j) = d^+(j)$ for all $j \in P^+$ and $v(j) = d^-(j)$ for all $j \in P^-$. Show also that this condition is equivalent to the compatibility of P with respect to a certain painting.

(*Hint*. Apply the max tension min path theorem. Look at the way the basic inequality in Section 6A was derived.)

6.10. (*Separation Problem*). A *separation* of N^+ from N^-, where N^+ and N^- are disjoint nonempty node sets, is a collection M of elementary cuts Q: $N^+ \downarrow N^-$ such that no two of the cuts in M have any arcs in common. The separation problem consists of determining a separation M of maximum cardinality (number of cuts). Show how this can be formulated as a special max tension problem, making use of the integrality theorem in Section 6L. (This is analogous to Exercise 5.2.)

6.11. (*Separation Problem*). Prove that the maximum of M in the preceding exercise equals the minimum of $|P|$ over all paths $P: N^+ \to N^-$.

(*Hint*. Apply the max tension min path theorem.)

6.12. (*Max Tension*). Show how the max tension problem can always be reduced (by "restructuring" the networks) to the case where all span intervals are of the form $(-\infty, d^+(j)]$ with $d^+(j)$ finite (see Exercise 6.4).

6.13. (*Extended Max Tension Problem*). In a network G with span intervals, suppose that two nonempty disjoint node sets N^+ and N^- are given, but also two functions $a^+\colon N^+ \to R$ and $a^-\colon N^- \to R$. The following is the *extended* max tension problem: find the highest value of $\alpha^+ - \alpha^-$, where α^+ and α^- are real numbers such that there exists a potential u on G that is feasible with respect to spans and satisfies

$$u(i) = a^+(i) + \alpha^+ \quad \text{for all } i \in N^+,$$

$$u(i) = a^-(i) + \alpha^- \quad \text{for all } i \in N^-.$$

Show that this can be reduced to the ordinary max tension problem in Section 6C in two different ways:

(a) First, by adding to G a pair of nodes s and s' attached to N^+ and N^-, respectively, by arcs as in Figure 6.8, but with a different choice of span intervals.

(b) Second, by taking $\bar{u}$ to be any potential that agrees with a^+ on N^+ and a^- on N^-, and then reformulating everything in terms of $u' = u - \bar{u}$ and the shifted span intervals $\tilde{D}(j) = [d^-(j) - \bar{v}(j), d^+(j) - \bar{v}(j)]$, where $\bar{v} = \Delta\bar{u}$.

6.14. (*Extended Min Path Problem*). Prove that the maximum in the extended max tension problem of the preceding exercise equals (under certain mild assumptions) the minimum in the following *extended* min path problem: minimize $a^+(i) + d^+(P) - a^-(i')$ over all paths $P\colon N^+ \to N^-$, where i denotes the initial node of P and i' the terminal node.

(*Remark*. With $d^+(P)$ interpreted as the cost of transporting one unit of a certain commodity along P, one can think of $a^+(i)$ as the buying price for one unit at i and $a^-(i')$ as the selling price for one unit at i'. Then $a^+(i) + d^+(P) - a^-(i')$ is the net cost for the entire transaction.)

6.15. (*Min Path Algorithm*). Show that the extended max tension and min path problems in Exercises 6.13 and 6.14 can be solved directly (without actually going through the reformulations offered in Exercise 6.13) just by initiating the min path algorithm with a potential u_0 that satisfies the modified constraints, namely, feasibility with respect to spans, $u_0 = a^+$ on N^+ and $u_0 = a^- +$ const. on N^-.

6.16. (*Min Path Algorithm*). Solve the max tension and min path problems in Figure 6.22. Take $u_0 \equiv 0$ in the min path algorithm.

6.17. (*Max Tension Algorithm*). The following procedure is closely parallel to the max flow algorithm. Verify the (italicized) claims made in the statement.

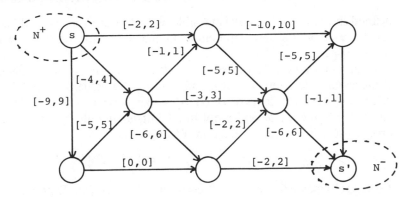

Figure 6.22

Given any potential u satisfying the constraints of the max tension problem, paint the arcs j of the network as follows in terms of $v = \Delta u$:

$$\begin{aligned}
\text{red} &\quad \text{if } d^-(j) < v(j) < d^+(j)\\
\text{black} &\quad \text{if } d^-(j) = v(j) < d^+(j)\\
\text{white} &\quad \text{if } d^-(j) < v(j) = d^+(j)\\
\text{green} &\quad \text{if } d^-(j) = v(j) = d^+(j)
\end{aligned}$$

Apply the painted network algorithm. If a compatible path $P: N^+ \to N^-$ is obtained, terminate; *u solves the max tension problem, and P solves the min path problem.* If a compatible cut $Q: N^+ \downarrow N^-$ is obtained, calculate

$$\alpha = \min\begin{cases} d^+(j) - v(j) & \text{for } j \in Q^+,\\ v(j) - d^-(j) & \text{for } j \in Q^-. \end{cases}$$

Then $\alpha > 0$. If $\alpha = \infty$, terminate; Q is of unlimited span. Otherwise set $u' = u + \alpha e_{N \setminus S}$ (where $[S, N \setminus S] = Q$). Then *u' still satisfies the constraints of the max tension problem and*

$$\left[\text{spread of } u' \text{ from } N^+ \text{ to } N^-\right] = \left[\text{spread of } u \text{ from } N^+ \text{ to } N^-\right] + \alpha.$$

Repeat the procedure with u'.

6.18. (*Max Tension Algorithm*). Show that in the algorithm of the preceding exercise the painted network subroutine need not be restarted from nothing in each iteration: after u has been replaced by u' and the painting altered accordingly, the previous routing θ is still compatible but can now be extended. Carrying the analysis further, show that the min path algorithm can be viewed simply as a way of implementing the max tension algorithm.

6.19. (*Initial Potentials*). Let G be an *acyclic* network with all span intervals of the form $(-\infty, d^+(j)]$, $d^+(j)$ finite. Show that the potential u_0

defined in the following way is feasible with respect to spans (and hence can be used as the initial potential in the min path algorithm when $N^+ = \{s\}$, $N^- = \{s'\}$).

First determine N^0, the set of nodes of G having no immediate predecessor (see Exercise 2.30), and give u_0 the value 0 on N^0. Then recursively, having defined u_0 on $S^k = N^0 \cup N^1 \cup \cdots \cup N^{k-1}$, determine N^k, the set of nodes not in S^k but having all their immediate predecessors in S^k, and give u_0 on N^k any value

$$\gamma_k \le \min\{d^+(j) + \gamma_{k-1} | j \in [S^k, N^k]^+\}.$$

Terminate when $S^k = N$, as will eventually happen (see Exercise 2.31). (*Remark.* The sets N^k are easy to determine from the *adjacency* matrix for G as follows. Add all the rows together to determine N^0, which consists of the nodes giving 0 entries in the sum; delete from the adjacency matrix the columns and rows corresponding to the nodes in N^0. Next do the same to the reduced matrix; this will yield N^1. Continuing in this way with smaller and smaller matrices get $N^2, \ldots, N^r$. This procedure has order of complexity $0(|N|^2)$.)

6.20. (*Min Path Algorithm*). Solve the max tension and min path problems in Figure 6.23 with $N^+ = \{s\}$, $N^- = \{s'\}$, by initiating the min path algorithm with a potential u_0 constructed by the method in Exercise 6.19.

6.21. (*Min Path Algorithm*). Show that if G is an acyclic network with all span intervals of the form $(-\infty, d^+(j)]$, $d^+(j)$ finite, a min path problem for $N^+ = \{s\}$, $N^- = \{s'\}$, can be solved as follows. Apply the min path algorithm as if $u_0 \equiv 0$ (even though this potential would not satisfy the constraints), but in the calculation of β and the determination of which nodes to add to S, consider only combinations i', j, i such that i has no immediate predecessor not in S and yet can be reached from S by a positive path (i.e., is a successor to something in S).

(*Hint.* Imagine that all nodes that are not successors to s have been deleted from G, along with the arcs incident to them. Investigate what would happen if the min path algorithm were applied to an initial

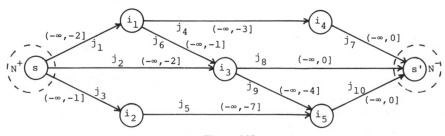

Figure 6.23

potential u constructed as in Exercise 6.19 but with the value of u on N_k chosen to be "vastly" lower than the minimum in the formula, the magnitude of "vast" growing vastly with k.)

(*Note.* This procedure generally will *not* simultaneously furnish a solution to the max tension problem, since one never even sees a potential defined on all of G.)

6.22. (*Min Path Problem*). Apply the method in the preceding exercise to the network in Figure 6.23. (The result may be compared with what happened in Exercise 6.20.)

6.23. (*Initial Potentials*). Let G be a network, *not* necessarily acyclic, whose span intervals are all of the form $(-\infty, d^+(j)]$ with $d^+(j)$ finite (as really involves no loss of generality; see Exercises 6.4 and 6.12). Let $G_1, \ldots, G_m$ be the strong components of G, and suppose for $k = 1, \ldots, m$ there is a known potential u^k on G_k that is feasible with respect to spans in G_k. Show how the procedure in Exercise 6.19 can be generalized to construct from these a potential u_0 on G that is feasible with respect to spans in G.

(*Hint.* This is related to Exercise 2.7.)

(*Remark.* If $d^+(j) \geq 0$ for all arcs in G_k, one can take $u^k \equiv 0$. The criterion in this exercise for the "easy" construction of a feasible potential therefore unifies the one for acyclic networks in Exercise 6.19 with the elementary one that $0 \in D(j)$ for all arcs j. It describes a general case where a feasible potential can be found by a method of complexity $0(|N|^2)$ rather than the higher order of complexity in the algorithms in Sections 6H, 6I, and 6K.)

6.24. (*Min Path Algorithm*). Suppose that $d^-(j) \leq 0 \leq d^+(j)$ for all arcs j *except* perhaps some in the cut $[N^+, N \setminus N^+]$. Show that if the min path algorithm is nevertheless applied with $u_0 \equiv 0$ (even though this potential might not be feasible with respect to spans), the path obtained will solve the restricted problem of minimizing $d^+(P)$ over all paths P: $N^+ \to N^-$ which have only initial node in N^+.

(*Hint.* Replace the given span interval by $(-\infty, d^+(j)]$ for j in $[N^+, N \setminus N^+]^+$ and by $[d^-(j), \infty)$ for j in $[N^+, N \setminus N^+]^-$. Consider what happens when the min path algorithm is applied to this modified network with u_0 taken to have the value 0 outside N^+ but a positive value on N^+, that is, large enough that u_0 becomes feasible.)

6.25. (*Arc Discrimination*). Demonstrate that the painted network algorithm with arc discrimination (Section 2G) can be regarded as a special case of the min path algorithm if span intervals are introduced in a certain way.

6.26. (*Breadth-First Search*). Show that the painted network algorithm with breadth-first search (Section 2G) can be identified with the min path algorithm (Dijkstra's method) in the case where span intervals are introduced as follows:

$$[-1,1] \quad \text{if } j \text{ is green}$$
$$(-\infty,1] \quad \text{if } j \text{ is white}$$
$$[-1,\infty) \quad \text{if } j \text{ is black}$$
$$(-\infty,\infty) \text{ if } j \text{ is red}$$

6.27. (*Dijkstra's Method*). Apply Dijkstra's method to solve the max tension and min path problems in Figure 6.22. (Take $u_0 \equiv 0$.)

6.28. (*Generalized Feasible Differential Theorem*). Show that the condition in the feasible differential theorem can be expressed equivalently as $0 \in D(P)$ for all elementary circuits P, where $D(P)$ is the interval defined in Exercise 6.3, and that in this form the theorem remains true even if the span intervals $D(j)$ are not all closed.

6.29. (*Feasible Potential Theorem*). Formulate and derive from the feasible differential theorem a necessary and sufficient condition for the existence of a solution to the "feasible potential problem" introduced in Section 6F (where there is also an interval $[d^-(i), d^+(i)]$ associated with each node i, and the constraint $d^-(i) \le u(i) \le d^+(i)$ must be satisfied).

6.30. (*Acyclic Networks*). Prove that a network is acyclic if and only if there exists a potential that is feasible with respect to the span intervals $D(j) = (-\infty, -1]$ for all $j \in A$.

(*Note.* This criterion can be used with the feasible differential algorithm to obtain a test for acyclicity that will produce a particular positive circuit if the network fails to be acyclic. This would involve more effort, of course, than the test in Exercise 2.31; see also the remark following Exercise 6.19.)

6.31. (*Acyclic Networks*). Prove that a network is acyclic if and only if its arc set can be expressed as the union of disjoint, positive cuts $Q_1, \ldots, Q_m$.

(*Hint.* The necessity can be deduced from the preceding exercise; consider the level sets of a potential u of the type in question.)

6.32. (*Infinite Spans*). The occurence of infinite values for $d^+(j)$ and $d^-(j)$ is generally an indication of special network structure that should be respected and even put to advantage in computations (e.g., the acyclic case), but it is also possible to treat infinite values as if they were very large finite values. Specifically, suppose one works instead with the *truncated spans*

$$d_M^+(j) = \begin{cases} d^+(j) & \text{if } d^+(j) < \infty, \\ 2M & \text{if } d^+(j) = \infty, \end{cases}$$

$$d_M^-(j) = \begin{cases} d^-(j) & \text{if } d^-(j) > -\infty, \\ -2M & \text{if } d^-(j) = -\infty, \end{cases}$$

where M is any number sufficiently large in the sense that

$$M > M_0 \triangleq \sum_{j \in A_+} |d^+(j)| + \sum_{j \in A_-} |d^-(j)|,$$

$$A_+ = \{ j \in A \mid d^+(j) \text{ finite} \},$$

$$A_- = \{ j \in A \mid d^-(j) \text{ finite} \}.$$

Show that for every path P without multiplicities, one has either $d^+(P) = \infty$ and $d_M^+(P) > M$, or $d^+(P) = d_M^+(P) < M$. Applying the feasible differential theorem, prove that the feasible differential problem with the truncated spans has a solution if and only if the original problem does. (Any solution to the truncated problem is also a solution to the original one.) Explain further how max tension problems could be solved in terms of the truncated spans.

6.33. (*Feasible Differential Algorithm*). Use the feasible differential algorithm to construct a potential that is feasible with respect to spans in the network of Figure 6.24. Start with the zero potential.

6.34. (*Feasible Differential Algorithm*). Use the feasible differential algorithm to detect in the network in Figure 6.25 a circuit P with $d^+(P) < 0$. Start with the zero potential.

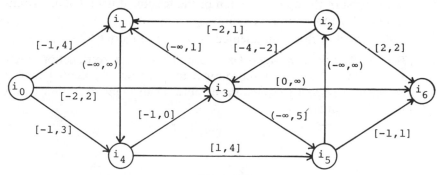

Figure 6.24

6.35. (*Feasible Differentials vs. Optimal Assignments*). Demonstrate that the feasible differential problem is equivalent to the *dual* of the optimal assignment problem given by the following data. Let K and L be two copies of the node set for G; for a node i, the copy in K is denoted by i^+, and the one in L is denoted by i^-. Declare i_1^+ compatible with i_2^- if there is an arc $j \sim (i_1, i_2)$ with $d^+(j) < \infty$ or an arc $j \sim (i_2, i_1)$ with $-d^-(j) < \infty$, in which case let the "cost" $q(i_1^+, i_2^-)$ be the lowest of all such values $d^+(j)$ or $-d^-(j)$; also declare every i^+ compatible with i^-, defining $q(i^+, i^-) = 0$.

Show that if [min] = [max] = 0 in the optimal assignment problem and its dual, the solution to the dual furnishes a solution to the feasible distribution problem in G. Otherwise [min] = [max] < 0; prove that in this case the solution to the optimal assignment problem provides in a certain way a circuit of negative upper span in G.

6.36. (*Double Feasible Differential Algorithm*). How can a "small" set $\overline{N} \subset N$ that is incident to every arc $j \in A$ be constructed?

(*Hint.* Consider the blocking problem in Section 5A where K and L are two copies of N, and a pair (k, l) is compatible if k and l represent the same node or adjacent nodes.)

6.37. (*Double Feasible Differential Algorithm*). Apply to the network in Figure 6.11 the double version of the feasible differential algorithm at the end of Section 6J, starting with $u \equiv 0$ and $\bar{i} = i_2$. (One iteration will suffice.)

6.38. (*Double Feasible Differential Algorithm*). Apply to the network in Figure 6.24 the double version of the feasible differential algorithm in Section 6H, starting with $u \equiv 0$.

6.39. (*Feasible Potential Problem*). In the case of general feasible potential problem mentioned in Section 6G, where there are span intervals associated with nodes as well as with arcs, suppose an initial potential u is known that satisfies the conditions in the arcs but not all the conditions at the nodes. Explain how a solution could be obtained in *one* iteration of the double version of the feasible differential algorithm (Section 6H) applied in a certain way.

(*Hint.* Work with the augmented network in Figure 6.9.)

6.40. (*Tension Rectification Algorithm*). Use the tension rectification algorithm (with the min path subroutine) to construct a potential that is feasible with respect to spans in the network in Figure 6.24. Start with the zero potential.

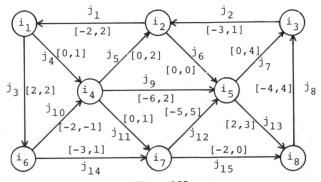

Figure 6.25

6.41. (*Tension Rectification Algorithm*). Use the tension rectification algorithm (with the min path subroutine) to detect in the network in Figure 6.25 a circuit P with $d^+(P) < 0$.

6.42. (*Tension Rectification Algorithm*). Verify the claims made in the statement of the tension rectification algorithm.

6.43. (*Tension Rectification Algorithm*). Work out and state in exact form the version of the tension rectification algorithm sketched at the end of Section 6I.

6.44. (*Triangle Inequality*). Prove that the span function for the network G, as defined in Section 6J, satisfies the triangle inequality.

6.45. (*Triangle Inequality*). Let N be any nonempty, finite set, and let d: $N \times N \to (-\infty, \infty]$ be a function that satisfies the triangle inequality 6J (1) and has $d(i, i) = 0$ for all i. Demonstrate that d is the span function for some network G with node set N.

6.46. (*Routing Optimization Procedure*). Construct a complete optimal routing with base s for the network in Figure 6.26. Initialize with the routing shown by the darkened arcs. From the answer, read off solutions to the max tension and min path problems for $N^+ = \{s\}$ and $N^- = \{s'\}$ (refer to the optimal routing theorem in Section 6J).

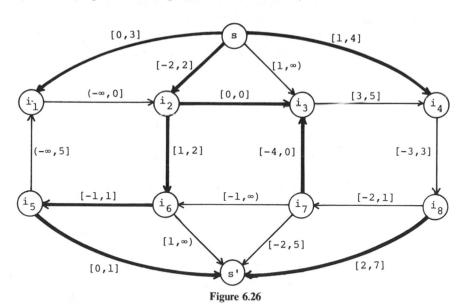

Figure 6.26

6.47. (*Elementary Optimal Routing Algorithm*). Prove that the following procedure has the properties claimed for it and is sure to terminate in a finite number of iterations.

Start with any complete compatible routing θ with base s and its associated potential u. Test the arcs j not used by θ to see if $d^-(j) \le$

$v(j) \le d^+(j)$. If this is true for all such arcs, θ *is optimal*. If not, select any arc j for which it is violated; let $(i', i) \sim j$ if $v(j) > d^+(j)$ and $(i, i') \sim j$ if $v(j) < d^-(j)$. If i lies on the θ-path from s to i', terminate; the portion of this path from i to i', followed by j, i, is *an elementary circuit P with* $d^+(P) < 0$. Otherwise redefine $\theta(i)$ to be j; then θ is still a *complete compatible routing* with base s, and its associated potential u differs from the previous one only in that *the values at i and all nodes whose θ-path passed through i have been lowered by* γ, where γ is the distance of $v(j)$ from $[d^-(j), d^+(j)]$. In this case repeat the step with the new θ.

(*Remark*. This algorithm is similar in many ways to the routing optimization procedure in Section 6K. One important difference, however, is that infeasibility in some arcs can be worsened while being improved in others. In fact feasibility can be lost and gained repeatedly in some arcs. Although there can only be a finite number of iterations, and this number may often turn out to be small, it is hard to *guarantee* more than that it is bounded by the total number of complete, compatible routings with base s. By contrast, the bound obtained for the procedure in Section 6K is much better.)

6.48. (*Elementary Optimal Routing Algorithm*). Apply the method in the preceding exercise to get a complete optimal routing with base s for the network in Figure 6.26. Start with the routing shown.

6.49. (*All-Node Min Path Algorithm*). Define the functions b^k on $N \times N$ recursively as follows. For $i \ne i'$, let

$$b^1(i, i') = \min \begin{cases} d^+(j) & \text{for arcs } j \sim (i, i') \\ -d^-(j) & \text{for arcs } j \sim (i', i) \end{cases}$$

(interpreted as $+\infty$ if there are no such arcs), whereas $b^1(i, i) = 0$. Let

$$b^{k+1}(i, i') = \min_{i'' \in N} \{ b^k(i, i'') + b^1(i'', i') \}.$$

Show that if there are no circuits P with $d^+(P) < 0$, then $b^n(i, i') = d(i, i')$ for all i, i', where $n = |N| - 1$.

(*Note*. This furnishes a finite algorithm for calculating all the values $d(i, i')$. It is interesting that if d and b^k are regarded as matrices of dimension $(n + 1) \times (n + 1)$, then b^{k+1} is a sort of generalized product $b^k * b^1$ in which the arithmetic operations $\cdot$ and $+$ of ordinary matrix multiplication are replaced by $+$ and "min," respectively. The calculation of such generalized products is "easy" for a computer. Of course, if only the values $d(s, i')$ are desired for some fixed s, this computational approach, which has order of complexity $0(|N|^3)$, is less efficient than the routing optimized procedure in Section 6L.)

6.50. (*All-Node Min Path Algorithm*). Augment the algorithm in the preceding exercise by a procedure that, on termination, furnishes a description of a system of optimal paths between all pairs of nodes.

6.51. (*All-Node Min Path Algorithm*). How could the presence of a circuit P with $d^+(P) < 0$ be detected by the algorithm of Exercise 6.49, if there were one?

(*Note.* The supplementary procedure in Exercise 6.50 could be employed in this case to produce a particular such circuit.)

6.52. (*Extreme Differentials*). How could one test (algorithmically) whether a given solution v to the feasible differential problem is an extreme differential or not?

6.53. (*Extreme Differentials*). Prove that the solutions to the feasible differential problem provided by the optimal routing theorem (Section 6J), in the case where N^+ consists of a single node, yield *extreme* differentials. (Thus an extreme differential can be determined, for example, by the routing optimization procedure in Section 6K.)

6.54. (*Extremal Representation Algorithm*). Develop and justify an algorithm, parallel to the one in Section 4K for flows, that constructs for any solution v to the feasible differential problem a representation as in Part 2 of the theorem in Section 6L.

6.55. (*Bottleneck Path Algorithm*). Justify the following algorithm for solving the bottleneck path problem as described in Exercise 5.46. At the start of each iteration there is a node set S with $N^+ \subset S$, $N^- \subset N \setminus S$, a routing θ of S with base N^+, and a function $w: S \to [-\infty, \infty]$. (Initially, $S = N^+$, θ is vacuous, and $w(i) = \infty$ for all $i \in S$.) Considering the cut $Q = [S, N \setminus S]$, let

$$
\alpha = \min \begin{cases} \min\{w(i'), r^+(j)\} & \text{for arcs } j \sim (i', i) \text{ in } Q^+, \\ \min\{w(i'), r^-(j)\} & \text{for arcs } j \sim (i, i') \text{ in } Q^-. \end{cases}
$$

Selecting any $\bar{i}', \bar{j}, \bar{i}$ for which the minimum is attained, add $\bar{i}$ to S, defining $\theta(\bar{i}) = \bar{j}$ and $w(\bar{i}) = \alpha$. Then repeat the procedure. Terminate as soon as a node $s' \in N^-$ is reached. The θ-path $P: N^+ \to s'$ then solves the problem, and $w(s')$ is the maximum.

6.56. (*Extreme Differentials as Extreme Points*). Show that the extreme differentials relative to a given system of span intervals $[d^-(j), d^+(j)]$ are the elements of the set

$$
V = \{v \in \mathcal{D} \mid d^-(j) \le v(j) \le d^+(j) \text{ for all } j \in A\}
$$

which *cannot* be expressed in the form

$$
v = (1 - \lambda)v' + \lambda v'' \quad \text{with } v' \in V, v'' \in V, v' \ne v'', 0 < \lambda < 1.
$$

(*Note.* V is a convex polyhedron, and this condition means that the extreme differentials are the extreme points of V.)

6N.*   COMMENTS AND REFERENCES

The min path problem has a very natural appeal, and the history of its applications can be traced from such questions as highway distances between cities and the number of steps required in passing from one state of a system to another (paths in dynamic networks), through the economic cost models developed in the early days of operations research, to abstract representations of situations in combinatorial optimization. An excellent example of the latter is the handling of the knapsack problem in Section 6B (see Fulkerson [1966]). One of the most celebrated applications, that of PERT and the critical path method in project scheduling, originated in 1958 with the United States Navy and has found widespread use in government and industry. The books of Muth and Thompson [1963], Miller [1963] and Kaufman and Desbazeille [1969] describe this kind of application in detail.

The systematic treatment of the min path problem here in terms of N^+ and N^- does not truly give any more generality than the traditional case of single nodes s and s'. However, it does aid in the development of computational methods, especially those that involve solving a sequence of min path problems for the same network, but with varying span intervals and choices of N^+ and N^-. (Such a method will be studied in Section 7J, for instance.) It would be awkward and conceptually debilitating in situations like this to have to replace the given network by an auxiliary one that changes from iteration to iteration.

The max tension problem and its duality with the min path problem are implicit in the work of Minty [1960], and this even goes for the generalization to nonclosed intervals (Exercise 6.8). Minty maintained a framework in which circuits and differentials were interchangeable with cuts and circulations, and he commented that one of his results contained the generalized max flow min cut theorem as a special case. He could also have commented that it contained the max tension min cut theorem, but Berge and Ghouila-Houri [1962] were actually the first to give an explicit statement of this theorem (with closed intervals and $N^+ = \{s\}$, $N^- = \{s'\}$).

In generalizing the max tension min cut theorem to node sets N^+ and N^-, there are two routes that could have been followed instead of the one in the text. The first corresponds to Exercise 6.7, and the second to Exercises 6.13 and 6.14. The first, where there are no constraints on u with respect to N^+ and N^- and the expression

$$\max\{u(i') - u(i) \,|\, i \in N^+, i \in N^-\}$$

is minimized, is in some ways very natural, but it cannot be solved in terms of a routing with base N^+ that is terminated as soon as N^- is reached. This case is therefore handled more appropriately by adjoining nodes s and s' to the network, as in Figure 6.3.

The second alternative replaces the constraints $u \equiv$ const. on N^+ and $u \equiv$ const. on N^- by $u \equiv a^+ +$ const. on N^+ and $u \equiv a^- +$ const. on N^-, where

a^+ and a^- are given functions. This turns out to be the right general formulation for the kind of problem that can be solved by routings in the manner just described, and an important application will be made in Section 7K. But it also involves specifying data in addition to N^+, N^-, and the span intervals for the network, which seems unpleasantly complicated, especially in view of the fact that in the many applications where N^+ and N^- do consist of single nodes, the specification of a^+ and a^- is superfluous. We prefer therefore to handle this second case in the manner suggested in (b) of Exercise 6.13.

The max tension min path theorem has not previously been made the foundation for computational developments. Functions that can be interpreted as potentials (defined on the whole network) have appeared in some of the techniques of dynamic programming that could be used to solve the min path problem, but not in formulations of the min path algorithm itself. This algorithm, whose polished form in Section 6E is due essentially to Dijkstra [1959], is customarily presented as valid only for networks with "non-negative costs," that is, having $d^-(j) \leq 0 \leq d^+(j)$ for all $j \in A$. The related method for acyclic networks (see Exercise 6.21) is regarded as a parallel but separate case. Whereas these two methods have order of complexity at most $0(|N|^2)$ in the digraph case, the ones usually offered for other kinds of networks are of order $0(|N|^3)$ and involve constructing a complete optimal routing with base s.

The advantage of approaching the subject by way of the max tension min path theorem lies in the realization of what truly matters, as far as ease of solvability of the min path problem is concerned. The real distinction should be made according to the availability of an initial potential u_0 that satisfies the constraints of the max tension problem. If such a potential could always be constructed by an $0(|N|^2)$ computation (an open question at present), then the min path problem could always be solved with complexity $0(|N|^2)$ (for digraphs). The feasible differential algorithm in Sections 6G and 6H, which is new and can be regarded as one of the fruits of the emphasis on max tension min path duality, shows at least that the general case can be covered without an $0(|N|^3)$ computation of a complete, optimal routing. One can get by with $0(|N_0||N|^2)$, where N_0 is the set of nodes "unbalanced" with respect to an initial choice of a potential that does not satisfy all the span constraints. In some situations (e.g., in Exercises 6.39 and 6.24) it can readily be arranged that $|N_0| = 1$.

Another advantage in the max tension min path approach, of course, is that it produces solutions to the max tension problem as well as to the min path problem. The fact that the min path algorithm is simply an expression of the "max tension algorithm," which in turn is naturally dual to the max flow algorithm (see Exercises 6.16 and 6.17), is especially interesting as a reflection of the profound unity of ideas in network programming. The max tension algorithm is introduced for the first time in this book.

The importance of the nonexistence of circuits of negative upper span is well recognized in studies of the min path problem and the construction of optimal routings, but its close connection with the existence of potentials satisfying span constraints is rarely mentioned. The feasible differential theorem, which ties the two existence questions together, was proved by Minty [1960] (again in the general form where the intervals need not be closed; see Exercise 6.28). Minty's constructive proof more or less boils down to the basic version of the tension rectification algorithm, although he did not present his algorithm as such nor offer refinements involving the min path algorithm as a subroutine, like the ones suggested in Section 6I.

A feature of both the feasible differential and tension rectification algorithms that deserves emphasis is the way they are oriented toward efficient detection of an explicit circuit P with $d^+(P) < 0$ when one exists. The same is true of the routing optimization procedure in Section 6L, which is an elaborated version of the feasible differential algorithm. This is important because in many applications it is precisely such circuits that are of interest. For example, the optimal distribution algorithm in Chapters 7 and 9 generalizes a well-known class of methods in which a feasible differential problem in effect must be tackled as a subproblem in each iteration. Only in the terminal iteration is a potential satisfying the span constraints actually obtained; all the other iterations are really aimed at constructing a circuit of negative upper span, which can be used in a certain improvement process.

In contrast, most of the methods in the existing literature which could be used to determine a circuit P with $d^+(P) < 0$ have not been designed with that as their primary focus. Chiefly they are algorithms for the computation of optimal routings which are based on the expectation that $d^+(P) \geq 0$ for all circuits P but contain checks on whether this condition is really satisfied. A partial exception, where circuits with negative upper span do receive direct attention, is the elementary optimal routing algorithm of Ford and Fulkerson [1962, p. 130] (described here in Exercise 6.47). This, however, suffers from an unsatisfactorily high bound on the number of steps that might be required. The new routing optimization procedure in Section 6L can be viewed as a closely related method that preserves the advantage of being able to start with a routing that is "nearly" optimal and use this to avoid a lot of unnecessary effort but whose order of complexity is only $0(|N_0||N|^2)$ (with $|N_0|$, as before, a measure of how far the initial routing is from being optimal).

Of course the new procedure in Section 6L, like all others for the optimal routing problem in cases where the possibility of circuits of negative upper span must be dealt with, requires a complete compatible routing for initiation. The feasible differential algorithm in Section 6G, on the other hand, does not require any such initial routing. It proceeds immediately with the job of finding circuit P with $d^+(P) < 0$, nevertheless employing min path techniques that can be implemented very efficiently through a modified form of Dijkstra's method (Section 6H). The same can be said of the tension rectification algorithm, once the refinements in Section 6I have been added.

For some networks certainly an initial complete routing that is compatible with respect to the given span intervals and based at a given node s (i.e., has base $N^+ = \{s\}$) is easy to come up with. For instance, if all the intervals

$[d^-(j), d^+(j)]$ are bounded (both $d^-(j)$ and $d^+(j)$ finite), any spanning tree in the network yields such a routing. Another case to be mentioned is the one where every span interval is of the form

$$[d^-(j), d^+(j)] = (-\infty, d(j)] \quad \text{with} \quad d(j) \quad \text{finite}, \tag{1}$$

and G is a *complete digraph*: for each pair $(i, i') \in N \times N$ with $i \neq i'$ there is a unique arc $j \sim (i, i')$. Then we get a complete compatible routing θ with base s by labeling each $i \neq s$ with the arc $j \sim (s, i)$. It must be borne in mind, however, that these cases are too restrictive for the kinds of applications where circuits P with $d^+(P) < 0$ must be detected as a subroutine in a general algorithm for optimizing flows or potentials, such as in Chapters 7 and 9. In these applications, even when quite elementary problems are involved, the span intervals need not all be bounded or of type (1), and they are likely to vary in form from iteration to iteration. Although it is true that the case of general span intervals can always be reduced to that of intervals all of the type (1) by restructuring the network (see Exercise 6.4 and Section 7I), having to do this repeatedly could be a nuisance. Clearly too, one must not lean too heavily on complete digraphs as models, since the networks occurring in economic applications are typically rather sparse (far from having an arc $j \sim (i, i')$ for every possible node pair), and this is just the sort of property that labeling methods (using "routings") may successfully exploit.

For an excellent analysis of the traditional methods for the optimal routing problem, consult Drefus [1969]. These methods include several clever refinements of the classical idea of "generalized matrix multiplication" in Exercise 6.49. Lawler [1976] likewise provides a fine overview and a discussion of related problems, like that of finding not only the best path but the second best, third best, and so forth. Another, more recent source for the latter problem is Shier [1979]. Numerical comparisons of optimal routing algorithms are given by Kelton and Law [1978], Golden [1976], and Pape [1974]. References to work on paring down the order of computational complexity are furnished by Klee [1978]. A particularly valuable article for its insights into the details of how such methods may be implemented on a computer is the one of Dial, Glover, Karney, and Klingman [1979]. Even this, however, is couched in terms of special assumptions on the network (in effect span intervals all of form (1)) and therefore is not immediately applicable to the variable situations which we have attempted to address here.

Shier and Witzgall [1980] discuss the sensitivity of solutions of the min path problem relative to changes in the spans, and Shogan [1979] studies a PERT model in which the spans can be random variables. Interesting applications of min path theory can be made to problems of location on networks; see the book of Handler and Murchandani [1979]. A dynamic min path model is treated by Halpern and Priess [1974]. Handler and Zang [1980] tackle a problem that can be formulated generally as follows: in a network with two systems of span intervals $[d_0^-(j), d_0^+(j)]$ and $[d_1^-(j), d_1^+(j)]$, minimize $d_0^+(P)$

over all paths $P: N^+ \to N^-$ satisfying $d_1^-(P) \le \alpha$ (for given α). The bottleneck path algorithm in Exercise 6.55, which closely resembles the min path algorithm, is essentially due to Fulkerson [1966]. It generalizes earlier methods of Hu [1961] and Pollack [1961] for finding a path of maximum capacity in the max flow problem.

More on the relationship between min path problems and network problems can be found in Frieze [1976].

7
NETWORKS WITH LINEAR COSTS

A feasible distribution problem typically has many different solutions, if it has any at all, so there is interest in looking for one that is best in some sense. In the optimal distribution problem the goal is to minimize a sum of costs given by convex functions associated with the arcs of the network. Analogously, in the optimal differential problem the goal is to optimize a certain cost expression over all solutions to a feasible differential problem.

This chapter treats such problems in the case of linear or piecewise linear costs. It is really just a prelude to the more general treatment in Chapters 8 and 9, which will largely subsume it. The main results could be derived as applications and refinements of the later ones, and that would be the most efficient method from a logical point of view. However, that would also place a lot of technical developments in the way of the reader who wants quick access to some of the simplest and most useful parts of the theory. It seems preferable therefore to start with a direct study of linear optimization, even though some aspects such as duality, and even some algorithms for the linear case, cannot be handled satisfactorily until a later stage.

First the linear optimal distribution problem is investigated by itself, and then the same pattern is followed for the linear optimal differential problem. The treatment of each includes theorems on the existence and characterization of solutions, a basic algorithm, and a version of it corresponding to the simplex method of linear programming. Although flows and potentials together enter the analysis of both problems, it is not until Section 7I that the deep connection between the problems really becomes apparent. There it is shown that they can be reduced to more special forms (the "elementary" problems) for which a powerful duality holds.

An important consequence of this duality is that any method for solving either the problem for flows or the one for potentials will automatically solve the other. This is illustrated in detail in Sections 7J and 7K for the "thrifty adjustment algorithm," which is actually just a form of the basic optimal differential algorithm but implemented and interpreted in terms natural to the optimal distribution problem. A completely symmetric method for solving a pair of problems in duality is the "out-of-kilter algorithm" in Section 7M.

7A.   LINEAR OPTIMAL DISTRIBUTION PROBLEM

Suppose each arc j of a connected network G has a capacity interval $[c^-(j), c^+(j)]$ and each node i has a supply $b(i)$, where $b(N) = 0$. A feasible distribution problem is defined by such data, but we will now be concerned also with the costs of flows. Suppose the cost of the flux $x(j) \in [c^-(j), c^+(j)]$ is given by a linear expression $d(j)x(j) + p(j)$, where $d(j)$ and $p(j)$ are constants associated with the arc j. The cost of a solution x to the feasible distribution problem is then the sum of these expressions. One would like it to be as low as possible.

Linear Optimal Distribution Problem.   *Minimize*

$$\sum_{j \in A} [d(j)x(j) + p(j)] = d \cdot x + \text{const.} \tag{1}$$

over all flows x such that

$$c^-(j) \le x(j) \le c^+(j) \quad \text{for all } j \in A$$
$$y(i) = b(i) \quad \text{for all } i \in N \ (\text{where } y = \text{div } x) \tag{2}$$

Some examples of this type of problem will be discussed in Section 7B. Of course the constants $p(j)$ could be dropped from (1) without any real loss of generality, since they contribute altogether only a constant term to the total cost. However, it is convenient to retain them because of the role they will later have in duality and the reformulation of piecewise linear problems as linear problems.

Note that the minimand (1) is a "linear" function of $x \in R^A$, and the constraints are a system of linear inequalities and equations in x (since $y(i) = \sum_{j \in A} e(i, j)x(j)$). Hence this is a *linear programming problem*. Certain methods of solution that will be discussed in this chapter turn out to be closely related to the general simplex method of linear programming, but they take advantage of the special properties of networks.

A flow x is called a *feasible solution* if it satisfies the constraints (2) and an *optimal solution* if, in addition, it minimizes the total cost (1) subject to (2). Linear programming problems are known to possess optimal solutions if they have feasible solutions and a finite optimal value. This fact will be established for the linear optimal distribution problem in a special form that will be useful later in computations.

The basic idea concerns the way that a feasible solution x can be improved by sending flux around a circuit P which is flow augmenting (in the sense used in Section 3F) and also satisfies

$$0 > \sum_{j \in P^+} d(j) - \sum_{j \in P^-} d(j) = d \cdot e_P \qquad (3)$$

The flow-augmenting property means that for $t > 0$ sufficiently small, at least the flow $x' = x + te_P$ will still satisfy the capacity constraints. (Since x is feasible, this will be true of x' if and only if $x(j) + t \le c^+(j)$ for all $j \in P^+$ and $x(j) - t \ge c^-(j)$ for all $j \in P^-$.) Then also $\text{div } x' = \text{div } x = b$, because $\text{div } e_P = 0$ for circuits. In other words, x' is another feasible solution, and by (3)

$$\begin{bmatrix} \text{cost} \\ \text{of } x' \end{bmatrix} = \begin{bmatrix} \text{cost} \\ \text{of } x \end{bmatrix} + td \cdot e_P < \begin{bmatrix} \text{cost} \\ \text{of } x \end{bmatrix}.$$

This idea will be exploited in Section 7C.

For now it is important merely to observe that if P is also of unlimited capacity (in the sense used in Section 3F), namely, if $c^+(j) = \infty$ for all $j \in P^+$ and $c^-(j) = -\infty$ for all $j \in P^-$, then t can be chosen arbitrarily high, and x' will be a feasible solution whose cost is arbitrarily near $-\infty$. An elementary circuit P that satisfies (3) and is also of unlimited capacity is called an *unbalanced circuit*. (Such circuits can be detected by the methods of Chapter 6; see Exercise 7.1).

Linear Existence Theorem for Optimal Flows. *Suppose the linear optimal distribution problem has at least one feasible solution. Then the infimum in the problem is finite if and only if no elementary circuit is unbalanced. In that event at least one optimal solution exists.*

The necessity of the circuit condition has just been observed, so we can concentrate on the sufficiency. This will be proved here by the theory of extreme flows in Section 4J, which will yield, as a bonus, corollaries peculiar to the present case of *linear* optimization and suggestive of the computational approach that will be explained in Section 7D (the simplex method). Termination arguments for all the applicable algorithms in the chapter will furnish alternative proofs. The theorem will eventually be generalized by results in Sections 8G and 8H that have another mode of proof.

Proof of Sufficiency. Assuming there are no unbalanced circuits, we shall proceed in effect by induction on the number of circuits of doubly unlimited capacity (i.e., circuits P having both $c^+(j) = +\infty$ and $c^-(j) = -\infty$ for all arcs $j \in P$). If there are no circuits of doubly unlimited capacity, the extremal representation theorem in Section 4J yields a parametric expression for the most general feasible solution:

$$x = \sum_{k=1}^{r} \lambda_k x_k + \sum_{l=1}^{q} \mu_l e_{P_l}, \qquad (4)$$

where each x_k is an extreme flow for the underlying feasible distribution problem and each P_l is an elementary circuit of unlimited capacity. The coefficients λ_k and μ_l range over all non-negative values, subject only to $\sum_{k=1}^r \lambda_k = 1$. Aside from the contribution of the constants $p(j)$, the cost of the flow (4) is

$$d \cdot x = \sum_{k=1}^r \lambda_k d \cdot x_k + \sum_{l=1}^q \mu_l d \cdot e_{P_l}, \tag{5}$$

where $d \cdot e_{P_l} \geq 0$ because P_l is not unbalanced. The problem therefore reduces to minimizing (5) over all choices of $\lambda_k \geq 0$ and $\mu_k \geq 0$ with $\sum_{k=1}^r \lambda_k = 1$. This minimum is obviously attained by choosing k_0 such that $d \cdot x_{k_0}$ is the smallest of the values $d \cdot x_k$, $k = 1, \ldots, r$, and then setting $\lambda_{k_0} = 1$, $\lambda_k = 0$ for all other k, and $\mu_l = 0$ for all l. Thus a certain extreme flow x_{k_0} is an optimal solution to the problem.

Suppose now that circuits of double unlimited capacity do exist in the problem, and let P be one of them. Since both P and its reverse P' are of unlimited capacity, but neither is unbalanced, one must have $d \cdot e_P = 0$. Thus if x is any feasible solution, the flow $x + te_P$ for arbitrary $t \in R$ is also feasible and has the same cost as x. It follows that nothing is lost in the problem if one selects any $j \in P$ and imposes the additional constraint $x(j) = 0$, for the optimal (respectively, feasible) solutions to the original problem would simply be the flows of the form $x + te_P$ with x an optimal (respectively, feasible) solution to the restricted problem. This operation amounts to redefining $c^+(j)$ and $c^-(j)$ both to be 0. It eliminates P from the list of circuits of doubly unlimited capacity without adding to the list and thereby reduces the problem to the case where the existence of an optimal solution has already been established.

Extreme Solutions

A feasible or optimal solution x is said to be *extreme* if it is an extreme flow for the underlying feasible distribution problem, that is, if the arc set $F_x = \{j \mid c^-(j) < x(j) < c^+(j)\}$ forms a forest (contains no elementary circuits). Such flows are useful in computations, as will be seen in Section 7D. The preceding proof shows that if the problem has at least one optimal solution x and is not "ill posed" in the sense of possessing an elementary circuit P of doubly infinite capacity with $d \cdot e_P = 0$ (in which case $x + te_P$ would be optimal for all $t \in R$), *then it has an optimal solution that is extreme.*

The final part of the proof reduces the "ill posed" case to this one by replacing some capacity intervals of the form $(-\infty, \infty)$ by $[0, 0]$ if necessary. Since this procedure does not destroy integrality, and since the integrality of $c^-(j)$, $c^+(j)$ and $b(i)$, implies the integrality of all extreme flows (see Section 4J), the following result is obtained as a by-product.

Integrality Theorem for Optimal Flows. *If the linear optimal distribution problem has an optimal solution, and if the capacities $c^-(j), c^+(j)$ and supplies $b(i)$ are all integral, then there is an optimal solution that is integral.*

An alternative proof of this theorem will be given in Section 7C in terms of the optimal distribution algorithm.

Piecewise Linear Costs

Instead of the linear expression $d(j)x(j) + p(j)$, the cost function associated with the arc j may be piecewise linear as in Figure 7.1. Specifically, suppose that the capacity interval is partitioned into s consecutive closed subintervals $C_1,\ldots,C_s$ with associated cost expressions

$$d_k x(j) + p_k \quad \text{for } x(j) \in C_k$$

which match at the "breakpoints" between the intervals and have $d_1 < d_2 < \ldots < d_s$. (Notorious examples of such functions with progressively higher slopes are encountered in income tax tables!)

This case can be reduced to the one of merely linear costs through a modification of the network. The key is the recognition that the cost assigned to $x(j)$ is the minimum of the expression $p_1 + d_1\xi_1 + \cdots + d_s\xi_s$ subject to $\xi_1 + \cdots + \xi_s = x(j)$, $\xi_1 \in C_1$, and $0 \le \xi_k \le l_k$ for $k = 2,\ldots,s$, where l_k is the length of the interval C_k (possibly ∞ for $k = s$). Intuitively, the flux is divided into a basic amount ξ_1 and "surpluses" $\xi_2,\ldots,\xi_s$ sent at higher rates of cost. If $x(j) \in C_1$ the minimum is attained simply by setting $\xi_1 = x(j)$, $\xi_2 = \ldots = \xi_s = 0$. Otherwise the minimum is attained by letting ξ_1 be the highest value in C_1 and assigning the excess amount of flux first to ξ_2 until the bound l_2 is reached, then to ξ_3 until l_3 is reached, and so forth until the total $x(j)$ is accounted for.

In network terms the same representation is accomplished by replacing the arc j by arcs $j_1,\ldots,j_s$ in parallel as in Figure 7.2. The flux $x(j_k)$ corresponds to

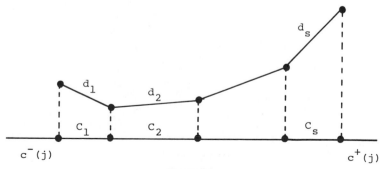

Figure 7.1

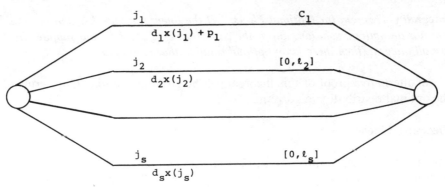

Figure 7.2

ξ_k and will automatically be adjusted in the course of optimization to reflect the original cost structure.

Although this kind of reformulation has its uses, it does suffer from the drawback that the number of arcs in the network may be increased greatly. A direct approach to piecewise linear problems, avoiding modifications of the network and therefore more suitable for successive approximation of more general, nonlinear problems, is also possible and will be covered by Chapters 8 and 9.

7B. EXAMPLES OF OPTIMIZATION OF FLOWS

Several applications of the optimal distribution model will now be described.

EXAMPLE 1.   (*Classical Transportation Problem*)

This problem was introduced in Example 10 in Section 5G in terms of integral flows, but the integrality requirement will now be relaxed. In simplest form the

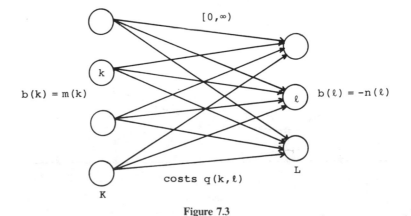

Figure 7.3

problem consists of minimizing

$$q(x) = \sum_{k \in K} \sum_{l \in L} q(k, l) x(k, l)$$

over all functions $x\colon K \times L \to [0, \infty)$ satisfying

$$\sum_{l \in L} x(k, l) = m(k) \quad \text{for all } k \in K,$$

$$\sum_{k \in K} x(k, l) = n(l) \quad \text{for all } l \in L.$$

Here K can be interpreted as a finite set of warehouses k with supplies $m(k)$, whereas L is a finite set of customers l with demands $n(l)$. The amount transported from k to l is $x(k, l) \geq 0$, and the cost of this per unit is $q(k, l)$.

To set this up as a linear optimal distribution problem, let G be the "complete bipartite" network having $N = K \cup L$ and $A = K \times L$ as in Figure 7.3. The supplies are $b(i) = m(k)$ for $i = k \in K$ and $b(i) = -n(l)$ for $i = l \in L$. For the arcs $j = (k, l)$ the capacity intervals are all $[0, \infty)$, and the cost expressions are $q(k, l)x(k, l)$; in other words, $c^-(j) = 0$, $c^+(j) = +\infty$, $d(j) = q(k, l)$, and $p(j) = 0$.

The integrality theorem in Section 7A implies that when $m(k)$ and $n(l)$ are all integral, there will be an integral optimal solution if there is an optimal solution at all. Thus there is no real loss of generality in restricting attention to integral flows x when $m(k)$ and $n(l)$ are integral, and this provides the bridge to the version of the problem in Section 5G. Upper bounds $x(k, l) \leq c(k, l)$ could easily be included.

EXAMPLE 2.   (*Optimal Assignments*)

The optimal assignment problem in Section 5E might be generalized as follows. Let $x(k, l)$ be the proportion of time that element $k \in K$ is assigned to element $l \in L$ (where (k, l) belongs to the compatible set $H \subset K \times L$ and $|K| = |L|$). The problem is to maximize

$$\sum_{(k, l) \in H} q(k, l) x(k, l)$$

subject to the constraints

$$x(k, l) \geq 0 \quad \text{for all } (k, l) \in H$$

$$\sum_{k \in K} x(k, l) = 1 \quad \text{for all } l \in L,$$

$$\sum_{l \in L} x(k, l) = 1 \quad \text{for all } k \in K,$$

where in the latter sums only pairs (k, l) in H are admitted.

This is a linear optimal distribution problem for a bipartite network like the one in Figure 7.3, but with arcs corresponding to pairs (k, l) not in H deleted. One has $b(k) = 1$, $b(l) = -1$.

The existence and integrality theorems imply there is an integral optimal solution if the constraints can be satisfied at all. (No circuit can be unbalanced, since it must cross back and forth between K and L.) But an integral flow x satisfying these constraints must have $x(k, l) = 0$ or $x(k, l) = 1$ for all k, l, and thus it corresponds to an assignment

$$M = \{(k, l) \in H \mid x(k, l) \neq 0\}.$$

Despite the apparent generality this example of optimal distribution therefore is equivalent to the purely combinatorial problem of optimal assignment.

EXAMPLE 3.   (*General Linear Transportation Problem*)

Example 1 may be extended by considering an arbitrary network of "transportation links" between various localities. Some nodes i represent warehouses $(b(i) > 0)$, some represent customers $(b(i) < 0)$, and others are transfer points where the flow is conserved $(b(i) = 0)$. Each link j has a capacity interval $[c^-(j), c^+(j)]$ and cost expression $d(j)x(j) + p(j)$, which might be identically zero in some instances. One seeks to distribute the supplies to meet the demands in the cheapest way possible.

The condition of fixed supplies and demands could be generalized by associating with each node i a capacity interval $[c^-(i), c^+(i)]$ and cost expression $d(i)y(i) + p(i)$, where $y(i)$ is the divergence at i. (For a node representing a customer and thus involving only negative values of $y(i)$, $d(i)$ could be the profit per unit delivered, and then the term $d(i)y(i)$ would be a negative cost.) This model may be regarded as a linear optimal distribution problem in the augmented network. Storage costs, costs of default on orders not filled, and so on, could also be handled in the manner explained in Example 2 in Section 3H.

EXAMPLE 4.   (*Dynamic Transportation Problem*)

The transportation problem can also be extended over a period of time in terms of flows in a dynamic network of the kind described in Section 1H. Besides covering the possible need to schedule shipments so as not to overload facilities, and in response to time-dependent costs and supplies, this allows for the fact that certain "warehouses" may be quicker at supplying certain "customers." By adopting the model in the preceding example with capacity intervals and cost functions also associated with the nodes, one could handle the case of "sales profits" that diminish the longer an order goes unfilled. There could also be "holdover" costs (and constraints) for quantities kept in a location from one time period to the next.

Many problems fit this kind of mold but involve elements of uncertainty in supply or demand and therefore require techniques beyond the scope of this book. An understanding of the corresponding deterministic case is nevertheless essential to their analysis.

EXAMPLE 5. (*Warehousing Problem*)

Here is a model similar in some respects to the preceding one, but it centers on the most profitable use of storage space.

Certain commodities X_k ($k = 1,\ldots,K$) are to be bought and sold at times $t = 1,\ldots,T$ subject to buying and selling prices that may vary with time but are known in advance. A warehouse of fixed total capacity is to be used to store the various amounts from the time they are bought till the time they are sold. The commodities also have different storage costs that may vary with time. There may be limitations on how much can be bought, sold, or stored. The problem is to determine a schedule of buying, selling, and storing that minimizes total cost (maximizes total profit).

This can be set up as an optimal distribution problem for a sort of dynamic network in which the flows represent transfers of storage space. Figure 7.4 illustrates the case of $K = 3$, $T = 3$. All commodities are measured in *units of storage space*, and all costs are likewise in such terms (e.g., cubic meters and dollars per cubic meter, if storage space is most appropriately measured in cubic meters).

The flux in an arc of the form $j_{k,t}$ represents the amount of commodity X_k sold at time t, whereas the flux in $j'_{k,t}$ is the amount bought. The flux in $\bar{j}_{k,t}$ is

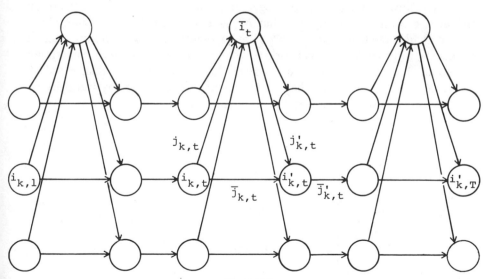

Figure 7.4

the quantity of X_k left in storage at time t, and the flux in $\bar{j}'_{k,t}$ is the quantity stored from t to $t + 1$. All these arcs have capacity intervals $[c^-(j), c^+(j)] \subset [0, \infty]$ and associated costs, the cost of selling being the negative of the profit of selling. Thus only non-negative flows are candidates for feasibility.

All nodes are conservative except those of the form $i_{k,1}$, where the supply is the amount of X_k initially in the warehouse, and those of the form $i_{k,T}$, where the supply is the negative of the amount of X_k which must be left over at the final time T. These requirements, which seem to imply the warehouse is already full at the beginning, stays full at all intermediate times, and is left full at the finish, involve no real loss of generality. A fictitious commodity representing unused storage space can always be introduced so that all the available space in the warehouse is accounted for. If no material goods are present initially, this can be represented by starting with the whole warehouse devoted to the "empty" commodity. The situation where the warehouse must be vacant at the end can be handled similarly. The cost of "selling" the empty commodity can be interpreted as the cost of converting unused space to active storage (perhaps a zero cost). The storage costs represent costs that are present even when space is not used.

Other interpretations of this kind of model are possible. For instance, the "warehouse" can be transmuted into a "work force." The "commodities" are different types of work, and units of storage space are replaced by worker-hours. The problem is how to allocate the available worker-hours to the different types of work in an optimal fashion over time.

EXAMPLE 6.　(*Max Flow Problem*)

The max flow problem can be regarded as a linear optimal distribution problem. For simplicity suppose that N^+ and N^- consist of solitary nodes s and s'. Add a feedback arc $\bar{j} \sim (s', s)$ as in Figure 7.5 with capacity interval $(-\infty, \infty)$, in order to get a network G' whose feasible circulations correspond to the feasible flows in G that are conserved at all nodes other than s and s'. The problem of maximizing the flux from s to s' in G is thereby transformed into the problem of maximizing $x(\bar{j})$ over all feasible circulations in G'.

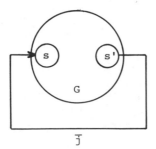

Figure 7.5

This may be regarded as the optimal distribution problem in G' having $b(i) = 0$ for all nodes i, cost coefficients $d(\bar{j}) = -1$ and $p(\bar{j}) = 0$ for the feedback arc, and zero costs ($d(j) = 0$ and $p(j) = 0$) for all other arcs.

Although the max flow problem gains nothing by this reformulation, a source of insight is provided in comparisons between the max flow algorithm and the more general optimization methods described later in this chapter and in Chapter 9.

7C.   OPTIMAL DISTRIBUTION ALGORITHM

The method about to be presented depends on a basic test for optimality. This is in terms of certain "curves" Γ_j associated with the arcs j in a linear optimal distribution problem, namely,

$$\Gamma_j = \begin{bmatrix} \text{set of all pairs } (x(j), v(j)) \in R^2, \text{ satisfying} \\ c^-(j) \leq x(j) \leq c^+(j) \text{ and having } v(j) \leq d(j) \text{ if} \\ x(j) < c^+(j), \text{ and } v(j) \geq d(j) \text{ if } x(j) > c^-(j) \end{bmatrix}. \qquad (1)$$

Figure 7.6 depicts Γ_j in the case where $c^-(j)$ and $c^+(j)$ are both finite. Of course if $c^+(j) = +\infty$, the vertical piece on the right is absent, whereas if $c^-(j) = -\infty$, it is the vertical piece on the left that disappears. If both $c^+(j) = +\infty$ and $c^-(j) = -\infty$, Γ_j degenerates to a horizontal line, whereas if $c^-(j) = c^+(j)$, it degenerates to a vertical line.

Linear Optimality Theorem for Flows.   *A flow x is an optimal solution to the linear optimal distribution problem if and only if div x = b, and there is a*

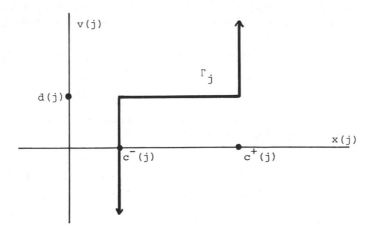

Figure 7.6

potential u whose differential v satisfies

$$(x(j), v(j)) \in \Gamma_j \quad \text{for all } j \in A. \tag{2}$$

Proof of Sufficiency. Suppose this condition is fulfilled by x and u. Then in particular, x is a feasible solution. Let x' be any other feasible solution. According to the definition of Γ_j,

$$c^-(j) < x(j) < c^+(j) \quad \Rightarrow \quad v(j) = d(j),$$

$$x(j) = c^+(j) \quad \Rightarrow \quad v(j) \geq d(j) \text{ and } x'(j) \leq x(j),$$

$$x(j) = c^-(j) \quad \Rightarrow \quad v(j) \leq d(j) \text{ and } x'(j) \geq x(j).$$

Hence $d(j)[x'(j) - x(j)] \geq v(j)[x'(j) - x(j)]$ in all cases. Adding this inequality over all $j \in A$, one obtains

$$d \cdot [x' - x] \geq v \cdot [x' - x] = -u \cdot \text{div}[x' - x] = -u \cdot [b - b] = 0.$$

Thus $d \cdot x' \geq d \cdot x$. In other words, x' cannot cost less than x, so x is optimal.

Proof of Necessity. The algorithm that follows, when applied to any feasible solution x, constructs either a potential u satisfying the condition in the theorem or a feasible solution x' "cheaper" than x. If x is optimal, only the first alternative is possible.

Interpretation

Think of x as a flow of goods and $u(i)$ as the price per unit at node i. Then $v(j)$ is the price increase (possibly negative) experienced in transporting goods through j, and $[d(j) - v(j)]x(j)$ is the *net* cost in transporting $x(j)$ units. The relations $(x(j), v(j)) \in \Gamma_j$ mean that this quantity is minimized in each arc j: if $d(j) - v(j) > 0$, one sends as little as possible $[x(j) = c^-(j)]$, whereas if $d(j) - v(j) < 0$, one sends as much possible $[x(j) = c^+(j)]$.

Algorithm

Given any feasible solution x to the linear optimal distribution problem, define a corresponding system of span intervals by

$$[d_x^-(j), d_x^+(j)] = \begin{cases} [d(j), d(j)] & \text{if } c^-(j) < x(j) < c^+(j), \\ (-\infty, d(j)] & \text{if } c^-(j) = x(j) < c^+(j), \\ [d(j), \infty) & \text{if } c^-(j) < x(j) = c^+(j), \\ (-\infty, \infty) & \text{if } c^-(j) = x(j) = c^+(j). \end{cases} \tag{3}$$

Apply any algorithm that terminates in either a solution to the feasible differential problem for these intervals, that is, in a potential u such that $v = \Delta u$ satisfies

$$d_x^-(j) \le v(j) \le d_x^+(j) \quad \text{for all } j \in A, \tag{4}$$

or in an elementary circuit P with $d_x^+(P) < 0$. (The feasible differential algorithm is especially designed for this, and refinements based on its use will be explained later.)

In the first alternative, x is an optimal solution and the algorithm terminates. Indeed, (4) is identical to (2), a condition already demonstrated to be sufficient for optimality.

In the second alternative the condition

$$0 > d_x^+(P) = \sum_{j \in P^+} d_x^+(P) - \sum_{j \in P^-} d_x^-(P)$$

means from (3) that $x(j) < c^+(j)$ for all $j \in P^+$, $x(j) > c^-(j)$ for all $j \in P^-$ (i.e., P is flow-augmenting), and

$$0 > \sum_{j \in P^+} d(j) - \sum_{j \in P^-} d(j) = d \cdot e_P. \tag{5}$$

Let

$$\alpha = \min \begin{cases} c^+(j) - x(j) & \text{for } j \in P^+, \\ x(j) - c^-(j) & \text{for } j \in P^-. \end{cases}$$

Then $\alpha > 0$. If $\alpha = \infty$, P is an unbalanced circuit, and the algorithm terminates; the infimum in the problem is $-\infty$ by the existence theorem in Section 7A. Otherwise let $x' = x + \alpha e_P$. Then x' is another feasible solution and

$$d \cdot x' = d \cdot x + \alpha d \cdot e_P < d \cdot x$$

by (5), so that x' is "cheaper" than x. Iterate with the new flow x'.

Termination

Conceivably the algorithm might generate an infinite sequence of feasible solutions with decreasing costs and thus never terminate. This could result from failure to detect an unbalanced circuit, even though the infimum in the problem is $-\infty$, but it could also occur when the infimum is finite. As a matter of fact the max flow algorithm may be identified with a special case of the optimal distribution algorithm (Exercise 7.8), so the example furnished in Exercise 3.12 carries over the present context and demonstrates that the costs might converge to a value short of the infimum.

Whether or not an unbalanced circuit exists (or equivalently by the existence theorem in Section 7A, whether or not the infimum in the problem is $-\infty$) can be settled straightforwardly at the outset (see Exercise 7.1). The other kind of misbehavior can be ruled out by a commensurability condition or by refinements that will be discussed later in this section.

The algorithm must terminate if the infimum in the problem is finite, and the values $c^-(j)$, $c^+(j)$ and $b(j)$ are all commensurable along with the initial flow values $x(j)$. Indeed, if these values are all multiples of a certain $\delta > 0$, then so are α and the successor flow values $x'(j)$. All the flows generated by the algorithm therefore belong to the same commensurability class, and at every iteration the decrease in cost is at least $\delta\varepsilon$, where ε is the smallest of the values $|d \cdot e_P|$ corresponding to (the finitely many) elementary circuits P with $d \cdot e_P < 0$. Since costs are bounded below, there cannot be an infinite sequence of iterations.

Of course if $c^-(j)$, $c^+(j)$ and $b(i)$ are commensurable, an initial feasible solution x with values $x(j)$ in the same commensurability class can be produced by the feasible distribution algorithm (or the flow rectification algorithm).

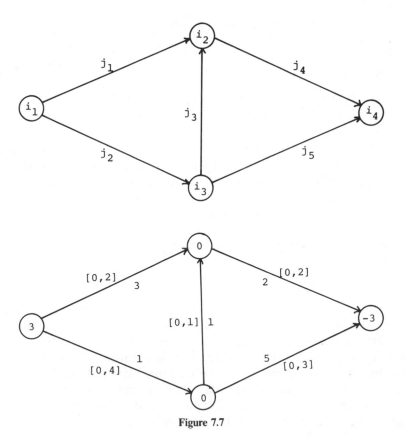

Figure 7.7

Integrality

Observe that the commensurability argument just given yields a second proof of the integrality theorem in Section 7A.

EXAMPLE 7

A network is shown in Figure 7.7 along with capacity intervals $[c^-(j), c^+(j)]$, supplies $b(i)$, and cost coefficients $d(j)$. (The constants $p(j)$ are all 0.) The

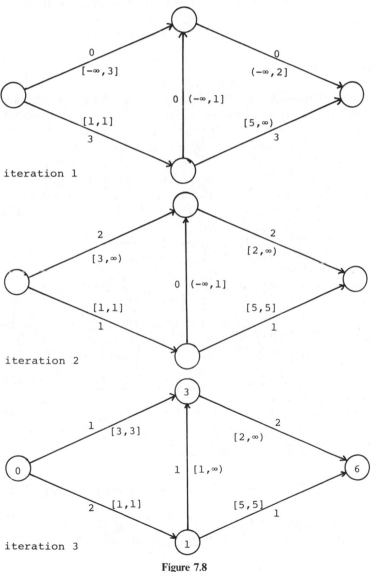

Figure 7.8

algorithm is initiated with an arbitrarily chosen feasible solution. Figure 7.8 displays the flows in successive iterations and their associated systems of *span* intervals $[d_x^-(j), d_x^+(j)]$. The corresponding circuits are

$$P_1 : i_1 \rightarrow i_2 \rightarrow i_4 \leftarrow i_3 \leftarrow i_1, \quad \alpha_1 = 2,$$

$$P_2 : i_1 \rightarrow i_3 \rightarrow i_2 \leftarrow i_1, \quad \alpha_2 = 1.$$

The flow tested in the third iteration turns out to be optimal; the corresponding potential values are indicated at the nodes.

Refinements

The feasible differential algorithm can be used as the main subroutine, and with the restart provision explained in Section 6H this is very advantageous. Each iteration of the optimal distribution algorithm except the last will begin with some potential u, work toward improving it node by node so as to achieve feasibility with respect to the span intervals $[d_x^-(j), d_x^+(j)]$, but will terminate by discovering a circuit P with $d_x^+(P) < 0$. The restart provision for the feasible differential algorithm modifies u in this case too: $w(i) - \beta$ is added to $u(i)$ for all $i \in S$, where $\beta = d_x^+(P)$. This modified u is taken as the potential at the beginning of the next iteration.

As observed in Section 6H, the terminal changes in u ensure that for every arc j, the distance of $v(j)$ form $[d_x^-(j), d_x^+(j)]$ is no greater than it was at the start of the subroutine: none of the quantities $\rho(i)$ has been decreased, and $\rho(\bar{i})$ (at the node treated in the final step of the subroutine, where the circuit P was detected) has been raised by the amount $|\rho(\bar{i})| + \beta \geq 0$. We now claim something more: the distance of $v(j)$ from $[d_x^-(j), d_x^+(j)]$ is not increased either when x is replaced by $x + \alpha e_P$. Thus *when the optimal distribution algorithm is implemented with this choice of subroutine, the distance of $(x(j), v(j))$ from Γ_j never increases for any arc j during the procedure.* In particular, once the condition $(x(j), v(j)) \in \Gamma_j$ is achieved for some arc, it is maintained forever after.

The proof of the claim is as follows. We know from Section 6H that at the stage just before x is replaced by $x + \alpha e_P$, the potential u (with which the feasible differential subroutine terminated) satisfies

$$v(j) \geq d_x^+(j) \quad \text{for all } j \in P^+, \qquad v(j) \leq d_x^-(j) \quad \text{for all } j \in P^-.$$

Thus for $j \in P^+$, where we know $c^-(j) \leq x(j) < x(j) + \alpha \leq c^+(j)$, the interval $[d_x^-(j), d_x^+(j)]$ is $(-\infty, d(j)]$ or $[d(j), d(j)]$, and its distance from $v(j)$ is

$v(j) - d(j)$; after replacement of x by $x + \alpha e_P$, the interval will be $[d(j), d(j)]$ or $[d(j), \infty)$, so the distance from $v(j)$ will either still be $v(j) - d(j)$, or it will have dropped to 0. Similarly for $j \in P^-$, the distance either stays the same or drops to 0, and of course for $j \notin P$ the distance is not affected because there is no change in $x(j)$.

As a further refinement a form of multipath implementation can be used to make more extensive improvements in x in each iteration, not just along a single circuit P. This is discussed toward the end of Section 7L and is shown to guarantee finite termination for the optimal distribution algorithm, provided that in the feasible differential subroutine one keeps with the same node $\bar{i}$ as long as $\rho(\bar{i}) < 0$.

7D.*   SIMPLEX METHOD FOR FLOWS

In each iteration of the optimal distribution algorithm in Section 7C, the attempt to find a potential u whose differential is feasible with respect to the span intervals $[d_x^-(j), d_x^+(j)]$ could be carried out as the construction of a complete optimal routing with base s, an arbitrarily selected node. (See the optimal routing theorem in Section 6J and the remarks near the end of Section 6K.) This would amount to solving the linear optimal distribution problem by a method that generates a sequence of complete routings and their associated potentials, each routing differing from its predecessor by only one arc.

Such an approach can be developed more directly in terms of the theory of spanning trees and pivoting in Chapter 4, as will now be explained. The procedure that is obtained corresponds to the general technique in linear programming known as the *simplex method with upper bounding*. It can be viewed essentially as an implementation of the optimal differential algorithm in which one source of trouble over termination (incommensurability of the data) is traded for another ("degeneracy"). The main computational virtue lies in the simplicity of the individual steps and the ease with which they can be executed. In worst-case situations the number of such steps could be forbiddingly high; nevertheless, the method has a reputation for good behavior in practice, at least when it is implemented with the spanning trees represented by routings in the manner suggested in the refinement at the end of this section.

One can assume without real loss of generality that the network G is connected. Designate some node s as a reference point. Then the maximal forests in G are the same as the spanning trees in G, which can in turn be identified with the routings of all of N with base s (see Section 4F). Each spanning tree F corresponds to a pair of Tucker representations of the circulation space and differential space, as explained in Section 4G.

In particular, for each choice of tension values $v(j)$ for $j \in F$, there are uniquely determined values $v(j)$ for $j \notin F$ such that v is a differential. There is a unique potential u with $\Delta u = v$ and $u(s) = 0$, namely, the one obtained by integrating v along the paths from s in the routing corresponding to F. On the other hand, for each choice of flux values $x(j)$ for $j \notin F$, there are uniquely determined values $x(j)$ for $j \in F$ such that x is a flow with div $x = b$ (under the assumption of course that $b(N) = 0$). This follows from the fact that at least one flow with divergence b exists (because the network is connected), and the difference between two such flows is a circulation. (A circulation that vanishes at all arcs outside F must be the zero circulation; see Section 4G.)

These relationships could be described in terms of a tableau of coefficients, which would bring out more clearly the connection between the procedure described next and the customary techniques in linear programming. The calculations that follow could then be carried out in terms of pivoting; see Section 4H and 4I.

Algorithm

Let x be any *extreme* feasible solution to the linear optimal distribution problem such that the set $F_x = \{j \in A | c^-(j) < x(j) < c^+(j)\}$ forms a forest, and let F be any spanning tree containing F_x. Let u be the unique potential with $u(s) = 0$ such that $v = \Delta u$ satisfies $v(j) = d(j)$ for all $j \in F$. Check whether

$$v(k) \geq d(k) \quad \text{for all } k \in A \setminus F \text{ with } x(k) = c^-(k) < c^+(k), \quad (1)$$

$$v(k) \leq d(k) \quad \text{for all } k \in A \setminus F \text{ with } x(k) = c^+(k) > c^-(k). \quad (2)$$

If yes, terminate; x is an optimal solution. If no, let $\bar{k}$ be any arc for which (1) or (2) is violated. (A rule that seems generally to improve the effectiveness is to choose $\bar{j}$ for which the violation is greatest.) There is an elementary circuit P, unique up to orientation, such that P uses $\bar{k}$ and otherwise only arcs of F (see Section 4H); if $\bar{k}$ violates (1), choose the orientation so that $\bar{k} \in P^+$, but if $\bar{k}$ violates (2) choose it so that $\bar{k} \in P^-$ (see Figure 7.9). Then $d \cdot e_P < 0$. Define

$$\alpha = \min \begin{cases} c^+(j) - x(j) & \text{for } j \in P^+. \\ x(j) - c^-(j) & \text{for } j \in P^-. \end{cases}$$

Then $\alpha \geq 0$. If $\alpha = \infty$, terminate; P is an unbalanced circuit, and the infimum in the problem is $-\infty$. If $\alpha < \infty$, let $\tilde{j}$ be any arc in P for which the minimum defining α is attained (possibly $\tilde{k} = \bar{k}$, although not if $\alpha = 0$). Define

$$x' = x + \alpha e_P, \qquad \tilde{F} = [F \setminus \{\tilde{j}\}] \cup \{\bar{k}\}.$$

Then x' is an extreme feasible solution, $\tilde{F}$ is a spanning tree containing $F_{x'} = \{j \in A | c^-(j) < x'(j) < c^+(j)\}$, and

$$d \cdot x' = d \cdot x - \alpha |v(\bar{k}) - d(\bar{k})| \leq d \cdot x.$$

Repeat the procedure with x', $\tilde{F}$.

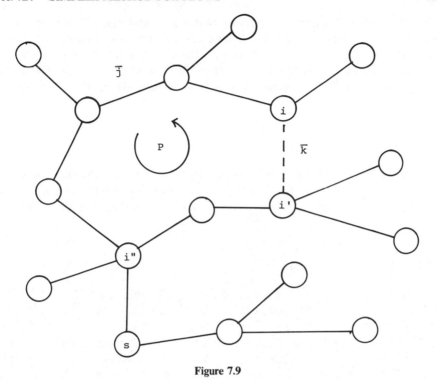

Figure 7.9

Justification

For every $k \in A \setminus F$, either $x(k) = c^-(k)$ or $x(k) = c^+(k)$, or both. The test performed on the potential u is therefore just a check of whether for all $j \in A$, $v(j)$ belongs to the associated span interval $[d_x^-(j), d_x^+(j)]$ in 7C(3), or equivalently, $(x(j), v(j)) \in \Gamma_j$. If so, x is optimal by the theorem in Section 7C.

On the other hand, if (1) is violated by an arc $\bar{k} \in F$ and P is the corresponding elementary circuit with $\bar{k} \in P^+$ but with all other arcs in F, then $v(j) = d(j)$ for all arcs of P except $\bar{k}$, and $v(\bar{k}) > d(\bar{k})$, so that

$$ d \cdot e_P = \sum_{j \in P^+} d(j) - \sum_{j \in P^-} d(j) < \sum_{j \in P^+} v(j) - \sum_{j \in P^-} v(j) $$

$$ = v \cdot e_P = -u \cdot \operatorname{div} e_P = 0. $$

(This shows in fact that $d \cdot e_P = d(\bar{k}) - v(\bar{k})$.) A similar argument shows that $d \cdot e_P < 0$ in the case where (2) is violated. If $\alpha = \infty$, then $c^+(j) = +\infty$ for all $j \in P^+$ and $c^-(j) = -\infty$ for all $j \in P^-$, and P is indeed an unbalanced circuit. (The infimum in the problem is $-\infty$ in this case according to the existence theorem in Section 7A.)

The feasibility of x implies otherwise that $\alpha \geq 0$, and that x' is feasible. For all arcs $k \in A \setminus F$ except possibly $\bar{k}$, one has $x'(k) = x(k)$ and consequently either $x'(k) = c^+(k)$ or $x'(k) = c^-(k)$ (because $F_x \subset F$). On the other hand, the choice of $\bar{j}$ ensures that either $x'(\bar{j}) = c^+(\bar{j})$ or $x'(\bar{j}) = c^-(\bar{j})$. Therefore $F_{x'} \subset \tilde{F}$. But $\tilde{F}$ is again a spanning tree (see Section 4I), so this shows x' is extreme.

Initiation

An extreme feasible solution with which to start the algorithm can be computed by the feasible distribution algorithm using arc discrimination in the painted network subroutine (see Exercise 4.42) or by the method in Section 4K.

Termination

The algorithm generates a sequence of extreme feasible solutions and associated spanning trees. In iterations having $\alpha = 0$, the flow stays the same, and only the spanning tree changes; however, in iterations having $\alpha > 0$, the flow is replaced by a cheaper one. Thus the algorithm cannot return to any extreme feasible solution that it has previously considered and abandoned, but there is danger that it might get stuck indefinitely on a particular one with an infinite sequence of iterations having $\alpha = 0$. If the latter possibility can be excluded somehow, the algorithm is sure to terminate eventually because there are only finitely many extreme feasible solutions altogether (see Section 4J).

In an iteration with $\alpha = 0$, the arc $\bar{j}$ either belongs to P^+ with $x(\bar{j}) = c^+(\bar{j})$, or it belongs to P^- with $x(\bar{j}) = c^-(\bar{j})$. Either way, it follows that $\bar{k} \neq \bar{j} \in F \setminus F_x$. Let us say that the extreme feasible solution x is *nondegenerate* if F_x is itself a spanning tree. Then F must coincide with F_x, so that $\bar{j}$ cannot belong to $F \setminus F_x$ and α therefore cannot be 0. This leads to the conclusion that *if every extreme feasible solution generated by the algorithm is nondegenerate, then the algorithm must terminate sooner or later.*

The trouble with such a nondegeneracy assumption, however, is that it is usually impossible to verify. Although this is not necessarily a serious matter in computational practice, one would like to refine the algorithm to get around it. This can be accomplished by taking extra care in the choice of $\bar{k}$ and $\bar{j}$ when there are ties.

Let the arcs $j \in A$ all be numbered in some arbitrary order; arcs lower in the numbering will be said to have higher priority. Suppose that in any iteration where more than one arc could serve as $\bar{k}$ (the arc entering the spanning tree). The one with the highest priority is selected; similarly with the choice of $\bar{j}$ (the arc leaving the tree) when $\alpha = 0$. This is *Bland's priority rule*. It will be demonstrated that *under Bland's priority rule there cannot be any infinite sequence of iterations with $\alpha = 0$, and the algorithm must therefore terminate.*

(Another way of ensuring termination, by a restriction to "strong" spanning trees, will be explained later in this section.)

In an infinite sequence of iterations with $\alpha = 0$, the extreme feasible solution x stays the same, while the spanning tree changes in a finite cycle that is repeated over and over. (Recall that the number of spanning trees is finite.) Consider such a cycle, and let B be the set of all arcs affected by it, or in other words, which are selected as $\bar{k}$ at some stage but as $\bar{j}$ at some other stage. Through selection as $\bar{k}$, it is clear that

$$j \in B \quad \text{implies } x(j) = c^-(j) < c^+(j) \text{ or } x(j) = c^+(j) > c^-(j). \quad (3)$$

Let j_0 denote the arc of lowest priority in B, and let F_0 be the spanning tree in the iteration when j_0 is chosen as $\bar{k}$. Let v_0 be the corresponding differential; thus either

$$x(j_0) = c^-(j_0) < c^+(j_0) \quad \text{and} \quad d(j_0) - v_0(j_0) < 0 \qquad (4a)$$

holds or

$$x(j_0) = c^+(j_0) > c^-(j_0) \quad \text{and} \quad d(j_0) - v_0(j_0) > 0. \qquad (4b)$$

Now consider the spanning tree F in an iteration when j_0 is chosen as $\bar{j}$. Let P be the circuit used in that iteration. Then $0 > d \cdot e_P$ and $v_0 \cdot e_P = 0$ (since v_0 is a differential and e_P a circulation), so that

$$0 > (d - v_0) \cdot e_P = \sum_{j \in A} [d(j) - v_0(j)] e_P(j). \qquad (5)$$

The selection of j_0 as $\bar{j}$ implies that $j_0 \in P^-$ in the case of (4a) but $j_0 \in P^+$ in the case of (4b). Hence

$$[d(j_0) - v_0(j)] e_P(j_0) > 0. \qquad (6)$$

Comparing (5) and (6), one sees there must be an arc $j_1 \neq j_0$ such that

$$[d(j_1) - v_0(j_1)] e_P(j_1) < 0. \qquad (7)$$

On the one hand, (7) implies $d(j_1) - v_0(j_1) \neq 0$, so $j_1 \notin F_0$ (because all arcs $j \in F_0$ have $v_0(j) = d(j)$ by the definition of v_0). On the other hand, (7) implies $j_1 \in P$, so either $j_1 \in F$ or j_1 is the arc about to enter F in place of j_0. This shows that j_1 must be one of the arcs affected by the cycle, that is, $j_1 \in B$. From (4) and the fact that j_0 was chosen over j_1 in the F_0 iteration, even though $j_1 \notin F_0$ and j_0 had the lowest priority, one is led to only two possibilities:

$$x(j_1) = c^-(j_1) < c^+(j_1) \quad \text{and} \quad d(j_1) - v_0(j_1) \leq 0, \quad \text{or} \qquad (8a)$$

$$x(j_1) = c^+(j_1) > c^-(j_1) \quad \text{and} \quad d(j_1) - v_0(j_1) \geq 0. \qquad (8b)$$

In (8a) one must have $j_1 \in P^+$ by (7), whereas in (8b) one must have $j_1 \in P^-$. Either way, j_1 is one of the arcs for which the minimum defining α is attained and is thus a candidate for selection as j. Since j_1 has higher priority than j_0, it is impossible for this to be an iteration when j_0 is chosen as j. The contradiction establishes the assertion that the algorithm must terminate under Bland's priority rule.

Refinement Using Strong Spanning Trees

Termination in a finite number of iterations can also be guaranteed by limiting attention to spanning trees that correspond to certain kinds of routings. Let us call F a *strong* spanning tree for s (the chosen reference node) and the flow x if it not only contains F_x but also has the property that the corresponding routing θ with base s is compatible (in the sense of Section 6J) with respect to the span intervals $[d_x^-(j), d_x^+(j)]$ in Section 7C. In view of the definition of these intervals compatibility of θ is equivalent to the condition that for each θ-path P, the arcs in P^+ have $x(j) < c^+(j)$, whereas those in P^- have $x(j) > c^-(j)$; in other words, P is flow augmenting for x. The potential u associated with F in the algorithm is then the same as the one associated with θ with respect to the intervals $[d_x^-(j), d_x^+(j)]$ (in the sense of Section 6J).

Suppose the algorithm can be initiated with F a strong spanning tree for s and x. The following refinement of the simplex method will guarantee that this property is maintained at every iteration and moreover in such a way that an infinite sequence of iterations with $\alpha = 0$ is impossible. Termination of the method is then certain, as already argued earlier. The refinement is formulated in terms of the routing θ.

Having picked an arc $\bar{k} \in A \setminus F$ that violates (1) or (2), let $(i', i) \sim \bar{k}$ in the case of (1) but $(i, i') \sim \bar{k}$ in the case of (2). Let i'' be the last node common to the θ-paths $P_i: s \to i$ and $P_{i'}: s \to i'$. The circuit P is obtained by following P_i from i'' to i', then $\bar{k}$ from i' to i, and finally $P_{i'}$ backward from i to i''; see Figure 7.8. (Possibly $i'' = i$, in which case the last portion of this description of P is vacuous.) Next, having calculated the α value corresponding to P, select as $\bar{j}$, from among the arcs of P furnishing the minimum in the definition of α, the *first* such arc that is encountered as P is traced starting at i''. Then proceed as before.

The effect of the pivot step on the routing will be that the nodes of P prior to $\bar{j}$ (when P is traced from i'') will be reached by following P from i'', whereas those after $\bar{j}$ will be reached by following the reverse of P. The claim that the new spanning tree $\tilde{F}$ is strong for s and the new flow x' reduces to the assertion that these two path segments are sure to be flow augmenting for x'. This is true by the choice of $\bar{j}$ for the portion of P from i'' up to $\bar{j}$ (because $\bar{j}$ marks the first obstacle encountered in trying to send additional flow along P). As for the reverse of P from i'' up to $\bar{j}$, there are two cases. If $\alpha = 0$, this must be part of $P_{i'}$ and hence flow augmenting for x' (because $x' = x$, and both P_i and $P_{i'}$) were flow augmenting for x inasmuch as F was a strong spanning tree for s and

x). If $\alpha > 0$, the entire reverse of P is actually flow augmenting for x', since one could subtract from x' the elementary flow just added to get x' from x. Thus the rule for choosing $\bar{j}$ does ensure that at all stages the spanning tree will still be strong for s and the flow on hand.

In programmable terms the effect of the iteration on the routing is as follows. If the arc $\bar{j}$ comes *after* the arc $\bar{k}$ (in tracing P from i''), θ is redefined to label each node of P between $\bar{k}$ and $\bar{j}$ with the arc of P that *precedes* it, and the number $\gamma = |v(\bar{k}) - d(\bar{k})| > 0$ is *subtracted* from the value of u at each such node and all nodes reached by the routing by passing through such nodes. If $\bar{j}$ comes *before* $\bar{k}$, θ is redefined to label each node of P between $\bar{j}$ and $\bar{k}$ with the arc of P that *succeeds* it, and γ is *added* to the value of u at each such node and all nodes reached by the routing by passing through such nodes. If $\bar{j} = \bar{k}$, there is no change in θ or u.

In iterations with $\alpha = 0$, we have seen that only the first of the three cases is possible. Since potential values are not increased in such iterations, and at least one is lowered each time, the spanning trees in any sequence of such iterations must all be different (the potential function being uniquely determined by the spanning tree and the choice of s). An infinite sequence of degenerate iterations is therefore impossible.

7E. LINEAR OPTIMAL DIFFERENTIAL PROBLEM

Consider now the problem of finding a "best" solution to a feasible differential problem with span intervals $D(j) = [d^-(j), d^+(j)]$. Suppose the cost of the tension $v(j) \in D(j)$ is given by a linear expression $c(j)v(j) + q(j)$, where $c(j)$ and $q(j)$ are constants associated with the arc j. The sum of these over all $j \in A$ is the *cost* of the differential v. For reasons that will ripen later in the theory of duality, it is desirable also to introduce terms with coefficients $b(i)$ for the potential values $u(i)$ at the nodes of the network and to cast the problem in the mold of maximization. *It will be assumed that the network is connected with* $b(N) = 0$.

Linear Optimal Differential Problem. *Maximize*

$$- \sum_{i \in N} b(i)u(i) - \sum_{j \in A} [c(j)v(j) + q(j)] = -b \cdot u - c \cdot v + \text{const.} \quad (1)$$

over all potentials u such that

$$d^-(j) \le v(j) \le d^+(j) \quad \text{for all } j \in A \quad (\text{where } v = \Delta u) \quad (2)$$

Examples will be described in Section 7F. The reason for assuming $b(N) = 0$ is that otherwise the supremum in the problem could not possibly be finite: adding an arbitrary real number t to all the potentials $u(i)$ would add $-tb(N)$

to the maximand without affecting the constraints. As for the assumption that the network is connected, there is no loss of generality because otherwise the problem would decompose into a separate one for each component.

The constants $q(j)$ do not really influence the task of optimization in any way, but they help in the derivation and comparison of certain models. For that matter even the coefficients $c(j)$ are redundant, since for $j \sim (i, i')$ one could replace $v(j)$ by $u(i') - u(i)$ and thereby absorb the second sum in (1) into the first. But this approach would not be illuminating in the long run, since it does not carry over to nonlinear cost expressions. At any rate it is clear that (1) amounts to a linear function of $u \in R^N$ and that (2) represents a system of linear inequality constraints on u. The problem at hand is therefore one of linear programming.

A *feasible solution* is a potential u that satisfies the constraints (2), and an *optimal solution* is one that yields the maximum in the problem. Results corresponding to those for the optimal distribution problem in Section 7A will now be established.

A feasible solution u can be improved by increasing tension across certain kinds of cuts $Q = [S, N \setminus S]$ that satisfy

$$0 > - \sum_{i \in S} b(i) + \sum_{j \in Q^+} c(j) - \sum_{j \in Q^-} c(j) = b \cdot e_{N \setminus S} + c \cdot e_Q \qquad (3)$$

(where $-b \cdot e_S$ has been written as $b \cdot e_{N \setminus S}$ for convenience because $\Delta e_{N \setminus S} = e_Q$; recall that $0 = b(N) = b(S) + b(N \setminus S) = b \cdot e_S + b \cdot e_{N \setminus S}$). If for $t > 0$ the potential $u' = u + t e_{N \setminus S}$ (with differential $v' = v + t e_Q$) is still a feasible solution, one has by (3) that

$$\begin{bmatrix} \text{cost} \\ \text{of } u' \end{bmatrix} = \begin{bmatrix} \text{cost} \\ \text{of } u \end{bmatrix} + t \begin{bmatrix} b \cdot e_{N \setminus S} + c \cdot e_Q \end{bmatrix} < \begin{bmatrix} \text{cost} \\ \text{of } u \end{bmatrix}.$$

Thus in the way the problem has been formulated, u' gives a higher value to the maximand than u does.

If besides satisfying (3), Q is also of unlimited span (as in Chapter 6), that is, has $d^+(j) = +\infty$ for all $j \in Q^+$ and $d^-(j) = -\infty$ for all $j \in Q^-$, then no matter how high t is chosen, u' will be a feasible solution. As t increases to $+\infty$, so will the value of the maximand in this case. A cut Q that satisfies (3) and is also of unlimited span is therefore called an *unbalanced cut*.

Linear Existence Theorem for Optimal Potentials.   *Suppose the linear optimal differential problem has at least one feasible solution. Then the supremum in the problem is finite if and only if no cut is unbalanced. In that event at least one optimal solution exists.*

The necessity of the condition is clear from fact cited just before the statement of the theorem. The sufficiency will be proved here in a manner that will bring to light additional properties of the linear optimal differential problem. Alternative proof is to be provided by termination arguments for the refined versions of the algorithms in Sections 7G and 7H and by the general existence theory in Sections 8G and 8H.

Proof of Sufficiency.   The assumption is that there are no unbalanced cuts in the problem. If it is true that also there are no nonempty cuts of doubly unlimited span (i.e., with all arcs j having $d^+(j) = \infty$ and $d^-(j) = -\infty$), the extremal representation theorem in Section 6L furnishes an expression of the most general feasible solution as

$$u = \sum_{k=1}^{r} \lambda_k u_k + \sum_{l=1}^{q} \mu_l e_{N \setminus S_l} + \text{const.}$$

where each $v_k = \Delta u_k$ is an extreme differential, each cut $Q_l = [S_l, N \setminus S_l]$ is elementary and of unlimited span, $\lambda_k \geq 0$, $\sum_{k=1}^{r} \lambda_k = 1$, $\mu_l \geq 0$. (This is obtained by integrating the representation for $v = \Delta u$; since the network is connected, v determines u up to an additive constant.) The problem can then be viewed as that of maximizing the expression

$$- \sum_{k=1}^{r} \lambda_k [b \cdot u_k + c \cdot v_k] - \sum_{l=1}^{q} \mu_l [b \cdot e_{N \setminus S_l} + c \cdot e_{Q_l}]$$

over all coefficients $\lambda_k \geq 0$, $\sum_{k=1}^{r} \lambda_k = 1$, $\mu_l \geq 0$. Here $b \cdot e_{N \setminus S_l} + c \cdot e_{Q_l} \geq 0$, because there are no unbalanced cuts, so the maximum is attained by choosing the index k_0 which yields the highest of the quantities $-b \cdot u_k - c \cdot v_k$ and setting $\lambda_{k_0} = 1$, $\lambda_k = 0$ for allother k, $\mu_l = 0$ for all l. In other words, a certain potential u_{k_0} is optimal.

If there is a nonempty cut $Q = [S, N \setminus S]$ of doubly unlimited span, it must have $b \cdot e_{N \setminus S} + c \cdot e_Q = 0$ (since neither Q nor its reverse are unbalanced, by assumption). Then for every feasible solution u the potential $u + te_{N/S}$ is feasible for all $t \in R$ and gives the same value to the expression being maximized. No generality would be lost therefore if any $j \in Q$ were selected and the constraint $v(j) = 0$ added to the problem. The optimal (respectively, feasible) solutions to the original problem would be the potentials of the form $u + te_{N \setminus S}$ such that u is an optimal (respectively, feasible) solution to the restricted problem. Thus we could pass to an equivalent problem in which the arc j has span interval $[0, 0]$ and consequently belongs to no nonempty cuts of doubly unlimited span. A finite sequence of such alterations reduces the problem to one without any nonempty cuts of unlimited span, and such a problem does have an optimal solution by the argument already given.

Extreme Solutions

A feasible (or optimal) solution u is said to be *extreme* if $v = \Delta u$ is an extreme differential for the span intervals $[d^-(j), d^+(j)]$ (see Section 6L). If all the spans $d^-(j)$, $d^+(j)$ are integral, this implies v is integral, but u itself might not be integral due to the arbitrariness of constants of integration. Of course in the

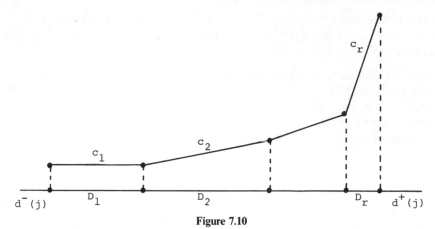

Figure 7.10

latter case there would be an integral potential u' with $\Delta u' = v$, and this would likewise be a feasible (or optimal) solution.

The argument for the existence theorem demonstrates that if the problem has at least one optimal solution x and does not possess any elementary cut $Q = [S, N \setminus S]$ of doubly unlimited span with $b \cdot e_{N \setminus S} + c \cdot e_Q = 0$ (in which event $u + t e_{N \setminus S}$ would also be optimal for all $t \in R$), then it has an extreme solution. The argument assumed connectedness of the network, but it could also be applied component by component. It also shows how a problem with cuts of the troublesome type just mentioned could be reduced to one without. It therefore yields the following result.

Integrality Theorem for Optimal Potentials.   *If the linear optimal differential problem has an optimal solution, and if the spans $d^-(j)$, $d^+(j)$, are all integral, then there is an optimal potential that is integral.*

This will be proved constructively in another manner in Section 7G.

Piecewise Linear Costs

As was seen in the case of flows, problems with piecewise linear cost functions for tension values, like the one in Figure 7.10, can be reduced to problems with linear costs. Suppose the span interval $[d^-(j), d^+(j)]$ can be partitioned into consecutive closed subintervals $D_1, \ldots, D_r$ with cost expressions

$$c_k v(j) + q_k \text{ for } v(j) \in D_k$$

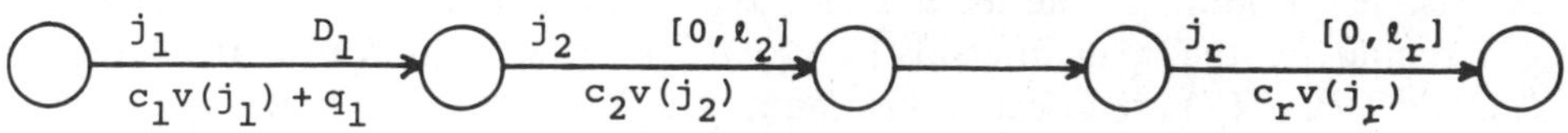

Figure 7.11

which match at the "breakpoints" between the intervals and have $c_1 < c_2 < \ldots < c_r$. The cost assigned to $v(j)$ is then the same as the maximum of $q_1 + c_1\eta_1 + \cdots + c_r\eta_r$ subject to $\eta_1 + \cdots + \eta_r = v(j)$, $\eta_1 \in D_1$ and $0 \leq \eta_k \leq l_k$ for $k = 2,\ldots,r$, where l_k is the length of D_k. Thus j can be replaced by arcs $j_1,\ldots,j_s$ in series as in Figure 7.11 to get the same effective cost structure.

7F. EXAMPLES OF OPTIMIZATION OF POTENTIALS

Optimal differential problems are encountered most typically as the "duals" of optimal distribution problems. The theory of problems dual to each other will be explained in Section 7I in a special case and in Chapter 8 in the general case. However, problems in terms of potentials do arise also in their own right.

EXAMPLE 8. (*Timing of Events*)

Consider as in Example 5 in Section 6C a network G that corresponds to a large-scale project. The nodes of G represent "events," whereas the arcs represent relationships between events. A potential u is a *schedule* for the project: $u(i)$ is the time when event i is to take place. For an arc $j \sim (i, i')$, the interval $[d^-(j), d^+(j)]$ constrains the time difference $v(j) = u(i') - u(i)$. However, in such a setting there may also be costs to take into account. For instance, the elapsed time $v(j)$ between a certain pair of events might be decreased by hiring extra personnel or increased at the expense of renting storage facilities. Such costs might well have a piecewise linear structure involving various "breakpoints" in rates. In particular, the time difference between the start and finish of the project may contribute costs of this type; if the project takes too long, a penalty proportional to the excess time may have to be paid.

The problem is to find the schedule that minimizes total cost subject to feasibility. Using the trick at the end of Section 7E for reformulating piecewise linear costs, this could be cast as a linear optimal differential problem.

EXAMPLE 9. (*Location of Facilities*)

A sequence of n facilities of some sort is to be located along a line (not necessarily straight) representing a highway, pipeline, transmission route, and so forth. The line is parameterized by the distance s measured along it from some reference point (positive in one direction, negative in the other). The coordinate s_i assigned to facility F_i must belong to a specified interval $[d^-(i), d^+(i)]$, and subject to this, one wants to minimize a total cost expression of the form

$$h(s_1,\ldots,s_n) = \sum_{1 \leq k < i \leq n} a(k, i)|s_k - s_i|,$$

where $0 \le a(k, i) < \infty$. (For certain i one might have $d^-(i) = d^+(i)$; this would mean that the location of F_i is already fixed, although it affects the costs associated with facilities yet to be located.)

Let G be the network whose node set is $N = \{0, 1, \ldots, n\}$ and whose arc set A consists of the pairs $(k, i) \in N \times N$ such that $k < i$. A potential u on G assigns facility F_i to position $s_i = u(i) - u(0)$, so that the distance $|s_k - s_i|$ can be identified with the tension magnitude $|v(j)|$ for $j = (k, i)$. Thus for arcs $j = (k, i)$ with $k > 0$ the span interval is $(-\infty, \infty)$ and the cost function is $\eta \to a(k, i)|\eta|$ (piecewise linear), whereas for arcs $j = (0, i)$ the span interval is $[d^-(i), d^+(i)]$ with zero cost.

EXAMPLE 10. (*Dual of Transportation Problem*)

Linear programming theory assigns the following dual problem to the classical transportation model in Example 1 in Section 7B:

$$\text{maximize} \quad \sum_{l \in L} n(l)u(l) - \sum_{k \in K} m(k)u(k)$$

$$\text{subject to} \quad u(l) - u(k) \le q(k, l) \quad \text{for all } (k, l) \in K \times L.$$

Here $u: K \times L \to R$ can be interpreted as a potential in the network of Figure 7.3. This is then a linear optimal differential problem having for the arc $j = (k, l)$ the span interval $[-\infty, q(k, l)]$ and cost rate $c(j) = 0$. The same pair of problems will be dual to each other in the sense explained in Section 7I and Chapter 8.

EXAMPLE 11. (*Duality for Optimal Assignments*)

The linear programming dual of the optimal assignment problem, when cast in the optimal distribution mode as in Example 2 in Section 7B, is

$$\text{maximize} \quad \sum_{l \in L} u(l) - \sum_{k \in K} u(k) \text{.}$$

$$\text{subject to} \quad u(l) - u(k) \le q(k, l) \quad \text{for all } (k, l) \in H.$$

This is a linear optimal differential problem for the network obtained from the one in Figure 7.3 by deleting arcs that correspond to pairs $(k, l) \notin H$. It is the dual problem that was introduced in Section 5E.

EXAMPLE 12. (*Max Tension*)

The max tension problem in Section 6C can be expressed as an optimal differential problem. For simplicity, suppose $N^+ = \{s\}$ and $N^- = \{s'\}$, so the problem consists of maximizing $u(s') - u(s)$ subject to $d^-(j) \le v(j) \le d^+(j)$

for all $j \in A$, where $v = \Delta u$. This corresponds to $b(s) = 1$, $b(s') = -1$, $b(i) = 0$ for all other nodes i, and $c(j) = 0 = q(j)$ for all arcs j.

An alternative reformulation, paralleling the one for the max flow problem in Example 6 in Section 7B, is to add a feedback arc $\bar{j} \sim (s', s)$ to the network. The span interval for $\bar{j}$ is $(-\infty, \infty)$, and the cost coefficients are $c(\bar{j}) = -1$, $q(\bar{j}) = 0$. In this case one takes $b(i) = 0$ for all $i \in N$. The expression to be maximized in the optimal differential problem works out to $-(-v(\bar{j})) = u(s') - u(s)$.

EXAMPLE 13.   (*Shared Cost Problem*)

This is interesting as a purely combinatorial model that can be formulated as an optimal differential problem because of integrality, much in a manner reminiscent of the matching and assignment theory in Chapter 5.

We are given two nonempty (disjoint) finite sets K and L and a subset H of $K \times L$. Two subsets S of K and T of L must be selected optimally in a certain sense. A cost $f(k)$ must be paid for each element k that is included in S, but a benefit $g(l)$ is received for each element l that is included in T. The difficulty lies in the fact that l cannot be placed in T unless all its *prerequisites*, the elements k such that $(k, l) \in H$, are in S. Let us call S and T admissible if every $l \in T$ has all its prerequisites included in S. The problem is then to maximize

$$h(S, T) = \sum_{l \in T} g(l) - \sum_{k \in S} f(k)$$

over all admissible pairs of subsets $S \subset K$ and $T \subset L$.

A network representation is furnished in Figure 7.12; each arc j has an indicated span interval $[d^-(j), d^+(j)]$ and cost coefficient $c(j)$. The corresponding optimal differential problem (with $b(i) = 0$, $p(j) = 0$) consists of maximizing

$$\sum_{l \in L} g(l)v(j_l) - \sum_{k \in K} f(k)v(j_k)$$

over all differentials v that are feasible with respect to spans. Since $v(j_k) = u(k) - u(s)$ and $v(j_l) = u(l) - u(s)$, the feasibility condition is simply this: the numbers $v(j_k)$ and $v(j_l)$ can be chosen arbitrarily subject to the constraints:

$$0 \le v(j_k) \le 1 \quad \text{and} \quad 0 \le v(j_l) \le 1,$$

$$v(j_k) \ge v(j_l) \quad \text{for all } (k, l) \in H.$$

It remains only to invoke integrality. We know there is no actual loss of generality in restricting attention to differentials v that are integral and hence

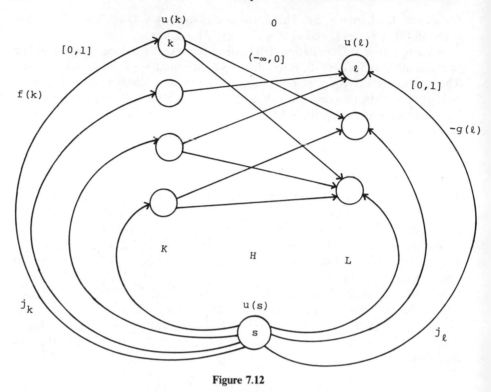

Figure 7.12

have all values $v(j_k)$ and $v(j_l)$ equal to 0 or 1. Identify S with the set of $k \in K$ having $v(j_k) = 1$ and T with the set of $l \in L$ having $v(j_l) = 1$. The constraint $v(j_k) \geq v(j_l)$ for all $(k, l) \in H$ ensures admissibility for S and T.

7G.   OPTIMAL DIFFERENTIAL ALGORITHM

Optimality of potentials can be characterized in a manner similar to that used for flows in Section 7C. Each arc j has assigned to it a certain "curve" Γ_j, where

$$
\Gamma_j = \begin{bmatrix} \text{set of all pairs } (x(j), v(j)) \in R^2 \text{ satisfying} \\ d^-(j) \leq v(j) \leq d^+(j) \text{ and having} \\ x(j) \leq c(j) \text{ if } v(j) < d^+(j), \\ \text{and } x(j) \geq c(j) \text{ if } v(j) > d^-(j). \end{bmatrix} \tag{1}
$$

The shape of Γ_j for finite $d^-(j)$, $d^+(j)$, is shown in Figure 7.13. If $d^+(j) = +\infty$, the horizontal piece on the right is absent, whereas if $d^-(j) = -\infty$, the

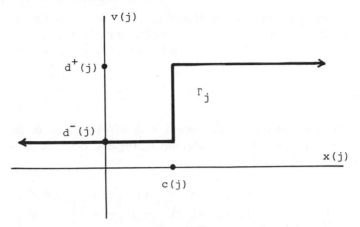

Figure 7.13

horizontal piece on the left is absent. If both $d^+(j) = +\infty$ and $d^-(j) = -\infty$, Γ_j is just a vertical line, whereas if $d^-(j) = d^+(j)$, it is a horizontal line.

Linear Optimality Theorem for Potentials. *A potential u is an optimal solution to the linear optimal differential problem if and only if there is a flow x with* div $x = b$ *that satisfies (along with $v = \Delta u$)*

$$(x(j), v(j)) \in \Gamma_j \quad \text{for all } j \in A. \tag{2}$$

Proof of Sufficiency. Certainly (2) implies $d^-(j) \leq v(j) \leq d^+(j)$ for all $j \in A$. In fact

$$d^-(j) < v(j) < d^+(j) \quad \Rightarrow \quad x(j) = c(j)$$
$$v(j) = d^+(j) \quad \Rightarrow \quad x(j) \geq c(j)$$
$$v(j) = d^-(j) \quad \Rightarrow \quad x(j) \leq c(j).$$

If $v' = \Delta u'$ is any differential satisfying $d^-(j) \leq v'(j) \leq d^+(j)$ for all arcs j, these implications show that $c(j)[v'(j) - v(j)] \geq x(j)[v'(j) - v(j)]$ for all arcs j. Therefore

$$c \cdot [v' - v] \geq x \cdot [v' - v] = -[\text{div } x] \cdot [u' - u] = -b \cdot [u' - u],$$

or equivalently,

$$-b \cdot u - c \cdot v \geq -b \cdot u' - c \cdot v'.$$

This means that no feasible solution u' can give a higher value to the maximand than u does.

Proof of Necessity. As explained in Section 7E, there is no real loss of generality in supposing that the network is connected, with $b(N) = 0$. In this setting a procedure will now be furnished that constructs for any feasible potential u either a flow x satisfying the condition in the theorem or a feasible potential u' which is "better" than u. If u is optimal, only the first alternative is possible.

Algorithm

The network is assumed to be connected, with $b(N) = 0$. Given any feasible solution u to the linear optimal differential problem, define a corresponding system of capacity intervals by

$$[c_u^-(j), c_u^+(j)] = \begin{cases} [c(j), c(j)] & \text{if } d^-(j) < v(j) < d^+(j), \\ (-\infty, c(j)] & \text{if } d^-(j) = v(j) < d^+(j), \\ [c(j), \infty) & \text{if } d^-(j) < v(j) = d^+(j), \\ (-\infty, \infty) & \text{if } d^-(j) = v(j) = d^+(j). \end{cases} \quad (3)$$

Apply any algorithm that terminates in either a solution to the feasible distribution problem for these intervals and the values $b(i)$ as supplies, that is, in a flow x that satisfies div $x = b$ and

$$c_u^-(j) \le x(j) \le c_u^+(j) \quad \text{for all } j \in A \quad (4)$$

or in a cut $Q = [S, N \setminus S]$ with $b(S) > c_u^+(Q)$. (The feasible distribution algorithm in Chapter 3 is ideal for this; see the refinement described at the end of this section.)

If the flow is obtained, terminate; u is optimal. In fact (4) is equivalent to (2), so the sufficient condition for optimality is fulfilled.

If the cut is obtained, the condition

$$b(S) > c_u^+(Q) = \sum_{j \in Q^+} c_u^+(j) - \sum_{j \in Q^-} c_u^-(j)$$

translates by way of (3), and the equation $-b(S) = b(N \setminus S)$, into

$$v(j) < d^+(j) \quad \text{for all } j \in Q^+, \qquad v(j) > d^-(j) \quad \text{for all } j \in Q^-,$$

$$0 > \sum_{i \in N \setminus S} b(i) + \sum_{j \in Q^+} c(j) - \sum_{j \in Q^-} c(j) = b \cdot e_{N \setminus S} + c \cdot e_Q. \quad (5)$$

Here Q cannot be empty because the network is connected and $b(N) = 0$.

Let

$$\alpha = \min \begin{cases} d^+(j) - v(j) & \text{for } j \in Q^+, \\ v(j) - d^-(j) & \text{for } j \in Q^-. \end{cases}$$

Then $\alpha > 0$. If $\alpha = \infty$, terminate; Q is unbalanced and the problem has [sup] $= +\infty$ (see the existence theorem in Section 7E). If $\alpha < \infty$, let $u' = u + \alpha e_{N \setminus S}$. Then u' is another feasible solution, and for $v' = \Delta u'$ one has

$$-b \cdot u' - c \cdot v' = -b \cdot u - c \cdot v - \alpha \big[b \cdot e_{N \setminus S} + c \cdot e_Q \big] > -b \cdot u - c \cdot v$$

by (5), so that u' is better than u. Iterate with the new potential u'.

Termination

The argument is similar to the one for the feasible distribution algorithm in Section 7C. By checking the presence of unbalanced cuts at the outset (see Exercise 7.16), one can eliminate the possibility that [sup] $= +\infty$ (see the existence theorem in Section 7E). A condition of commensurability then does the trick. *The algorithm must terminate if the supremum in the problem is finite and the spans $d^-(j)$, $d^+(j)$, are all commensurable along with the initial potential values $u(i)$.* For if these numbers are all multiples of $\delta > 0$, then so are α and the modified potential values $u'(i)$. All the feasible solutions generated by the algorithm therefore belong to the same commensurability class, and at every iteration the increase in the maximand is at least $\delta \varepsilon$, where ε is the smallest of the values $|b \cdot e_{N \setminus S} + c \cdot e_Q|$ corresponding to cuts $Q = [S, N \setminus S]$ with $b \cdot e_{N \setminus S} + c \cdot e_Q < 0$. It follows that only finitely many iterations can take place, because the maximand is bounded above.

If the spans are commensurable, an initial feasible solution produced by the feasible differential algorithm (starting from the zero potential, say, or from the potential associated with some complete, compatible routing with a base s, as in Section 6K) will have its values $u(i)$ in the same commensurability class.

Integrality

The commensurability argument for termination yields a second proof of the integrality theorem in Section 7E.

EXAMPLE 14

Figure 7.14 gives a network with span intervals $[d^-(j), d^+(j)]$, and coefficients $c(j)$ for the arcs and coefficients $b(i)$ for the nodes. The node numbers in Figure 7.15 are the *potential* values $u(i)$ in successive iterations, starting from the zero potential (which happens to be feasible); the intervals in Figure 7.15 are the associated *capacity* intervals $[c_u^-(j), c_u^+(j)]$. In the third iteration a flow is found that solves the feasible distribution problem for these capacity intervals and the numbers $b(i)$ in Figure 7.14 as supplies; the flow is indicated

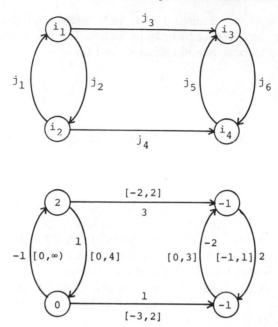

Figure 7.14

along the arcs. The node numbers in this iteration thus give a potential that solves the linear optimal differential problem.

Refinements

If the feasible distribution algorithm is used as the subroutine in each iteration, a very appealing property emerges. It turns out that the flow x at the end of the subroutine can be used to restart it the next time. At all stages one will have $(x(j), v(j)) \in \Gamma_j$, despite modifications in x and v.

Indeed, in applying the feasible distribution algorithm, one can begin with arbitrary $x(j)$ in $[c_u^-(j), c_u^+(j)]$, and the flux values will stay in these intervals to the end of the subroutine. Thus one will begin with $(x(j), v(j)) \in \Gamma_j$, and this will still hold for the modified x when the time comes to modify v. At that stage the feasible distribution algorithm will have produced a cut $Q = [S, N \setminus S]$ that separates $N^+ = \{i | y(i) < b(i)\}$ from $N^- = \{i | y(i) > b(i)\}$ and has

$$[\text{flux of } x \text{ across } Q] = c_u^+(Q) < b(S).$$

The equality here implies $x(j) = c_u^+(j)$ for all $j \in Q^+$, and $x(j) = c_u^-(j)$ for

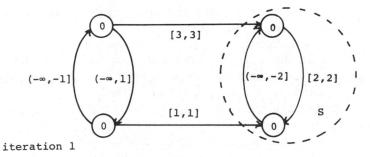

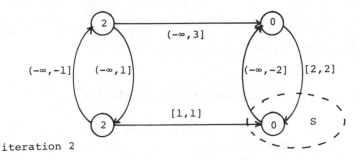

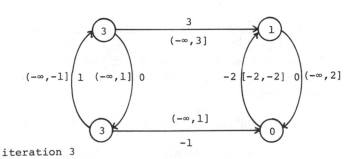

Figure 7.15

all $j \in Q^-$; hence according to the definition of $c_u^+(j)$ and $c_u^-(j)$, one will have $x(j) = c(j)$ for all $j \in Q$, as well as the properties (5) noted earlier. Thus for every $j \in Q$, the point $(x(j), v(j))$ will lie on the *vertical* segment of Γ_j in Figure 7.12, and moreover in such a position that, without leaving the segment, it can be raised by α for $j \in Q^+$ and lowered by α for $j \in Q^-$. This is exactly what happens in passing to the new potential. Therefore at the start of the next iteration of the optimal distribution algorithm one will still have $(x(j), v(j)) \in \Gamma_j$ for all $j \in A$. In particular, $x(j)$ will belong to the intervals $[c_u^-(j), c_u^+(j)]$

corresponding to the new potential u and consequently be suitable for restarting the feasible distribution subroutine.

Note that each modification of x has the effect of reducing the discrepancies $|y(i) - b(i)|$ at certain nodes i without increasing them at any other nodes. When the discrepancies have all been brought to zero, the optimality conditions div $x = b$ and $(x(j), v(j)) \in \Gamma_j$ are all satisfied.

A further refinement can be introduced in the way u and v are modified. In the following the procedure that has been given, it may happen that several iterations of the optimal distribution go by without any change in x because a cut for improving u is detected each time on the first round of the feasible distribution subroutine. Analysis of this situation brings the realization that the same sort of thing goes on as in the max tension algorithm (see Exercises 6.17 and 6.18), and it can be handled very efficiently, in the same way, by a min path implementation. This results in the "thrifty adjustment algorithm" explained in Sections 7J, 7K, and 7L (see Exercise 7.34).

7H.*   SIMPLEX ALGORITHM FOR POTENTIALS

The linear optimal differential algorithm can be realized in slightly modified detail as a spanning tree method like the one in Section 7D for flows. (Assume still that the network is connected and $b(N) = 0$.)

Algorithm

Let u be any *extreme* feasible solution to the linear optimal differential algorithm, that is, such that the set $F_v' = \{k \in A|\ d^-(k) < v(k) < d^+(k)\}$ forms a coforest, and let F be any spanning tree disjoint from F_v'. Let x be the unique flow satisfying $x(k) = c(k)$ for all $k \in A \setminus F$ and div $x = b$ (see Section 4G). Check whether

$$x(j) \le c(j) \quad \text{for all } j \in F \text{ with } v(j) = d^-(j) < d^+(j), \tag{1}$$

$$x(j) \ge c(j) \quad \text{for all } j \in F \text{ with } v(j) = d^+(j) > d^-(j). \tag{2}$$

If yes, terminate; x is an optimal solution. If no, let $\bar{j}$ be any arc for which (1) or (2) is violated. The deletion of $\bar{j}$ from F would partition the nodes of the network into two sets: the nodes reachable from the initial node of $\bar{j}$ using only arcs of $F \setminus \{\bar{j}\}$ and those reachable similarly from the terminal node of $\bar{j}$. Let S denote the first set if $\bar{j}$ violates (1) and the second if $\bar{j}$ violates (2). Then $Q = [S, N \setminus S]$ is a cut having $b \cdot e_{N \setminus S} + c \cdot e_Q < 0$ and using no arc of F besides $\bar{j}$. Define

$$\alpha = \min \begin{cases} d^+(k) - v(k) & \text{for } j \in Q^+, \\ v(k) - d^-(k) & \text{for } j \in Q^-. \end{cases}$$

Then $\alpha \geq 0$. If $\alpha = \infty$, terminate; Q is an unbalanced cut, and the supremum in the problem is $+\infty$. If $\alpha < \infty$, let $\bar{k}$ be any arc in Q for which the minimum defining α is attained (possibly $\bar{k} = \bar{j}$, although not if $\alpha = 0$). Define

$$u' = u + \alpha e_{N \setminus S}, \qquad \tilde{F} = [F \setminus \{\bar{j}\}] \cup \{\bar{k}\}.$$

Then u' is an extreme feasible solution, $\tilde{F}$ is a spanning tree disjoint from $F_{v'} = \{k \in A \mid d^-(k) < v'(k) < d^+(k)\}$, and

$$-b \cdot u' - c \cdot v' = -b \cdot u - c \cdot v + \alpha |x(\bar{j}) - c(\bar{j})| \geq -b \cdot u - c \cdot v.$$

Repeat the procedure with u', $\tilde{F}$.

Justification

The spanning tree F is chosen with the property that either $v(j) = d^+(j)$ or $v(j) = d^-(j)$ for every $j \in F$. If (1) and (2) are satisfied, then one has $x(j) \in [c_u^-(j), c_u^+(j)]$ for every $j \in A$ (see 7G(3)), and the optimality condition for u in the theorem in Section 7G is fulfilled. If (1) is violated by $\bar{j}$ and the cut $Q = [S, N \setminus S]$ is chosen in the manner described, then $\bar{j} \in Q^+$ with $x(\bar{j}) > c(\bar{j})$, but all other arcs k in Q are not in F and consequently have $x(k) = c(k)$. This implies that

$$\sum_{k \in Q^+} x(k) - \sum_{k \in Q^-} x(k) > \sum_{k \in Q^+} c(k) - \sum_{k \in Q^-} c(k),$$

or to put it another way,

$$b \cdot e_{N \setminus S} + c \cdot e_Q < b \cdot e_{N \setminus S} + x \cdot e_Q = (\operatorname{div} x) \cdot e_{N \setminus S} + x \cdot \Delta e_{N \setminus S} = 0,$$

where the difference between the two sides is $|x(\bar{j}) - c(\bar{j})|$. The same reasoning applies when (2) is violated. If $\alpha = \infty$, then $d^+(k) = +\infty$ for all $k \in Q^+$ and $d^-(k) = -\infty$ for all $k \in Q^-$, so Q is unbalanced; the supremum in the problem is then $+\infty$ by the existence theorem in Section 7E. If $\alpha < \infty$, one has $\alpha \geq 0$ because $d^-(j) \leq v(j) \leq d^+(j)$ for all $j \in A$. The choice of α ensures that the same inequalities are satisfied by $v'(j) = v(j) + \alpha e_Q(j)$. Since the minimum in the definition of α is attained for $\bar{k}$, one has either $v'(\bar{k}) = d^+(\bar{k})$ or $v'(\bar{k}) = d^-(\bar{k})$. Hence no arc in $\tilde{F} = [F \setminus \{\bar{j}\}] \cup \{\bar{k}\}$ can have $d^-(j) < v'(j) < d^+(j)$, inasmuch as no arc in F has $d^-(j) < v(j) < d^+(j)$, although $v'(j) = v(j)$ for all arcs in $F \setminus \{\bar{j}\}$ (such arcs being disjoint from Q). In other words, $\tilde{F}$ is disjoint from $F_{v'}$ as claimed. Since $\tilde{F}$ is another spanning tree (see Exercise 4.34), this implies u' is again extreme.

Initiation

To generate an extreme feasible solution u with which to start the algorithm, one can apply any method for constructing an optimal routing with base s (an arbitrary node); see Exercise 6.53.

Termination

Iterations with $\alpha = 0$ may cause trouble, because they do not involve any increase in the maximand or even any change in the feasible solution at hand. In such an iteration either $\bar{k} \in Q^+$ and $v(\bar{k}) = d^+(\bar{k})$, or instead $\bar{k} \in Q^-$ and $v(\bar{k}) = d^-(\bar{k})$. Therefore $\bar{k} \notin F'_v \cup F$. This is impossible if the extreme feasible solution u is *nondegenerate* in the sense that the complement of F'_v is itself a spanning tree (hence the only candidate for F). It follows that *if every extreme feasible solution generated by the algorithm is nondegenerate, then the algorithm must terminate in finitely many iterations*. Indeed, every iteration then has $\alpha > 0$ and leads to a higher value of the maximand. Thus none of the extreme feasible solutions generated by the algorithm can be encountered a second time. Since there are only a finite number of extreme feasible solutions, as seen in Section 6J (disregarding the arbitrary constant of integration which has no role in the maximand), the procedure cannot be repeated infinitely often.

An alternative approach is to invoke *Bland's priority rule*: number the arcs in some order ("priority"), and whenever there is more than one candidate for $\bar{j}$ or for $\bar{k}$ choose the one with highest priority. *Under Bland's priority rule there cannot be an infinite sequence of iterations with $\alpha = 0$, and the algorithm must therefore terminate*. This can be proved by arguments like those in Section 7D (Exercise 7.23).

7I.   DUALITY AND THE ELEMENTARY PROBLEMS

In solving either of the optimization problems in this chapter, one ends up with both a flow and a potential. This hints at some kind of duality. A full theory of duality will be presented in Chapter 8, but an interesting and important example is covered by the results already obtained. The trick is to reduce the linear models to the seemingly more special case where all the capacity intervals have the form $[c(j), \infty)$, and all the span intervals have the form $(-\infty, d(j)]$. The way to do this will be explained later in this section.

Let G be a connected network with supplies $b(i)$ satisfying $b(N) = 0$. Let each arc j have associated numbers $c(j)$ and $d(j)$ (finite), and let $p(j)$ and $q(j)$ be constants related by

$$q(j) = -c(j)d(j) - p(j), \qquad p(j) = -c(j)d(j) - q(j). \qquad (1)$$

Elementary Optimal Distribution Problem.   Minimize

$$\sum_{j \in A} [d(j)x(j) + p(j)]$$

over all flows x satisfying $x(j) \geq c(j)$ for all $j \in A$ and $y(i) = b(i)$ for all $i \in N$, where $y = \operatorname{div} x$.

Elementary Optimal Differential Problem.   Maximize

$$- \sum_{i \in N} b(i)u(i) - \sum_{j \in A} [c(j)v(j) + q(j)]$$

over all potentials u satisfying $v(j) \le d(j)$ for all $j \in A$, where $v = \Delta u$.

For these problems the characteristic curves Γ_j defined in 7C(1) and 7G(1) coincide:

$$\Gamma_j = \{(x(j), v(j)) \in R^2 \mid x(j) \ge c(j), v(j) \le d(j),$$

$$[x(j) - c(j)] \cdot [v(j) - d(j)] = 0\}. \quad (2)$$

(The form of Γ_j is illustrated in Figure 7.16.) Therefore the optimality theorems in Sections 7C and 7G concern the same set of conditions on x and u:

$$\text{div } x = b \quad \text{and} \quad (x(j), v(j)) \in \Gamma_j \quad \text{in (2) for all } j \in A. \quad (3)$$

These conditions thus characterize both x and u as optimal solutions to the respective problems. The problems are said to be *dual* to each other. (An economic interpretation of the duality will be furnished toward the end of Section 7K.)

This leads to an important conclusion, based on the way the algorithms in this chapter use the optimality conditions. If an elementary optimal distribution problem is solved by either the algorithms in 7C or 7D, the potential on hand at termination solves the corresponding elementary optimal differential problem. Likewise, if an elementary optimal differential problem is solved by either of the algorithms in 7G or 7H, the flow x on hand at termination solves the corresponding elementary optimal distribution problem.

Such duality is intriguing, but also very useful because it makes possible a *dual approach to computation*: an elementary problem of either type can be solved by solving its dual. The number of computational methods that can be

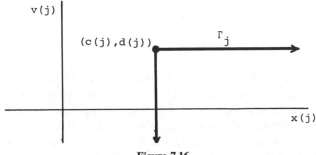

Figure 7.16

used to solve a given problem is thereby doubled. The dual approach is often quite rewarding. For instance, the algorithm given for the optimal assignment problem in 5F is really just the linear optimal differential algorithm as applied to the dual problem (cf. Example 2 in 7B, Example 11 in 7F, and Exercise 7.22). The same pattern can be followed with the classical transportation problem and its dual (cf. Example 1 in 7B, Example 10 in 7F, and Exercise 7.26). Duality will be utilized further in the general algorithms in 7J and 7M.

Another aspect of duality is seen in the relationship between the minimum and maximum.

Elementary Duality Theorem. *If both of the elementary problems possess feasible solutions, then*

$$\begin{bmatrix} minimum\ in\ the\ elementary \\ optimal\ distribution\ problem \end{bmatrix} = \begin{bmatrix} maximum\ in\ the\ elementary \\ optimal\ differential\ problem \end{bmatrix}$$

(*both finite*). *If only the first problem possesses a feasible solution, it has* [inf] = $-\infty$. *If only the second problem possesses a feasible solution, it has* [sup] = $+\infty$.

Proof.   Let x and u be feasible solutions to the respective problems. Then $x \cdot v = -b \cdot u$. Using this and (1), one can express the difference between the minimand and the maximand as

$$\sum_{j\in A} [d(j)x(j) + p(j)] - \left(-\sum_{i\in N} b(i)u(i) - \sum_{j\in A} [c(j)v(j) + q(j)] \right)$$

$$= \sum_{j\in A} [d(j)x(j) + c(j)v(j) - x(j)v(j) + p(j) + q(j)]$$

$$= \sum_{j\in A} [x(j) - c(j)][d(j) - v(j)] \geq 0.$$

Thus one has

$$\sum_{j\in A} [d(j)x(j) + p(j)] \geq -\sum_{i\in N} b(i)u(i) - \sum_{j\in A} [c(j)v(j) + q(j)],$$

$$(4)$$

and equality holds if and only if (3) is satisfied. In particular, (4) implies that $\infty > $ [inf] $\geq$ [sup] $> -\infty$, and moreover that [min] = [max] if there exist x and u satisfying (3). But the finiteness of [inf] or [sup] ensures that the problem in question has an optimal solution (cf. the existence theorems in 7A and 7E). Since optimal solutions are characterized by (3) (cf. the optimality theorems in 7C and 7G), the desired conclusion is apparent.

The reader who is familiar with linear programming will want to take note of the case where $c(j) = p(j) = q(j) = 0$. The elementary optimal distribution problem in vector notation then takes the form

$$\text{minimize } d \cdot x \text{ subject to } x \geq 0, \, Ex = b, \tag{5}$$

where E is the incidence matrix. The elementary optimal differential problem becomes

$$\text{maximize } -u \cdot b \text{ subject to } -uE \leq d. \tag{6}$$

This is a standard dual pair of linear programming problems, and the duality theorem above reduces to the standard one that would be valid for an arbitrary matrix E. The conditions $(x(j), v(j)) \in \Gamma_j$ reduce to the complementary slackness property of linear programming:

$$x \geq 0, \, uE + d \geq 0, \, [uE + d] \cdot x = 0.$$

As a matter of fact, the elementary problems can always be reformulated as (5) and (6). In the flow problem, make a change of variables $x'(j) = x(j) - c(j)$ and drop the resulting constant term from the minimand. In the potential problem, rewrite the maximand in the form $-b' \cdot u + \text{const.}$ for a new vector b' by substituting $v(j) = u(i') - u(i)$ in the term $c(j)v(j)$ when $j \sim (i', i)$. The results above therefore represent a special case of linear programming duality which has been developed in independent terms convenient for problems in networks. The formulation with possibly nonzero $c(j), p(j), q(j)$, facilitates transformations between various types of models (elementary, linear, piecewise linear, etc.) and also puts the problems in a better perspective for the general duality theory in Chapter 8.

It will now be demonstrated that any linear problem can be recast as an elementary problem. Since piecewise linear problems can always be recast as linear problems (see Sections 7A and 7E), this opens up a vast class of problems to the duality results and in particular to the dual computational approach. Two disadvantages should not be forgotten, however: the size of the network may be increased enormously and rather redundantly, and a direct interpretation for the duality may be lost. These observations provide some of the motivation for the alternative setting in Chapter 8.

Reduction of the Linear Optimal Distribution Problem to the Elementary Case

An arc j with a linear cost expression $d(j)x(j) + p(j)$ over a general capacity interval $[c^-(j), c^+(j)]$ can be replaced by one or more arcs with linear cost expressions but capacity intervals that are finite on the left, infinite on the right. There are four modes of treatment, depending on whether $c^-(j)$ and $c^+(j)$ are finite or infinite.

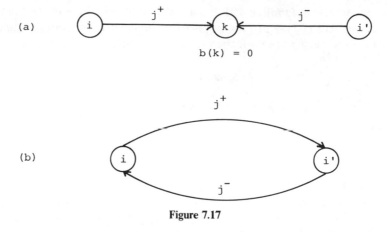

Figure 7.17

If $c^-(j)$ is finite but $c^+(j) = \infty$, the interval already has the form $[c(j), \infty)$, and there is nothing to do. Much the same is true if $c^+(j)$ is finite but $c^-(j) = -\infty$, except that j must be replaced by its reverse j'; this reverses the sign of the flux, so the capacity interval becomes $[c(j'), \infty)$ for $c(j') = -c^+(j)$, whereas the cost expression becomes $d(j')x(j') + p(j')$ for $d(j') = -d(j)$, $p(j') = p(j)$.

If both $c^+(j)$ and $c^-(j)$ are finite for an arc $j \sim (i, i')$, replace j by a pair of arcs j^+ and j^- in series as in Figure 7.17a. Note that $x(j^+) = -x(j^-)$; this number can be identified with $x(j)$. For j^+ the interval is $[c^-(j), \infty)$ and the cost is $d(j)x(j^+) + p(j)$, whereas for j^- the interval is $[-c^+(j), \infty)$ and the cost is 0.

Finally, if $c^+(j)$ and $c^-(j)$ are both infinite, replace j by a pair of arcs j^+ and j^- in opposition as in Figure 7.17b. The original flux corresponds to $x(j^+) - x(j^-)$. For j^+ the interval is $[0, \infty)$ and the cost is $d(j)x(j^+) + p(j)$; also for j^- the interval is $[0, \infty)$, but the cost is $-d(j)x(j^-)$.

Reduction of the Linear Optimal Differential Problem to the Elementary Case

The procedure is similar for replacing arcs with linear costs $c(j)v(j) + q(j)$ over span intervals $[d^-(j), d^+(j)]$ by arcs with span intervals that are infinite on the left, finite on the right.

If $d^-(j) = \infty$ but $d^+(j)$ is finite, the interval already has the form $(-\infty, d(j)]$. If $d^-(j)$ is finite and $d^+(j) = \infty$, replace j by its inverse j' with interval $(-\infty, -d^-(j)]$ and cost expression $-c(j)v(j') + q(j)$.

If $d^+(j)$ and $d^-(j)$ are both finite for an arc $j \sim (i, i')$, replace j by a pair of arcs j^+ and j^- in opposition as in Figure 7.17b. Then $v(j^+) = -v(j^-)$, and this value corresponds to the original tension $v(j)$. For j^+ the span interval is $(-\infty, d^+(j)]$ and the cost expression is $c(j)v(j^+) + q(j)$, whereas for j^- the interval is $(-\infty, -d^-(j)]$ and the cost is 0.

If $d^+(j)$ and $d^-(j)$ are both infinite, replace $j \sim (i, i')$ by arcs j^+ and j^- in series as in Figure 7.17a. This represents the tension $v(j)$ by the difference $v(j^+) - v(j^-)$. For both j^+ and j^- the interval is $(-\infty, 0]$; the cost expression for j^+ is $c(j)v(j^+) + q(j)$, but for j^- it is $-c(j)v(j^-)$.

7J.   THRIFTY ADJUSTMENT ALGORITHM

Duality makes it possible to solve the linear optimal distribution problem by means of the optimal differential algorithm, and with a number of refinements this becomes a very attractive and efficient approach. Here we look at the resulting procedure in polished form as it applies to a pair of *elementary* problems dual to each other in the sense of Section 7I. The generalization to nonelementary linear problems is covered in Exercises 7.36 and 7.41 and to piecewise linear problems in Section 9D.

The procedure can essentially be obtained just by using the feasible distribution algorithm as the subroutine in the optimal differential algorithm, as already discussed in the last part of Section 7F. The flow x produced by the subroutine in one round is taken as the initial flow in the next round, when the subroutine is restarted. It turns out then that the rule for improving the potential bears a close resemblance to the pattern in the max tension algorithm, and like the latter it can be implemented as the min path algorithm (for greatest efficiency, in the form of the node-scanning refinement in Section 6E). The verification of these connections, important as they are to understanding the proper place of the procedure in the theory, is left to Exercise 7.34, however.

Instead, the algorithm is presented and justified in a direct manner that leads to an illuminating interpretation in Section 7K. In this interpretation the potential values represent "prices" that rise in response to market forces, and the paths used to adjust surpluses and shortages in supply are always chosen to be as "thrifty" as possible. Refinements in implementation are described in Section 7L.

Recall that the capacity intervals in the elementary optimal distribution problem are

$$[c^-(j), c^+(j)] = [c(j), \infty) \quad \text{for all } j \in A, \tag{1}$$

whereas the span intervals in the elementary optimal differential problem are

$$[d^-(j), d^+(j)] = (-\infty, d(j)] \quad \text{for all } j \in A. \tag{2}$$

Corresponding to any potential u whose differential v is feasible with respect to the spans (2), we introduce the modified capacity intervals

$$[c_u^-(j), c_u^+(j)] = \begin{cases} [c(j), \infty) & \text{if } v(j) = d(j), \\ [c(j), c(j)] & \text{if } v(j) < d(j). \end{cases} \tag{3}$$

Likewise, corresponding to any flow x that is feasible with respect to the capacities (1), we introduce the modified span intervals

$$[d_x^-(j), d_x^+(j)] = \begin{cases} (-\infty, d(j)] & \text{if } x(j) = c(j), \\ [d(j), d(j)] & \text{if } x(j) > c(j). \end{cases} \tag{4}$$

These modified intervals specialize the ones used in the optimal distribution and differential algorithms of Sections 7C and 7G to the case of the "elementary" curves Γ_j in 7I(2) (Figure 7.16). We have

$$(x(j), v(j)) \in \Gamma_j \Leftrightarrow x(j) \in [c_u^-(j), c_u^+(j)] \Leftrightarrow v(j) \in [d_x^-(j), d_x^+(j)]. \tag{5}$$

Statement of the Algorithm

At the start of the general iteration, one has a flow x and a potential u such that the equivalent conditions (5) hold for all $j \in A$. (Then u is a feasible solution to the elementary optimal differential problem, but x is not necessarily a feasible solution to the elementary optimal distribution problem, since it is not necessarily true that $y(i) = b(i)$ for all i, where $y = \text{div } x$.)

Let $N^+ = \{i|b(i) > y(i)\}$ and $N^- = \{i|b(i) < y(i)\}$. If $N^+ = \emptyset = N^-$, terminate; one has $\text{div } x = b$, and it follows from $(x(j), v(j)) \in \Gamma_j$ for all $j \in A$ that x solves the elementary optimal differential problem (see Section 7I). Otherwise $N^+ \neq \emptyset \neq N^-$ (under the blanket assumption that $b(N) = 0$).

Apply the min path algorithm to N^+, N^-, and the span intervals (4) with starting potential $u_0 = u$. (Do this even though u_0 may not under this choice be constant on N^+ and N^-. The min path algorithm operates entirely in terms of the shifted span values obtained by subtracting off $v_0 = \Delta u_0$, and by our assumption that the conditions in (5) hold for $v_0 = v$, these will have the form

$$[d_x^-(j) - v_0(j), d_x^+(j) - v_0(j)]$$

$$= \begin{cases} (-\infty, d(j) - v_0(j)] & \text{if } x(j) = c(j), \\ [0,0] & \text{if } x(j) > c(j), \end{cases} \tag{6}$$

where $d(j) - v_0(j) \geq 0$. In particular, these intervals will contain 0, just as always in the execution of the min path algorithm.)

If the min path subroutine ends with a cut $Q: N^+ \downarrow N^-$ of infinite span with respect to (4), terminate; Q is an unbalanced cut. Then the elementary optimal differential problem has $[\sup] = +\infty$ (see the existence theorem in Section 7E), whereas the elementary optimal distribution problem has no feasible solutions (see the duality theorem in Section 7I).

Otherwise the min path subroutine ends with a path $P: N^+ \to N^-$ and a potential $u_0 + w$. Replace $u = u_0$ by $u = u_0 + w$, and for the new intervals (3) calculate

$$\alpha = \min \begin{cases} x(j) - c(j) & \text{for } j \in P^-, \\ b(i) - y(i) & \text{for the initial node } i \text{ of } P, \\ y(i') - b(i') & \text{for the terminal node } i' \text{ of } P. \end{cases} \tag{7}$$

Then $0 < \alpha < \infty$. Replace x by $x + \alpha e_P$, thereby reducing the discrepancy between the values of y and b at the initial and terminal nodes of P by α (and deleting the corresponding node from N^+ or N^- if the discrepancy has been removed entirely). Iterate with this new x and u, which will still satisfy the conditions in (5) for all $j \in A$.

Initiation

One can take u initially to be any potential whose differential is feasible with respect to the intervals (2), that is, any potential that is a feasible solution to the elementary optimal differential problem, and then choose $x(j)$ arbitrarily in the intervals (3); see (5). Such a potential can be determined by the general algorithms of Chapter 6 if necessary, but often one is immediately available. For instance, $u \equiv 0$ works if $d(j) \geq 0$ for all j, as is common in some types of applications. Another easy case is encountered when the elementary optimal distribution problem has been obtained by reformulating (as in Section 7I) a more general linear or piecewise linear problem with capacity intervals $[c^-(j), c^+(j)]$ that are all bounded (see Exercise 7.39).

Observe that regardless of the construction of u, the intervals (3) always contain $c(j)$, so that it is always possible to choose $x(j) = c(j)$ for all $j \in A$. However, other choices of x sometimes seem more natural (see Example 15 in this section).

Justification of the Algorithm

First we verify that when the subroutine terminates with a cut, the cut is unbalanced. A cut $Q: N^+ \downarrow N^-$ that is of infinite span with respect to the intervals (4) is also of infinite span with respect to the intervals (2) and has $x(j) = c(j)$ for all $j \in Q^- = Q$. Furthermore it is of the form $[S, N \setminus S]$ with $b(i) \leq y(i)$ for all $i \in N \setminus S$, at least one of the inequalities being strict. Hence

$$b \cdot e_{N \setminus S} + c \cdot e_Q = b \cdot e_{N \setminus S} + x \cdot \Delta e_{N \setminus S}$$

$$= b \cdot e_{N \setminus S} - (\operatorname{div} x) \cdot e_{N \setminus S}$$

$$= (b - y) \cdot e_{N \setminus S} < 0,$$

and this means Q is an unbalanced cut for the elementary optimal differential problem.

The other thing we must check is that the conditions in (5) are preserved in nonterminal iterations.

In applying the min path algorithm with u_0 not necessarily constant on N^+ or on N^-, we are certainly still solving the max tension and min path problems for the shifted intervals (6) (because everything is the same as if we had started with these intervals, which all contain 0, and the zero potential). The solutions obtained are w and P, respectively, and the common maximum and minimum is a certain $\beta \geq 0$, which is also the constant value assumed by w on N^-. In particular, we have

$$d_x^-(j) - v_0(j) \leq \Delta w(j) \leq d_x^+(j) - v_0(j) \quad \text{for all } j \in A, \tag{8}$$

and the expressions

$$\begin{bmatrix} \text{spread of } \Delta w \\ \text{from } N^+ \text{ to } N^- \end{bmatrix} = \begin{bmatrix} \text{spread of } \Delta w \\ \text{relative to } P \end{bmatrix} = \sum_{j \in P^+} \Delta w(j) - \sum_{j \in P^-} \Delta w(j),$$

and

$$\begin{bmatrix} \text{upper span of } P \\ \text{with respect to the intervals (6)} \end{bmatrix}$$

$$= \sum_{j \in P^+} [d_x^+(j) - v_0(j)] - \sum_{j \in P^-} [d_x^-(j) - v_0(j)]$$

$$= d^+(P) - [\text{spread of } v_0 \text{ relative to } P] \tag{9}$$

have the same value β. This equality implies by (8) that $\Delta w(j) = d_x^+(j) - v_0(j)$ for all $j \in P^+$ and $\Delta w(j) = d_x^-(j) - v_0(j)$ for all $j \in P^-$. Thus in terms of the new potential $u = u_0 + w$ and its differential $v = v_0 + \Delta w$, we have

$$d_x^-(j) \leq v(j) \leq d_x^+(j) \quad \text{for all } j \in A, \tag{10}$$

$$v(j) = \begin{cases} d_x^+(j) & \text{for all } j \in P^+, \\ d_x^-(j) & \text{for all } j \in P^-. \end{cases} \tag{11}$$

From (10) we conclude that the conditions (5) are still satisfied by the old x and new u; in particular, $c_u^-(j) \leq x(j) \leq c_u^+(j)$ for the new intervals (3). On the other hand, (11) tells us that $v(j) = d(j)$ for all $j \in P$, and hence $[c_u^-(j), c_u^+(j)] = [c(j), \infty)$ for all $j \in P$. It also says $x(j) > c(j)$ for all $j \in P^-$. Thus $0 < \alpha < \infty$ in (7) as claimed, and

$$c_u^-(j) \leq x(j) + \alpha e_P(j) \leq c_u^+(j) \quad \text{for all } j \in A.$$

The conditions in (5) therefore will continue to hold also when x is replaced by $x + \alpha e_P$.

Termination

If numbers $b(i)$, $c(j)$ and initial flux values $x(j)$ are commensurable, finite termination is assured, exactly as in the feasible distribution algorithm. However, under certain refinements of the procedure it is possible to establish finite termination without resorting to any commensurability assumption; see Section 7L.

EXAMPLE 15

Figure 7.18 shows a network with supplies $b(i)$, costs $d(j)$, and capacity intervals $[c(j), \infty)$. These define the elementary optimal distribution problem obtained by the reduction procedure in Section 7I from the linear problem in Example 7 in Section 7C (Figure 7.7). The thrifty adjustment algorithm can be applied with $u \equiv 0$ and $x \equiv 0$ initially. (*Note:* This illustrates what was said earlier about the choice $x \equiv c$ sometimes not being very "natural." The choice $x \equiv 0$ corresponds here to a flow in the original problem in Example 7, whereas $x \equiv c$ does not.)

The first diagram in Figure 7.19 shows the shifted span intervals 7J(6) corresponding to $x \equiv 0$ and $u_0 = u \equiv 0$, as well as the routing θ and $w(i)$ values calculated from these by the min path algorithm. An elementary flow along the θ-path $P: N^+ \to N^-$ with flux $\alpha = 1$ is added to the zero flow, and w is added to the zero potential, to get the new x and u in the second diagram. This u now becomes u_0; the new shifted span intervals are displayed in the third diagram along with the corresponding new θ and w. The potential in the fourth diagram is obtained by adding this w to the potential in the second diagram, and the flow is obtained by adding a flux of $\alpha = 1$ along the θ-path P. One more iteration produces the flow x and potential u in the last diagram. Now $N^+ = \emptyset = N^-$, so x and u are optimal solutions to the respective problems.

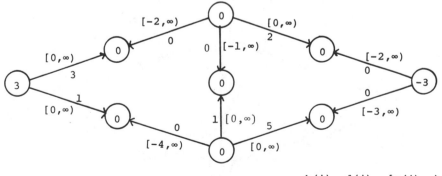

$$b(i), \ d(j), \ [c(j), \infty)$$

Figure 7.18

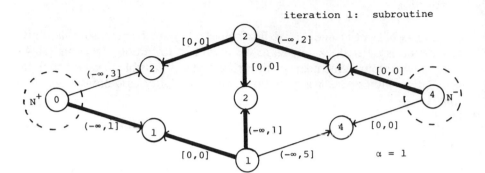

iteration 1:   subroutine

$\alpha = 1$

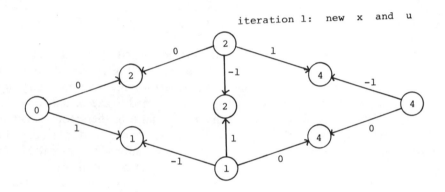

iteration 1:   new   x   and   u

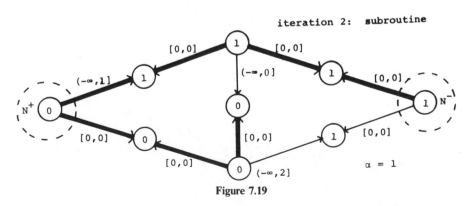

iteration 2:   subroutine

$\alpha = 1$

Figure 7.19

iteration 2: new x and u

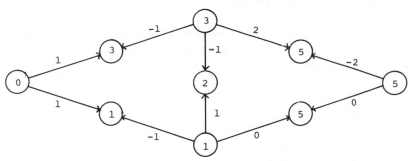

iteration 3: subroutine

N^+

N^-

$\alpha = 1$

iteration 3: new x and u

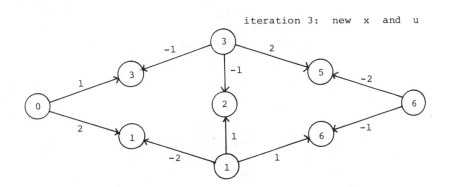

7K.*   INTERPRETATIONS

Behind the name of the thrifty adjustment algorithm is a very appealing economic interpretation. To get at this, we first need to look more closely at the nature of the min path problem solved in each iteration.

Let $\bar{u}$ denote the potential at the very start of the algorithm. The u_0 in the general iteration satisfies $u_0 \equiv \bar{u}$ on N^+ and $u_0 \equiv \bar{u} + \bar{\beta}$ on N^- for a certain $\bar{\beta} \geq 0$ because u_0 is obtained by a series of steps in which a non-negative function w that vanishes on N^+ and is constant on N^- is added to the potential at hand. (The sets N^+ and N^- in earlier iterations have to include the ones in the iteration under discussion, since nodes are never added to them, only deleted.) In terms of $\bar{u}$, one has in 7J(9)

$$[\text{spread of } v_0 \text{ relative to } P] = u_0(i') - u_0(i)$$

$$= \bar{u}(i') - \bar{u}(i) + \bar{\beta},$$

where i still denotes the initial node of P and i' the terminal node. Furthermore from 7J(4)

$$d_x^+(P) = \begin{cases} d \cdot e_P & \text{if } x(j) > c(j) \text{ for all } j \in P^-, \\ +\infty & \text{otherwise.} \end{cases}$$

Therefore in minimizing 7J(9), we are actually minimizing the expression

$$m(P) = \bar{u}(i) + \sum_{j \in P^+} d^+(j) - \sum_{j \in P^-} d^-(j) - \bar{u}(i') \quad \text{for } P\colon i \to i' \quad (1)$$

over all the paths $P\colon N^+ \to N^-$ that are *flow augmenting* for x with respect to the given capacity intervals 7J(1) in the elementary optimal distribution problem. The paths in this class are just the ones that would be of interest if we were simply applying the feasible distribution algorithm, or in other words, the available *adjustment paths*. The α corresponding to such a P in 7J(7) is also the value that would be specified by that algorithm: the greatest flux along P that can be added to x without violating feasibility with respect to capacities or overshooting the goal of reducing the discrepancy between y and b.

The only difference then between the thrifty adjustment algorithm and the feasible distribution algorithm, as far as the flows are concerned, is that *instead of accepting any of the adjustment paths that could be used at a particular stage, we insist on one that is "best" in the sense of minimizing $m(P)$ in* (1). Note that the minimum of $m(P)$ is $\mu = \bar{\beta} + \beta$, where $\bar{\beta}$ is the minimum of 7J(9) as earlier. Of course $\bar{\beta}$ is itself the sum of the β values in previous iterations (if any). Thus in each iteration the minimum of $m(P)$ increases by $\beta \geq 0$ over what it was in the preceding iteration.

The meaning of the quantity (1) is the key to an economic interpretation. As in Section 7C, let us think of x as representing a flow of a certain commodity.

At each node i, $b(i)$ is the given supply and $u(i)$ the current market price enjoyed by the commodity. A price system $\bar{u}$ has been specified at the start of the procedure. In these terms $m(P)$ is the *marginal cost* associated with buying another unit of the commodity at the initial node of P (where there is a surplus), transporting it along P, and then selling it at the terminal node of P (where there is a shortage). In minimizing this marginal cost, we are looking among the available adjustment paths for one that is as *thrifty* as possible. We then send as much along it as feasible. If this does not finish erasing the surpluses and shortages, we repeat the procedure for the updated collection of available adjustment paths. The marginal costs μ of the paths used from iteration to iteration form a *nondecreasing* sequence (the increment in the minimum being $\beta \geq 0$). This means that adjustments are ever more expensive; the cheapest possibilities are used first.

The discussion up to this point demonstrates that the thrifty adjustment algorithm could actually be formulated without reference to any changes in potential values, namely, as a form of the feasible distribution algorithm having an extra criterion for path selection in terms of a fixed $\bar{u}$. But limiting ourselves to such a framework of flows could be tantamount to throwing away information about the corresponding optimal differential problem and its solution. Such information will somehow have to be generated anyway, at least implicitly, because of the way potentials enter into the min path subroutine. The duality between potential changes and flow changes has an interesting interpretation too.

Remember that the flow $\bar{x}$ at the start of the algorithm cannot be chosen arbitrarily but must satisfy $(\bar{x}(j), \bar{v}(j)) \in \Gamma_j$, where $\bar{v} = \Delta\bar{u}$. As we have seen in Section 7C, this means that $\bar{x}(j)$ minimizes the net cost $[d(j) - \bar{v}(j)]x(j)$ that can be incurred in transporting through j a quantity $x(j)$ which is feasible with respect to the capacity interval $[c(j), \infty)$ assigned to j. From the fact that this holds for all $j \in A$, we know $\bar{x}$ and $\bar{u}$ would solve the respective problems if the vector $\bar{b} = \operatorname{div}\bar{x}$ agreed with b. In providing us at the end of each iteration with a new price system u such that $(x(j), v(j)) \in \Gamma_j$ for all $j \in A$, the algorithm ensures that this property is maintained at all times: the flow x is always optimal for the modified optimal distribution problem in which b is replaced by the current $y = \operatorname{div} x$.

As for the price system u, recall that it is constructed from the flow x and potential u_0 at the *start* of the iteration as $u_0 + w$, where w is a solution to the max tension problem for N^+, N^-, and the shifted span intervals 7J(6). Since $\Delta w = \Delta u - \Delta u_0 = v - v_0$, this can be translated as saying that u solves the following "extended" problem (see Exercise 6.13): over all potentials u with differential v feasible with respect to the span intervals $[d_x^-(j), d_x^+(j)]$ and with $u - u_0$ constant on N^+ and on N^-, maximize

$$h(u) = \begin{bmatrix} \text{constant value} \\ \text{of } u - u_0 \text{ on } N^- \end{bmatrix} - \begin{bmatrix} \text{constant value} \\ \text{of } u - u_0 \text{ on } N^+ \end{bmatrix}.$$

Of course u_0 can be replaced by $\bar{u}$ here, since $u_0 - \bar{u}$ is itself constant on N^+ and N^-, and the maximum in question is then the value $\mu = \bar{\beta} + \beta$ instead of β.

Thus for the flow x at the start of the iteration, we seek the maximum μ for which there is a price system u agreeing with $\bar{u}$ on N^+ and $\bar{u} + \mu$ on N^- and satisfying the equivalent conditions in 7J(5). This gives the highest price difference between N^+ and N^- that is compatible with local optimality of x relative to the net cost of the flux in each arc. In passing to a new u of such type in the course of each iteration, we are merely allowing the market price at each node i to rise by a natural amount $w(i) \geq 0$ which reflects the increasing expense of tapping the surpluses in N^+ to transfer an additional quantity of the commodity to i.

The elementary optimal differential problem can itself be given meaning in this setting. Each potential u handled by the algorithm is a feasible solution, since $v(j) \leq d(j)$ certainly holds at all times for all arcs. The inequalities $v(j) \leq d(j)$ can be interpreted as in Example 6 in Section 6C. They are the natural conditions that must be satisfied by any system of market prices consistent with the given transportation costs ($d(j)$ for using j forward, ∞ for using j backward).

Let us place ourselves in the position of a monopoly that operates the market for the commodity in question at every node and also receives all payments for transportation. We are free to set the prices $u(i)$, subject to the fundamental constraints just mentioned, and then reap the benefits of the economic response in sales and transport requisitions. We imagine this response to be based on the decisions of one or many individuals in the business of buying quantities of the commodity at various locations, shipping to adjacent locations, and then selling. These individuals refuse to lose money; they must at least break even. They are further restrained by the requirement that the flux $x(j)$ in the arc j must be at least $c(j)$, an amount prescribed by circumstances or exterior agreements. The possible responses are thus the flows x that satisfy $x(j) \geq c(j)$ for all arcs with net cost coefficient $d(j) - v(j) = 0$ but have $x(j) = c(j)$ for all arcs with $d(j) - v(j) > 0$. The divergence of x does not matter because the entrepeneurs are not constrained by the total stocks $b(i)$ they possess at each node. The market allows them to take care of any differences through sales and purchases.

For any flow x and its divergence y, we have $u \cdot y + v \cdot x = 0$; adding this to the maximand in the elementary optimal differential problem and using the relation $-q(j) = p(j) + c(j)d(j)$ in 7I(1), we can rewrite the maximand as

$$\sum_{i \in N} u(i)[y(i) - b(i)] + \sum_{j \in A} [d(j)x(j) + p(j)]$$

$$- \sum_{j \in A} [d(j) - v(j)][x(j) - c(j)]. \qquad (2)$$

For the flows that are the possible responses to a price system u, the third sum vanishes. The rest then gives the benefits we, the monopoly, could reap by choosing u (remember that the value of this expression is actually independent of the *particular* response x): the first sum is the net profit in market transactions, and the second is the income in transportation payments. Thus in this problem we are interested in choosing a (feasible) price system that maximizes the economic advantage that can be gained from the situation.

The effect of an iteration of the thrifty adjustment algorithm is to add to the maximand expression (2) the quantity

$$\sum_{i \in N} w(i)[y(i) - b(i)] = \beta \sum_{i \in N^-} [y(i) - b(i)] > 0$$

when $\beta > 0$. (Here x denotes the flow at the start of the iteration. Recall that $w(i) = 0$ for all $i \in N^+$ and $w(i) = \beta$ for all $i \in N^-$, whereas $y(i) - b(i) = 0$ for all $i \notin N^+ \cup N^-$. The third sum in (2) stays at zero through all stages of the procedure.) Iterations with $\beta = 0$ have $w \equiv 0$ and do not change u. All changes in the price system are therefore in the direction of greater economic advantage.

7L.*   MULTIPATH IMPLEMENTATION

The min path subroutine in the thrifty adjustment algorithm can be executed in multipath form (see the end of Section 6E). This will produce a multirouting Θ that represents *all* the paths suitable for use in a given iteration, or with a lexicographical device (likewise explained at the end of Section 6E), all the suitable paths that satisfy the secondary criterion of employing as few arcs as possible. Such an approach can lead to a sharp improvement in theoretical efficiency.

We have seen in the preceding section that the thrifty adjustment algorithm is a version of the feasible distribution algorithm in which not every flow-augmenting path $P: N^+ \to N^-$ is acceptable but only the ones that actually minimize the quantity $m(P)$ in 7J(1) (where $\bar{u}$ is the starting potential). The minimum μ of this quantity never goes down, but there might be a sequence of iterations in which it stays at the same level. In such iterations one has $\beta = 0$ and $w \equiv 0$ in the subroutine; the potential u remains unchanged. In trying to estimate the number of steps the algorithm could require to reach termination, a lot turns out to depend on how this situation is handled.

Another observation about the nature of the acceptable paths will reveal the way to proceed. In the justification of the algorithm in Section 7J, it was seen that a path $P: N^+ \to N^-$ solves the subproblem tackled by the min path subroutine if and only if $v(j) = d_x^+(j)$ for all $j \in P^+$ and $v(j) = d_x^-(j)$ for all $j \in P^-$. Here x is the flow at the start of the iteration, but v is the differential of the potential generated by the iteration. Recalling the definition of the intervals $[d_x^-(j), d_x^+(j)]$ in 7J(4), we see that this condition can be expressed as

$$x(j) < c_u^+(j) \quad \text{for all } j \in P^+ \quad \text{and} \quad x(j) > c_u^-(j) \quad \text{for all } j \in P^-.$$

In other words, *the paths eligible for use in any iteration are precisely the paths* $P: N^+ \to N^-$ *that are flow augmenting for the given flow* x *with respect to the capacity intervals* $[c_u^-(j), c_u^+(j)]$ *corresponding to the potential* u *generated by the iteration.*

In a sequence of iterations at the same μ level, the intervals $[c_u^-(j), c_u^+(j)]$, like u, remain unchanged. The thrifty adjustment algorithm thus reduces temporarily to a phase of the basic feasible distribution algorithm and could suffer from the same foibles. We know from Section 3J that these foibles can be eliminated by the Dinic-Karzanov method, and it remains only to investigate the best way of adapting this to the present context.

The trick is to implement the min path subroutine in every iteration in the multipath, two-criterion form (described near the end of Section 6E) that constructs a multirouting Θ with the following property: the Θ-paths from N^+ to N^- are exactly the paths, among the ones already characterized as acceptable, that also use the fewest arcs possible. Then equivalently according to what has just been observed, a path $P: N^+ \to N^-$ will be a Θ-path if and only if it is a flow-augmenting path for the (given) flow x relative to the (currently generated) intervals $[c_u^-(j), c_u^+(j)]$, and no other such path uses fewer arcs. As a matter of fact this property will hold not just for paths terminating in N^-, but for $P: N^+ \to i$ with i arbitrary, since after all, the contraction of Θ would not be affected (although possibly truncated) if $\{i\}$ were taken as N^- instead.

Such a multirouting Θ provides the framework for applying the Dinic-Karzanov procedure in the form given in Section 3J. (The set N_k still consists of the nodes reached from N^+ by Θ-paths employing k arcs.) This procedure has the effect of perhaps lumping together several iterations of the thrifty adjustment algorithm, all at the same μ level, into a single "grand iteration." The possibility remains of a sequence of "grand iterations" at the same μ level. But in this case, because we would really be back entirely in the context of the feasible distribution algorithm for fixed intervals $[c_u^-(j), c_u^+(j)]$, we know from Section 3J that the number of arcs used by the paths in successive iterations must increase. Hence the number of "grand iterations" at any μ level is certainly bounded by $|N| - 1$.

Termination and Order of Complexity

Suppose for simplicity that the network in question is a digraph, so $|A| \leq |N|^2$. The min path subroutine is then of complexity $0(|N|^2)$, even if carried out in special multipath form specified earlier (see Section 6E). Each "grand iteration" consists of this subroutine followed by an application of the Dinic-Karzanov method, which is likewise $0(|N|^2)$, as seen in Section 3J. Since there are at most

$|N| - 1$ "grand iterations" at any μ level, we are able to estimate the complexity of calculations at any μ level as $O(|N|^3)$.

In order to obtain bounds on the algorithm as a whole, it is necessary to estimate how many times μ can increase in the course of the procedure. Each μ value encountered is in particular of the form $m(P)$ in 7K(1) for a certain elementary path $P: \overline{N}^+ \to \overline{N}^-$, where $\overline{N}^+$ and $\overline{N}^-$ are the initial N^+ and N^- (and $\overline{u}$ is the initial potential). The number of such paths is finite, so this does at least show that only finitely many increases in μ are possible. Thus the algorithm is sure to terminate in a finite number of steps when implemented in the multipath form.

Unfortunately, the number of elementary paths $P: \overline{N}^+ \to \overline{N}^-$ can be very large, and the $m(P)$ values associated with them can be all different (e.g., if the numbers $d(j)$ are all mutually incommensurable and $\overline{u} \equiv 0$). This does not necessarily mean that the number of μ levels actually encountered will be equally large because the paths yielding μ are also required to be flow augmenting with respect to the flow x that is on hand. It does show that a more subtle argument would be needed to establish a better overall bound, if such is possible at all.

Application to the Optimal Distribution Algorithm

Similar ideas can be invoked when the optimal distribution algorithm is implemented in terms of the feasible differential algorithm (and its special restart provision), as described near the end of Section 7C. Recall that the feasible differential algorithm uses a min path subroutine (with respect to certain modified span intervals) to find a circuit $P: \overline{i} \to \overline{i}$ with $d_x^+(P) < 0$, and then it replaces x by $x + \alpha e_P$ for a certain $\alpha > 0$. This subroutine can be executed in multipath form just as before to get a multirouting Θ that represents *all* the eligible circuits that could be used to improve x.

Actually Θ is not quite a multirouting in the usual sense, since the node $\overline{i}$ has a double role; ordinary routings and multiroutings do not give circuits. But this is a minor matter of terminology. It is easy to adapt the method to this situation in order to make larger-scale improvements in x that do not involve just one circuit at a time. Then the number of "grand iterations" that can take place at the same node $\overline{i}$ before a definite rise in $\rho(\overline{i})$ occurs is bounded by $|N|$. Arguing as before, one can show that $\rho(\overline{i})$ will be lifted to 0 in finitely many steps. A new choice of $\overline{i}$ can then be made. The procedure terminates when $\rho \equiv 0$. (Remember from Section 7C that ρ values never rise during the procedure; this carries over of course to the multipath implementation, see Exercise 7.44.) Thus when the optimal distribution algorithm is refined in this manner, it always terminates in a finite number of iterations, regardless of commensurability.

The resulting algorithm resembles the thrifty adjustment algorithm in many ways. The kind of thing that goes on in the latter for paths $P: N^+ \to N^-$ occurs instead for paths $P: \overline{i} \to \overline{i}$, but for one choice of $\overline{i}$ after another. This feature may seem to reflect badly on the ultimate prospects for efficiency of the optimal distribution algorithm: it has to do over and over again what the thrifty adjustment algorithm (actually a refinement of the optimal differential algorithm) does just once. But in practice the picture is not so clear; the phases for individual i may turn out to be relatively brief. Another fact to take into account is that in the optimal distribution algorithm, the flow x is always a feasible solution to the corresponding problem, whereas in the thrifty adjustment algorithm it is feasible only at the end, when it is also optimal. In some applications, especially large problems where one might be tempted to stop short of optimality, this could be an important consideration.

7M.*   OUT-OF-KILTER ALGORITHM

Yet another method can be used to solve an elementary pair of problems in duality. This is the out-of-kilter algorithm, which has the interesting feature that the two problems are treated with complete symmetry. Although in the long run it does not seem to offer any advantage over the methods already described, even when loaded with refinements, it is nevertheless important for theoretical as well as esthetic reasons. An expanded form for piecewise linear problems in Section 9F will be shown in Chapter 10 to be valid for a much more general kind of optimization, not restricted to network models. The general out-of-kilter algorithm will also be seen to embody the flow rectification algorithm in Section 3K and the tension rectification algorithm in Section 6I as special cases (Exercises 9.14 and 9.15).

The method can be applied to any flow x satisfying $\operatorname{div} x = b$ and any potential u. There is no initial requirement of feasibility of either x or u. As in all the other approaches, the curves Γ_j play a crucial role. An arc j is said to be *out of kilter* if $(x(j), v(j)) \notin \Gamma_j$, where $v = \Delta u$.

Algorithm

If no arc is out of kilter, terminate; x and u are optimal solutions to the two problems (see Section 7I). Otherwise select any out-of-kilter arc $\bar{j}$. Paint the arcs of the network as follows, according to the location of each point $(x(j), v(j))$ relative to Γ_j (see Figure 7.20):

$$
\begin{array}{ll}
\text{black} & \text{if } x(j) > c(j) \text{ and } v(j) < d(j) \\
\text{red} & \text{if } x(j) = c(j) \text{ and } v(j) < d(j) \\
\text{green} & \text{if } x(j) > c(j) \text{ and } v(j) = d(j) \\
\text{white} & \text{if } x(j) = c(j) \text{ and } v(j) = d(j), \text{ or} \\
& \text{if } x(j) < c(j), \text{ or if } v(j) > d(j)
\end{array}
$$

(Note that all out-of-kilter arcs, including $\bar{j}$, will be black or white.) Apply

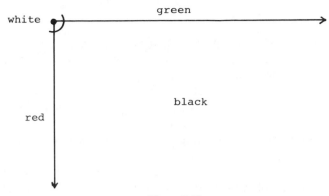

Figure 7.20

Minty's algorithm (the special form of the painted network algorithm in Section 2H) to determine either a compatible elementary circuit P containing $\bar{j}$ or a compatible cut Q containing $\bar{j}$.

In the case of the circuit P, let

$$\alpha = \min \begin{cases} x(j) - c(j) & \text{for } j \in P^- \\ c(j) - x(j) & \text{for } j \in P^+ \text{ with } v(j) < d(j) \\ c(\bar{j}) - x(\bar{j}) & \text{if } \bar{j} \in P^+ \text{ with } v(\bar{j}) \le d(\bar{j}). \end{cases} \tag{1}$$

Then $\alpha > 0$. If $\alpha = \infty$ (i.e., the minimum is vacuous), terminate; P is an unbalanced circuit for the elementary optimal distribution problem. (Therefore by the theorem in Section 7I this problem either has no feasible solution or has [inf] $= -\infty$; the dual problem has no feasible solution.) Otherwise replace x by $x + \alpha e_P$, and return to the beginning with this new flow (and the same u).

In the case of a cut $Q = [S, N \setminus S]$, let

$$\alpha = \min \begin{cases} d(j) - v(j) & \text{for } j \in Q^+ \\ v(j) - d(j) & \text{for } j \in Q^- \text{ with } x(j) > c(j) \\ v(\bar{j}) - d(\bar{j}) & \text{if } \bar{j} \in Q^- \text{ with } x(\bar{j}) \ge c(\bar{j}). \end{cases} \tag{2}$$

Then $\alpha > 0$. If $\alpha = \infty$ (i.e., the minimum is vacuous), terminate; Q is an unbalanced cut for the elementary optimal differential problem. (Therefore by the theorem in Section 7I, this problem either has no feasible solution or has [sup] $= \infty$; the dual problem has no feasible solution.) Otherwise replace u by $u + \alpha e_{N \setminus S}$ (and v by $v + \alpha e_Q$), and return to the beginning with this new potential (and the same x).

In either case the distance of each new point $(x(j), v(j))$ from the corresponding Γ_j is no greater than for the previous point. In particular, the arcs that were previously in kilter are still in kilter.

Justification

Black arcs correspond to points $(x(j), v(j))$ that in Figure 7.20 could be moved either "north" or "west," at least a little bit, without the distance from Γ_j being increased. For red arcs it is "north" or "south"; for green arcs, "east" or "west"; and for white arcs, "east" or "south." Motions "north" or "south" correspond to the circuit case, with $j \in P^+$ or $j \in P^-$, respectively. Motions "east" or "west" correspond to the cut case, with $j \in Q^+$ or $j \in Q^-$, respectively. Compatibility with the painting therefore ensures in either case that the change from x to $x + \alpha e_P$ or from v to $v + \alpha e_Q$ will not move any point $(x(j), v(j))$ farther from Γ_j. The amount of motion (if there is motion at all) will be α units in one of the directions in question. The formulas for α in the two cases give the greatest shift possible without the distance from some Γ_j beginning to increase and without "overshooting" (i.e., moving farther than would be necessary to bring the selected out-of-kilter arc $\bar{j}$ into kilter, if this can indeed be accomplished by the choice of α).

If the minimum (1) is vacuous, P must be a *positive* circuit ($e_P \geq 0$) consisting of green arcs with $v(j) = d(j)$ and white arcs with $v(j) \geq d(j)$; the arc $\bar{j}$ in P must be white with $v(\bar{j}) > d(\bar{j})$. These conditions imply $d \cdot e_P < v \cdot e_P = 0$. It follows that P is an unbalanced circuit, the capacity intervals in the elementary optimal distribution problem being of the form $[c(j), \infty)$ for all j.

Similarly, if the minimum in (2) is vacuous, Q must be a *negative* cut ($e_Q \geq 0$) consisting of red arcs with $x(j) = c(j)$ and white arcs with $x(j) \leq c(j)$; the arc $\bar{j}$ in Q must be white with $x(\bar{j}) < c(\bar{j})$. Then $c \cdot e_Q < x \cdot e_Q = x \cdot \Delta e_{N \setminus S} = -(\text{div } x) \cdot e_{N \setminus S} = -b \cdot e_{N \setminus S}$. Thus $c \cdot e_Q + b \cdot e_{N \setminus S} < 0$, and since the span intervals in the elementary optimal differential problem are all of the form $(-\infty, d(j)]$, this says Q is an unbalanced cut.

Termination

Very mild conditions will guarantee that only a finite number of iterations can take place before the algorithm terminates in one of the three possible ways. The general argument will be postponed, however, until the expanded version of the algorithm is treated in Section 9G. Here we shall look only at a particularly noteworthy case.

Let us suppose that $x(j) \geq c(j)$ and $v(j) \geq d(j)$ for all $j \in A$ initially; that is, x and u start out as feasible solutions to the respective problems. (We have already supposed div $x = b$.) From what has been observed, this property will be maintained throughout the computation. Out-of-kilter arcs can only be black. For such an arc j, the differences $x(j) - c(j)$ and $d(j) - v(j)$ are both positive; they remain unchanged in iterations where j does not belong to the circuit or cut that is produced, but if j does belong (as is true in particular for the selected out-of-kilter arc $\bar{j}$) exactly one of them decreases by α.

Thus every iteration furnishes a definite improvement in "out-of-kilterness" in this situation. A simple commensurability condition is enough, for example, to imply that only a finite number of such improvements is possible. Suppose the numbers $c(j)$, $d(j)$, and initial values $x(j)$, $v(j)$, are all commensurable. These are all multiples then of a certain $\delta > 0$, and so must be α as defined by (1) or (2). The commensurability class is therefore preserved in each iteration, and the decrease in $x(\bar{j}) - c(\bar{j})$ or $d(\bar{j}) - v(\bar{j})$ is in every instance at least δ. Hence in this case there can be only a finite number of iterations before all arcs are in kilter.

Comparison with Other Methods

In the case where x starts out feasible (but u is not necessarily feasible) the out-of-kilter algorithm can be regarded simply as the version of the optimal distribution algorithm in Section 7C (applied to an *elementary* pair of problems in duality) that uses the tension rectification algorithm in Section 6I, rather than the feasible differential algorithm, as the main subroutine (Exercise 7.46; see Exercise 7.47 for a dual characterization). This fact, besides placing the method in perspective, points the way to a number of refinements. First there are the refinements of the tension rectification algorithm, already explained in Section 6I, that are capable of putting it on a par with the feasible differential algorithm in terms of efficiency. (These can be adapted even to the case where x is not feasible initially.) Then there is the possibility of multipath implementation, which can be introduced very much in the manner described in the last part of Section 7L (see Exercise 7.50).

The method produced by these refinements is closely akin to the sharpened form of the optimal distribution algorithm with feasible differential subroutine, which as will be recalled from the end of Section 7C, likewise works under the assumption that $\operatorname{div} x = b$ and never increases the distance of $(x(j), v(j))$ from Γ_j. The two approaches are nevertheless distinct and in a certain limited sense "opposite" (see the remarks in Section 6I). The out-of-kilter algorithm does of course have the advantage that *neither x nor u* need be feasible initially. (When x is not feasible, something like the flow rectification algorithm in Section 3K comes into play; see Exercise 7.47.) However, this is not especially important because the construction of a feasible x or u generally appears to involve a computation of lower order of complexity than the construction of optimal solutions, at least for network problems. (This will not necessarily hold for the generalizations in Chapter 10.)

There are also parallels between the out-of-kilter algorithm and the version of the optimal differential algorithm that uses the basic feasible distribution algorithm as the main subroutine. The latter method too can be formulated in terms of the painting that corresponds to Figure 7.20, but with $(x(j), v(j)) \in \Gamma_j$ for $j \in A$ and the painted network algorithm (for paths from a set N^+ to a set N^-) employed in place of Minty's algorithm (see Exercise 7.33). In effect this method relaxes the constraint div $x = b$ in order to have all arcs always in kilter instead. When refinements are added, we eventually end up with the sharp form of the thrifty adjustment algorithm.

Anyway it seems inescapable that the simplicity and symmetry of formulations in terms of paintings, compatible paths and cuts, must be abandoned sooner or later in the quest for highly efficient implementation. Of all the optimization methods described, only the thrifty adjustment algorithm remains really intuitive and capable of natural interpretation, even when brought to the peak of efficiency.

7N.*   EXERCISES

7.1. (*Unbalanced Circuits*). Show that the unbalanced circuits for the linear optimal distribution problem are the circuits of negative upper span for the following system of span intervals: the arc j is assigned

$$
\begin{aligned}
&[d(j), d(j)] && \text{if } c^-(j) = -\infty \text{ and } c^+(j) = +\infty, \\
&(-\infty, d(j)] && \text{if } c^-(j) > -\infty \text{ and } c^+(j) = +\infty, \\
&[d(j), \infty) && \text{if } c^-(j) = -\infty \text{ and } c^+(j) < +\infty, \\
&(-\infty, \infty) && \text{if } c^-(j) > -\infty \text{ and } c^+(j) < +\infty.
\end{aligned}
$$

(The presence of such a circuit can therefore be detected by trying to solve the feasible differential problem for these intervals by the methods in Chapter 6.)

7.2. (*Extreme Solutions*). Assume in the linear distribution problem that there are no circuits of doubly unlimited capacity, so that extreme feasible solutions exist. Let $x_1, \ldots, x_q$ be the optimal solutions which are extreme flows, and let $P_1, \ldots, P_t$ be the elementary circuits of unlimited capacity (if any) such that $d \cdot e_{P_l} = 0$. Prove that x is an optimal solution if and only if it can be expressed in the form

$$
x = \sum_{k=1}^{q} \lambda_k x_k + \sum_{l=1}^{t} \mu_l e_{P_l},
$$

where $\lambda_k \geq 0, \sum_{k=1}^{q} \lambda_k = 1, \mu_l \geq 0$.

(*Hint.* Milk this out of the proof of the existence theorem in Section 7A.)

7.3. (*Piecewise Linear Costs*). Reformulate as a linear optimal distribution problem the problem of minimizing

$$
\|x\|_1 = \sum_{j \in A} |x(j)|
$$

over all flows x that satisfy $|x(j)| \leq c(j)$ for all $j \in A$ and div $x = b$.

7.4. (*Integrality in Piecewise Linear Problems*). Prove that if the costs in an optimal distribution problem are piecewise linear with all breakpoints (including $c^-(j)$ and $c^+(j)$, when finite) integral, then an integral optimal solution exists if the problem has an optimal solution at all.

(*Hint.* Reformulate as a linear problem.)

7.5. (*Transportation Problem*). Show that the classical transportation problem in Section 7B has an optimal solution if it has a feasible solution, regardless of the signs of the coefficients $q(k, l)$.

(*Note.* It has a feasible solution if and only if total supply equals total demand.)

7.6. (*Max Flow*). Demonstrate that the max flow problem with general node sets N^+ and N^- can be formulated as a linear optimal flow problem (an optimal distribution problem in the augmented network).

7.7. (*Optimality Conditions and Duality*). Suppose in the linear optimal distribution problem that every arc j has both $c^+(j)$ and $c^-(j)$ finite and $p(j) = 0$. Show that for all values $x(j) \in [c^-(j), c^+(j)]$ and $v(j) \in (-\infty, \infty)$ one has

$$x(j)[d(j) - v(j)]$$
$$\geq \min\{c^+(j)[d(j) - v(j)], c^-(j)[d(j) - v(j)]\},$$

and equality holds if and only if $(x(j), v(j))$ belongs to the set Γ_j in 7C(1). Prove from this that the inequality

$$\sum_{j \in A} x(j)d(j) \geq - \sum_{i \in N} b(i)u(i)$$
$$+ \sum_{j \in A} \min\{c^+(j)[d(j) - v(j)],$$
$$c^-(j)[d(j) - v(j)]\} \tag{1}$$

is valid for every feasible solution x and every potential u and its differential v, and equality holds if and only if $(x(j), v(j)) \in \Gamma_j$ for every $j \in A$. Deduce by way of the existence and optimality theorems in Sections 7A and 7C that if the minimum in the linear optimal distribution problem is finite, it equals the maximum in the problem

$$\text{maximize} - \sum_{i \in N} b(i)u(i)$$
$$+ \sum_{j \in A} \min\{c^+(j)[d(j) - v(j)], c^-(j)[d(j) - v(j)]\} \tag{2}$$

over all potentials u, where $v = \Delta u$; in fact if x is an optimal solution, the potentials that solve (2) are precisely the ones such that $(x(j), v(j))$ $\in \Gamma_j$ for all $j \in A$.

(*Remark.*   This demonstrates that the potential with which the optimal distribution algorithm (or the simplex algorithm for flows) terminates solves (2). The dual problem (2) is the one assigned by the theory in Chapter 8 in a framework that does not require $c^+(j)$ and $c^-(j)$ to be finite.)

7.8.   (*Max Flow Algorithm*). Show that when the optimal distribution algorithm is applied to the max flow problem, reformulated as in Example 6 in Section 7B, it reduces to the max flow algorithm.

7.9.   (*Optimal Distribution Algorithm*). Solve the linear optimal distribution problem in Figure 7.21, following the pattern in Example 7 in Section 7C. The figure indicates the capacity intervals $[c^-(j), c^+(j)]$ and cost coefficients $d(j)$; one has $p(j) = 0$ for all $j \in A$ and $b(i) = 0$ for all $i \in N$. Start with the zero flow.

7.10.   (*Primal Transportation Algorithm*). Describe in direct terms the method for solving the classical transportation problem in Section 7B that is obtained by applying the optimal distribution algorithm. (Much as was the case when the max flow algorithm was specialized to the matching algorithm in Section 5B, this can be done in terms of a rectangular tableau.)

7.11.   (*Optimal Assignment Algorithm*). Show that the optimal distribution algorithm, when applied to the optimal assignment problem as formulated in Example 2 in Section 7B, yields the following method as an alternative to the Hungarian assignment algorithm.

Given any assignment M, either there exist numbers $u(k)$ for $k \in K$ satisfying

$$u(k') \leq u(k) + q(k, M(k)) \quad \text{for all } k, k', \text{ in } K$$

or there is a sequence $k^1, \ldots, k^m$ in K such that

$$q(k^1, M(k^1)) + q(k^2, M(k^2)) + \cdots + q(k^m, M(k^m))$$

$$> q(k^m, M(k^1)) + q(k^1, M(k^2)) + \cdots + q(k^{m-1}, M(k^m)).$$

(A special constructive procedure can be distilled for deciding between these alternatives.) in the first case, M is an optimal assignment. In the second case, an assignment M' which is better than m is obtained by setting

$$M'(k^m) = M(k^1), \ M'(k^1) = M(k^2), \ldots, M'(k^{m-1}) = M(k^m),$$

$$M'(k) = M(k) \quad \text{for all other } k \in K.$$

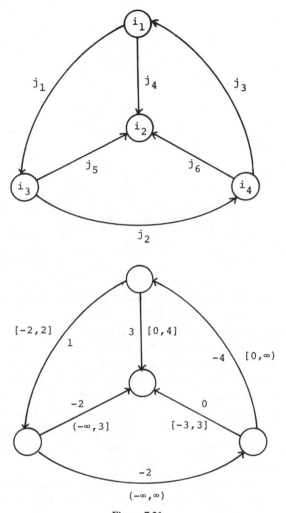

Figure 7.21

7.12. (*Linear Optimal Circulation Problem*). Show that any linear optimal distribution problem can be reduced to one in which all supplies $b(i)$ are zero by passing to the augmented network. Thus there would be no real loss of generality in assuming $b = 0$.

7.13. (*Simplex Algorithm for Flows*). Use the simplex method in Section 7D to solve the linear optimal circulation problem in Figure 7.21 (see Exercise 7.9 for explanation). Start with the extreme flow x and associated spanning tree F in Figure 7.22. (One iteration will suffice.)

7.14. (*Simplex Algorithm for Flows*). Prove that the algorithm in Section 7D reduces to a particular "realization" of the linear optimal distribution

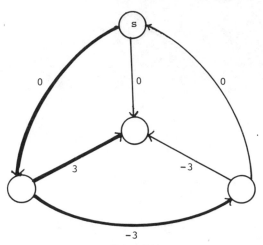

Figure 7.22

algorithm in the case where all extreme feasible solutions are nondegenerate (as defined in Section 7D).

7.15. (*Degeneracy*). Show that for the optimal assignment problem formulated as a linear optimal distribution problem as in Section 7B and with all pairs (k, l) compatible, *none* of the extreme feasible solutions is nondegenerate (as defined in Section 7D). Argue from the fact that extreme feasible solutions must be integral.

(*Remark.* Much the same phenomenon can be expected, although not with such inevitability, in the case of the classical transportation problem and other models involving bipartite networks, and this mars the effectiveness of the simplex algorithm.)

7.16. (*Unbalanced Cuts*). Show that the unbalanced cuts for the linear optimal differential problem are the cuts $Q = [S, N \setminus S]$ with the property that $b(S) > c^+(Q)$ for the following system of capacity intervals $[c^-(j), c^+(j)]$:

$$
\begin{array}{ll}
[c(j), c(j)] & \text{if } d^-(j) = -\infty \text{ and } d^+(j) = +\infty, \\
(-\infty, c(j)] & \text{if } d^-(j) > -\infty \text{ and } d^+(j) = +\infty, \\
[c(j), \infty) & \text{if } d^-(j) = -\infty \text{ and } d^+(j) < +\infty, \\
(-\infty, \infty) & \text{if } d^-(j) > -\infty \text{ and } d^+(j) < +\infty.
\end{array}
$$

Such a cut, if one exists, can therefore be constructed by the feasible distribution algorithm or the flow rectification algorithm.

7.17. (*Max Tension*). Formulate the max tension problem with general node sets N^+ and N^- as a linear optimal differential problem in the augmented network.

7.18. (*Piecewise Linear Costs*). Reformulate as a linear optimal differential problem the problem of minimizing

$$\|u\|_1 = \sum_{i \in N} |u(i)|$$

over all potentials u such that $v = \Delta u$ satisfies $d^-(j) \le v(j) \le d^+(j)$ for all $j \in A$.

(*Hint.* Pass to the augmented network, and use the trick at the end of Section 7E to reduce piecewise linear costs to linear costs.)

7.19. (*Integrality in Piecewise Linear Problems*). Prove that if the costs in an optimal differential problem are piecewise linear with all breakpoints (including $d^-(j)$ and $d^+(j)$, when finite) integral, then an integral optimal solution exists if the problem has an optimal solution at all.

(*Hint.* Reformulate in terms of linear costs.)

7.20. (*Optimality Conditions and Duality*). Suppose in the linear optimal differential problem that every arc j has both $d^-(j)$ and $d^+(j)$ finite and $q(j) = 0$. Show that for all values $x(j) \in (-\infty, \infty)$ and $v(j) \in [d^-(j), d^+(j)]$ one has

$$v(j)[x(j) - c(j)]$$
$$\le \max\{d^+(j)[x(j) - c(j)], d^-(j)[x(j) - c(j)]\},$$

and equality holds if and only if $(x(j), v(j))$ belongs to the set Γ_j in 7G(1). Prove from this that the inequality

$$\sum_{j \in A} \max\{d^+(j)[x(j) - c(j)], d^-(j)[x(j) - c(j)]\}$$

$$\ge -\sum_{i \in N} b(i)u(i) - \sum_{j \in A} c(j)v(j) \qquad (3)$$

holds for every feasible solution u and every flow x satisfying $\operatorname{div} x = b$, and it holds as an equation if and only if $(x(j), v(j)) \in \Gamma_j$ for every $j \in A$. Deduce by way of the existence and optimality theorems in Sections 7E and 7G that if the maximum in the linear optimal differential problem is finite, it equals the minimum in the problem.

$$\text{minimize} \sum_{j \in A} \max\{d^+(j)[x(j) - c(j)], d^-(j)[x(j) - c(j)]\} \qquad (4)$$

over all flows x satisfying $\operatorname{div} x = b$; in fact if u is an optimal solution, the flows that solve (4) are precisely the ones such that $(x(j), v(j)) \in \Gamma_j$ for all $j \in A$.

(*Remark.* This demonstrates that the flow with which the optimal differential algorithm (or the simplex algorithm for potentials) terminates solves (4). The dual problem (4) is the one assigned to this case by Chapter 8.)

7.21. (*Optimal Differential Algorithm*). Solve the linear optimal differential problem in Figure 7.23 by the algorithm of Section 7G. (Use the notation of Example 14 in Section 7G.) Figure 7.23 gives the intervals $[d^-(j), d^+(j)]$ and coefficients $c(j)$ along the arcs and the coefficients $b(i)$ at the nodes.

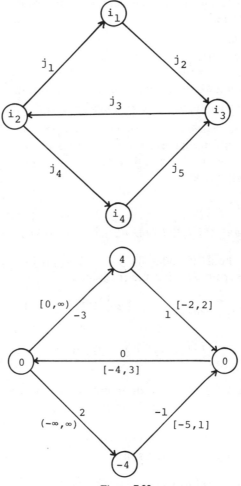

Figure 7.23

7.22. (*Hungarian Assignment Algorithm*). Show that the optimal differential algorithm, when applied to the dual of the optimal assignment problem (see Example 11 in Section 7F), reduces in effect to the algorithm in Section 5F.

(*Note.* This illustrates the dual approach explained in Section 7I for computing solutions to elementary optimal distribution problems, here the optimal assignment problem as formulated in Example 2 in Section 7B.)

7.23. (*Simplex Algorithm for Potentials*). Prove the assertion at the end of Section 7H, that Bland's priority rule ensures termination.

(*Hint.* Imitate the argument for Bland's rule in Section 7D.)

7.24. (*Simplex Algorithm for Potentials*). Solve the optimal differential problem in Figure 7.23 (see Exercise 7.21 for explanation) by the simplex method in Section 7H. Start with the extreme potential u and associated spanning tree F indicated in Figure 7.24.

7.25. (*Elementary Algorithms*). Write out the special forms of the optimal distribution and differential algorithms that solve the dual pair of elementary problems in Section 7I.

7.26. (*Elementary Algorithms*). Figure 7.25 gives the data for an elementary optimal distribution problem and an elementary optimal differential problem having $c(j) = p(j) = q(j) = 0$ for all j; the numbers at the nodes are $b(i)$, and the numbers at the arcs are $d(j)$. First solve the two problems by the optimal distribution algorithm, and then solve them by the optimal differential algorithm.

7.27. (*Dual Transportation Algorithm*). Describe in detail the dual approach to solving the classical transportation problem.

(*Hint.* The problem in question is "elementary" in the sense of Section 7I. Apply the optimal differential algorithm to its dual; see Example 10 in Section 7F.)

7.28. (*Elementary Problems*). Reformulate the linear optimal distribution problem in Exercise 7.9 as an elementary problem in the sense of Section 7I.

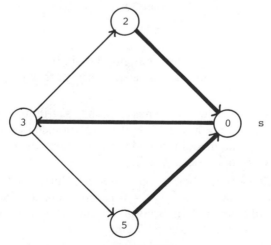

Figure 7.24

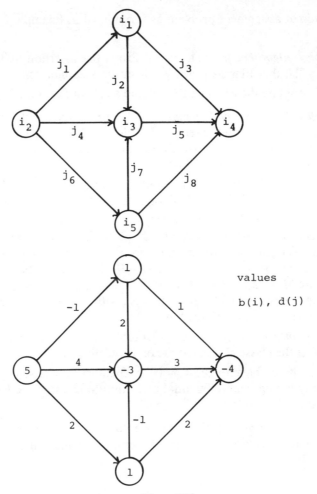

Figure 7.25

7.29. (*Elementary Problems*). Reformulate the linear optimal differential problem in Exercise 7.21 as an elementary problem in the sense of Section 7I.

7.30. (*Duality*). Show that when a linear optimal distribution problem with $c^+(j)$ and $c^-(j)$ finite and $p(j) = 0$ is reformulated as an elementary problem by the method in Section 7I, the dual thereby obtained is equivalent to the one in Exercise 7.7.

7.31. (*Duality*). Show that when a linear optimal differential problem with $d^+(j)$ and $d^-(j)$ finite and $q(j) = 0$ is reformulated as an elementary problem by the method in Section 7I, the dual thereby obtained is equivalent to the one in Exercise 7.20.

7.32. (*Optimal Assignment Duality*). Show that the duality theorem for optimal assignments in Section 5E is a special case of the elementary duality theorem in Section 7I (see Example 11 in Section 7F).

7.33. (*Optimal Differential Algorithm*). Suppose that the optimal differential algorithm is executed with the feasible distribution algorithm as the main subroutine (the flow left over from one iteration being used to start the next; see the end of Section 7G). Show that the resulting procedure is equivalent to the following.

　　The general step begins with a flow x and a potential u such that $(x(j), v(j)) \in \Gamma_j$ for all $j \in A$. If the sets $N^+ = \{i \mid b(i) > y(i)\}$ and $N^- = \{i \mid b(i) < y(i)\}$ satisfying $N^+ = \varnothing = N^-$, terminate; the optimality conditions are satisfied. Otherwise $N^+ \ne \varnothing \ne N^-$. In this event paint the arcs of the network as indicated in Figure 7.26, and apply the painted network algorithm. If a compatible path $P: N^+ \to N^-$ is obtained, let

$$\alpha = \min \begin{cases} c_u^+(j) - x(j) & \text{for } j \in P^+, \\ x(j) - c_u^-(j) & \text{for } j \in P^-, \\ b(i) - y(i) & \text{for the initial node } i \text{ or } P, \\ y(i') - b(i') & \text{for the terminal node } i' \text{ of } P, \end{cases}$$

where $[c_u^-(j), c_u^+(j)]$ is the interval defined in 7E(3). Then $0 < \alpha < \infty$. Replace x by $x + \alpha e_P$, and return to the beginning. If a cut $Q: N^+ \downarrow N^-$ is obtained, let

$$\alpha = \min \begin{cases} d^+(j) - v(j) & \text{for } j \in Q^+, \\ v(j) - d^-(j) & \text{for } j \in Q^-. \end{cases}$$

Then $\alpha > 0$. If $\alpha = \infty$, terminate; Q is unbalanced. Otherwise reduce u by $u + \alpha e_{N \setminus S}$ (and v by $v + \alpha e_Q$), and return to the beginning.

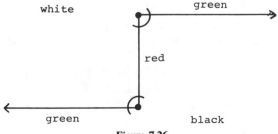

Figure 7.26

7.34. (*Thrifty Adjustment Algorithm*). Demonstrate that the thrifty adjustment can be regarded simply as a method of implementing the optimal differential algorithm with the feasible distribution algorithm as the main subroutine.

(*Hint.* Specialize the formulation of the algorithm given in the preceding exercise to the case of an elementary pair of problems, and investigate what happens in a succession of iterations where only u changes and x stays the same. The situation will be just like the one in Exercise 6.18, where the max tension algorithm was seen to be implementable as the min path algorithm).

(*Remark.* It should be clear from this how to extend the thrifty adjustment algorithm to solve a general linear optimal differential problem, not necessarily elementary.)

7.35. (*Thrifty Adjustment Algorithm*). Use the thrifty adjustment algorithm to solve the elementary pair of problems in Exercise 7.26 (Figure 7.25). Start with the potential u and flow x indicated in Figure 7.27. (Verify first that these do satisfy the requirements for initiation.)

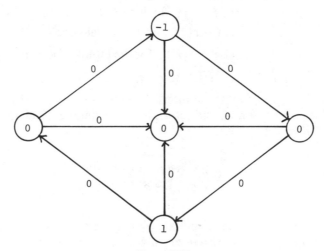

Figure 7.27

7.36. (*Thrifty Adjustment Algorithm for Linear Optimal Distribution Problem*) Justify the following extended version of the thrifty adjustment algorithm as a means of solving linear optimal distribution problems that are not necessarily elementary. (Give arguments parallel to the ones in Section 7J.)

In the general step, x and u satisfy $(x(j), v(j)) \in \Gamma_j$ for every $j \in A$, where Γ_j given by 7C(1) (see Figure 7.6). If the sets $N^+ = \{i \mid b(i) > y(i)\}$ and $N^- = \{i \mid b(i) < y(i)\}$ are empty, terminate; the optimality conditions are fulfilled. Otherwise $N^+ \neq \emptyset \neq N^-$. In this case apply the min path algorithm to N^+, N^-, and the span intervals in 7C(3), taking u as the initial potential u_0. If a cut $Q: N^+ \downarrow N^-$ of unlimited span for these intervals is obtained, terminate; the optimal distribution problem has no feasible solution. Otherwise replace $u = u_0$ by $u = u_0 + w$ (the potential generated by the min path algorithm), and for the Θ-path $P: N^+ \to N^-$ calculate

$$
\alpha = \min \begin{cases} c^+(j) - x(j) & \text{for } j \in P^+, \\ x(j) - c^-(j) & \text{for } j \in P^-, \\ b(i) - y(i) & \text{for the initial node } i \text{ of } P, \\ y(i') - b(i') & \text{for the terminal node } i' \text{ of } P. \end{cases}
$$

Then $0 < \alpha < \infty$. Replace x by $x + \alpha e_P$, and return to the beginning of the procedure with the new x and u.

7.37. (*Thrifty Adjustment Algorithm for Linear Optimal Distribution Problems.*) Demonstrate that the procedure in Exercise 7.36 can be derived by reformulating the linear optimal distribution problem as an elementary problem (in the manner of Section 7I), applying the version of the algorithm already available for the elementary problem and then translating what happens back into the original setting.

7.38. (*Thrifty Adjustment Algorithm for Linear Optimal Distribution Problems*). Use the algorithm in Exercise 7.36 to solve the linear optimal distribution problem in Example 7 in Section 7C directly.

(*Note.* It is interesting to compare the result with the solution of the same problem, recast in elementary form, in Example 15 in Section 7J. This illustrates the idea in Exercise 7.37.)

7.39. (*Thrifty Adjustment Algorithm*). Show that when the thrifty adjustment algorithm is applied to an elementary pair of problems obtained by reformulating a linear optimal differential problem whose intervals $[c^-(j), c^+(j)]$ are all bounded, an initial feasible potential u is immediately available.

(*Remark.* When such a linear problem is solved *directly* by generalized form of the algorithm in Exercise 7.36, one can in fact start with *any* potential u. For each $v(j)$ it will be possible to choose $x(j)$ such that $(x(j), v(j)) \in \Gamma_j$, because the curves Γ_j in 7C(1) and Figure 7.6 will not be missing either of the vertical segments.)

7.40. (*Thrifty Adjustment Algorithm*). What procedure, studied earlier, does the thrifty adjustment algorithm reduce to when $d(j) = 0$ for all $j \in A$ and the initial potential is taken to be the zero potential?

7.41. (*Thrifty Adjustment Algorithm for Linear Optimal Differential Problems*). Justify the following version of the thrifty adjustment algorithm that can be applied directly to any linear differential problem, not necessarily elementary.

(*Hint.* Extend the arguments in Section 7J.)

In the general step, x and u satisfy $(x(j), v(j)) \in \Gamma_j$ for every $j \in A$, where Γ_j is given by 7G(1) (see Figure 7.12). If the sets $N^+ = \{i|\ b(i) > y(i)\}$ and $N^- = \{i|\ b(i) < y(i)\}$ are empty, terminate; the optimality conditions are fulfilled. Otherwise $N^+ \neq \emptyset \neq N^-$. In this case apply the min path algorithm to the span intervals

$$
[d_x^-(j), d_x^+(j)] = \begin{cases} [d^+(j), d^+(j)] & \text{if } x(j) > c(j), \\ [d^-(j), d^+(j)] & \text{if } x(j) = c(j), \\ [d^-(j), d^-(j)] & \text{if } x(j) < c(j), \end{cases}
$$

taking u as the initial potential u_0. If a cut $Q: N^+ \downarrow N^-$ of unlimited span for these intervals is obtained, terminate; the optimal differential problem has $[\sup] = \infty$. Otherwise replace $u = u_0$ by $u = u_0 + w$ (the potential generated by the min path algorithm), and for the Θ-path P: $N^+ \rightarrow N^-$, and calculate

$$
\alpha = \min \begin{cases} c(j) - x(j) & \text{for } j \in P^+ \text{ with } x(j) < c(j), \\ x(j) - c(j) & \text{for } j \in P^- \text{ with } x(j) > c(j), \\ b(i) - y(i) & \text{for the initial node } i \text{ of } P, \\ y(i') - b(i') & \text{for the terminal node } i' \text{ of } P. \end{cases}
$$

Then $0 < \alpha < \infty$. Replace x by $x + \alpha e_P$, and return to the beginning of the procedure with the new x and u.

7.42. (*Thrifty Adjustment Algorithm for Linear Optimal Differential Problems*). Show that the procedure in the preceding exercise can be derived by reformulating the linear optimal differential problem (in the manner of Section 7I), applying the version of the algorithm already available for the elementary problem, and then translating what happens back into the original setting.

7.43. (*Thrifty Adjustment Algorithm for Linear Optimal Differential Problems*). Use the procedure in Exercise 7.41 to solve the problem in Example 14 in Section 7G.

7.44. (*Multipath Implementation*). Verify that when the optimal distribution algorithm (with the feasible differential algorithm as subroutine) is implemented in multipath form (see the last part of Section 7L), it remains true that the values $\rho(i)$ never sink.

7.45. (*Out-of-Kilter Algorithm*). Solve the elementary pair of problems in Exercise 7.26 (Figure 7.25) by the out-of-kilter algorithm in Section 7M. Start with $u \equiv 0$ and $x(j_1) = x(j_2) = x(j_6) = x(j_7) = 0$, $x(j_3) = x(j_8) = 1$, $x(j_4) = 5$, $x(j_5) = 2$.

7.46. (*Out-of-Kilter Algorithm*). Show that when the out-of-kilter algorithm is applied with initial x and u such that x is a feasible solution to the elementary optimal distribution problem, it can be identified with the version of the optimal distribution algorithm that uses the tension rectification algorithm (Section 3K) as the main subroutine.

7.47. (*Out-of-Kilter Algorithm*). Show that when the out-of-kilter algorithm is applied with initial x and u such that u is a feasible solution to the elementary optimal differential problem, it can be identified with the version of the optimal differential algorithm that uses the flow rectification algorithm (Section 6I) as the main subroutine.

7.48. (*Out-of-Kilter Algorithm*). Extend the out-of-kilter algorithm to handle the linear (not necessarily elementary) optimal distribution problem directly.

(*Hint.*   Follow the lead of Exercise 7.46.)

7.49. (*Out-of-Kilter Algorithm*). Extend the out-of-kilter algorithm to handle the linear (not necessarily elementary) optimal differential problem directly.

(*Hint.*   Follow the lead of Exercise 7.47.)

7.50. (*Multipath Implementation*). Describe how the out-of-kilter algorithm could be refined to enhance its efficiency.

(*Hint.*   A min path subroutine with multipath implementation can be used; see the discussion of the optimal distribution algorithm toward the end of Section 7L and the refinements of the tension rectification algorithm in Section 6I.)

7.51. (*Elementary Simplex Methods*). Specialize the statement of the simplex method for flows (Section 7D) to the case of solving an elementary pair of problems in duality (Section 7I).

7.52. (*Elementary Simplex Methods*). Specialize the statement of the simplex method for potentials (Section 7H) to the case of solving an elementary pair of problems in duality (Section 7I).

7P.*   COMMENTS AND REFERENCES

Linear optimal flow models are suited to a great many applications and often lead to very large-scale problems; see the survey of Glover and Klingman [1975]. The subject also has a rich history. The classical transportation problem in Example 1 of Section 7B was formulated by Hitchcock [1941]. A similar problem was studied around the same time by Kantorovitch [1942]. However, the theory did not really get rolling until the advent of linear programming, with the simplex method of Dantzig [1949] and the duality theorem of Gale, Kuhn, and Tucker [1951].

The linear programming duality theorem made it possible to characterize optimal solutions to the transportation problem in terms of dual variables. The transportation problem was subsequently extended to general networks with upper as well as lower bounds on flows. Charnes and Cooper [1958] showed how piecewise linear costs could also be treated. Characterizations of optimality in these cases had to be based, however, on reduction to what we have called *elementary* optimal distribution problems, actually with $c(j) = 0$ for all $j \in A$, since this was the standard form for duality in linear programming (see remarks in Section 7I). As a matter of fact any elementary optimal distribution problem can be reformulated even as a classical problem of the Hitchcock type, although not without unpleasant consequences in dimensionality; see Ford and Fulkerson [1962, pp. 127–130] for a discussion. The trouble with such reformulations, besides increases in dimensionality of course, is that they do violence to the given network structure and render the conditions that are obtained harder to interpret. To avoid this, one has to abandon linear programming duality in favor of a more general theory like the one in Chapter 8.

Nevertheless, there are some results about optimal flows that are best understood in the linear programming framework, and the theorems in Sections 7A and 7E about extreme solutions are among them. These theorems expose properties that are crucial to the formulation and validation of the simplex method in linear programming, and they provide a key to the adapted versions of that method in Sections 7D and 7H. The integrality theorems in Sections 7A and 7E, on the other hand, reflect features peculiar to networks and closely related structures. Another characteristic of the network case is the constructive ease with which an arbitrary solution can be represented as a combination of extreme solutions by the algorithm for flows in Section 4K, which has not previously been recorded (likewise for the one for differentials in Section 6L).

The simplex method was first applied to classical transportation problems by Dantzig [1951]. He observed the integrality property and the fact that it allowed the general pivoting transformations in the algorithm to be replaced by a construction of "stepping stone" paths in a rectangular array. Later he invented the "upper bounding" technique by which the simplex method could be applied directly to linear programs with variables constrained to intervals of the form $[0, c]$, rather than $[0, \infty)$. The desire to treat capacity constraints in network flow problems more efficiently was a major motivation for this. In his book [1962] Dantzig discusses at length the upper bounding technique and the specialization of the simplex method to networks. The only novelty of our version in Section 7D is the treatment of arbitrary intervals $[c^-(j), c^+(j)]$, which may or may not be bounded or even half-bounded, and the specialization to the network context of one of the termination rules of Bland [1978].

The refinement using "strong" trees is due to Cunningham [1976]. No comparable refinement is known for the dual simplex method in Section 7H.

It has been reported that in applications to large-scale flow problems, more than 90% of the pivoting steps in the simplex method can be degenerate (see Gavish and Schweitzer [1974]; Gavish, Schweitzer, and Shlifer [1977], Bradley, Brown, and Graves [1978]). In a sequence of degenerate steps the method more or less "flounders around" with little information to go on. Rules for finite termination, in the absence of good bounds on the number of iterations, do not offer much reassurance. Cunningham [1979] points out that "stalling," the possibility of arbitrarily long degenerate sequences, is just as dangerous to practicality as "cycling," the possibility of returning over and over to a previous state, and he offers some rules for avoiding it. See also the examples and suggestions of Zadeh [1973a, 1973b, 1980].

Computational experience with the simplex method, in networks and in general linear programming, has always turned out to be much better than could be accounted for in theory. For this reason the method has remained popular, despite its seeming drawbacks. At present it is still the subject of much research, especially in the network case. Breakthroughs in coding the method for classical transportation problem came with work of Srinivasan and Thompson [1973] and Glover, Karney, Klingman, and Napier [1974]. For this problem and its special case, the optimal assignment problem, Barr, Glover, and Klingman [1977, 1978], have shown it is enough to consider spanning trees of a limited class. This complements the results of Cunningham [1976] concerning "strong" trees (Section 7D). Adolphson [1979] proposes a version of the simplex method in which spanning trees are replaced by nonmaximal forests of arcs in response to degeneracy. For the state of the art in implementing the simplex algorithm in flow problems with linear costs, see Bradley, Brown, and Graves [1977], Ali, Helgason, Kennington, and Hall [1977], and the book of Kennington and Helgason [1980].

The fact that optimization problems for potentials can be important for their own sake, rather than just through duality with flow problems, has been slow in dawning. The PERT problem, which can be formulated in terms of max tension (Example 5 in Section 6C and Example 12 in Section 7F), and its natural extension in Example 8 in Section 7F, have been the most familiar examples, although the facilities location problem (Example 9) in Section 7F is also well known. The shared cost model in Example 13 in Section 7F (see Rhys [1970] and Balinski [1970]) has previously been formulated rather more complicatedly as a min cut problem. The direct expression in terms of potentials is new.

Ford and Fulkerson [1957] gave the first network flow algorithm that was not just an implementation of the simplex method. This was the natural extension to the classical transportation problem of the Hungarian assignment algorithm of Kuhn [1955] made possible by duality theory and the recognition that the optimal assignment problem was a kind of transportation problem, an insight gained from the integrality property. We can think of this method here as the specialized form of the optimal differential algorithm described in

Exercise 7.33 (but without mention of "paintings," an idea due to Minty) applied in the "elementary" case to the dual of the classical transportation problem (in Example 10 in Section 7F) with initial flow $x \equiv 0$.

The method of Ford and Fulkerson served as a prototype for the min cost flow algorithm of Busacker and Gowen [1961] (see also Jewell [1962]). As formulated by Lawler [1976] with the ideas of Klein [1967] included, so as to get around earlier restrictions to non-negative costs $d(j)$ and the like, the min cost flow algorithm combines features of what we have called the optimal distribution algorithm and the thrifty adjustment algorithm. Without significant loss of generality in practice, the capacity intervals are taken to have $0 = c^-(j) < c^+(j) < \infty$, and it is assumed that there is a single source s and a single sink s'. The method starts with any flow x satisfying $0 \leq y(s) \leq b(s), 0 \geq y(s') \geq b(s')$, and $0 = y(i) = b(i)$ for all other nodes i. First there is a phase consisting of iterations that search for improvement *circuits* of the kind in the optimal distribution algorithm, just as if $y(s) = b(s)$ and $y(s') = b(s')$. Such circuits are detected by trying to construct an optimal routing with base s for the intervals $[d_x^-(j), d_x^+(j)]$. This phase terminates when the construction succeeds and thereby yields a potential u whose differential v satisfies $(x(j), v(j)) \in \Gamma_j$ for all $j \in A$. The method then shifts over to iterations of the thrifty adjustment algorithm (in the "linear" form; see Exercise 7.36) for paths $P: s \to s'$ until the constraints $y(s) = b(s)$, $y(s') = b(s')$, are truly fulfilled. Iterations in the second phase have a lower order of complexity than in the first phase, because Dijkstra's method can be used (an observation of Edmonds and Karp [1972]) instead of a general method for constructing an optimal routing.

Actually the first phase in the min cost flow algorithm just explained is superfluous. This is demonstrated by our thrifty adjustment algorithm (in the extended form in Exercise 7.36 in the case of *nonelementary* linear problems). If the intervals $[c^-(j), c^+(j)]$ are all bounded, one can start with *any* potential u and choose $x(j)$ arbitrarily in the interval

$$[c_u^-(j), c_u^+(j)] = \begin{cases} [c^+(j), c^+(j)] & \text{if } v(j) > d(j), \\ [c^-(j), c^+(j)] & \text{if } v(j) = d(j), \\ [c^-(j), c^-(j)] & \text{if } v(j) > d(j), \end{cases}$$

so as to obtain a flow that satisfies $(x(j), v(j)) \in \Gamma_j$ for all $j \in A$. Of course y may differ from b at more than just two nodes, and it is not possible to insist *in advance* on which nodes they will be. But for the thrifty adjustment algorithm this does not matter; it works with paths $P: N^+ \to N^-$, whatever N^+ and N^- turn out to be, and the min path subroutine it employs can be executed as Dijkstra's method in *every* iteration. It appears then that by focusing on the "simpler" case of a single source and sink, the min cost flow algorithm really makes things harder than necessary.

Although the first phase of the min cost flow algorithm can be circumvented in the situation described, that improvement procedure in terms of circuits does have merit in the general context of the optimal distribution algorithm, despite the threat of a higher order of complexity in each iteration. One reason, already mentioned in Section 7M, is that it generates an ever-improving sequence of *feasible* solutions to the optimal distribution problem, rather than a sequence that just becomes feasible at the very end. Perhaps more important, this method can also be extended to the case of convex cost functions (Chapter 9) in terms of searching for a minimum along a ray in each iteration as in general nonlinear programming, the direction of the ray being given by a certain e_P. The thrifty adjustment algorithm can be extended only to piecewise linear costs, and the more "pieces," the slower it is likely to be. It is therefore not very promising in situations where a fine piecewise linearization of a more general cost function is suggested.

At all events the optimal distribution algorithm need not be implemented with a method of constructing an optimal routing with base s for the main subroutine, as in the first phase of the min cost flow algorithm. The tension rectification algorithm can be invoked instead, for instance, and as noted in Exercise 7.46, what one gets with this as the subroutine in the optimal distribution algorithm is the out-of-kilter method.

The out-of-kilter method was discovered by Fulkerson [1961] and Minty [1961] independently. Fulkerson was working in the framework of mathematical programming and Minty in terms of electrical analogies and what amounted to piecewise linear, rather than just linear cost functions. Both realized that the method can be initiated with neither x nor u feasible, and it will systematically bring arcs into kilter without ever increasing distances of the points $(x(j), v(j))$ from curves the Γ_j.

For work on computer implementations of the out-of-kilter method, see Barr, Glover, and Klingman [1974], Glover, Karney, and Klingman [1974], and Aashiani and Magnanti [1976]. A scaling technique of Edmonds and Karp [1971] provides an interesting theoretical improvement in efficiency which does not, however, appear particularly effective in practice. For other proposals see Zadeh [1979].

The optimal distribution algorithm can also be executed with the feasible differential algorithm (with its restart provision) as the subroutine, and this approach is new, because the feasible differential algorithm is new. As with the out-of-kilter method, the points $(x(j), v(j))$ never get farther from the curves Γ_j. Especially with the incorporation of a Dinic-Karzanov approach, as outlined in Section 7L, one can look for efficiency surpassing that of at least the forms of the out-of-kilter method used up till now. Multipath implementation has not previously been incorporated into network optimization algorithms, aside from the max flow case.

The thrifty adjustment algorithm with multipath implementation, viewed as a refined form of the optimal differential algorithm, should furnish improvements in case to which the latter can be specialized, for instance, in sharpening the Hungarian assignment method (see Exercise 7.22).

The bibliography of Golden and Magnanti [1977] lists over a thousand items under thirteen headings and could help in locating results on special topics in network optimization.

8

OPTIMAL FLOWS
AND POTENTIALS

In the optimization problems in Chapter 7 only linear cost expressions were allowed. However, piecewise linear cost functions with progressively increasing slopes could be reduced to that case by modifications of the network. The aim of this chapter is to develop the theory in direct terms, not only for such piecewise linear costs but for general convex costs.

One reason for the extension, of course, is that nonlinear cost functions do arise in applications; even if they can be approached through a series of piecewise linear approximations, it may be awkward computationally and conceptually to replace the given network by a series of more complicated and probably much larger networks. However, another important motivation is that the duality between optimization of flows and optimization of potentials, discussed in Section 7I for "elementary" problems, cannot be brought to full flower even in the linear or piecewise linear case except in the framework of convex functions and their conjugates. Duality is valuable for the insights it provides into the nature of a problem and the meaning of the conditions characterizing its solution. It greatly expands the range of computational techniques by showing how a problem for flows can be tackled by way of a problem for potentials, and vice versa.

For a function on a real interval, convexity is equivalent to the existence of nondecreasing left and right derivatives. Assigning to an arc j a particular capacity interval supplied with a convex cost function is actually equivalent to assigning to it a certain monotone marginal cost relation Γ_j that determines the function up to a constant of integration. Such relations play a role in characterizing optimality, much as in Chapter 7, but they can also be regarded as correspondences between flux and tension that generalize Ohm's law.

A curve Γ_j derived from a cost function for the flux in j can also be derived in dual fashion from a cost function for the tension in j. The two convex functions are "conjugate" to each other. This is the secret for passing from an optimal distribution problem to an optimal differential problem in the general case.

8A. CONVEX COST FUNCTIONS

From now on, each arc j of the network G has associated with it a capacity interval $C(j)$ (nonempty) and a function $f_j: C(j) \rightarrow R$ giving the *cost* $f_j(\xi)$ of each flux value $\xi \in C(j)$. It is assumed that f_j is *convex* in the sense that

$$\xi = (1 - \lambda)\xi' + \lambda\xi'', \qquad 0 < \lambda < 1$$

$$\Rightarrow f_j(\xi) \leq (1 - \lambda)f_j(\xi') + \lambda f_j(\xi'') \tag{1}$$

(see Figure 8.1).

The convention is adopted that $f_j(\xi) = +\infty$ for all $\xi \notin C(j)$. Then (1) holds for all ξ', ξ'', not just in $C(j)$, so f_j is a convex function on all of R with

$$\{\xi \in R | f_j(\xi) < \infty\} = C(j). \tag{2}$$

Since the set (2) is nonempty and f_j has only $+\infty$ as a possible value, not $-\infty$, one says f_j is *proper*. Note that for any function satisfying (1) the set (2) must be an interval. Thus, instead of starting with $C(j)$, an equivalent approach is to say that each arc j has assigned to it a *proper convex function* f_j *on the whole real line*. Then (2) defines the capacity interval corresponding to f_j.

The right endpoint of $C(j)$ is denoted by $c^+(j)$, and the left endpoint by $c^-(j)$. However, *it is not required that* $C(j)$ *be closed*, and therefore $C(j)$ might not contain $c^+(j)$ and $c^-(j)$ even if these values are finite. This may seem like a degree of generality not worth the effort, but it turns out that a restriction of the theory to the case of closed capacity intervals would appear quite awkward and unnatural in the light of duality. The closedness of the interval (2) for a proper convex function f_j is tied in with the asymptotic behavior of the

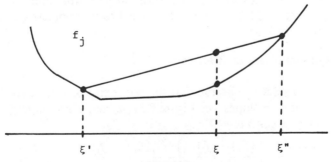

Figure 8.1

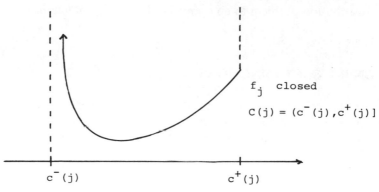

Figure 8.2

conjugate cost function for tension in the arc j, as will be evident in Section 8E. Thus one kind of restriction would lead to another, and through considerations of symmetry one would be forced into a framework more cumbersome than had been anticipated.

A weaker condition than the closedness of $C(j)$ serves perfectly well. The function f_j itself is assumed to be *closed* in the following sense.

1. If $-\infty < c^-(j) < c^+(j)$ and $f_j(\xi) \to \alpha < \infty$ as $\xi \downarrow c^-(j)$, then $c^-(j) \in C(j)$ and $f_j(c^-(j)) = \alpha$.
2. If $c^-(j) < c^+(j) < \infty$ and $f_j(\xi) \to \beta < \infty$ as $\xi \uparrow c^+(j)$, then $c^+(j) \in C(j)$ and $f_j(c^+(j)) = \beta$.

Because of the natural limit properties possessed by all convex functions, described later in this section, this is actually equivalent to the condition that f_j be continuous relative to the closure of $C(j)$. Thus if $c^+(j)$ is finite, one has $c^+(j) \notin C(j)$ only in the case where $C(j)$ has a nonempty interior and $f_j(\xi) \to +\infty$ as $\xi \uparrow c^+(j)$. Similarly, if $c^-(j)$ is finite, one has $c^-(j) \notin C(j)$ only if $C(j)$ has a nonempty interior and $f_j(\xi) \to +\infty$ as $\xi \downarrow c^-(j)$. See Figure 8.2.

Some basic properties of closed proper convex functions on R will be described in the rest of this section. These need to be appreciated for the sake of what will be done later.

Difference Quotients and Derivatives

First of all it is useful to observe that the convexity of f_j is equivalent to the following condition on slopes (see Figure 8.3): *for any three values $\xi_1 < \xi_2 < \xi_3$ in the interval $C(j)$, one has*

$$\frac{f_j(\xi_2) - f_j(\xi_1)}{\xi_2 - \xi_1} \le \frac{f_j(\xi_3) - f_j(\xi_1)}{\xi_3 - \xi_1} \le \frac{f_j(\xi_3) - f_j(\xi_2)}{\xi_3 - \xi_2} \tag{3}$$

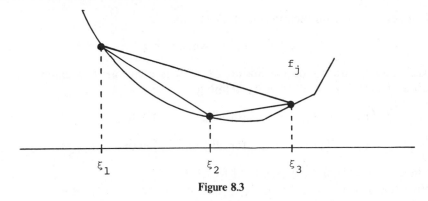

Figure 8.3

(Exercise 8.1). From this one sees that f_j is convex if and ony if for each $\xi \in C(j)$ the difference quotient $[f_j(\zeta) - f_j(\xi)]/[\zeta - \xi]$ is nondecreasing as a function of $\zeta \in R \setminus \{\xi\}$.

It follows that for each $\xi \in C(j)$ both the left and the right derivatives (left and right *marginal costs*)

$$f'_{j-}(\xi) = \lim_{\zeta \uparrow \xi} \frac{f_j(\zeta) - f_j(\xi)}{\zeta - \xi}, \qquad f'_{j+}(\xi) = \lim_{\zeta \downarrow \xi} \frac{f_j(\zeta) - f_j(\xi)}{\zeta - \xi},$$

exist (possibly as $+\infty$ or $-\infty$) and satisfy

$$f'_{j-}(\xi) \leq f'_{j+}(\xi), \tag{4}$$

the two values being equal and finite if and only if f_j is actually differentiable at ξ; see Figure 8.4. In fact it follows from (3) that both of the one-sided

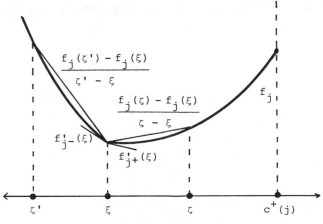

Figure 8.4

derivatives are finite on the interior of $C(j)$ with

$$f'_{j+}(\xi') \leq f'_{j-}(\xi'') \quad \text{when } \xi' < \xi''. \tag{5}$$

Thus f'_{j-} and f'_{j+} are nondecreasing on $C(j)$. It is convenient to extend these functions from $C(j)$ to all of R by defining

$$f'_{j-}(\xi) = f'_{j+}(\xi) = +\infty \quad \text{for all } \xi \notin C(j) \text{ with } \xi \geq c^+(j),$$

$$f'_{j-}(\xi) = f'_{j+}(\xi) = -\infty \quad \text{for all } \xi \notin C(j) \text{ with } \xi \leq c^-(j). \tag{6}$$

The inequalities (4) and (5) then hold for all $\xi \in R$.

Note that if $c^-(j) < c^+(j) \in C(j)$, one always has $f'_{j+}(c^+(j)) = +\infty$, but $f'_{j-}(c^+(j))$ can be finite or $+\infty$. Similarly, if $c^+(j) > c^-(j) \in C(j)$, one always has $f'_{j-}(c^-(j)) = -\infty$, but $f'_{j+}(c^-(j))$ can be finite or $-\infty$. These possibilities are illustrated in Figure 8.5. In the degenerate case where the interval $C(j)$ consists of a single point $\bar{\xi}$ (so $c^-(j) = c^+(j) = \bar{\xi}$), one has $f'_{j+}(\bar{\xi}) = +\infty$ and $f'_{j-}(\bar{\xi}) = -\infty$, so that f'_{j+} and f'_{j-} have only infinite values.

Nevertheless, whether $C(j)$ has a nonempty interior or not, the set

$$\tilde{C}(j) = \left\{ \xi \in R | f'_{j-}(\xi) < +\infty \quad \text{and } f'_{j+}(\xi) > -\infty \right\}$$

$$= \left\{ \xi \in R | \exists \eta \in R, f'_{j-}(\xi) \leq \eta \leq f'_{j+}(\xi) \right\} \subset C(j) \tag{7}$$

is a *nonempty* interval with the same endpoints $c^+(j)$ and $c^-(j)$ as $C(j)$. The points lying to the right of $\tilde{C}(j)$ are those with $f'_{j-}(\xi) = f'_{j+}(\xi) = +\infty$, whereas those to the left have $f'_{j-}(\xi) = f'_{j+}(\xi) = -\infty$. The interval $\tilde{C}(j)$ will play as important a role as $C(j)$ in some of what follows.

Continuity

Until now nothing has been made of the closedness assumption on f_j, and this has been in order not to obscure the limit properties that are natural conse-

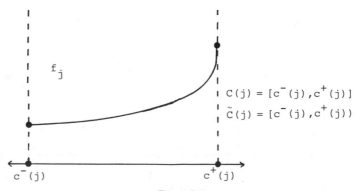

$$C(j) = [c^-(j), c^+(j)]$$
$$\tilde{C}(j) = [c^-(j), c^+(j))$$

Figure 8.5

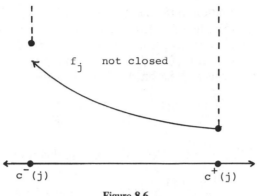

Figure 8.6

quences of convexity alone. Suppose that $c^-(j) < c^+(j)$ (since the case where $C(j)$ reduces to a single point—the only point where f_j is finite—is not of interest in a discussion of continuity). At all points ξ strictly between $c^-(j)$ and $c^+(j)$, f_j has been seen to possess finite left and right derivatives. This property implies continuity at ξ both from the left and from the right. Thus f_j is certainly continuous at ξ when $c^-(j) < \xi < c^+(j)$, as well as when $\xi > c^+(j)$ or $\xi < c^-(j)$ (f_j then being identically $+\infty$).

To see what happens at the endpoints, consider the difference quotient $[f_j(\xi) - f_j(\bar{\xi})]/[\xi - \bar{\xi}]$ as a function of ξ for fixed $\bar{\xi}$ satisfying $c^-(j) < \bar{\xi} < c^+(j)$. Since the quotient is nondecreasing it has a right limit at $c^-(j)$ and a left limit at $c^+(j)$ (either limit possibly $+\infty$). The same is therefore true of f_j itself (since the only term of the difference quotient which could bring the existence of a limit into question is $f_j(\xi)$).

Thus if f_j were not closed, this could only be because its value at a finite endpoint $c^+(j)$ or $c^-(j)$ did not agree with the natural limit value as the endpoint is approached from the interior of $C(j)$, being too high as in Figure 8.6. The discrepancy could be remedied by redefining f_j at the endpoint in question. Then f_j would be a continuous function relative to the closure of $C(j)$.

Limits of Derivatives

One-sided limits of the left and right derivative functions obey a very simple law by virtue of the closedness assumption on f_j. Namely, for all $\xi \in R$ one has

$$f_{j-}'(\xi +) = f_{j+}'(\xi) = f_{j+}'(\xi +),$$

$$f_{j-}'(\xi -) = f_{j-}'(\xi) = f_{j+}'(\xi -). \tag{8}$$

In other words, f_{j+}' is the right continuous "regularization" of f_{j-}', whereas f_{j-}' is the left continuous "regularization" of f_{j+}'.

The first double equation in (8) is established as follows. Since $f'_{j+}(\xi) \le f'_{j-}(\zeta)$ $\le f'_{j+}(\zeta)$ by (4) and (5) when $\xi < \zeta$, one has $f'_{j+}(\xi) \le f'_{j-}(\xi +) \le f'_{j+}(\xi +)$. All three quantities equal $+\infty$ if $\xi \ge c^+(j)$ or $\xi < c^-(j)$, so the proof can be completed by showing that $f'_{j+}(\xi) \ge f'_{j+}(\xi +)$ when $c^-(j) \le \xi < c^+(j)$. At such a point ξ, f_j is continuous at least on the right, so that

$$\frac{f_j(\eta) - f_j(\xi)}{\eta - \xi} = \lim_{\zeta \downarrow \xi} \frac{f_j(\eta) - f_j(\zeta)}{\eta - \zeta}.$$

Since $[f_j(\eta) - f_j(\zeta)]/[\eta - \zeta] \ge f'_{j+}(\zeta)$ when $\eta > \zeta$, this implies

$$\frac{f_j(\eta) - f_j(\xi)}{\eta - \xi} \ge f'_{j+}(\xi +) \quad \text{for all } \eta > \xi.$$

Therefore

$$f'_{j+}(\xi) = \lim_{\eta \downarrow \xi} \frac{f_j(\eta) - f_j(\xi)}{\eta - \xi} \ge f'_{j+}(\xi +)$$

as claimed. The proof of the other double equation in (8) is similar.

Convex Costs as Integrals of Nondecreasing Marginal Costs

In many applications the function f_j is not given directly but is constructed by integrating a marginal cost function ϕ_j defined on a certain interval and nondecreasing there. This construction will now be explored in a framework adequate for the present level of generality.

For notational convenience ϕ_j can always be extended beyond its interval of definition, whatever that may be, by assigning it the value $+\infty$ at points lying to the right and $-\infty$ at points to the left. Thus one can suppose that ϕ_j is a nondecreasing extended-real-valued function on the whole real line R which is not the constant function $+\infty$ nor the constant function $-\infty$. Using notation whose appropriateness will be seen shortly, define

$$\tilde{C}(j) = \{\xi \in R | \phi(\xi -) < +\infty \text{ and } \phi(\xi +) > -\infty\}$$

$$= \{\xi \in R | \exists \eta \in R, \varphi(\xi -) \le \eta \le \varphi(\xi +)\}. \tag{9}$$

This is a nonempty interval whose left and right endpoints will be denoted by $c^-(j)$ and $c^+(j)$. (Certainly $\tilde{C}(j)$ contains every value ξ for which $\phi(\xi)$ is finite; in the degenerate case where there is no such value, a point $\bar{\xi}$ exists such that $\phi(\xi) = +\infty$ for all $\xi > \bar{\xi}$ and $\phi(\xi) = -\infty$ for all $\xi < \bar{\xi}$, and then $\tilde{C}(j)$ consists just of $\bar{\xi}$.)

Fix any point $\bar{\xi} \in \tilde{C}(j)$ and any number $\alpha \in R$, and define

$$f_j(\xi) = \int_{\bar{\xi}}^{\xi} \phi_j(t)\, dt + \alpha \quad \text{for all } \xi \in R. \tag{10}$$

Then f_j is a closed proper convex function with

$$f_j^-(\xi) = \phi_j(\xi -) \quad \text{and} \quad f_j^+(\xi) = \phi_j(\xi +) \quad \text{for all } \xi \in R. \qquad (11)$$

(See Figure 8.7.) An outline of the proof of this fact will now be given. Note that (11) reconciles the use of the symbol $\tilde{C}(j)$ in (9) as well as in (7).

   Although ϕ_j might be continuous or even finite everywhere, the integral is well defined (always finite or $+\infty$) because ϕ_j is nondecreasing, and because the choice of $\bar{\xi}$ excludes the possibility that ϕ_j might take on both $+\infty$ and $-\infty$ at points interior to the interval of integration. It is clear that $f_j(\bar{\xi}) = \alpha$, $f_j(\xi) = +\infty$ for $\xi > c^+(j)$ and for $\xi < c^-(j)$. Morever $f_j(\xi) < \infty$ for all $\xi \in \tilde{C}(j)$. The set

$$C(j) = \{\xi \in R | f_j(\xi) < +\infty\}$$

is an interval containing $\tilde{C}(j)$ and having the same endpoints $c^-(j)$ and $c^+(j)$. The continuity properties of the integral in (9) ensure that f_j is continuous relative to the corresponding closed interval.

   For arbitrary $\xi \in C(j)$ and $\zeta \in R$ with $\zeta \neq \xi$, one has

$$\frac{f_j(\zeta) - f_j(\xi)}{\zeta - \xi} = \frac{1}{\zeta - \xi} \int_\xi^\zeta \phi_j(t)\, dt. \qquad (12)$$

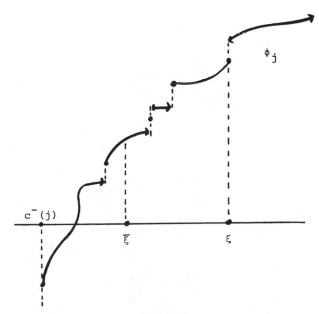

Figure 8.7

Since ϕ_j is nondecreasing, the expression on the right of (12) (which is by definition the average value of ϕ_j over the interval between ξ and ζ) is nondecreasing as a function of ζ and hence so is the difference quotient on the left of (12). Therefore f_j is convex. The derivative relation (11) is likewise apparent from (12).

8B.   CHARACTERISTIC CURVES

Although f_j generally has two separate derivative functions f'_{j+} and f'_{j-}, these are so closely connected by the limit formulas as to constitute different aspects of the same mathematical object. The underlying simplicity can be captured in geometric form through the introduction of the set

$$\Gamma_j = \left\{ (\xi, \eta) \in R^2 \,\middle|\, f'_{j-}(\xi) \le \eta \le f'_{j+}(\xi) \right\}. \tag{1}$$

This is called the *characteristic curve* associated with the arc j. It is very near to being the graph of a nondecreasing function, but it may contain vertical segments, as illustrated in Figure 8.8. In applications the variable η in (1) turns out to measure tension, so that Γ_j expresses a certain relationship between flux

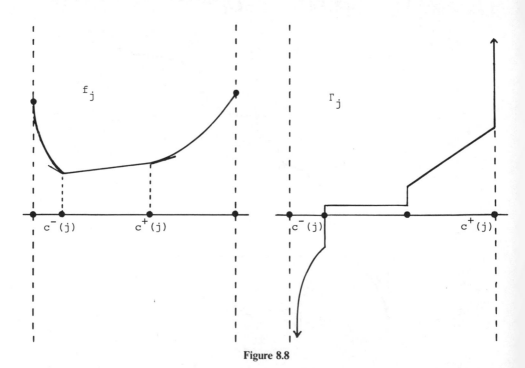

Figure 8.8

and tension peculiar to the arc j. Observe that

$$\tilde{C}(j) = \{\xi \in R | \exists \eta \in R \quad \text{with} f_{j-}'(\xi) \leq \eta \leq f_{j+}'(\xi)\}$$

$$= [\text{projection of } \Gamma_j \text{ on the horizontal axis}]. \tag{2}$$

What conditions on a set $\Gamma_j \subset R^2$ are necessary and sufficient for Γ_j to be representable in the form (1) for some closed proper convex function f_j? An answer is provided by the foregoing results on differentiation and integration: there should exist a nondecreasing function $\phi_j: R \to R \cup \{\pm\infty\}$, not identically $+\infty$ or identically $-\infty$, such that

$$\Gamma_j = \{(\xi, \eta) \in R^2 | \phi_j(\xi -) \leq \eta \leq \phi_j(\xi +)\}. \tag{3}$$

(Starting from f_j, one can take $\phi_j = f_{j+}'$ or $\phi_j = f_{j-}'$, and then (2) turns into (3) by virtue of 8A(8). In the reverse direction, one can define f_j and ϕ_j by 8A(10) and invoke 8A(12). This characterization leads to another which is more symmetric (Exercise 8.6): Γ_j has the form described if and only if it is a *maximal monotone relation* in R^2 in the following sense:

1.   For arbitrary $(\xi, \eta) \in \Gamma_j$ and $(\xi', \eta') \in \Gamma_j$, one has either $(\xi, \eta) \leq (\xi', \eta')$ or $(\xi', \eta') \leq (\xi, \eta)$.
2.   For arbitrary $(\xi, \eta) \notin \Gamma_j$, there exists $(\xi', \eta') \in \Gamma_j$ such that neither $(\xi, \eta) \leq (\xi', \eta')$ nor $(\xi', \eta') \geq (\xi, \eta)$.

This means that Γ_j is totally ordered with respect to the usual coordinatewise partial ordering of R^2 but is not properly included in any other totally ordered set.

The cost function f_j is actually determined by the characteristic curve Γ_j uniquely up to an additive constant. Indeed, the one-sided derivatives f_{j-}' and f_{j+}' are uniquely determined by Γ_j by (1) and hence so is the interval $C(j)$ (and its endpoints $c^-(j)$ and $c^+(j)$). In the degenerate case where $c^-(j) = c^+(j)$ (so that Γ_j is just a vertical line), f_j is trivially unique up to an additive constant, since there is only one point where the value of f_j is not $+\infty$. If $c^-(j) < c^+(j)$, both f_{j-}' and f_{j+}' are finite on the open interval between $c^-(j)$ and $c^+(j)$. If another function on this interval has these same one-sided derivatives, its difference with f_j has zero everywhere as both its left and right derivative (hence as its two-sided derivative) and as such is constant. Therefore f_j is unique on the interior of $\tilde{C}(j)$, up to an additive constant. The overall uniqueness then follows from the expression of the values of f_j at the endpoints of $\tilde{C}(j)$ as limits of values on the interior (which is possible due to "closedness").

In summary, although we started out by assigning to each arc j a closed proper convex function f_j (and capacity interval $C(j)$), the domain of finiteness of f_j), we could just as well have begun with a system of maximal monotone relations Γ_j (and their projection intervals $\tilde{C}(j)$). Although Γ_j can be obtained from f_j by differentiation, it is also possible to obtain f_j from Γ_j by integration. The choice of the constant of integration has no real effect on optimization problems like those in this chapter, where a sum of costs is minimized.

8C.   PIECEWISE LINEAR OR QUADRATIC COSTS

An important case that will deserve much special attention occurs when the capacity interval $C(j)$ is closed and can be divided into a finite number of subintervals on which the cost function f_j is linear, as already considered in Section 7A, see Figure 7.1. Then f_j is said to be *piecewise linear*, and the finite endpoints of the subintervals where f_j has different slopes are called the *breakpoints* for f_j. (Thus $c^-(j)$ and $c^+(j)$ are breakpoints if and only if finite.) Of course the slopes of the consecutive linear pieces (if there are more than one) must form an ascending sequence, or f_j would not be convex. The one-sided limits at the breakpoints must also agree, so that f_j is continuous relative to $[c^-(j), c^+(j)]$.

In similar fashion one describes the case where f_j is *piecewise quadratic*. The constant second derivatives of the quadratic pieces must be non-negative, so that the slope is nondecreasing as required by convexity. The breakpoints in this case are defined as the (finite) values where there is a discontinuity in the slope or in the second derivative or both. (Again $c^+(j)$ and $c^-(j)$ are included as breakpoints if and only if finite.) The piecewise quadratic class includes the piecewise linear class.

Functions of these types are often seen in applications as pieced-together approximations of functions that are more difficult to treat or incompletely known. But they also arise naturally when costs change abruptly due to penalties or bonuses when the flux exceeds or falls short of certain "quotas." An example already mentioned is the kind of function in income tax tables. Duality is another source of such functions: in passing to conjugate costs in

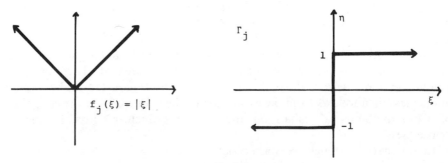

Figure 8.9

Section 8E, a function with one "piece" may well be replaced by a function with several pieces. A specific, useful example of piecewise linearity is the absolute value function; see Figure 8.9.

Before considering piecewise linear or quadratic functions in general form, it is instructive to take a close look at the case where there is just a single linear (or constant) piece:

$$f_j(\xi) = \begin{cases} d(j)\xi + p(j) & \text{if } \xi \in [c^-(j), c^+(j)] \\ +\infty & \text{if } \xi \notin [c^-(j), c^+(j)] \end{cases} \tag{1}$$

Then f_j is called a *linear* cost function, although the linearity is only relative to the interval $C(j) = [c^-(j), c^+(j)]$. (Perhaps this is a good time to recall the rule that a capacity interval $C(j)$ contains only real numbers, never $+\infty$ or $-\infty$. This seems belied by the symbolism $C(j) = [c^-(j), c^+(j)]$ when $c^-(j)$ and $c^+(j)$ might be infinite, but the notational alternatives would be too cumbersome. Of course in a formula like (1), ξ always denotes a real number; there is no question of defining f_j at $-\infty$ or $+\infty$.)

There are many occasions where it is useful to take f_j simply as the *indicator* of the interval $C(j)$:

$$f_j(\xi) = \begin{cases} 0 & \text{if } \xi \in [c^-(j), c^+(j)], \\ +\infty & \text{if } \xi \notin [c^-(j), c^+(j)]. \end{cases} \tag{2}$$

This means that a capacity constraint is associated with the arc j, but there is no contribution from flux in j to the cost expression one wants to optimize. Clearly (2) is the case of (1) with $d(j) = 0$ and $p(j) = 0$.

Also included in (1) is the degenerate case where $c^-(j) = c^+(j)$. This arises mainly in the context of (2), or, in other words, through the introduction of constraints that require the flux in an arc j to equal a certain given value.

The characteristic curve Γ_j corresponding to a linear cost function (1) has an especially simple form, which, however, depends on the nature of the interval $C(j)$. The five possibilities are displayed in Figures 8.10 through 8.14.

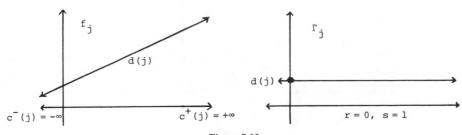

Figure 8.10

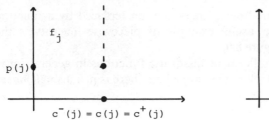
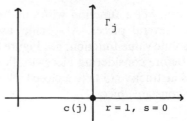

Figure 8.11

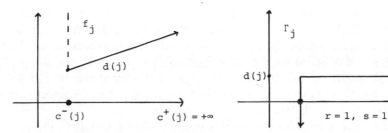

Figure 8.12

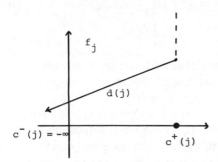
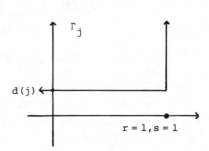

Figure 8.13

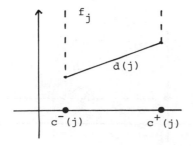
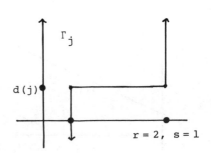

Figure 8.14

In the general piecewise linear case, f_j has r breakpoint values $c_1(j) < c_2(j)$
$< \cdots < c_r(j)$, and s consecutive slope values $d_1(j) < d_2(j) < \cdots < d_s(j)$.
Although r and s can differ by at most 1, their relationship will be seen to
depend on the finiteness of $c^-(j)$ and $c^+(j)$. The values of r and s associated
with the five "linear" cases already considered are given in the corresponding
figures. The remaining four cases are illustrated by Figures 8.15 through 8.18.

From this analysis it is evident that *the class of piecewise linear functions f_j
corresponds to the class of characteristic curves Γ_j which are step relations in the
sense of being composed of a finite number of horizontal and vertical line
segments.*

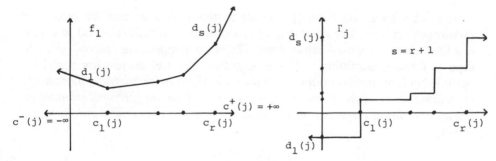

Figure 8.15

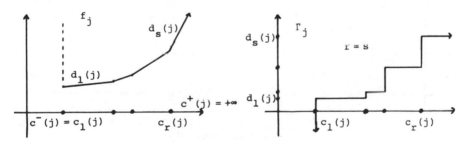

Figure 8.16

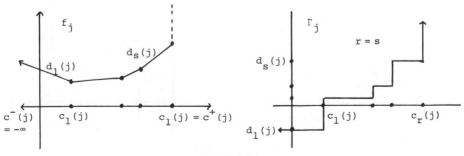

Figure 8.17

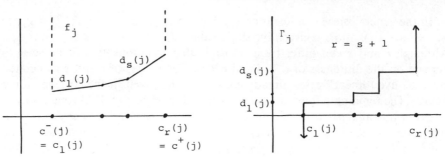

Figure 8.18

It is easy to go on from this result to the analogous one for piecewise quadratic functions. The derivative of a nonlinear quadratic (convex) function is a linear function with positive slope. Therefore in admitting "pieces" of such type to f_j one is admitting to Γ_j line segments that are not just horizontal or vertical but have positive slope. *Thus the class of piecewise quadratic functions f_j corresponds to the class of characteristic curves Γ_j which are polygonal relations in the sense of being composed of finitely many line segments (none with negative slopes).*

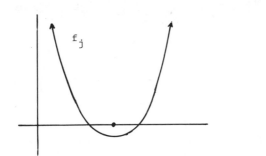

Figure 8.19

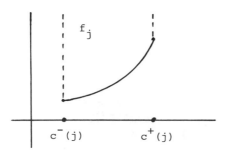

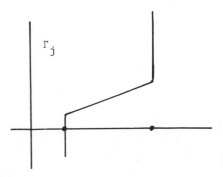

Figure 8.20

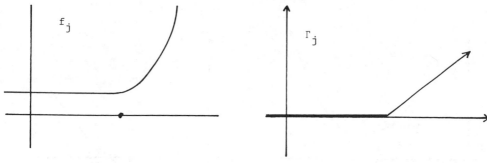

Figure 8.21

Some examples are shown in Figures 8.19 through 8.24. These and the earlier ones for linear and piecewise linear costs should give the reader a feeling of how easy it is in many important cases to recover f_j from Γ_j by integration. Such a procedure will be of practical interest later in constructing the cost function g_j for tensions that is conjugate to f_j, since this can be accomplished by passing to the inverse relation Γ_j^{-1} and then integrating (see Section 8F).

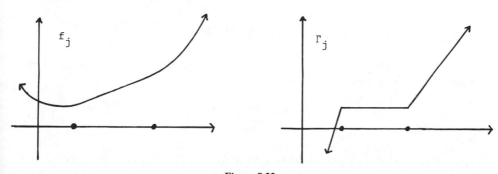

Figure 8.22

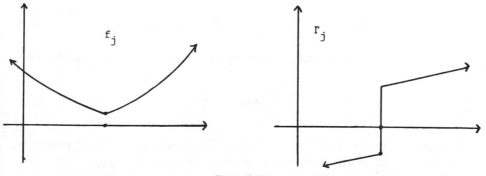

Figure 8.23

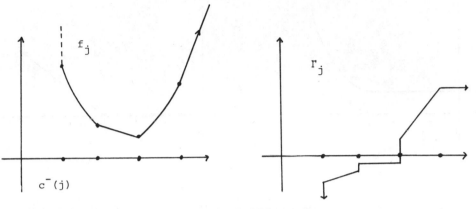

Figure 8.24

An obvious but important fact deserves to be noted in conclusion: when f_j is piecewise linear or quadratic, then $\tilde{C}(j) = C(j) = [c^-(j), c^+(j)]$. This simplifies many of the results in this chapter when they are applied to problems whose costs are of such type.

8D. OPTIMAL DISTRIBUTION PROBLEM

With a cost function f_j assigned to each arc j of the network G, it is natural to define the *cost of a flux x in G* as

$$\Phi(x) = \sum_{j \in A} f_j(x(j)). \tag{1}$$

Only flows satisfying the capacity constraints $x(j) \in C(j)$ will really be of interest, but the conventions in Section 8A about ∞ play a useful role in this respect. The closed proper convex function f_j is finite on $C(j)$ but $+\infty$ everywhere else. Therefore

$$\Phi(x) < +\infty \Leftrightarrow x(j) \in C(j) \quad \text{for all } j \in A. \tag{2}$$

Suppose from now on that a supply $b(i)$ is associated with each node i and that

$$0 = \sum_{i \in N} b(i) = b(N). \tag{3}$$

Suppose also that the network G is connected. This assumption and Equation (3) are no real restriction in our main concern, the study of the problem stated next. They ensure the existence of at least one flow x satisfying $\operatorname{div} x = b$ (as can be seen by applying the feasible distribution theorem to the case where all capacity intervals are $(-\infty, \infty)$).

Optimal Distribution Problem. *Minimize the cost* $\Phi(x)$ *over all flows x satisfying*

$$x(j) \in C(j) \quad \text{for all } j \in A \quad \text{and} \quad y(i) = b(i) \quad \text{for all } i \in N, \qquad (4)$$

where $y = \text{div } x$.

Examples of this type of problem have been discussed in Section 7B in the case of linear costs, and these can be generalized to nonlinear costs. Other examples will be described in Section 8I.

Flows satisfying (4) are called *feasible solutions*. A feasible solution at which the minimum is attained is an *optimal solution*, or simply a *solution*. A necessary and sufficient condition for the existence of a feasible solution is provided of course by the feasible distribution theorem in Section 3H if the capacity intervals are all of the closed form $C(j) = [c^-(j), c^+(j)]$: besides (3), one must have

$$b(S) \leq c^+(Q) \quad \text{for all cuts } Q = [S, N \setminus S]. \qquad (5)$$

When $C(j)$ is not necessarily closed, the more general condition furnished in Exercise 3.14 (with $C(i) = [b(i), b(i)]$) substitutes for (5):

$$b(S) \in C(Q) \quad \text{for all cuts } Q = [S, N \setminus S], \qquad (6)$$

where

$$C(Q) = \sum_{j \in Q^+} C(j) - \sum_{j \in Q^-} C(j) = \left\{ \sum_{j \in Q^+} \xi_j - \sum_{j \in Q^-} \xi_j \mid \xi_j \in C(j) \right\}, \qquad (7)$$

The question of the existence of an optimal solution is more difficult, but a necessary and sufficient condition will be furnished in Section 8L.

In some contexts it will be important to replace $C(j)$ by the interval $\tilde{C}(j)$ defined in 8A(7). A flow x is said to be a *regularly feasible solution* if

$$x(j) \in \tilde{C}(j) \quad \text{for all } j \in A \quad \text{and } y(i) = b(i) \quad \text{for all } i \in N. \qquad (8)$$

Of course, "regular feasibility" coincides with "feasibility" for classes of problems having $\tilde{C}(j) = C(j)$ for all arcs j. Classes with this property that deserve a name are the *linear problems*, with every f_j of the form

$$f_j(x(j)) = \begin{cases} d(j)x(j) + p(j) & \text{if } x(j) \in [c^-(j), c^+(j)], \\ +\infty & \text{if } x(j) \notin [c^-(j), c^+(j)], \end{cases}$$

the *piecewise linear problems* (every f_j piecewise linear) and the *piecewise quadratic problems* (every f_j piecewise quadratic). These classes all actually have $\tilde{C}(j) = C(j) = [c^-(j), c^+(j)]$, so that the simpler criterion (5) can be invoked for feasibility.

The optimal distribution problem can also be expressed more completely as

$$\text{minimize } \Phi(x) \text{ over all flows } x \text{ satisfying div } x = b. \tag{9}$$

The two forms are equivalent because of (2). The feasible solutions to the problem are the flows satisfying $\Phi(x) < \infty$ and div $x = b$. If such flows exist, they will be the only ones worthy of consideration in (9). If not, then $+\infty$ is the minimum in (9). Thus one can write

$$\left[\begin{array}{c} \text{inf in optimal} \\ \text{distribution problem} \end{array} \right] = \inf\{\Phi(x) \mid x \in R^A \text{ and div } x = b\} \tag{10}$$

with the understanding that "min $= +\infty$" is another way of asserting the nonexistence of feasible solutions.

A special case of the optimal distribution problem is the *optimal circulation problem*, where $b(i) = 0$ for all $i \in N$. A seemingly more general model, on the other hand, is the *optimal flow problem*. In this there is also a closed proper convex function f_i (and interval $C(i)$) assigned to each i, and one minimizes the cost

$$\sum_{j \in A} f_j(x(j)) + \sum_{i \in N} f_i(y(i)) \qquad (y = \text{div } x) \tag{11}$$

over all flows x satisfying

$$x(j) \in C(j) \quad \text{for all } j \in A, \qquad y(i) \in C(i) \quad \text{for all } i \in N. \tag{12}$$

The optimal distribution problem is the case where

$$f_i(y(i)) = \begin{cases} 0 & \text{if } y(i) = b(i), \\ +\infty & \text{if } y(i) \neq b(i). \end{cases} \tag{13}$$

(This, incidentally, illustrates the expediency of admitting to the general theory degenerate cost functions that are finite at one point only.)

However, any optimal flow problem in G can be regarded as an optimal circulation problem in the augmented network $\overline{G}$, just as was the case with the corresponding feasibility problems in Section 3H. Thus the three models are, in truth, equivalent. The choice of the optimal distribution problem as the basic one is dictated by convenience.

8E.   CONJUGATE COSTS

The problem dual to the optimal distribution problem will be formulated in Section 8G. This requires that each arc j have assigned to it a span interval

$D(j)$ and a cost $g_j(\eta)$ for each tension value $\eta \in D(j)$. The same properties of convexity and continuity are again crucial.

Adopting the approach in Section 8A, what we need for each arc j is a closed proper convex function g_j. The corresponding span interval is then

$$D(j) = \{\eta \in R \mid g_j(\eta) < +\infty\}. \tag{1}$$

The left and right derivatives g'_{j-} and g'_{j+} describe a certain maximal monotone relation whose integral is g_j and whose projection on the η-axis is the interval

$$\tilde{D}(j) = \{\eta \in R \mid g'_{j-}(\eta) < +\infty \quad \text{and} \quad g'_{j+}(\eta) > -\infty\} \subset D(j). \tag{2}$$

The intervals $D(j)$ and $\tilde{D}(j)$ are not necessarily closed, but they have the same endpoints $d^-(j)$ and $d^+(j)$, and g_j is continuous relative to $[d^-(j), d^+(j)]$.

It is a remarkable fact that the specification of a closed proper convex cost function f_j for flows is equivalent to the specification of a closed proper convex cost function g_j for tensions. The two functions are said to be *conjugate* to each other. They will turn out to be related to each other by the formulas

$$g_j(\eta) = \sup_{\xi \in R} \{\xi\eta - f_j(\xi)\} = -\inf_{\xi \in C(j)} \{f_j(\xi) - \xi\eta\}, \tag{3}$$

$$f_j(\xi) = \sup_{\eta \in R} \{\xi\eta - g_j(\eta)\} = -\inf_{\eta \in D(j)} \{g_j(\eta) - \xi\eta\}. \tag{4}$$

Important consequences of these formulas will be derived later in this section. Other formulas, frequently more convenient than these for constructing g_j from f_j, or vice versa, will also be furnished.

The reader may have had some inkling of the duality correspondence among cost functions from the discussion of characteristic curves in Section 8B. The function f_j is determined up to an additive constant as the integral of a maximal monotone relation Γ_j. But Γ_j has a natural *inverse*, defined by

$$\Gamma_j^{-1} = \{(\eta, \xi) \in R^2 \mid (\xi, \eta) \in \Gamma_j\} \tag{5}$$

(see Figure 8.25). This too is a maximal monotone relation, since the definition of that notion is symmetric in ξ and η. The integral of Γ_j^{-1} is a certain closed proper convex function of η, unique up to an additive constant. It will be demonstrated that with the right choice of the constant, this function turns out to be g_j. In other words, *it will be shown that g_j, as defined by (3), has Γ_j^{-1} as its marginal cost curve*:

$$\Gamma_j^{-1} = \{(\eta, \xi) \in R^2 \mid g'_{j-}(\eta) \le \xi \le g'_{j+}(\eta)\}. \tag{6}$$

Thus

$$g'_{j-}(\eta) \le \xi \le g'_{j+}(\eta) \Leftrightarrow (\xi, \eta) \in \Gamma_j \Leftrightarrow f'_{j-}(\xi) \le \eta \le f'_{j+}(\xi) \tag{7}$$

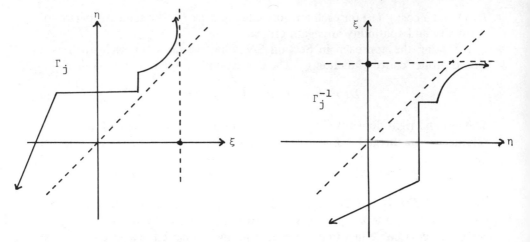

Figure 8.25

and

$$\tilde{D}(j) = \left\{ \eta \in R \mid \exists \xi \in R \quad \text{with } (\eta, \xi) \in \Gamma_j^{-1} \right\} \tag{8}$$

$$= \left[\text{projection of } \Gamma_j \text{ on the } \textit{vertical } \text{axis} \right]$$

(see Figure 8.26 and formula 8B(2) for $\tilde{C}(j)$).

The fact that $\tilde{C}(j)$ and $\tilde{D}(j)$ are the projections of the characteristic curves Γ_j on the horizontal and vertical axes gives much insight into the dual roles played by the capacity bounds and span bounds. It yields at once the formulas

$$d^+(j) = \lim_{\xi \to +\infty} f_{j-}'(\xi) = \lim_{\xi \to +\infty} f_{j+}'(\xi),$$

$$d^-(j) = \lim_{\xi \to -\infty} f_{j-}'(\xi) = \lim_{\xi \to -\infty} f_{j+}'(\xi), \tag{9}$$

as well as

$$c^+(j) = \lim_{\eta \to +\infty} g_{j-}'(\eta) = \lim_{\eta \to +\infty} g_{j+}'(\eta),$$

$$c^-(j) = \lim_{\eta \to -\infty} g_{j-}'(\eta) = \lim_{\eta \to -\infty} g_{j+}'(\eta). \tag{10}$$

Thus $d^+(j)$ and $d^-(j)$ are the extreme slope values for the graph of f_j, and whether or not they are actually contained in $\tilde{D}(j)$ depends on the *asymptotic behavior* of f_j. At the same time $c^+(j)$ and $c^-(j)$ are the extreme slope values

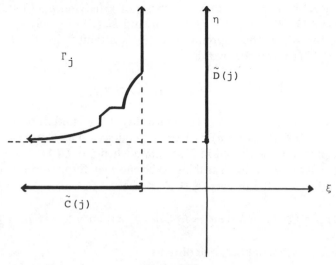

Figure 8.26

for g_j, and their membership in $\tilde{C}(j)$ depends on the asymptotic behavior of g_j.

Another fact that will be established about conjugate cost functions is that

$$f_j(\xi) + g_j(\eta) \geq \xi\eta \quad \text{for all } \xi, \eta, \tag{11}$$

with

$$f_j(\xi) + g_j(\eta) = \xi\eta \Leftrightarrow (\xi, \eta) \in \Gamma_j. \tag{12}$$

This will play a central role in the duality theory for the optimal distribution problem.

Proof of the Conjugacy Properties.   Given a closed proper convex function f_j (and the associated relation Γ_j), let g_j be defined by (3). Then (11) holds. It must be demonstrated that (12) holds too, and that g_j is a closed proper convex function satisfying (6). This will prove that g_j can be obtained alternatively as the integral of Γ_j^{-1} for which the constant of integration is chosen to validate (12). But then f_j can be obtained likewise from g_j as the integral of $(\Gamma_j^{-1})^{-1} = \Gamma_j$ for which the constant of integration is chosen to validate (12). It will follow therefore that f_j is given by (4), and the verification of the asserted properties of conjugacy will be complete.

As a start, let us consider a fixed value of η and try to characterize the corresponding values of ξ, if any, for which the supremum in (3) is attained. These are the values that minimize the closed proper convex function $h(\xi) =$

$f_j(\xi) - \xi\eta$ over R. Therefore they are the ones satisfying $h'_-(\xi) \leq 0 \leq h'_+(\xi)$ (see Exercise 8.11). But $h'_-(\xi) = f'_j(\xi) - \eta$, and $h'_+(\xi) = f'_{j+}(\xi) - \eta$. Thus the values of ξ for which the supremum in (3) is attained are those such that $f'_{j-}(\xi) \leq \eta \leq f'_{j+}(\xi)$. In other words,

$$g_j(\eta) = \xi\eta - f_j(\xi) \Leftrightarrow f'_{j-}(\xi) \leq \eta \leq f'_{j+}(\xi). \qquad (13)$$

This equivalence is the same as (12). Hence (12) is true, and in particular there are values of η for which $g_j(\eta) < +\infty$. Of course, $g_j(\eta)$ can never be $-\infty$, since the expression maximized in (3) is finite when $\xi \in C(j)$.

The convexity of g_j is the next thing to be checked. Supposed $\eta = (1 - \lambda)\eta' + \lambda\eta''$ with $0 < \lambda < 1$. For every $\xi \in R$ one has

$$(1 - \lambda)g_j(\eta') \geq (1 - \lambda)\big[\xi\eta' - f_j(\xi)\big], \qquad \lambda g_j(\eta'') \geq \lambda\big[\xi\eta'' - f_j(\xi)\big].$$

Adding these two inequalities, one obtains

$$(1 - \lambda)g_j(\eta') + \lambda g_j(\eta'') \geq \xi\eta - f_j(\xi) \quad \text{for all } \xi \in R.$$

The supremum in ξ then yields

$$(1 - \lambda)g_j(\eta') + \lambda g_j(\eta'') \geq g_j(\eta).$$

Therefore g_j is a proper convex function.

Furthermore g_j is closed, for if not, the value of g_j at a certain finite endpoint of $D(j)$ would exceed the limit value β of g_j as the endpoint was approached from the interior of $D(j)$ (recall the discussion of continuity properties of convex functions in Section 8A). Then the interval $\{\eta \in R|g_j(\eta) \geq \beta\}$ would not be closed, a property contradictory to the representation furnished by (3):

$$\{\eta \in R| g_j(\eta) \leq \beta\} = \bigcap_{\xi \in C(j)} \{\eta \in R| \xi\eta \leq \beta + f_j(\xi)\}.$$

Relations (11) and (12), already established, imply for any fixed ξ that the closed proper convex function $k(\eta) = g_j(\eta) - \xi\eta$ satisfies $k(\eta) \geq -f_j(\xi)$ for all $\eta \in R$, and that $k(\eta) = -f_j(\xi)$ when $(\xi, \eta) \in \Gamma_j$. In the latter case $-f_j(\xi)$ must be the minimum value of k, and since it is attained at η one has $k'_-(\eta) \leq 0 \leq k'_+(\eta)$, or in other words, $g'_{j-}(\eta) - \xi \leq 0 \leq g'_{j+}(\eta) - \xi$. Therefore

$$\{(\eta, \xi) \in R^2| g'_{j-}(\eta) \leq \xi \leq g'_{j+}(\eta)\} \supset \{(\eta, \xi)| (\xi, \eta) \in \Gamma_j\} = \Gamma_j^{-1}. \quad (14)$$

The first set in (14) is the maximal monotone relation associated with g_j. But Γ_j^{-1} is also a maximal monotone relation. Strict inclusion is therefore impossible in (14), and (6) is proved.

8F.   EXAMPLES OF CONJUGATE FUNCTIONS

How can g_j be constructed from f_j (or the reverse)? Besides applying the definition 8E(3) directly, which is sometimes difficult, there is the possibility of using integration. The curve Γ_j must be determined precisely along with its inverse Γ_j^{-1}, and then Γ_j^{-1} is integrated with the right choice of the additive constant. Specifically, let $(\bar{\xi}, \bar{\eta})$ be any point of Γ_j (so that $g_j(\bar{\eta}) = \bar{\xi}\bar{\eta} - f_j(\bar{\xi})$ by 8E(12)), and let Ψ_j be any nondecreasing function that represents Γ_j^{-1} in the sense of

$$\{(\eta, \xi) \in R^2 | \Psi_j(\eta -) \le \xi \le \Psi_j(\eta +)\} = \Gamma_j^{-1}$$

Then

$$g_j(\eta) = \bar{\xi}\bar{\eta} - f_j(\bar{\eta}) + \int_{\bar{\eta}}^{\eta} \Psi_j(t)\, dt \quad \text{for all } \eta \in R. \tag{1}$$

There is still another method of obtaining g_j which is usually easier, although it again requires complete knowledge of Γ_j. First determine the interval $\tilde{D}(j)$ as the projection of Γ_j on the vertical axis. For $\eta \in \tilde{D}(j)$, use any corresponding ξ with $(\xi, \eta) \in \Gamma_j$ to get the value of g_j by the formula $g_j(\eta) = \xi\eta - f_j(\xi)$ (see 8E(12)). Next extend g_j to the closure of $D(j)$ by continuity, and define $g_j(\eta) = +\infty$ for all η outside the closure of $\tilde{D}(j)$. This yields the whole function g_j, and in particular the span interval $D(j)$. Of course the endpoints $d^-(j)$ and $d^+(j)$ of $D(j)$ and $\tilde{D}(j)$ can also be obtained directly from 8E(9).

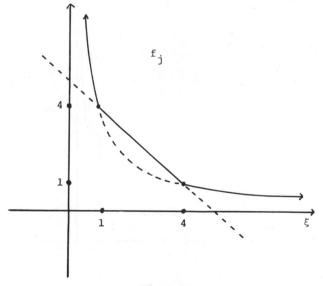

Figure 8.27

EXAMPLE 1

The procedure just described will be illustrated by a specific calculation. (Simpler examples of conjugacy will be discussed later.) Let

$$f_j(\xi) = \begin{cases} \max\{5 - \xi, 4/\xi\} & \text{if } \xi > 0, \\ +\infty & \text{if } \xi \le 0, \end{cases}$$

as in Figure 8.27. From an inspection of what happens to the slope of f_j in the

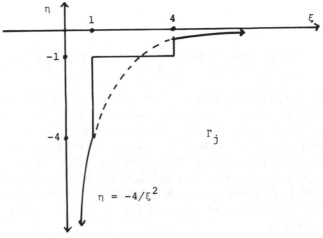

Figure 8.28

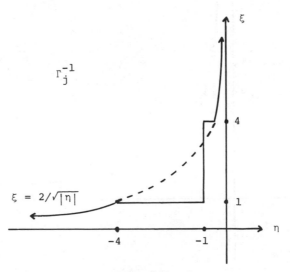

Figure 8.29

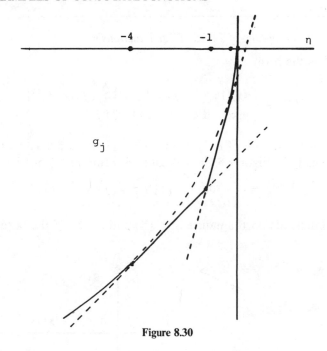

Figure 8.30

extreme, one sees immediately that $d^-(j) = -\infty$ and $d^+(j) = 0$ (see 8E(9)). Note that f_{j-}' and f_{j+}' differ only at $\xi = 1$ and $\xi = 4$. The corresponding characteristic curve $\tilde{\Gamma}_j$ is shown in Figure 8.28 and its inverse in Figure 8.29. Observe from the projections on the η-axis that $\tilde{D}(j) = (0, \infty)$. The function g_j is given on $\tilde{D}(j)$ by

$$g_j(\eta) = \begin{cases} \xi\eta - (4/\xi) & \text{for } \eta \le -4 \quad \text{and} \quad \xi = 2/\sqrt{|\eta|}, \\ 1 \cdot \eta - f_j(1) & \text{for } -4 \le \eta \le -1, \\ 4 \cdot \eta - f_j(4) & \text{for } -1 \le \eta \le -1/4, \\ \xi\eta - (4/\xi) & \text{for } -1/4 \le \eta < 0 \quad \text{and} \quad \xi = 2/\sqrt{|\eta|}. \end{cases}$$

But when $\xi = 2/\sqrt{|\eta|}$, one has $\xi\eta - (4/\xi) = -4\sqrt{|\eta|}$. Taking the limit of this expression as $\eta \downarrow 0$ one obtains $g_j(0) = 0$. Thus $D(j) = [0, \infty)$ and

$$g_j(\eta) = \begin{cases} -4\sqrt{|\eta|} & \text{for } -\infty < \eta \le -4, \\ \eta - 4 & \text{for } -4 \le \eta \le -1, \\ 4\eta - 1 & \text{for } -1 \le \eta \le -1/4, \\ -4\sqrt{|\eta|} & \text{for } -1/4 \le \eta \le 0, \\ +\infty & \text{for } \eta > 0. \end{cases}$$

This function is depicted in Figure 8.30.

EXAMPLE 2.   (*Conjugates of Linear Cost Functions*)

Suppose f_j has the form

$$f_j(\xi) = \begin{cases} d(j)\xi + p(j) & \text{if } \xi \in [c^-(j), c^+(j)] \\ +\infty & \text{if } \xi \notin [c^-(j), c^+(j)]. \end{cases}$$

Depending on the nature of $c^-(j)$ and $c^+(j)$, there are five cases to treat, as displayed already in Figures 8.10 through 8.14. From 8E(3) one has

$$d_j(\eta) = \sup_{c^-(j) \le \xi \le c^+(j)} \{\xi[\eta - d(j)] - p(j)\}$$

and this too depends on the nature of $c^-(j)$ and $c^+(j)$. If the latter are both

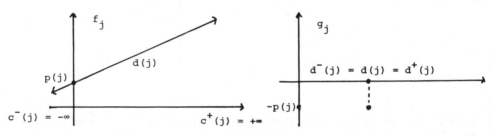

Figure 8.31

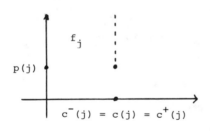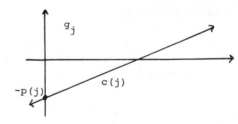

Figure 8.32

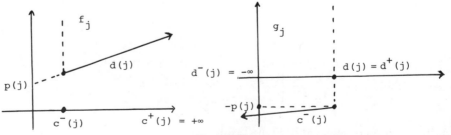

Figure 8.33

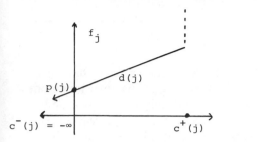

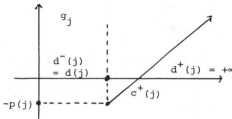

Figure 8.34

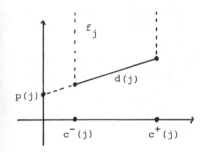

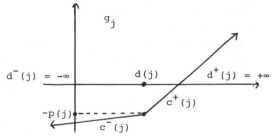

Figure 8.35

finite, it is obvious that

$$g_j(\eta) = \begin{cases} c^+(j)[\eta - d(j)] - p(j) & \text{if } \eta \ge d(j), \\ c^-(j)[\eta - d(j)] - p(j) & \text{if } \eta \le d(j). \end{cases}$$

This function is not linear but piecewise linear. In the general case it turns out that the same formula works for g_j, provided the right arithmetic is invoked for $c^+(j)$ for $c^-(j)$ when these are infinite:

$$(\pm\infty)\cdot\alpha = \begin{cases} \pm\infty & \text{if } \alpha > 0, \\ 0 & \text{if } \alpha = 0, \\ \mp\infty & \text{if } \alpha < 0. \end{cases}$$

The five possibilities are shown in Figures 8.31 through 8.35. Note that Figure 8.33 gives the kind of cost function pairs that were encountered in the "elementary" problems in Section 7I.

Especially important is the case where f_j is the indicator of the interval $[c^-(j), c^+(j)]$, that is, where $d(j) = 0$ and $p(j) = 0$. Then with the same arithmetic for $\pm\infty$ one has

$$g_j(\eta) = \begin{cases} c^+(j)\eta & \text{if } \eta \ge 0, \\ c^-(j)\eta & \text{if } \eta \le 0. \end{cases}$$

EXAMPLE 3.   (*Piecewise Linear Costs*)

The closed proper convex functions f_j which are piecewise linear have been characterized in Section 8B as the ones corresponding to maximal monotone relations Γ_j of step form. But Γ_j has this form if and only if Γ_j^{-1} does, and g_j corresponds to Γ_j^{-1}. *Thus f_j is piecewise linear if and only if g_j is piecewise linear.*

It is hard to find an expression for f_j and g_j in terms of linear pieces that is satisfactory in representing conjugacy in all situations. However, this is not really necessary, as a simple prescription works for passing from one function to the other and covers not only the four general cases exemplified in Section 8B by Figures 8.15 through 8.18, but also the five "linear" cases of the preceding example.

Given f_j piecewise linear, let $c_1(j) < \cdots < c_r(j)$ be the breakpoints and $d_1(j) < \cdots < d_s(j)$ the successive slope values. Determine the interval $D(j)$

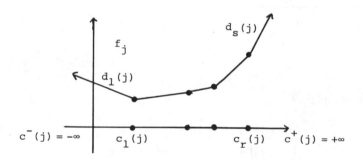

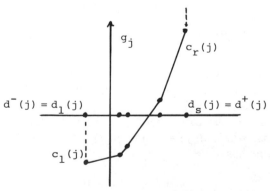

Figure 8.36

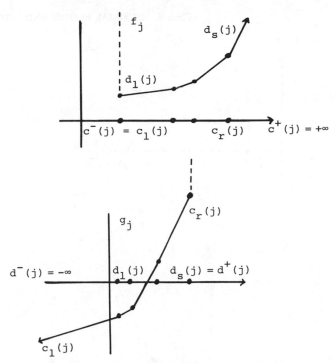

Figure 8.37

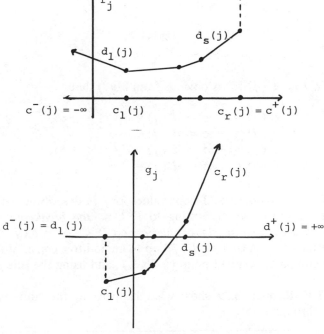

Figure 8.38

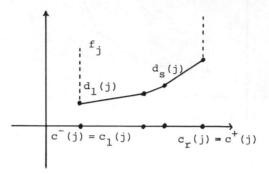

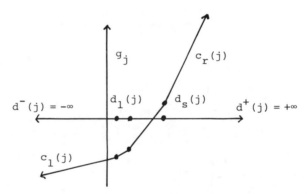

Figure 8.39

$$= \tilde{D}(j) = [d^-(j), d^+(j)];$$ it is obvious from 8E(9) that

$$
\begin{aligned}
c^-(j) &= -\infty \Rightarrow d^-(j) = d_1(j) > -\infty, \\
c^-(j) &> -\infty \Rightarrow d^-(j) = -\infty, \\
c^+(j) &= +\infty \Rightarrow d^+(j) = d_s(j) < +\infty, \\
c^+(j) &< +\infty \Rightarrow d^+(j) = +\infty.
\end{aligned}
$$

The roles of the breakpoints and slope values for f_j in describing the "steps" in the curve Γ_j are reversed in passing to Γ_j^{-1}, so *the breakpoints of* g_j *are* $d_1(j) < \cdots < d_s(j)$ *and the slope values are* $c_1(j) < \cdots < c_r(j)$. This information determines g_j on $[d^-(j), d^+(j)]$ up to an additive constant that can be fixed by selecting a convenient point $(\bar{\xi}, \bar{\eta}) \in \Gamma_j$ and using the rule $g_j(\bar{\eta}) = \bar{\xi}\bar{\eta} - f_j(\bar{\xi})$.

Figures 8.36 through 8.39 show what happens in the four examples in Figures 8.15 through 8.18.

EXAMPLE 4.   (*Piecewise Quadratic Costs*)

The class of piecewise quadratic functions can be handled much as in the preceding example. It corresponds to the maximal monotone relations which are polygonal, a class preserved under taking inverses. Therefore *f_j is piecewise quadratic if and only if g_j is piecewise quadratic.*

In constructing g_j from f_j, it is simplest to work with the fact that *a line segment* in Γ_j *with positive slope* α *corresponds to a line segment in* Γ_j^{-1} *with slope*

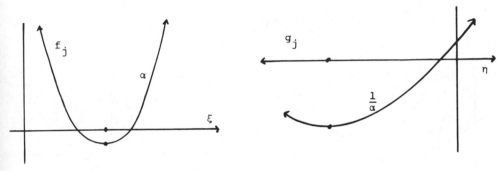

Figure 8.40

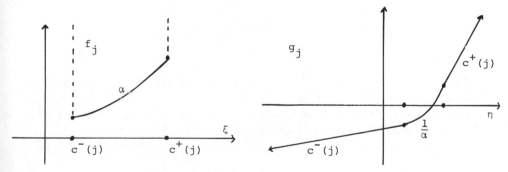

Figure 8.41

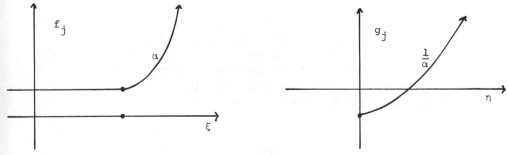

Figure 8.42

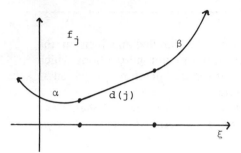
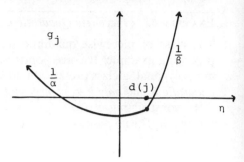

Figure 8.43

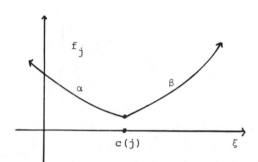
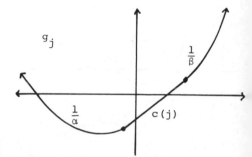

Figure 8.44

α^{-1}. Thus a nonlinear piece of f_j where the second derivative is the constant α corresponds to a nonlinear piece of g_j where the second derivative is α^{-1}. For example, the conjugates of the functions f_j depicted in Section 8C in Figures 8.10 through 8.23 are given in Figures 8.40 through 8.44.

8G.   OPTIMAL DIFFERENTIAL PROBLEM

On the basis of the relations explained in Section 8E, a dual description can be given of the data introduced in setting up the optimal distribution problem. There is a connected network G with supplies $b(i)$ that satisfy $b(N) = 0$. Each arc j has a span interval $D(j)$, the domain of finiteness of a closed proper convex function g_j giving the costs associated with tension in j.

For such data it will be fruitful to study the expression

$$\Psi(u) = - \sum_{i \in N} u(i)b(i) - \sum_{j \in A} g_j(v(j)), \qquad (1)$$

where u is an arbitrary potential and $v = \Delta u$. Clearly $\Psi(u)$ can be a real number or $-\infty$, and

$$\Psi(u) > -\infty \Leftrightarrow v(j) \in D(j) \quad \text{for all } j \in A. \tag{2}$$

Notice that $\Psi(u)$ actually depends only on the differential v. Indeed if u' is another potential with $\Delta u' = v$, then u' and u differ only by a constant (since G is connected): $u'(i) = u(i) + \alpha$ for all $i \in N$. Hence

$$\sum_{i \in N} u'(i)b(i) = \sum_{i \in N} u(i)b(i) + \alpha b(N) = \sum_{i \in N} u(i)b(i),$$

because $b(N) = 0$.

In the case of an elementary optimal distribution problem, each function f_j has the form

$$f_j(\xi) = \begin{cases} d(j)\xi + p(j) & \text{if } \xi \geq c(j), \\ +\infty & \text{if } \xi < c(j), \end{cases} \tag{3}$$

and the conjugate is then

$$g_j(\eta) = \begin{cases} c(j)\eta + q(j) & \text{if } \eta \leq d(j), \\ +\infty & \text{if } \eta > d(j), \end{cases} \tag{4}$$

where $q(j) = -c(j)d(j) - p(j)$, as seen in Example 2 in Section 8F (see Figure 8.33). Thus (1) reduces to the expression maximized in the corresponding elementary optimal differential problem in Section 7I. The dual of the optimal distribution problem in the general case is the following.

Optimal Differential Problem. *Maximize $\Psi(u)$ over all potentials u whose differential v satisfies*

$$v(j) \in D(j) \quad \text{for all } j \in A. \tag{5}$$

The objective here is to seek from among the solutions to the feasible differential problem with span intervals $D(j)$ one that is best in the sense of maximizing the expression (1). Since $\Psi(u)$ really depends only on $v = \Delta u$, everything could be expressed entirely in terms of "differentials," and this is the justification for the name of the problem. However, in applications and algorithms it is generally more convenient to work with the corresponding potentials, despite the arbitrariness of the constant of integration.

A potential u whose differential v satisfies (5) is called a *feasible solution*. If it also furnishes the maximum of Ψ, it is called an *optimal solution*, or simply a *solution* to the optimal differential problem. The feasible differential theorem in Section 6B furnishes a necessary and sufficient condition for the existence of a feasible solution in the case where all the span intervals are closed, $D(j) = [d^-(j), d^+(j)]$: one must have

$$d^+(P) \geq 0 \quad \text{for all elementary circuits } P. \tag{6}$$

The form of the condition that serves even when $D(j)$ may not be closed is

$$0 \in D(P) \quad \text{for all elementary circuits } P, \tag{7}$$

where

$$D(P) = \sum_{j \in P+} D(j) - \sum_{j \in P-} D(j) = \left\{ \sum_{j \in P+} \eta_j - \sum_{j \in P-} \eta_j \mid \eta_j \in D(j) \right\} \tag{8}$$

(see Exercise 6.5). A potential u is a *regularly feasible solution* if

$$v(j) \in \tilde{D}(j) \quad \text{for all } j \in A \tag{9}$$

(see 8E(2) for the definition of $\tilde{D}(j)$).

The optimal differential problem is *linear* if every g_j has the form

$$g_j(v(j)) = \begin{cases} c(j)v(j) + q(j) & \text{if } v(j) \in [d^-(j), d^+(j)], \\ +\infty & \text{if } v(j) \notin [d^-(j), d^+(j)]. \end{cases}$$

It is *piecewise linear* if every g_j is piecewise linear, and *piecewise quadratic* if every g_j is piecewise quadratic. In all these cases one has $\tilde{D}(j) = D(j) = [d^-(j), d^+(j)]$, so the notions of feasibility and regular feasibility coincide. Examples 3 and 4 in Section 8F show how such classes of problems behave under duality. To wit: *the optimal differential problem is piecewise linear (respectively, piecewise quadratic) if and only if the optimal distribution problem is piecewise linear (respectively, piecewise quadratic)*. But the dual of a linear problem is not necessarily linear (just piecewise linear, except in special cases); see Exercises 8.31 and 8.32.

In view of (2) the optimal differential problem is equivalent to maximizing $\Psi(u)$ over *all* potentials u. Thus

$$\begin{bmatrix} \text{sup in optimal} \\ \text{distribution problem} \end{bmatrix} = \sup\{ \Psi(u) \mid u \in R^N \}, \tag{10}$$

where "$\sup = -\infty$" means that feasible solutions do not exist.

The optimal differential problem is dual to the optimal circulation problem with $b(i) = 0$ for all $i \in N$, so that the first sum drops out of (1). It is sometimes worthwhile to consider a generalized model, the *optimal potential problem*, where each node i has a closed proper convex function g_i with associated interval $D(i)$, and one maximizes

$$- \sum_{i \in N} g_i(u(i)) - \sum_{i \in A} g_j(v(j)) \quad (v = \Delta u) \tag{11}$$

subject to

$$u(i) \in D(i) \quad \text{for all } i \in N, \qquad v(j) \in D(j) \quad \text{for all } j \in A. \qquad (12)$$

However, the latter can just as well be viewed as an optimal differential problem in the augmented network $\overline{G}$ with all supplies equal 0. It is dual to the general optimal flow problem in which 8D(11) is minimized subject to 8D(12) (see Exercise 8.36). For such duality, f_i and g_i are conjugate to each other, as well as f_j and g_j.

8H.   DUALITY THEOREM AND EQUILIBRIUM CONDITIONS

The duality between the optimal distribution problem and the optimal differential problem is tied in with a joint characterization of solutions to the two problems in terms of the curves Γ_j, just as it was in the elementary case treated in Section 7I. But these curves now come to life and become a focus of interest in their own right. The idea is that Γ_j may be given directly as an "equilibrium relation" that must hold between flux and tension in the arc j.

Network Equilibrium Problem.   *Find a flow x and a potential u such that*

$$\operatorname{div} x = b \quad and \quad (x(j), v(j)) \in \Gamma_j \quad for \ all \ arcs \ j. \qquad (1)$$

Henceforth (1) will be referred to as the *network equilibrium conditions*.

The fact that each of the three elements f_j, g_j, and Γ_j determines the other two leads to a similar conclusion about the optimal distribution problem, the optimal differential problem, and the network equilibrium problem. Any one of the three problems carries the other two along with it. In particular, any network equilibrium problem concerning maximal monotone relations Γ_j corresponds to a dual pair of optimization problems. The implications of this will be savored later in this section. Of course Γ_j only determines the conjugate pair f_j, g_j, up to an arbitrary constant of integration, so the objective functions Φ and Ψ in the corresponding optimal distribution and differential problems are likewise determined only up to an additive constant. But the choice of the constant has no real effect on minimizing or maximizing.

A more general form of the network equilibrium problem would be to seek a flow x and potential u such that

$$(x(i), v(j)) \in \Gamma_j \quad \text{for all } j \in A \quad and \quad (y(i), u(i)) \in \Gamma_i \quad \text{for all } i \in N,$$

$$(2)$$

where Γ_i is a maximal monotone relation associated with the node i and $y = \operatorname{div} x$. In (2), Γ_i is the vertical line in R^2 which intersects the horizontal axis at the point $b(i)$ (so that $y(i)$ must equal $b(i)$, but $u(i)$ can be anything). However, this is a generalization more in name than in fact. The conditions (2) can be cast back into the form (1) with $b = 0$ simply by passing to the augmented network with a distribution node $\bar{i}$ and letting Γ_i be the characteristic curve of the arc $j_i \sim (\bar{i}, i)$.

The network equilibrium conditions enter the study of optimality through the fact that for any flow x and potential u one has, for all arcs j,

$$f_j(x(j)) \geq x(j)v(j) - g_j(v(j)), \qquad \text{with equality} \Leftrightarrow (x(j), v(j)) \in \Gamma_j$$

(3)

(see 8E(11) (12)). Suppose $\operatorname{div} x = b$, so that

$$\sum_{j \in A} x(j)v(j) = - \sum_{i \in N} u(i)b(i).$$

When (3) is added over all $j \in A$, one gets

$$\sum_{j \in A} f_j(x(j)) \geq - \sum_{i \in N} u(i)b(i) - \sum_{j \in A} g_j(v(j)),$$

with equality if and only if $(x(j), v(j)) \in \Gamma_j$ for all $j \in A$. This proves that

$$\Phi(x) \geq \Psi(u) \quad \text{for all } x \text{ and } u \text{ such that } \operatorname{div} x = b,$$

(4)

with equality holding if and only if x and u actually satisfy the network equilibrium conditions.

One thing apparent from (4) is the inequality

$$\inf\{\Phi(x) \mid x \in R^A, \operatorname{div} x = b\} \geq \sup\{\Psi(u) \mid u \in R^N\},$$

which says that

$$\begin{bmatrix} \text{inf in optimal} \\ \text{distribution problem} \end{bmatrix} \geq \begin{bmatrix} \text{sup in optimal} \\ \text{differential problem} \end{bmatrix}.$$

Furthermore from the case of equality in (4) one can conclude that *x and u satisfy the network equilibrium conditions if and only if* $\operatorname{div} x = b$ *and*

$$\begin{bmatrix} \text{inf in optimal} \\ \text{distribution problem} \end{bmatrix} = \Phi(x) = \Psi(u) = \begin{bmatrix} \text{sup in optimal} \\ \text{differential problem} \end{bmatrix}.$$

These conditions therefore imply that x and u are optimal.

The following, stronger results actually hold.

Network Duality Theorem.   *If at least one of the two problems has a feasible solution, then*

$$\begin{bmatrix} \text{inf } in \; optimal \\ distribution \; problem \end{bmatrix} = \begin{bmatrix} \text{sup } in \; optimal \\ differential \; problem \end{bmatrix}.$$

Network Equilibrium Theorem.   *A flow x and potential u satisfy the network equilibrium conditions* (1) *if and only if x solves the optimal distribution problem and u solves the optimal differential problem.*

The second theorem is an immediate corollary of the first and the conclusions just drawn about (4). It reduces the network equilibrium problem to solving a pair of optimization problems. Existence and uniqueness theorems for solutions to the optimization problems will therefore be applicable to the existence and uniqueness of x and u satisfying the network equilibrium conditions (see Section 8L and Exercises 8.48 through 8.50 and 8.55 through 8.57).

In the duality theorem one should remember that "inf $= +\infty$" means there are no feasible solutions to the first problem, whereas "sup $= -\infty$" means there are no feasible solutions to the second. Thus the equation includes the assertion that *if the optimal distribution problem has feasible solutions, then its infimum is $-\infty$ if and only if the optimal differential problem has no feasible solutions. Similarly, if the optimal differential problem has feasible solutions, then its supremum is $+\infty$ if and only if the optimal distribution problem has no feasible solutions.* In the exceptional case not covered by the equation, one has both inf $= +\infty$ and sup $= -\infty$. That this can indeed occur is demonstrated by the example in Exercise 8.38(a).

When $b(i) = 0$ for all $i \in N$, the duality theorem reduces to an assertion about the circulation space $\mathscr{C}$ and its orthogonal complement, the differential space $\mathscr{D}$:

$$\inf\left\{ \sum_{j \in A} f_j(x(j)) \mid x \in \mathscr{C} \right\} = \sup\left\{ -\sum_{j \in A} g_j(v(j)) \mid v \in \mathscr{D} \right\}, \qquad (5)$$

unless the infimum is $+\infty$ and the supremum is $-\infty$ (i.e., unless there exists neither a feasible circulation nor a feasible differential). For the version corresponding to the general optimal flow problem and optimal potential problem, see Exercise 8.36.

The elementary duality theorem in Section 7I is of course the special case of the preceding theorem corresponding to the conjugacy of the functions 8G(3)(4). But many other results are also special cases. This includes max flow min cut, max tension min path, and the duality theorem for optimal assignments (Exercises 8.40, 8.41, and also 7.32). Here we are speaking only of the parts of these results that assert [inf] = [sup], since the question of the attainment of the infimum and supremum is relegated in the general theory to the existence theorems in Section 8L.

An algorithmic proof of the duality theorem will be provided in Section 9H. To avoid the impression of logical circularity, it should be kept in mind, however, that the inequality [inf] ≥ [sup] has already been established along with the *sufficiency* of the network equilibrium conditions for the optimality of x and u.

The equilibrium conditions can be interpreted economically in the case of a general transportation much as for the linear problem in Section 7C. We know that $(x(j), v(j)) \in \Gamma_j$ if and only if the maximum of $\xi_j v(j) - f_j(\xi_j)$ over all $\xi_j \in C_j$ is attained at $\xi_j = x(j)$. Thinking of ξ_j as the amount of a certain commodity being sent through j, and $v(j)$ as the rise in value (market price) from the initial to the terminal node of j, we see that $\xi_j v(j) - f_j(\xi_j)$ represents net profit (i.e., return from sales minus purchase cost minus transportation cost for ξ_j units). Equilibrium requires this to be maximized, subject to capacity constraints, in every arc.

Prices are discussed further in Example 7 in the next section.

8I. EQUILIBRIUM MODELS

Some applications where the network equilibrium problem arises on its own will now be described.

EXAMPLE 5.   (*Electrical Networks*)

Think of each arc $j \sim (i, i')$ as an electrical component with two terminals joined to the nodes i and i'. For instance, it may be a battery (voltage source), a generator (current source), a diode (one-way switch or gate), a linear or nonlinear resistor, or even a "black box" obtained by lumping together a two-terminal subnetwork comprised of elements such as those already mentioned. Each component j has a characteristic curve Γ_j, a maximal monotone relation describing the states of current $x(j)$, and voltage difference $v(j)$ that are possible for j. The following four cases are depicted in Figure 8.45:

1. If j is a classical resistor with resistance $r(j) > 0$, then Γ_j is a line with equations $v(j) = r(j)x(j)$ (Ohm's law).
2. If j is an ideal generator that supplies current in the amount $c(j)$, that is, pumps this amount of current from i to i' regardless of the voltages at i', then Γ_j is a vertical line with equation $x(j) = c(j)$.
3. If j is an ideal battery that maintains a voltage difference of $d(j)$ between i and i', regardless of how much current flows from i to i', then Γ_j is a horizontal line with equation $v(j) = d(j)$.

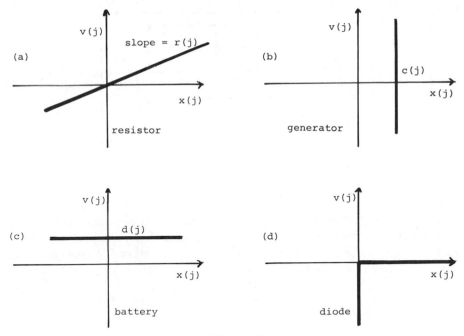

Figure 8.45

4. If j is an ideal diode, it has two possible states: either the potential difference is negative and no current can flow through j at all (the gate is closed), or the potential difference is zero and an arbitrary amount of current can flow through j in the positive direction (the gate is open). Then Γ_j is the union of the nonpositive $v(j)$-axis and the nonnegative $x(j)$-axis.

As another illustration, the "black box" in Figure 8.46 has as its characteristic curve the maximal monotone relation in Figure 7.16 that typifies "elementary" duality. The topic of "black boxes" will be taken up in Section 8N. It turns out that any maximal monotone relation that is polygonal can be represented by a "black box" made from only these four types of components. If "nonlinear resistors" are admitted, the curves approach general maximal monotone relations in appearance.

If the electrical network is supplied with current at its various nodes, the amount at node i being $b(i)$ (perhaps 0), the result ought to be a certain flow x and potential u satisfying the network equilibrium conditions, unless the relations are incompatible in some way (e.g., there is a "short circuit"). In other words, determining the "internal" currents and voltages corresponding to "external" supplies $b(i)$ amounts to solving a network equilibrium problem. Theorems about the existence and uniqueness of solutions to such problems

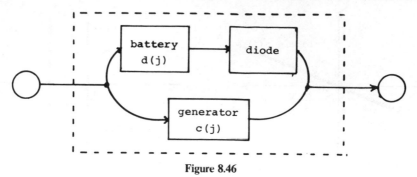

Figure 8.46

are applicable in this setting. The correspondence with the optimal distribution problem characterizes the resultant flow x as one that minimizes a certain quantity (see Exercises 8.42). In the classical case of every Γ_j linear, at least, this is the "rate of energy dissipation."

EXAMPLE 6. (*Hydraulic Networks*)

The same concepts that make sense for electricity are valid for water, or any incompressible fluid for that matter. For instance, a city water supply system can be represented by a network whose arcs correspond to pipes, aquaducts, and pumping stations, and each of these has a corresponding "resistance relation."

    In models of this type, $u(i)$ can be interpreted as the *pressure head*, that is, the level to which water would rise in an open vertical pipe inserted at node i. Certain nodes may represent reservoirs, and then there would be constraints imposed on $u(i)$, or more generally, relations Γ_i associated with $u(i)$ and $y(i)$. This would place us in the context of 8H(2).

EXAMPLE 7. (*Price Equilibrium*)

Consider a modified feasible distribution problem in which the supplies $b(i)$ are not fixed in advance but must be determined by equilibrium considerations. Specifically, suppose one is concerned with the flow of a certain good whose *demand* at node i is a *nonincreasing* function of the price $u(i)$ at this node. In notation convenient in the present setting, $b(i) = \psi_i(u(i))$, where ψ_i: $R \to R$ is continuous and *nondecreasing*. (As a special case ψ_i could be constant or even identically zero for some nodes.) The prices $u(i)$ are variables whose values must be determined from the demand relations and the economic fact that if the price difference $\Delta u(j)$ in an arc is positive, the flux $x(j)$ will be the maximum possible; if, however, $\Delta u(j)$ is negative, it will be the minimum possible.

The task is therefore to determine a flow x and a potential u such that for $y = \operatorname{div} x$ and $v = \Delta u$ one has

$$y(i) = \psi_i(u(i)) \quad \text{for all } i \in N$$

$$c^-(j) \le x(j) \le c^+(j) \quad \text{for all } j \in A,$$

$$x(j) = c^+(j) \quad \text{for } j \text{ such that } v(j) > 0,$$

$$x(j) = c^-(j) \quad \text{for } j \text{ such that } v(j) < 0.$$

But these are relations of the form 8H(2) with Γ_i and Γ_j maximal monotone (Exercise 8.42). One is therefore confronted with solving a certain equilibrium problem in the augmented network (see Exercise 8.36).

This example can be generalized by supposing merely that for each arc j there is a maximal monotone relation Γ_j expressing the dependence of the flux $x(j)$ on the price differential $v(j)$.

8J.   IMPROVEMENT OF FLOWS

Several properties of the cost function Φ in the optimal distribution problem are critical in the development of algorithms and will now be explored. Of particular interest are questions related to replacing a feasible solution x by another feasible solution $x + te_P$ that satisfies $\Phi(x + te_P) < \Phi(x)$.

Let us observe first of all, from the continuity of each f_j relative to the closed interval $[c^-(j), c^+(j)]$, that Φ *is continuous relative to the set*

$$\{x \in R^A \mid c^-(j) \le x(j) \le c^+(j) \quad \text{for all } j \in A\},$$

which is the closure of the set of flows satisfying $x(j) \in C(j)$ for all arcs j (and therefore contains all flows whose cost is finite). All level sets of the form $\{x \in R^A \mid \Phi(x) \le \alpha\}$ or $\{x \in R^A \mid \Phi(x) = \alpha\}$ with α an arbitrary real number are therefore closed. It follows by 8D(2) and the continuity of the linear mapping $x \to \operatorname{div} x$ that all sets of the form

$$\{x \in R^A \mid x \text{ is a feasible solution with } \Phi(x) \le \alpha\} \tag{1}$$

are closed. The question of when such sets are also bounded will be taken up in Section 8L, and the answer will be used in ascertaining the existence of optimal solutions. At all events, note that the *set of optimal solutions is closed in* R^A, since it coincides with (1) for α equal to the infimum in the problem.

The cost function Φ is also *convex* on R^A in the sense that if $x = (1 - \lambda)x' + \lambda x''$ for arbitrary flows x, x' and $0 < \lambda < 1$, then

$$\Phi(x) \le (1 - \lambda)\Phi(x') + \lambda\Phi(x'').$$

This is immediate from the fact that Φ is the sum of functions f_j satisfying for each flux $x(j) = (1 - \lambda)x'(j) + \lambda x''(j)$ the inequality

$$f_j(x(j)) \leq (1 - \lambda)f_j(x'(j)) + \lambda f_j(x''(j)).$$

It implies that the sets (1), and in particular the set of all optimal solutions, are *convex* in the sense of containing along with arbitrary x' and x'' everything of the form $(1 - \lambda)x' + \lambda x''$ with $0 < \lambda < 1$.

The convexity of Φ has the consequence that for an arbitrary feasible solution x and any flow z the function $f(t) = \Phi(x + tz)$ is convex in $t \in R$; in fact f is a closed proper convex function (Exercise 8.15). Geometrically, the set of flows $x + tz$ as t ranges over $[0, \infty)$ forms a "ray" in the space R^A emanating from x in the direction of z. Especially important in the development of algorithms is the case where z is the incident vector e_P for an elementary path P, because some methods iteratively replace a feasible solution x by one of the form $x' = x + \alpha e_P$. In this context one has

$$f(t) = \Phi(x + te_P) = \sum_{j \in P^+} f_j(x(j) + t) + \sum_{j \in P^-} f_j(x(j) - t) + \sum_{j \notin P} f_j(x(j)).$$

$$(2)$$

If P is a circuit, the flow $x + te_P$ will have the same divergence b as x and therefore will be another feasible solution to the optimal distribution problem if and only if $f(t) < \infty$ (see 8D(2)). Of course the values of t for which $f(t) < \infty$ are the ones for which $x(j) + t \in C(j)$ for all $j \in P^+$ and $x(j) - t \in C(j)$ for all $j \in P^-$.

The right derivative $f'_+(0)$ in (2) is of great interest, since it bears on whether $x + te_P$ is a "cheaper" flow than x for small, positive values of t. The derivative $f'_+(t)$, for t such that $x + te_P$ is feasible, is the limit as $s \downarrow t$ of the difference quotient

$$\frac{f(s) - f(t)}{s - t} = \sum_{j \in P^+} \frac{f_j(x(j) + s) - f_j(x(j) + t)}{s - t}$$

$$+ \sum_{j \in P^-} \frac{f_j(x(j) - s) - f_j(x(j) - t)}{s - t}.$$

Therefore

$$f'_+(t) = \sum_{j \in P^+} f'_{j+}(x(j) + t) - \sum_{j \in P^-} f'_{j-}(x(j) - t), \qquad (3)$$

at least if $x + te_P$ is regularly feasible so that $f'_{j+}(x(j) + t) > -\infty$ and $f'_{j-}(x(j) - t) < +\infty$ in (3). If $x + te_P$ is feasible but not regularly feasible, the right side of (3) might involve adding terms that are oppositely infinite and thus be undefined. However, the formula remains valid if in such cases one follows the rule $\infty - \infty = +\infty$. This can be seen from the fact that arcs j yielding $+\infty$ in the sum already do so for the difference quotients, whereas those yielding $-\infty$ only do so in the limit. With $\infty - \infty = +\infty$, (3) actually is valid for all $t \in R$, even if $x + te_P$ is not feasible (but x is). Clearly $f'_+(t) = +\infty$ if and only if, for all $s > t$, $x + se_P$ is not feasible. In similar fashion one obtains

$$f'_-(t) = \sum_{j \in P^+} f'_{j-}(x(j) + t) - \sum_{j \in P^-} f'_{j+}(x(j) - t) \tag{4}$$

for all t, but this time interpreting $\infty - \infty$ as $-\infty$.

Especially important, of course, is the case where $f'_+(0) < 0$, since then $\Phi(x + te_P) < \Phi(x)$ for $t > 0$ sufficiently small. This case corresponds to having $d_x^+(P) < 0$ for

$$\left[d_x^-(j), d_x^+(j) \right] = \left[f'_{j-}(x(j)), f'_{j+}(x(j)) \right].$$

It will be pursued in Section 9A.

Recall now that in the optimal distribution problem every arc j has associated with it not only capacity bounds $c^+(j)$ and $c^-(j)$ but also span bounds $d^+(j)$ and $d^-(j)$ that are implicit in the duality in Section 8E. Hence for every circuit P, the quantities $d^+(P)$ and $d^-(P)$ are well defined. Our next goal is to relate these quantities and the condition $d^+(P) < 0$ to the derivatives (3) and (4), and to formulate the appropriate generalization of the concept of "unbalanced circuit" that played such an important role in the existence theorem for the linear problem in Section 7A.

The feasible differential theorem says that the existence of an elementary circuit P with $d^+(P) < 0$ is equivalent to the nonexistence of a potential u whose differential v satisfies $v(j) \in [d^-(j), d^+(j)]$ for all arcs j. This is not quite the same as the nonexistence of a feasible solution to the optimal differential problem, since $D(j)$ might not be closed. The sharpened condition for that, as pointed out in Section 8G as an application of the generalized feasible differential theorem in Exercise 6.28, is the existence of an elementary circuit P with $0 \notin D(P)$, where $D(P)$ is the interval in 8G(8). Passing to the reverse of P if necessary, one can write this condition as $D(P) \subset (-\infty, 0)$. Of course $d^-(P)$ and $d^+(P)$ are the endpoints of $D(P)$, and it is possible to have $D(P) \subset (-\infty, 0)$ even though $d^+(P) = 0$.

It is useful at this stage to introduce the corresponding concepts for the intervals $\tilde{D}(j)$. Thus for an elementary circuit P, let

$$\tilde{D}(P) = \sum_{j \in P+} \tilde{D}(j) - \sum_{j \in P-} \tilde{D}(j) = \left\{ \sum_{j \in P+} \eta_j - \sum_{j \in P-} \eta_j \,\middle|\, \eta_j \in \tilde{D}(j) \right\}. \tag{5}$$

This is an interval contained in $D(P)$ and, like the latter, having $d^-(P)$ and $d^+(P)$ as its endpoints.

An elementary circuit P will be said to be *unbalanced* if $\tilde{D}(P) \subset (-\infty, 0)$. The existence of an unbalanced elementary circuit is therefore equivalent, by the generalized feasible differential theorem just cited, to the nonexistence of a potential u whose differential v satisfies $v(j) \in \tilde{D}(j)$ for all arcs j, or in other words, to the nonexistence of a *regularly* feasible solution to the optimal differential problem.

Obviously, for problems with $\tilde{D}(j) = D(j) = [d^-(j), d^+(j)]$, such as piecewise linear or quadratic problems, the unbalanced circuits are simply the ones with $d^+(P) < 0$, and their presence can be determined by an application of the feasible differential algorithm or tension rectification algorithm (see Chapter 6). In particular, this notation reduces to what was introduced in Section 7A in connection with the linear optimal distribution problem: if $f_j(\xi) = d(j)\xi + p(j)$ on $C(j) = [c^-(j), c^+(j)]$, one has

$$\tilde{D}(j) = D(j) = \begin{cases} [d(j), d(j)] & \text{if } c^-(j) = -\infty, c^+(j) = +\infty, \\ [d(j), \infty) & \text{if } c^-(j) = -\infty, c^+(j) < +\infty, \\ (-\infty, d(j)] & \text{if } c^-(j) > -\infty, c^+(j) = +\infty, \\ (-\infty, \infty) & \text{if } c^-(j) > -\infty, c^+(j) < \infty \end{cases}$$

(see 8E(9)), so that $\tilde{D}(j) \subset (-\infty, 0)$ if and only if P is a circuit of unlimited capacity with $d \cdot e_P < 0$.

The connection with feasibility explains why unbalanced circuits are important in the study of the optimal differential problem, but their role in the optimal distribution problem is equally important. As pointed out earlier, for any feasible solution x to the optimal distribution problem and any elementary circuit P, the function $f(t) = \Phi(x + te_P)$ is closed proper convex in $t \in R$ with derivatives given by (3) and (4) (the rule $\infty - \infty = +\infty$ being invoked in (3) if necessary when x is not *regularly* feasible but $\infty - \infty = -\infty$ in (4)). According to the limit formulas 8E(9), one has

$$\lim_{t \uparrow +\infty} f'_{j-}(x(j) + t) = \lim_{t \uparrow +\infty} f'_{j+}(x(j) + t) = d^+(j),$$

$$\lim_{t \uparrow +\infty} f'_{j-}(x(j) - t) = \lim_{t \uparrow +\infty} f'_{j+}(x(j) - t) = d^-(j).$$

Therefore

$$\lim_{t \uparrow +\infty} f'_-(t) = \lim_{t \uparrow +\infty} f'_+(t) = d^+(P), \tag{6}$$

and similarly

$$\lim_{t \downarrow -\infty} f'_-(t) = \lim_{t \downarrow -\infty} f'_+(t) = d^-(P). \tag{7}$$

By the same kind of reasoning based on 8E(8), one has $\eta \in \tilde{D}(P)$ if and only if $\eta \leq f'_+(t)$ for all t sufficiently high, and $\eta \geq f'_-(t)$ for all t sufficiently low. Hence

$$\tilde{D}(P) = \left[\begin{array}{c} \text{projection of the maximal monotone relation } \Gamma \\ \text{corresponding to } f \text{ onto the vertical axis.} \end{array} \right] \quad (8)$$

Thus P is an unbalanced circuit if and only if $f'_+(t) < 0$ for all t. This may be compared with the fact, evident from (6), that $d^+(P) \leq 0$ if and only if $f'_+(t) \leq 0$ for all t, whereas $d^+(P) < 0$ if and only if for some $\varepsilon > 0$ one has $f'_+(t) \leq -\varepsilon$ for all t.

Recall now that a function f on R is *decreasing* if $f(t_2) < f(t_1)$ whenever $t_1 < t_2$ and $f(t_1) < \infty$. It is *nonincreasing* if $f(t_2) \leq f(t_1)$ whenever $t_1 < t_2$, and *strongly decreasing* if for some $\varepsilon > 0$ one has $f(t_2) \leq f(t_1) - \varepsilon(t_2 - t_1)$ whenever $t_1 < t_2$ and $f(t_1) < \infty$. Combining the derivative forms of these properties (see Exercise 8.10) with the facts just recorded, one obtains the following result.

Descent Theorem for Flows. *Let P be any elementary circuit, and let $f(t) = \Phi(x + te_P)$, where x is any feasible solution to the optimal distribution problem. Then:*

1. *f is nonincreasing $\Leftrightarrow d^+(P) \leq 0$.*
2. *f is decreasing $\Leftrightarrow P$ is unbalanced, i.e., $\tilde{D}(P) \subset (-\infty, 0)$.*
3. *f is strongly decreasing $\Leftrightarrow d^+(P) < 0$.*

Corollary. *If $f(t) = \Phi(x + te_P)$ is nonincreasing, decreasing, or strongly decreasing for one choice for the feasible solution x, then it has the same property for every x.*

8K.   IMPROVEMENT OF POTENTIALS

Corresponding properties of the function Ψ maximized in the optimal differential problem will now be derived.

Convexity of g_j implies Ψ is *concave* on R^N in the sense that if $u = (1 - \lambda)u' + \lambda u''$ with $0 < \lambda < 1$ (and hence $v = (1 - \lambda)v' + \lambda v''$ for the corresponding differentials), then

$$\Psi(u) \geq (1 - \lambda)\Psi(u') + \lambda\Psi(u'').$$

Furthermore the continuity of g_j relative to the closure $[d^-(j), d^+(j)]$ of $D(j)$ implies Ψ is *continuous relative to the set*

$$\{u \in R^N \mid d^-(j) \leq v(j) \leq d^+(j) \text{ for all } j \in A\}.$$

Then the set

$$\{u \in R^N | v(j) \in D(j) \text{ for all } j \in A \text{ and } \Psi(u) \geq \beta\} \tag{1}$$

is convex and closed (see 8G(2)). In particular, the set of all optimal solutions is convex and closed, since it is the set (1) for β taken to be the maximum value in the problem.

Maximizing Ψ is equivalent to minimizing the convex function $-\Psi$. The formulation of the problem in terms of maximization is dictated by the goal of duality with the optimal distribution problem, but it is sometimes simpler to work with $-\Psi$ in the fine details of algorithmic development, since otherwise one would be obliged, for example, to refer to a parallel theory of "closed proper concave functions" on R and their one-sided derivatives.

For arbitrary potentials u and w such that u is feasible, $f(t) = -\Psi(u + tw)$ is a closed proper convex function of $t \in R$ (Exercise 8.29). Note that the set $\{u + tw | t \geq 0\}$ is a ray in the potential space R^N emanating from u in the direction of w. The case of greatest interest occurs when a cut $Q = [S, N \setminus S]$ is under consideration and w is the potential $e_{N \setminus S}$, where

$$e_{N \setminus S}(i) = \begin{cases} 0 & \text{if } i \in S, \\ 1 & \text{if } i \in N \setminus S. \end{cases}$$

The differential of w is then $\Delta e_{N \setminus S} = e_Q$, so that

$$f(t) = -\Psi(u + te_{N \setminus S}) = -tb(S) + \sum_{i \in N} u(i)b(i)$$

$$+ \sum_{j \in Q^+} g_j(v(j) + t) + \sum_{j \in Q^-} g_j(v(j) - t) + \sum_{j \notin Q} g_j(v(j)) \tag{2}$$

by the rule

$$b(N \setminus S) = -b(S) \tag{3}$$

(a consequence of $0 = b(N) = b(S) + b(N \setminus S)$). Taking the right derivative in (2), one obtains

$$f'_+(t) = -b(S) - \sum_{j \in Q^+} g'_{j+}(v(j) + t) - \sum_{j \in Q^-} g'_{j-}(v(j) - t). \tag{4}$$

As with the derivative formulas in Section 8J, this is valid even for t such that $u + te_{N \setminus S}$ is not regularly feasible, provided one interprets $\infty - \infty$ as $+\infty$. Similarly, one has

$$f'_-(t) = -b(S) + \sum_{j \in Q^+} g'_{j-}(v(j) + t) - \sum_{j \in Q^-} g'_{j+}(v(j) - t), \tag{5}$$

but with $\infty - \infty$ interpreted as $-\infty$.

Note that the condition $f'_+(0) < 0$, which implies $\Psi(u + \alpha e_{N\setminus S}) > \Psi(u)$ for $\alpha > 0$ sufficiently small, can be expressed as $c_u^+(Q) < b(S)$ with respect to the intervals

$$\left[c_u^-(j), c_u^+(j) \right] = \left[g'_{j-}(v(j)), g'_{j+}(v(j)) \right].$$

This fact is the basis of the general optimal differential algorithm in Section 9C.

At all events the quantities $c^+(Q)$ and $c^-(Q)$ are well defined for every cut Q in terms of the capacity bounds $c^-(j)$ and $c^+(j)$ associated implicitly (by the duality in Section 8E) with each arc j. The existence of a cut $Q = [S, N\setminus S]$ with $c^+(Q) < b(S)$ is equivalent to the nonexistence of a flow x that satisfies $x(j) \in [c^-(j), c^+(j)]$ for all arcs j and $\operatorname{div} x = b$ (feasible distribution theorem). Note that such a cut must be nonempty, because of the blanket assumption in this chapter that the network G is connected and $b(N) = 0$. To cover the general case where $C(j)$ might not be closed, the condition $c^+(Q) < b(S)$ must be refined to $b(S) \notin C(Q)$ in accordance with the generalized feasible differential theorem in Exercise 3.14 (with $C(i)$ taken to be the degenerate interval $[b(i), b(i)]$). However, this can be reduced to $C(Q) \subset (-\infty, b(S))$ by passing to the reverse cut, if necessary. Thus the existence of a cut with $C(Q) \subset (-\infty, b(S))$ is equivalent to the nonexistence of a feasible solution to the optimal distribution problem.

A cut $Q = [S, N\setminus S]$ will be called *unbalanced* if actually $\tilde{C}(Q) \subset (-\infty, b(S))$, where

$$\tilde{C}(Q) = \sum_{j \in Q^+} \tilde{C}(j) - \sum_{j \in Q^-} \tilde{C}(j) = \left\{ \sum_{j \in Q^+} \xi_j - \sum_{j \in Q^-} \xi_j \,\middle|\, \xi_j \in \tilde{C}(j) \right\}. \quad (6)$$

By the theorem just cited, the existence of an unbalanced cut is equivalent to the nonexistence of a *regularly* feasible solution to the optimal distribution problem. Clearly $\tilde{C}(Q)$ is an interval having $c^-(Q)$ and $c^+(Q)$ as its endpoints, and $\tilde{C}(Q) \subset C(Q)$. Thus for problems satisfying $\tilde{C}(j) = C(j) = [c^-(j), c^+(j)]$, as is true in the piecewise linear or quadratic case, the unbalanced cuts are just the ones with $c^+(Q) < b(S)$, and their existence can be detected by the feasible distribution algorithm or the flow rectification algorithm. In the case of a linear optimal differential problem, they are the same as the unbalanced cuts introduced in Section 7E (Exercise 8.44).

Consider next the closed proper convex function $f(t) = -\Psi(u + te_{N\setminus S})$ in (2), where u is a feasible solution to the optimal differential problem. The one-sided derivatives of f are expressed in terms of the cut $Q = [S, N\setminus S]$ by (4) and (5) (where again the rule $\infty - \infty = +\infty$ should be invoked if necessary in case u is not *regularly* feasible). An argument like the one in

Section 8J, in terms of 8E(10), establishes that

$$\lim_{t \uparrow +\infty} f'_-(t) = \lim_{t \uparrow +\infty} f'_+(t) = -b(S) + c^+(Q),$$

$$\lim_{t \downarrow -\infty} f'_-(t) = \lim_{t \downarrow -\infty} f'_+(t) = -b(S) + c^-(Q), \tag{7}$$

and furthermore

$$-b(S) + \tilde{C}(Q) = \left[\begin{array}{c} \text{projection of the maximal monotone relation} \\ \text{corresponding to } f \text{ on the vertical axis} \end{array} \right] \tag{8}$$

(Exercise 8.46). *Thus Q is an unbalanced cut if and only if $f'_+(t) < 0$ for all t.* On the other hand $c^+(Q) \le b(S)$ if and only if $f'_+(t) \le 0$ for all t, whereas $f'_+(Q) < b(S)$ if and only if for some $\varepsilon > 0$ one has $f'_+(t) \le -\varepsilon$ for all t.
    This yields the following result (by Exercise 8.10).

Descent Theorem for Potentials. *Let $Q = [S, N \setminus S]$ be any cut, and let $f(t) = -\Psi(u + te_{N\setminus S})$, where u is any feasible solution to the optimal differential problem. Then:*
   1. *f is nonincreasing $\Leftrightarrow c^+(Q) \le b(S)$.*
   2. *f is decreasing $\Leftrightarrow Q$ is unbalanced, that is, $\tilde{C}(Q) \subset (-\infty, b(S))$.*
   3. *f is strongly decreasing $\Leftrightarrow c^+(Q) < b(S)$.*

Corollary. *If $f(t) = -\Psi(u + te_{N\setminus S})$ is nonincreasing, decreasing, or strongly decreasing for one choice of the feasible solution u, then it has the same property for every u.*

8L. EXISTENCE OF SOLUTIONS

The facts about unbalanced cuts and circuits lead to the following theorems, whose proof is the main goal of this section and the next.

Existence Theorem for Optimal Flows. *Suppose the optimal distribution problem has a feasible solution. Then the following are equivalent:*
   1. *The optimal distribution problem has an optimal solution.*
   2. *The optimal differential problem has a regularly feasible solution.*
   3. *No elementary circuit P is unbalanced.*
   4. *For every feasible solution x to the optimal distribution problem and every elementary circuit P, the function $f(t) = \Phi(x + te_P)$ attains its minimum at some $t \in R$.*

Existence Theorem for Optimal Potentials. *Suppose the optimal differential problem has a feasible solution. Then the following are equivalent:*
   1. *The optimal differential problem has an optimal solution.*

2.   *The optimal distribution problem has a regularly feasible solution.*

3.   *No cut Q is unbalanced.*

4.   *For every feasible solution u to the optimal differential problem and every cut $Q = [S, N \setminus S]$, the function $f(t) = -\Psi(u + te_{N \setminus S})$ attains its minimum at some $t \in R$.*

Corollary.   *The optimal distribution problem and the optimal differential problem both have optimal solutions if and only if both have regularly feasible solutions.*

For the specialization of these results to piecewise linear or quadratic problems, see Exercise 8.47.

The proof of the existence theorems will not depend on the part of the proof of the network duality theorem that has been postponed until Section 9H (i.e., the proof that equality holds, rather than just $\leq$). But when that result and its consequence, the network equilibrium theorem, are invoked, one obtains another existence result as a corollary.

Existence Theorem for Network Equilibrium.   *The following are equivalent:*

1.   *The network equilibrium relations are satisfied by some flow x and potential u.*

2.   *No elementary circuit P is unbalanced, and no cut Q is unbalanced.*

3.   *Both the optimal distribution problem and the optimal differential problem have regularly feasible solutions.*

The equivalence of Assertions 2 and 3 of all three theorems is already known, of course, from Section 8H. The equivalence of Assertions 3 and 4 of the first two theorems is immediate from the descent theorems of Sections 8J and 8K and the fact that a closed proper convex function that is neither decreasing nor increasing attains its minimum somewhere (see Exercise 8.12). (If there is an f of the given form that is increasing, then by reversing directions one gets an f that is decreasing.) It is clear also from the descent theorems and their corollaries that, for either problem, Assertion 1 implies 4. (For if $\Phi(x + te_P)$ were decreasing in t for every feasible solution x, then no flow could be optimal. Likewise, if $-\Psi(u + te_{N \setminus S})$ were decreasing in t for every feasible solution u, then no potential could be optimal.)

The remaining argument, that the existence of optimal solutions follows in each case from Assertion 4, will be completed in the next section after some groundwork.

Uniqueness of Solutions

Results on the uniqueness of optimal solutions are outlined in Exercises 8.55, 8.56, and 8.57.

8M.*   BOUNDEDNESS OF OPTIMIZING SEQUENCES

The optimal distribution problem is said to have the *boundedness property* if for every $\alpha \in R$ the closed convex set

$$\{ x \in R^A | \operatorname{div} x = b \text{ and } \Phi(x) \leq \alpha \} \tag{1}$$

is bounded. This property is equivalent to one of direct computational interest —the boundedness of every minimizing sequence of flows, or in other words, of every sequence of feasible solutions $\{ x^k \}_{k=1}^{\infty}$ such that $\Phi(x^k)$ is finite and tends to the infimum in the problem (Exercise 8.51). When this property holds, then by the compactness of closed bounded sets in R^A and the continuity properties of Φ (see Section 8J), every minimizing sequence has at least one cluster point, and any such is an optimal solution (Exercise 8.52). Thus in particular, at least one optimal solution exists.

The corresponding notion for the optimal differential problem is affected in its formulation by the fact that the feasibility of a potential u and the associated value $\Psi(u)$ are unaltered if a constant t is added to $u(i)$ at every node i. To put it another way, potentials are of interest only up to an arbitrary additive constant, and questions of boundedness should not hang on the choice of the constant. The property needed is therefore defined in terms of differentials. The optimal differential problem is thus said to have the *boundedness property* if for every $\beta \in R$ the (closed convex) set

$$\{ v \in R^A | \exists u \text{ with } \Delta u = v \text{ and } \Psi(u) \geq \beta \} \tag{2}$$

is bounded. This is equivalent to the boundedness of every *maximizing sequence of differentials*, or in other words, of every differential sequence $\{ v^k \}_{k=1}^{\infty}$ corresponding to a sequence of feasible solutions $\{ u^k \}_{k=1}^{\infty}$ with $\Psi(u^k)$ finite and tending to the supremum in the problem. When the latter holds, there is at least one cluster point for $\{ v^k \}_{k=1}^{\infty}$, and any such is the differential of a potential u that is an optimal solution. [This follows from the closedness of the differential space $\mathscr{D}$ in R^A (see Exercise 2.12), the continuity properties of Ψ in Section 8K, and the fact observed in Section 8G that $\Psi(u) = \Psi(u')$ if $\Delta u = \Delta u'$.]

Boundedness Theorems for Flows.   *Suppose there is at least one feasible solution to the optimal distribution problem. Then the following are equivalent:*

1. *The optimal distribution problem has the boundedness property.*
2. $d^+(P) > 0$ *for all elementary circuits P.*
3. *The arc set $F = \{ j | d^-(j) = d^+(j) \}$ contains no elementary circuits, and there is a feasible solution u to the optimal differential problem such that $d^-(j) < v(j) < d^+(j)$ for all $j \notin F$.*

Boundedness Theorem for Potentials. *Suppose there is at least one feasible solution to the optimal differential problem. Then the following are equivalent:*

1. *The optimal differential problem has the boundedness property.*
2. $b(S) < c^+(Q)$ *for all nonempty cuts Q.*
3. *The arc set* $F' = \{ j| \, c^-(j) = c^+(j)\}$ *contains no nonempty cuts, and there is a feasible solution x to the optimal distribution problem such that* $c^-(j) < x(j) < c^+(j)$ *for all* $j \notin F'$.

Note that in these theorems F forms a *forest* in the network (see Section 4F), while F' is a *coforest* (see Exercise 4.28).

Proof of the Boundedness Theorem for Flows. Condition 1 $\Rightarrow$ Condition 2. If there were an elementary circuit P with $d^+(P) \le 0$, then by the descent theorem for flows in Section 8J the expression $\Phi(x + te_P)$ would be nonincreasing in t. A set of the form (1) would therefore contain along with each of its elements x the entire ray $\{x + te_P| \, t \ge 0\}$ and could not be bounded.

Condition 2 $\Rightarrow$ Condition 3. For $j \notin F$, let $\hat{D}(j)$ be the open interval $(d^-(j), d^+(j))$, whereas for $j \in F$ let $\hat{D}(j)$ be the degenerate interval consisting of the point $d^-(j) = d^+(j)$. For an elementary circuit P, the interval

$$\hat{D}(P) = \sum_{j \in P^+} \hat{D}(j) - \sum_{j \in P^-} \hat{D}(j) = \left\{ \sum_{j \in P^+} \eta_j - \sum_{j \in P^-} \eta_j | \, \eta_j \in \hat{D}(j) \right\} \quad (3)$$

reduces to $(d^-(P), d^+(P))$ if $P \not\subset F$, but to the single point $d^-(P) = d^+(P)$ if $P \subset F$. The reverse circuit P', of course, has $d^+(P') = -d^-(P)$. Therefore Condition 2 is equivalent to the condition that for every elementary circuit P one has $P \not\subset F$ and $0 \in \hat{D}(P)$. But if $0 \in \hat{D}(P)$ for every elementary circuit P, then by the generalized feasible differential theorem (see Exercise 6.5) there is a potential u such that $v(j) \in \hat{D}(j)$ for all $j \in A$. This is the property asserted in Condition 3.

Condition 3 $\Rightarrow$ Condition 1. Let u be a potential with $d^-(j) < v(j) < d^+(j)$ for all $j \notin F$, and choose $\varepsilon > 0$ small enough that $v(j) + \varepsilon < d^+(j)$ and $v(j) - \varepsilon > d^-(j)$ for all $j \notin F$. Then by the conjugacy formula

$$f_j(\xi) = \sup_{d^-(j) \le \eta \le d^+(j)} \{\xi\eta - g_j(\eta)\}$$

one has for all $j \notin F$

$$f_j(\xi) \ge \xi[v(j) + \varepsilon] - g_j(v(j) + \varepsilon),$$

$$f_j(\xi) \ge \xi[v(j) - \varepsilon] - g_j(v(j) - \varepsilon),$$

whereas for $j \in F$

$$f_j(\xi) = \xi v(j) - g_j(v(j)).$$

Choosing the constant μ_j to be $g_j(v(j))$ for $j \in F$ and the larger of $g_j(v(j) + \varepsilon)$ and $g_j(v(j) - \varepsilon)$ for $j \notin F$, one therefore has

$$f_j(\xi) \geq \xi v(j) - \mu_j + \varepsilon|\xi| \quad \text{for } j \notin F,$$

$$f_j(\xi) = \xi v(j) - \mu_j \quad \text{for } j \in F.$$

Any feasible solution x to the optimal distribution problem satisfies div $x = b$, so that $x \cdot v = -b \cdot u$; hence

$$\Phi(x) \geq \sum_{j \in F} \left[x(j)v(j) - \mu_j + \varepsilon|x(j)| \right] + \sum_{j \in F} \left[x(j)v(j) - \mu_j \right]$$

$$= \varepsilon \sum_{j \notin F} |x(j)| - \mu \quad \left(\mu = \sum_{j \in A} \mu_j + b \cdot u \right).$$

Every element of the set (1) thus has

$$|x(j)| \leq (\alpha + \mu)/\varepsilon \quad \text{for all } j \notin F. \tag{4}$$

Now fix any $\bar{x}$ in (1); for every other x in (1) the flow $x - \bar{x}$ is a circulation, because div$(x - \bar{x}) = b - b = 0$. Since F is a forest, there is a spanning tree F_0 such that $F_0 \supset F$. In terms of the corresponding Tucker representation of the circulation space $\mathscr{C}$ (see Section 4G), one has for certain coefficients $a(j, k)$:

$$x(j) - \bar{x}(j) = \sum_{k \notin F_0} a(j, k)[x(k) - \bar{x}(k)] \quad \text{for all } j \in F_0.$$

But (4) implies $|x(k) - \bar{x}(k)| \leq 2(\alpha + \mu)/\varepsilon$ for all $k \notin F_0$, so the boundedness of the set (1) follows from this representation.

Proof of the Boundedness Theorem for Potentials. This is parallel to the argument just furnished and is left to the reader as Exercise 8.53.

Remaining Proof of the Existence Theorem for Optimal Flows. It has already been observed in Section 8L that in this theorem Condition 1 $\Rightarrow$ Condition 2 $\Leftrightarrow$ Condition 3 $\Leftrightarrow$ Condition 4. The task now is to derive Condition 1 from the other conditions. This will be accomplished by an inductive argument in two stages which reduces the problem to an equivalent one having the boundedness property.

Assume the equivalent conditions of Conditions 2, 3, and 4 hold. Then in particular from Condition 3, every elementary circuit P has $0 \in \check{D}(P)$, and consequently $d^-(P) \leq 0 \leq d^+(P)$. Let us say that P is of *Type 1* if $d^-(P) < 0 < d^+(P)$, of *Type 2* if $d^-(P) = 0 = d^+(P)$, and *Type 3* if either $d^-(P) < 0 = d^+(P)$ or $d^-(P) = 0 < d^+(P)$. Note that the circuits of Type 2 are thus precisely the ones such that, for all $j \in P$, $d^-(j) = d^+(j)$ (i.e., f_j is linear).

If all elementary circuits are of Type 1, the boundedness theorem just proved implies the existence of an optimal solution. (Every minimizing sequence for the optimal distribution problem has a cluster point that is optimal.)

Consider now the case where all elementary circuits are either of Type 1 or of Type 2. Suppose $\bar{P}$ is of Type 2, and let $\bar{j}$ be one of the arcs in $\bar{P}$. For every flow x, the expression $\Phi(x + te_{\bar{P}})$ is linear in t with slope $d^-(P) = d^+(P) = 0$; hence it is constant in t. Therefore the problem has an optimal solution with flux 0 in $\bar{j}$ if it has an optimal solution at all. The question of existence is not affected, therefore, if the constraint $x(\bar{j}) = 0$ is added to the problem, or what amounts to the same thing, if $f_{\bar{j}}$ is redefined to be the function

$$f_{\bar{j}}(\xi) = \begin{cases} 0 & \text{if } \xi = 0, \\ \infty & \text{if } \xi \neq 0. \end{cases}$$

In that event one has $d^+(\bar{j}) = \infty$ and $d^-(\bar{j}) = -\infty$, so that the altered problem cannot have any circuit of Type 2 containing $\bar{j}$. A finite sequence of such modifications reduces the problem to one of having only circuits of Type 1 (but still satisfying Condition 4, hence Conditions 2 and 3), and this is the case for which existence of an optimal solution was verified in the preceding paragraph.

To complete the proof, it will be shown that circuits of Type 3 can be transformed into circuits of Type 2 without sacrificing properties given by Conditions 2, 3, and 4. Let $\bar{P}$ be of Type 3. Passing to the reverse of $\bar{P}$ if necessary, one has $d^-(\bar{P}) < d^+(\bar{P}) = 0 \in \check{D}(\bar{P})$, so that $d^+(j) \in \check{D}(j)$ for all $j \in \bar{P}^+$ and $d^-(j) \in \check{D}(j)$ for all $j \in \bar{P}^-$. This means that $f_{j+}'(\xi) = f_{j-}'(\xi) = d^+(j)$ for all ξ sufficiently high when $j \in \bar{P}^+$, whereas $f_{j+}'(\xi) = f_{j-}'(\xi) = d^-(j)$ for all ξ sufficiently low when $j \in \bar{P}^-$. In other words, there are constants α_j, ξ_j, such that

$$f_j(\xi) = d^+(j)\xi + \alpha_j \quad \text{for all } \xi \geq \xi_j \text{ when } j \in \bar{P}^+,$$

$$f_j(\xi) = d^-(j)\xi + \alpha_j \quad \text{for all } \xi \leq \xi_j \text{ when } j \in \bar{P}^-. \tag{5}$$

Define

$$\hat{f}_j(\xi) = \begin{cases} d^+(j)\xi + \alpha_j & \text{if } j \in \bar{P}^+, \\ d^-(j)\xi + \alpha_j & \text{if } j \in \bar{P}^-, \\ f_j(\xi) & \text{if } j \notin \bar{P}, \end{cases}$$

and let $\hat{\Phi}$ be the function obtained from Φ when f_j is replaced by $\hat{f}_j$ for all arcs j. One has $f_j(\xi) \geq \hat{f}_j(\xi)$ for all ξ, j, by convexity, so $\Phi(x) \geq \hat{\Phi}(x)$. In fact for every flow x one has $\Phi(x + te_{\bar{P}}) = \hat{\Phi}(x + te_{\bar{P}})$ for all t sufficiently high by (5).

Therefore the modified problem with functions f_j is equivalent to the one with functions f_j as far as the existence of optimal solutions is concerned, and it still has the property given by Condition 4 (hence Conditions 2 and 3). In the modified problem the circuit $\overline{P}$ is of Type 2, not Type 3. Moreover every circuit of Type 2 for the original problem remains of Type 2 for the modified problem (since $\hat{f}_j = f_j$ when f_j is linear). It follows that a finite sequence of modifications of this sort will reduce the problem to one in which there are no elementary circuits of Type 3. (An infinite sequence is impossible, since the cost function for at least one more arc is made linear at each stage.) Then the existence of an optimal solution is ensured by the argument already given.

Remaining proof of the Existence Theorem for Optimal Potentials. This follows essentially the same lines as the preceding proof in reducing a general problem to one for which the boundedness property for potentials holds. For such a problem the existence of an optimal solution is assured. We omit the details.

8N.* BLACK BOXES

The treatment of piecewise linear cost functions in Chapter 7 hinged on the representation of an arc with such costs by a subnetwork comprised of arcs having a simpler cost structure. Similar ideas were touched on in the discussion of the network equilibrium problem in Section 8I: an arc might correspond to a "black box" made of various components, and its characteristic curve would then be the "resultant" in some sense of those of the components. The time has come to look more closely at the question of what happens when a subnetwork is lumped together as a single arc, partly for the sake of clarifying the

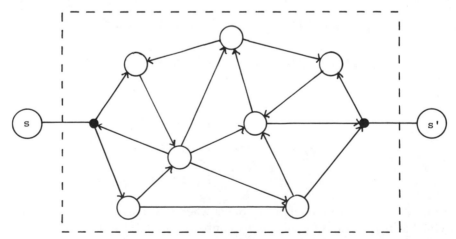

Figure 8.47

procedure and enhancing its usefulness but also as an illustration of the powerful theoretical consequences of the duality and existence theorems.

An application to project cost curves (PERT) is described in Example 10 at the end of the section.

The object of study is a connected network G with two distinguished nodes s and s' as in Figure 8.47. Each arc j of G has a characteristic curve Γ_j corresponding to a pair of conjugate convex cost functions f_j and g_j. We want to regard G as a "black box," that is, as a single arc with input node s and output node s'. The question is what characteristic curve and cost functions should be associated with the "black box," and whether they have the properties and relationships one might hope for. The first part of this question is easily answered in a natural way, but the second part requires the full weight of the theory.

Let ξ and η denote the flux and tension from s to s'. *The characteristic curve Γ of the "black box" is defined to be the set of all pairs $(\xi, \eta) \in R^2$ such that G has a flow x with divergence y and a potential u with differential v satisfying*

$$y(s) = \xi = -y(s'), \qquad y(i) = 0 \quad \text{for all nodes } i \neq s, s', \tag{1}$$

$$u(s') - u(s) = \eta, \tag{2}$$

$$(x(j), v(j)) \in \Gamma_j \quad \text{for all arcs } j \in A. \tag{3}$$

The cost function for flux through the "black box" is

$$f(\xi) = \inf\left\{ \sum_{j \in A} f_j(x(j)) \mid x \text{ is a flow satisfying (1)} \right\}, \tag{4}$$

whereas the cost function for tension is

$$g(\eta) = \inf\left\{ \sum_{j \in A} g_j(v(j)) \mid u \text{ is a potential satisfying (2)} \right\} \tag{5}$$

The precise relationship between Γ, f, and g is described in the "black box theorem" that follows the next two examples.

EXAMPLE 8. (*Arcs in Series*)

Suppose G consists merely of arcs j_1 and j_2 in series as in Figure 8.48a. Then $\xi = x(j_1) = x(j_2)$, whereas $\eta = v(j_1) + v(j_2)$. The characteristic curve is the *sum*:

$$\Gamma = \Gamma_{j_1} + \Gamma_{j_2}$$

$$= \left\{ (\xi, \eta) \mid \exists \eta_1, \eta_2 : (\xi, \eta_1) \in \Gamma_{j_1}, (\xi, \eta_2) \in \Gamma_{j_2}, \eta_1 + \eta_2 = \eta \right\}.$$

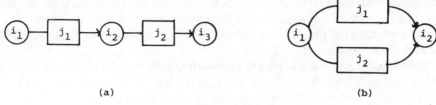

(a)   (b)

Figure 8.48

The cost functions are

$$f(\xi) = f_{j_1}(\xi) + f_{j_2}(\xi),$$

$$g(\eta) = \inf\{ g_{j_1}(\eta_1) + g_{j_2}(\eta_2) | \eta_1 + \eta_2 = \eta \}.$$

EXAMPLE 9. *(Arcs in Parallel)*

If G consists of arcs $j_1 j_2$ in parallel as in Figure 8.48*b*, the dual situation is encountered. Then $\xi = x(j_1) + x(j_2)$, and $\eta = v(j_1) = v(j_2)$. The characteristic curve is the *inverse sum*:

$$\Gamma = \left(\Gamma_{j_1}^{-1} + \Gamma_{j_2}^{-1} \right)^{-1}$$

$$= \{ (\xi, \eta) | \exists \xi_1, \xi_2 : (\xi_1, \eta) \in \Gamma_{j_1}, (\xi_2, \eta) \in \Gamma_{j_2}, \xi_1 + \xi_2 = \xi \}.$$

The cost function are

$$f(\xi) = \inf\{ f_{j_1}(\xi_1) + f_{j_2}(\xi_2) | \xi_1 + \xi_2 = \xi \},$$

$$g(\eta) = g_{j_1}(\eta) + g_{j_2}(\eta).$$

Black Box Theorem. *The functions f and g are closed proper convex and conjugate to each other if $f(\xi)$ is finite for some ξ, or if $g(\eta)$ is finite for some η, or if there exist ξ and η, such that $f(\xi) < \infty$ and $g(\eta) < \infty$. This is the case in particular if $\Gamma \neq \varnothing$, in which event Γ is the maximal monotone relation corresponding to f and g.*

   Moreover $\Gamma \neq \varnothing$ if and only if there is a flow x satisfying $x(j) \in \tilde{C}(j)$ for all arcs j and $y(i) = 0$ for all nodes $i \neq s, s'$, as well as a potential u satisfying $v(j) \in \tilde{D}(j)$ for all arcs j.

Proof. Let G' be the network obtained by adding to G a feedback arc $\bar{j} \sim (s', s)$. For any choice of conjugate functions f_j, g_j, one has

$$\inf\left\{ \sum_{j \in A} f_j(x(j)) + f_{\bar{j}}(x(\bar{j})) \mid x \text{ circulation in } G' \right\}$$

$$= \sup\left\{ - \sum_{j \in A} g_j(v(j)) - g_{\bar{j}}(v(\bar{j})) \mid v \text{ differential in } G' \right\} \qquad (6)$$

by the duality theorem in Section 8H, if either [inf] $< \infty$ or [sup] $> -\infty$. Apply this first to an arbitrary $\xi \in R$ and the conjugate pair

$$f_{\bar{j}}(x(\bar{j})) = \begin{cases} 0 & \text{if } x(\bar{j}) = \xi \\ \infty & \text{if } x(\bar{j}) \neq \xi \end{cases}$$

$$g_{\bar{j}}(v(\bar{j})) = \xi v(\bar{j}) \qquad (7)$$

The result is that the equation

$$f(\xi) = \sup_{\eta \in R} \{ \xi\eta - g(\eta) \} \qquad (8)$$

holds if either $f(\xi) < \infty$ or there is at least one η with $g(\eta) < \infty$, the second condition following by equation (8) from the first if also $f(\xi) > -\infty$.

Next apply (6) to an arbitrary $\eta \in R$ and the conjugate pair

$$f_{\bar{j}}(x(\bar{j})) = -x(\bar{j})\eta,$$

$$g_{\bar{j}}(v(\bar{j})) = \begin{cases} 0 & \text{if } v(\bar{j}) = -\eta, \\ \infty & \text{if } v(\bar{j}) \neq -\eta. \end{cases} \qquad (9)$$

This yields the fact that the equation

$$\inf_{\xi \in R} \{ f(\xi) - \xi\eta \} = -g(\eta),$$

or equivalently

$$g(\eta) = \sup_{\xi \in R} \{ \xi\eta - f(\xi) \}, \qquad (10)$$

holds if either $g(\eta) < \infty$ or there is at least one ξ with $f(\xi) < \infty$, the second condition following by equation (10) from the first if also $g(\eta) > -\infty$.

Putting the two results together, one sees the equivalence of the conditions: (1) $f(\xi)$ is finite for some ξ, (2) $g(\eta)$ is finite for some η, and (3) there exist ξ and η such that $f(\xi) < \infty$ and $g(\eta) < \infty$. Furthermore under these conditions equations (8) and (10) are valid for *all* ξ and η, and hence neither f nor g can have the value $-\infty$ anywhere (and thus are "proper"). It is easily verified that (8) implies f is convex with all the intervals of the form $\{\xi \in R | f(\xi) \leq \alpha\}$ closed. Therefore f is a closed proper convex function. Similarly, (10) implies g is a closed proper convex function. Of course f and g are conjugate to each other by (8) and (10).

If $\Gamma \neq \varnothing$, then any flow x and potential u corresponding to one of the pairs (ξ, η) in the definition of Γ satisfy the conditions in the last paragraph of the theorem, namely,

$$x(j) \in \tilde{C}(j) \quad \text{for all } j \quad \text{and} \quad y(i) = 0 \quad \text{for all } i \neq s, s', \qquad (11)$$

$$v(j) \in \tilde{D}(j) \quad \text{for all } j, \qquad (12)$$

because $\tilde{C}(j)$ and $\tilde{D}(j)$ are the projections of Γ_j on the horizontal and vertical axes. If u satisfies (12), one has $g(\eta) < \infty$ for $\eta = u(s') - u(s)$, and in fact u is a regularly feasible solution to the optimal differential problem in (6) for the functions (7), regardless of the value of ξ. The corresponding optimal distribution problem then has an optimal solution by the existence theorem for flows in Section 8L, if it has a feasible solution. In other words, if there is a potential satisfying (12), then $g(\eta) < \infty$ for a certain η, and for every ξ such that $f(\xi) < \infty$ the infimum in the definition (4) of $f(\xi)$ is attained by at least one flow x. By a dual argument in terms of the functions (9), if there is a flow x satisfying (11), then $f(\xi) < \infty$ for a certain ξ, and for every η such that $g(\eta) < \infty$ the infimum in the definition (5) of $g(\eta)$ is attained by at least one potential u.

It follows that if (11) and (12) can be satisfied, then f and g are closed proper convex functions conjugate to each other, and the pairs (ξ, η) with $f(\xi) + g(\eta) = \xi\eta$ are the ones such that there exist x and u in the network G satisfying (1), (2), and

$$\sum_{j \in A} f_j(x(j)) + \sum_{j \in A} g_j(v(j)) = - \sum_{i \in N} y(i)u(i).$$

This equation can also be written as

$$\sum_{j \in A} \left[f_j(x(j)) + g_j(v(j)) - x(j)v(j) \right] = 0,$$

and hence it is equivalent to (3) by 8E(11) (12). But the pairs (ξ, η) with $f(\xi) + g(\eta) = \xi\eta$ are the ones in the maximal monotone relation corresponding to f and g (see (8E(12))). This relation is therefore identical to Γ, and in particular $\Gamma \neq \varnothing$.

EXAMPLE 10.   (*Project Cost Curves*)

Returning to the situation in Example 5 in Section 7F, let us think of a PERT model where $u(i)$ is the time at which event i in a large project is scheduled to occur. Then $v(j)$ represents the elapsed time between the events corresponding to the nodes of j, and $g_j(v(j))$ is the associated cost. The start and finish of the project are represented by s and s' (the roles of these nodes need no longer be reversed, as was dictated earlier by the desire to conform to a max tension formulation). Therefore η is the total time taken by the project. For any specified value of η there is an associated minimal cost for executing the project in exactly η time units (possibly infinite cost if the execution is infeasible), and this is $g(\eta)$.

The theorem tells us that if there is at least one value of η for which the project can be executed with finite minimum cost, the function g is closed, proper, and convex, and its derivatives (the corresponding marginal costs) are described by the set Γ (actually Γ^{-1}).

EXAMPLE 11.   (*Sensitivity Analysis*)

Sometimes it is necessary to analyze how the minimum cost in an optimal distribution problem might depend on certain of the values $b(i)$, $c^+(j)$, or $c^-(j)$ as parameters. Taking advantage of the tricks of reformulating the problem in terms of circulations and capacity intervals that involve only a single bound (see Chapter 3), one can pose the question in the following form. Given an optimal *circulation* problem and an arc j_0 that represents a capacity constraint $x(j_0) \le c_0$ (i.e., such that the associated function f_{j_0} vanishes on $(-\infty, c_0]$ and is $+\infty$ everywhere else), how can the infimal value in the problem vary when c_0 varies?

To reduce this to a black box problem, pass from diagram a in Figure 8.49 to diagram b: replace the interval $(-\infty, c_0]$ by $(-\infty, 0]$, but think of the terminal node s of j_0 as an input and the initial node s' as an output. The feasible flows through the black box in Figure 8.49b that have $y(s) = -y(s') = c_0$ can be identified with the feasible circulations in Figure 8.49a. The object of inquiry is thus $f(c_0)$, where f is given by (4). The black box theorem yields the conclusion that $f(c_0)$ is a closed proper convex function of c_0 (if it is finite for some choice of c_0), and its derivatives are described by the black box curve Γ.

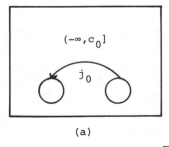

(a)

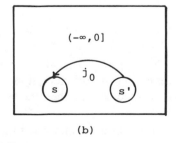

(b)

Figure 8.49

Sensitivity analysis relative to the spans $d^+(j)$ and $d^-(j)$ in an optimal differential problem can be approached similarly; see Exercise 8.62.

Computation

At the end of Section 9E, a procedure will be given for constructing f, g, and Γ in the case where each Γ_j is of staircase form (f_j and g_j piecewise linear). Then Γ will turn out likewise to be of staircase form (f and g piecewise linear). This procedure could be used in Example 10 to determine how the total cost of the project varies with the specified duration. It could also perform the sensitivity analysis in Example 11 in the piecewise linear case.

8P.* EXERCISES

8.1. (*Convex Costs*). Prove that f_j is convex on the interval $C(j)$ if and only if for every choice of values $\xi_1 < \xi_2 < \xi_3$ in $C(j)$ one has (see Figure 8.3):

$$\frac{f_j(\xi_2) - f_j(\xi_1)}{\xi_2 - \xi_1} \leq \frac{f_j(\xi_3) - f_j(\xi_1)}{\xi_3 - \xi_1} \leq \frac{f_j(\xi_3) - f_j(\xi_2)}{\xi_3 - \xi_2}.$$

8.2. (*Convex Costs*). Sketch an example of a closed proper convex function f_j such that $C(j)$ is a bounded interval of the form $[c^-(j), c^+(j))$ and $\tilde{C}(j) = (c^-(j), c^+(j))$.

8.3. (*Convex Costs*). Sketch the graph of the function f_j defined on R by

$$f_j(\xi) = \begin{cases} 2\xi - 1 & \text{if } \xi > 1, \\ \xi^2 & \text{if } 0 \leq \xi \leq 1, \\ -\xi & \text{if } -1 \leq \xi < 0, \\ +\infty & \text{if } \xi < -1 \end{cases}$$

What are $c^-(j)$, $c^+(j)$ and $C(j)$ in this case? What are the formulas for f'_{j-} and f'_{j+}? What is $\tilde{C}(j)$?

8.4. (*Continuity of Derivatives*). Prove that if the closed proper convex function f_j is differentiable on an interval $I \subset C(j)$ (in the ordinary two-sided sense), then its derivative function f'_j is continuous relative to I.

(*Hint.* Use the one-sided limit laws in Section 8A.)

8.5. (*Integration*). Let ϕ_j be the nondecreasing function on R defined by

$$\phi_j(\xi) = \begin{cases} \xi & \text{if } \xi < 0, \\ 1 & \text{if } 0 \le \xi < 1, \\ 2 & \text{if } \xi = 1, \\ 3 & \text{if } 1 < \xi < 2, \\ +\infty & \text{if } \xi \ge 2. \end{cases}$$

Sketch the graphs of ϕ_j and the function f_j which is the integral of ϕ_j in the sense of formula 8A(10) with $\bar{\xi} = 0$, $\alpha = 0$. What are the values of $\phi_j(\xi -)$ and $\phi_j(\xi +)$ at $\xi = 0$, 1 and 2?

8.6. (*Characteristic Curves*). Sketch the curve Γ_j that corresponds to the function f_j in Exercise 8.3.

8.7. (*Maximal Monotone Relations*). Prove that a subset Γ_j of R^2 can be expressed in the form

$$\Gamma_j = \left\{ (\xi, \eta) \in R^2 \,|\, \phi_j(\xi -) \le \eta \le \phi_j(\xi +) \right\}$$

for some nondecreasing function ϕ_j on R, not identically $+\infty$ or $-\infty$, if and only if Γ_j is a maximal monotone relation. (In particular, if ϕ_j is a finite continuous nondecreasing function on R, then the graph of ϕ_j is a maximal monotone relation.)

8.8. Sketch the maximal monotone relation Γ_j that corresponds to the function ϕ_j in Exercise 8.5 in the manner described in Exercise 8.7.

8.9. (*Integration*). Let f_j be a closed proper convex function, and let ϕ_j be an arbitrary function on R satisfying

$$f_j'(\xi) \le \phi_j(\xi) \le f_j'^+(\xi) \quad \text{for all } \xi \in R.$$

Show that ϕ_j is nondecreasing and not identically $+\infty$ or identically $-\infty$, and moreover

$$f_j(\xi) = \int_{\bar{\xi}}^{\xi} \phi_j(t)\, dt + f_j(\bar{\xi}) \quad \text{for all } \bar{\xi} \in \tilde{C}(j).$$

(Make use of the fact that Γ_j determines f_j uniquely up to an additive constant.)

8.10. (*Nonincreasing Convex Functions*). Let f be a closed proper convex function on R, let Γ be the corresponding maximal monotone relation in R^2, and let $\tilde{D} = \{ \eta \in R | \exists \xi \text{ with } (\xi, \eta) \in \Gamma \}$ (the interval that is the projection of Γ on the vertical axis). One says, of course, that f is *nonincreasing* (respectively, *decreasing*) if $\xi_1 < \xi_2$ implies $f(\xi_2) \le f(\xi_1)$

(respectively, $f(\xi_2) < f(\xi_1)$ unless $f(\xi_1) = +\infty = f(\xi_2)$); it is *strongly decreasing* if there is an $\varepsilon > 0$ such that $\xi_1 < \xi_2$ implies $f(\xi_2) \leq f(\xi_1) - \varepsilon(\xi_2 - \xi_1)$ unless $f(\xi_1) = +\infty = f(\xi_2)$. Prove that

(a) f is a nonincreasing $\Leftrightarrow f'_+(\xi) \leq 0$ for all $\xi \in R$
$$\Leftrightarrow \tilde{D} \subset (-\infty, 0].$$

(b) f is decreasing $\Leftrightarrow f'_+(\xi) < 0$ for all $\xi \in R$
$$\Leftrightarrow \tilde{D} \subset (-\infty, 0).$$

(c) f is strongly decreasing $\Leftrightarrow \exists \varepsilon > 0$ such that $f'_+(\xi) \leq -\varepsilon$ for all $\xi \in R$
$$\Leftrightarrow [\text{closure of } \tilde{D}] \subset (-\infty, 0).$$

(*Hint.*   Use difference quotients for necessity, integration for sufficiency.)

8.11.   (*Minimum of a Convex Function*). Prove that a closed proper convex function f on R has its minimum at ξ if and only if $f'_-(\xi) \leq 0 \leq f'_+(\xi)$.

(*Hint.*   Argue from the fact that the one-sided derivatives are monotone limits of difference quotients.)

8.12.   (*Minimum of a Convex Function*). Let f be a closed proper convex function on R. Prove that if f is neither a decreasing function nor an increasing function, then f attains its minimum at some point.

(*Hint.*   Invoke the results in the two preceding exercises in analyzing the condition $0 \in \tilde{D}$.)

8.13.   (*Minimum of a Convex Function*). Let f be a closed proper convex function on R that is piecewise quadratic. Prove that if f does not attain its minimum at any point, then f must be a strongly decreasing function or a strongly increasing function.

(*Hint.*   Apply the results of Exercises 8.10 and 8.12.)

8.14.   (*Elevator Problem*). This is an example involving the minimum of a piecewise linear convex function. An elevator in a building with n floors is being designed to wait at a fixed level $\xi \in [1, n]$ between calls. The probability of the elevator being called to floor k is a known number p_k, and the time the elevator takes to respond to such a call is proportional to the distance $|\xi - k|$. The problem is how to choose ξ to minimize the expected response time (assuming for simplicity that there is no overlap between calls, so the elevator is always able to return to its waiting level before the next call). This amounts to minimizing

$$f(\xi) = \sum_{k=1}^{N} p_k |\xi - k|$$

over all $\xi \in R$. Prove in fact that the minimum of f over R is attained for ξ equal to one of the integers k in $[1, N]$ (although it might also be attained for other values), and thus one of the floors is optimal as the waiting level. Give a condition in terms of the probabilities that characterizes the floor in question.

(*Hint.* Apply the results in Exercises 8.10 through 8.12.)

8.15. (*Optimal Distribution Problem*). Let $f(t) = \Phi(x + tz)$, where Φ is the cost function in Section 8D and x and z are arbitrary flows such that $\Phi(x) < \infty$. Prove that f is a closed proper convex function.

8.16. (*Optimal Distribution Problem*). For $f(t) = \Phi(x + tz)$ with x a feasible solution and z an arbitrary flow, derive a formula for $f'_+(t)$ that generalizes the one given for the case of $z = e_P$ in 8J(3). Use it to show that if $x + tz$ is regularly feasible for some $t > 0$, then x itself is regularly feasible if and only if $f'_+(0) > -\infty$.

8.17. (*Regular Feasibility*). Suppose the optimal distribution problem has at least one regularly feasible solution, and let x be an optimal solution. Show that x must be regularly feasible.

(*Hint.* Apply Exercises 8.15 and 8.16 with $z = x' - x$, where x' is a regularly feasible solution.)

8.18. (*Regular Feasibility*). Let $f(t) = \Phi(x + te_P)$, where x is a *regularly* feasible solution to the optimal distribution problem and P is an elementary circuit. Using the fact that f is a closed proper convex function with left and right derivatives given by 8J(3) and 8J(4), prove that the values of t for which the flow $x + te_P$ is regularly feasible are those satisfying $f'_+(t) > -\infty$ and $f'_-(t) < +\infty$.

(*Hint.* Consider separately the cases $t > 0$, $t < 0$, and $t = 0$, and make use of the fact that f_{j+}' and f_{j-}' are nondecreasing functions. The conditions $x(j) \in \tilde{C}(j)$ in the definition of regular feasibility means that $f_{j+}'(x(j)) > -\infty$ and $f_{j-}'(x(j)) < +\infty$ for all $j \in A$.)

8.19. (*Optimal Flows*). Let x be a feasible solution to the optimal distribution problem. Prove that x is optimal if and only if for every circulation z the function $f(t) = \Phi(x + tz)$ satisfies $f'_+(0) \geq 0$.

(*Hint.* Apply the condition in Exercise 8.11.)

8.20. (*Min Path versus Optimal Flow*). Consider the problem of sending one unit of flux from a node s to a node s' when the cost per unit of flux in arc j is $d^+(j)$ in the positive direction and $d^-(j)$ in the negative direction. This is an optimal distribution problem with $b(s) = 1$, $b(s') = -1$, and $b(i) = 0$ for all other nodes i, and

$$f_j(\xi) = \begin{cases} d^+(j)\xi & \text{if } \xi \geq 0, \\ d^-(j)\xi & \text{if } \xi \leq 0. \end{cases}$$

Regarding $d^+(j)$ and $d^-(j)$ as spans and assuming $d^+(P) > 0$ for all circuits P, prove that a flow x is an *integral* optimal solution if and only if x is of the form e_P for an elementary path P solving the min path problem for $N^+ = \{s\}$, $N^- = \{s'\}$. (For simplicity, assume $d^-(j)$ and $d^+(j)$ are finite. Use the conformal realization theorem in Section 4B to express x in terms of paths.)

(*Remark.* This flow problem has an integral optimal solution if it has an optimal solution at all (see the theorem in Exercise 7.4), so it is actually *equivalent* to the min path problem.)

8.21. (*Conjugate Costs*). What is the function g_j conjugate to $f_j(\xi) = |\xi|$? More generally, determine the conjugate of

$$f_j(\xi) = \begin{cases} d^+(j)\xi & \text{if } \xi \geq 0, \\ d^-(j)\xi & \text{if } \xi \leq 0. \end{cases}$$

8.22. (*Conjugate Costs*). Show that the conjugate of $f_j(\xi) = \frac{1}{2}\alpha\xi^2 + \beta\xi + \gamma$, where $\alpha > 0$, is $g_j(\eta) = \frac{1}{2}\alpha^*\eta^2 + \beta^*\eta + \gamma^*$, where $\alpha^* = 1/\alpha$, $\beta^* = -\beta/\alpha$, and $\gamma^* = (\beta^2/2\alpha) - \gamma$. More generally, show that if f_j is a finite differentiable function on R whose derivative f_j' is increasing, then

$$g_j(\eta) = \begin{cases} [f_j']^{-1}(\eta) - f_j([f_j']^{-1}(\eta)) & \text{if } d^-(j) < \eta < d^+(j), \\ +\infty & \text{if } \eta < d^-(j) \text{ or } \eta > d^+(j), \end{cases}$$

where

$$d^+(j) = \lim_{\xi \to +\infty} f_j'(\xi) \quad \text{and} \quad d^-(j) = \lim_{\xi \to -\infty} f_j'(\xi);$$

the values of g_j at the end points $d^-(j)$ and $d^+(j)$ (if these are finite) are the limits of the expression over $d^-(j) < \eta < d^+(j)$ as $\eta \downarrow d^-(j)$ and as $\eta \uparrow d^+(j)$. Apply this formula to $f_j(\xi) = (1 + \xi^2)^{1/2}$.

8.23. (*Conjugate Costs*). Determine the conjugate g_j of the function f_j in Exercise 8.3, and likewise for Exercise 8.5.

8.24. (*Conjugate Costs*). Determine the pair of functions f_j and g_j corresponding to $\Gamma_j = \{(\xi, \eta)| - \pi/2 < \xi < \pi/2, \eta = \tan \xi\}$ and having $f_j(0) = 0$.

8.25. (*Conjugate Costs*). For closed proper convex functions f_j and g_j conjugate to each other, show that $g_j(0) = -\inf f_j$ and $f_j(0) = -\inf g_j$. Furthermore show that the points $\xi \in R$ where f_j attains its minimum are those satisfying $g_{j-}'(0) \leq \xi \leq g_{j+}'(0)$, whereas the points η where g_j attains its minimum are those satisfying $f_{j-}'(0) \leq \eta \leq f_{j+}'(0)$.

In particular, f_j is bounded below if and only if $0 \in D(j)$, whereas f_j attains its minimum if and only if $0 \in \tilde{D}(j)$. Similarly, g_j is bounded below if and only if $0 \in C(j)$, whereas g_j attains its minimum if and only if $0 \in \tilde{C}(j)$.

8.26. (*Conjugate Costs*). A line in R^2 is said to be *asymptotic* to the graph of f_j if it does not meet the graph, yet there are points of the graph arbitrarily close to it. Prove that a nonvertical line of such type exists if and only if $\tilde{D}(j) \neq D(j)$. Dually, there is a nonvertical line asymptotic to the graph of g_j if and only if $\tilde{C}(j) \neq C(j)$. The existence of vertical asymptotes, on the other hand, corresponds to the conditions $D(j) \neq [d^-(j), d^+(j)]$ and $C(j) \neq [c^-(j), c^+(j)]$.

(*Remark.* This shows that the case where $\tilde{C}(j) = C(j) = [c^-(j), c^+(j)]$ and $\tilde{D}(j) = D(j) = [d^-(j), d^+(j)]$ is the one where neither f_j nor g_j has an asymptote.)

8.27. (*Conjugate Costs*). Prove that when f_j and g_j are piecewise linear, it is impossible for both to be finite on all of R (i.e., to have $C(j) = (-\infty, \infty) = D(j)$). Likewise, it is impossible for $C(j)$ and $D(j)$ both to be bounded.

8.28. (*Conjugate Costs*). Demonstrate that for every $\xi \in C(j)$ one has

$$f_j(\xi') \leq f_j(\xi) + (\xi' - \xi)d^+(j) \quad \text{for all } \xi \geq \xi',$$

$$f_j(\xi') \leq f_j(\xi) + (\xi' - \xi)d^-(j) \quad \text{for all } \xi \leq \xi'.$$

Similarly, for every $\eta \in D(j)$ one has

$$g_j(\eta') \leq g_j(\eta) + (\eta' - \eta)c^+(j) \quad \text{for all } \eta' \geq \eta,$$

$$g_j(\eta') \leq g_j(\eta) + (\eta' - \eta)c^-(j) \quad \text{for all } \eta' \leq \eta.$$

8.29. (*Optimal Differential Problem*). Let $f(t) = -\Psi(u + te_{N \setminus S})$ as in 8K(2). Prove that f is a closed proper convex function whose right and left derivatives are given by 8K(4) and 8K(5).

8.30. (*Linear Duality*). Demonstrate that the dual of a linear optimal distribution problem, as computed in terms of conjugate functions in the case where the bounds $c^+(j)$ and $c^-(j)$ are finite, is the piecewise linear problem described in Exercise 7.6. (The dual is the same even if $c^+(j)$ and $c^-(j)$ are not necessarily finite, but the formula must be interpreted according to the convention in Example 2 in Section 8F.)

8.31. (*Linear Duality*). Demonstrate that the dual of a linear optimal differential problem, as computed in terms of conjugate functions in the case where the bounds $d^+(j)$ and $d^-(j)$ are finite, is the piecewise linear problem described in Exercise 7.20.

8.32. (*Piecewise Linear Duality*). Consider the problem of finding a solution x to a feasible distribution problem (with supplies $b(i)$ and closed capacity intervals $[c^-(j), c^+(j)]$) which is nearest to a given flow $\bar{x}$ in the sense of minimizing the norm

$$\|x - \bar{x}\|_1 = \sum_{j \in A} |x(j) - \bar{x}(j)|.$$

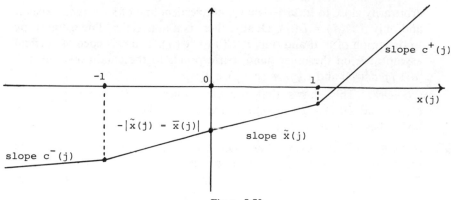

Figure 8.50

Set this up as an optimal distribution problem with piecewise linear cost functions f_j. Show that the dual is the piecewise linear optimal differential problem in which each function g_j has the form in Figure 8.50, where $\tilde{x}(j)$ is the element of $[c^-(j), c^+(j)]$ nearest to $\bar{x}(j)$. (Figure 8.50 only covers the case where $c^+(j)$ and $c^-(j)$ are finite, but the generalization should be obvious.)

8.33. (*Linear Programming Duality*). Formulate the problem of minimizing $d \cdot x$ over all flows $x \geq 0$ satisfying $Ex = b$ (where E is the incidence matrix) as an optimal differential problem. Demonstrate that the corresponding dual problem consists of maximizing $-u \cdot b$ over all potentials u satisfying $-uE \leq d$.

(*Note.* These are dual problems in the sense of linear programming theory.)

8.34. (*Piecewise Linear Duality*). Formulate as an optimal potential problem (i.e., an optimal differential problem in the augmented network) the problem of minimizing the norm

$$\|u\|_1 = \sum_{i \in N} |u(i)|$$

over all potentials u such that $d^-(j) \leq \Delta u(j) \leq d^-(j)$ for all j. What is the dual problem?

(*Hint.* This will be a certain optimal flow problem, i.e., an optimal distribution problem in the augmented network.)

8.35. (*Min Cut versus Optimal Potential*). Suppose in the optimal differential problem that the network is of the form in Figure 8.51 with $b(i) = 0$ for all nodes i, $g_j(\eta) = -\eta$, and for all arcs $j \neq \bar{j}$

$$g_j(\eta) = \begin{cases} c^+(j)\eta & \text{if } \eta \geq 0, \\ c^-(j)\eta & \text{if } \eta \leq 0. \end{cases}$$

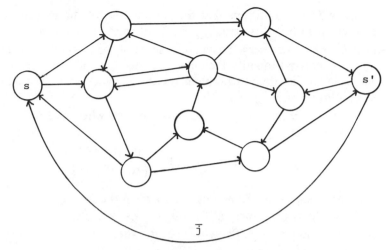

Figure 8.51

Assuming that $c^+(Q) > 0$ for all cuts $Q = [S, N \setminus S]$ with $\varnothing \neq S \neq N$, prove that a potential u is an *integral* optimal solution if and only if Δu is of the form e_Q for some cut Q solving the min cut problem for $N^+ = \{s\}$, $N^- = \{s'\}$ in the network with $\bar{j}$ deleted. (For simplicity, assume that $c^+(j)$ and $c^-(j)$ are finite. Use the conformal realization theorem for differentials in Exercise 4.16.)

(*Remark.* If this problem has an optimal solution at all, it has an integral optimal solution; see Exercise 7.19. Therefore it is *equivalent* to the min cut problem.)

8.36. (*Generalized Duality Theorem*). Show by reformulation in terms of the augmented network that the dual of the general optimal flow problem (minimize 8D(11) subject to 8D(12)) is the general optimal potential problem (maximize 8G(9) subject to 8G(10)). Deduce from the network duality theorem that for arbitrary closed proper convex functions f_i, f_j, and their conjugates g_i, g_j, one has (for $y = \operatorname{div} x$, $v = \Delta u$)

$$\inf\left\{ \sum_{j \in A} f_j(x(j)) + \sum_{i \in N} f_i(y(i)) \right\}$$

$$= \sup\left\{ - \sum_{j \in A} g_j(v(j)) - \sum_{i \in N} g_i(u(i)) \right\},$$

except for the possibility that the infimum might be $+\infty$ and the supremum $-\infty$. Finally, show that the network equilibrium conditions in this setting are

$$(x(j), v(j)) \in \Gamma_j \quad \text{for all } j \in A,$$

$$(y(i), u(i)) \in \Gamma_i \quad \text{for all } i \in N.$$

8.37. (*Duality Theorem*). Prove that the special case of the network duality theorem with $b = 0$ is actually equivalent to the general case.

8.38. (*Duality and Existence*). To illustrate some of the possibilities consistent with the network duality theorem, consider the following examples of the optimal distribution problem for the network in Figure 8.52 with $b(i_1) = b(i_2) = b(i_3) = 0$:

(a)   inf $= +\infty$, sup $= -\infty$. Show this occurs when $f_{j_1}(\xi) = -\xi$, $f_{j_4}(\xi) = 0$, and

$$f_{j_2}(\xi) = f_{j_3}(\xi) = \begin{cases} 0 & \text{if } \xi \geq 1, \\ \infty & \text{if } \xi < 1. \end{cases}$$

(b)   inf $=$ sup $= \infty$. Show this occurs when the functions are the same as in part (a), except $f_{j_1}(\xi) \equiv 0$.

(c)   inf $=$ sup $= -\infty$. Show this occurs when $f_{j_1}(\xi) = -\xi$ and $f_{j_2}(\xi) = f_{j_3}(\xi) = f_{j_4}(\xi) \equiv 0$.

(d)   inf and sup finite, neither attained. Show this occurs when $f_{j_1}(\xi) = e^{-\xi}$, $f_{j_4}(\xi) = 0$,

$$f_{j_2}(\xi) = f_{j_3}(\xi) = \begin{cases} -\sqrt{\xi} & \text{if } \xi \geq 0, \\ \infty & \text{if } \xi < 0. \end{cases}$$

(e)   inf and sup finite, only sup attained. Show this occurs when $f_{j_1}(\xi) = e^{-\xi}$ and $f_{j_2}(\xi) = f_{j_3}(\xi) = f_{j_4}(\xi) \equiv 0$.

(f)   inf and sup finite, only inf attained. Show this occurs when functions are as in case (d), but $f_{j_1}(\xi) \equiv 0$.

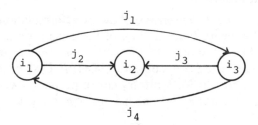

Figure 8.52

8.39. (*Feasibility*). One consequence of the network duality theorem in Section 8H is that if either problem fails to have a feasible solution, then the other problem cannot have a finite optimal value (inf or sup, as the case may be). Prove this directly from the inequalities in Exercise 8.28 under the assumption that $C(j)$ and $D(j)$ are closed.

8.40. (*Max Flow Min Cut*). Demonstrate that the max flow min cut theorem is a special case of the network duality theorem in Section 8H (except for the assertions about the existence of solutions, which are covered by the existence theorem in Section 8L).

(*Hint.* There is no loss of generality in assuming $N^+ = \{s\}$, $N^- = \{s'\}$. Use the result of Exercise 8.35.)

8.41. (*Max Tension Min Path*). Demonstrate that the max tension min path theorem is a special case of the network duality theorem in Section 8H (except for the assertions about the existence of solutions, which are covered by the existence theorem in Section 8L). (There is no loss of generality in assuming $N^+ = \{s\}$, $N^- = \{s'\}$. Use the result in Exercise 8.20.)

8.42. (*Equilibrium*). Suppose the network equilibrium problem is solved by the pair (x, u) and also by (x', u'). Prove that it is also solved then by (x', u) and by (x, u').

8.43. (*Electrical Equilibrium*). What are the functions f_j and g_j that correspond to the four basic types of electrical components depicted in Figure 8.45? What is the general form of the optimal distribution and optimal differential problems that one gets when the arc set is divided into four categories, each associated with one of these four types of components?

8.44. (*Price Equilibrium*). Show that the conditions in Example 7 in Section 8I are of the form $(y(i), u(i)) \in \Gamma_i$ and $(x(j), v(j)) \in \Gamma_j$ for certain maximal monotone relations Γ_i and Γ_j and thus correspond to an optimal flow problem and optimal potential problem dual to each other as in Exercise 8.36. Give expression to these problems by constructing cost function pairs f_i, g_i, and f_j, g_j, corresponding to Γ_i and Γ_j. (For simplicity in the latter, assume that $c^-(j)$ and $c^+(j)$ are finite and that $N = N_1 \cup N_2$, $N_1 \cap N_2 = \emptyset$, where Γ_i is a continuous increasing function with all of R as its domain and range for $i \in N_1$ but a constant function with value $b(i)$ for $i \in N_2$.)

8.45. (*Unbalanced Cuts*). Show that the definition of "unbalanced cut" in Section 8K agrees with the one given in Section 7E in the case where the optimal differential problem is linear.

8.46. (*Unbalanced Cuts*). Prove 8K(7) and 8K(8).

8.47. (*Existence Theorem for Piecewise Quadratic Problems*). Prove that for piecewise quadratic costs the following assertions are all equivalent:

(a) The optimal distribution problem has an optimal solution.

(b) The infimum in the optimal distribution problem is finite.

(c) The optimal distribution problem has a feasible solution, and there is no elementary circuit P with $d^+(P) < 0$.

(d) The optimal differential problem has a feasible solution.

(e) The supremum in the optimal differential problem is finite.

(f) The optimal differential problem has a feasible solution, and there is no cut Q with $c^+(Q) < b(S)$.

(In fact the same is true for any class of problems such that $\tilde{C}(j) = C(j) = [c^-(j), c^+(j)]$ and $\tilde{D}(j) = D(j) = [d^-(j), d^+(j)]$ for all arcs j. (*Hint.* Apply the duality theorem in Section 8H and the existence theorems in Section 8L.)

8.48. (*Equilibrium*). Suppose all the characteristic curves Γ_j are *polygonal* (i.e., composed of a finite number of line segments that are horizontal, vertical, or of positive slope). Prove that the network equilibrium problem has a solution if and only if there exists a flow $\tilde{x}$ satisfying $c^-(j) \le \tilde{x}(j) \le c^+(j)$ for all $j \in A$ and div $\tilde{x} = b$, and there exists a potential $\tilde{u}$ satisfying $d^-(j) \le \Delta\tilde{u}(j) \le d^+(j)$ for all $j \in A$.

8.49. (*Electrical Equilibrium*). Give a necessary and sufficient condition in terms of circuits and cuts for the equilibrium problem to have a solution in the case of an electrical network composed entirely of ideal batteries, generators, and linear resistors (Figures 8.45*a*, *b*, and *c*; see Example 5 in Section 8I).

8.50. (*Price Equilibrium*). Develop a necessary and sufficient condition for the network equilibrium problem to have a solution in the case of Example 7 in Section 8I with all the intervals $[c^-(j), c^+(j)]$ bounded.

8.51. (*Boundedness Property*). Prove that the optimal distribution problem has the boundedness property if and only if every minimizing sequence of flows is bounded.

8.52. (*Optimizing Sequences*). Prove that every cluster point of a minimizing sequence of flows solves the optimal distribution problem. Likewise, every cluster point of a maximizing sequence of potentials solves the optimal differential problem.

8.53. (*Boundedness Property*). Prove the boundedness theorem for potentials by an argument analogous to the one for flows.

8.54. (*Optimality*). Suppose the optimal distribution problem has at least one regularly feasible solution. Deduce from the theorems in Sections 8H and 8L that x is an optimal solution if and only if there is a potential u which together with x satisfies the network equilibrium conditions. (Dually, supposing that the optimal differential problem has at least one regularly feasible solution, it can be shown that u is an optimal solution if and only if there is a flow x which together with u satisfies the network equilibrium conditions.)

8.55. (*Uniqueness of Solutions*). Prove the following uniqueness theorem:

(a)   The optimal distribution problem has at most one optimal solution if the optimal differential problem has an optimal solution u such that the arc set $\{ j \in A | g'_{j-}(v(j)) < g'_{j+}(v(j))\}$ is a forest (i.e., contains no elementary circuits).

(b)   The optimal differential problem has at most one optimal solution (up to the arbitrary constant of integration) if the optimal distribution problem has an optimal solution x such that the arc set

$\{j \in A | f_j'(x(j)) < f_j'^+(x(j))\}$ is a coforest (i.e., contains no non-empty cuts).

(*Hint.* Argue by way of the theorems in Section 8H and the representations of flows and differentials in terms of spanning trees; see Sections 4F and 4G.)

8.56. (*Uniqueness of Solutions*). Derive the following corollary of the theorem in Exercise 8.55 and the optimality condition in Exercise 8.54.

(a) The optimal distribution problem has at most one optimal solution if it has a regularly feasible solution and there is a spanning tree F such that for every $j \notin F$ the relation Γ_j includes no horizontal segments.

(b) The optimal differential problem has at most one optimal solution (up to the arbitrary constant of integration) if it has a regularly feasible solution and there is a spanning tree F such that for every $j \in F$ the relation Γ_j includes no vertical segments.

(*Remark.* The condition that Γ_j contain no horizontal segments means that f_j is strictly convex on $C(j)$, or equivalently that $\tilde{D}(j)$ is open and g_j is differentiable throughout $\tilde{D}(j)$. Dually, the condition that Γ_j contain no vertical segments means that g_j is strictly convex on $D(j)$, or equivalently that $\tilde{C}(j)$ is open and f_j is differentiable throughout $\tilde{C}(j)$. Strict convexity means that strict inequality holds in 7A(1).)

8.57. (*Uniqueness of Solutions*). State and prove a sufficient condition in terms of the relations Γ_j for the network equilibrium problem to have at most one solution (x, u) (up to the arbitrary constant of integration in u).

(*Hint.* See the preceding exercise. Note that such a condition is very far from being *necessary*.)

8.58. (*Black Boxes*). Devise an electrical "black box" whose characteristic curve is the one in Figure 8.53*a*. Generalize to the cases in Figures 8.53*b* and *c*.

(*Hint.* Build up the desired curve from those of the basic components in Figure 8.45 by means of the operations in Examples 8 and 9 in Section 8N and "reversal" of arcs, which reflects their characteristic curves through the origin.)

8.59. (*Black Boxes*). Devise an electrical "black box" made of basic components in Figure 8.45 and having as its characteristic curve the maximal monotone relation corresponding to the function in Exercise 8.3.

(*Hint.* See the preceding exercise.)

8.60. (*Black Boxes*). Demonstrate that any maximal monotone relation Γ of the polygonal variety associated with piecewise quadratic functions can be represented as the characteristic curve of an electrical "black box" consisting only of the basic components in Figure 8.45.

(*Hint.* Extend the argument in Exercise 8.58.)

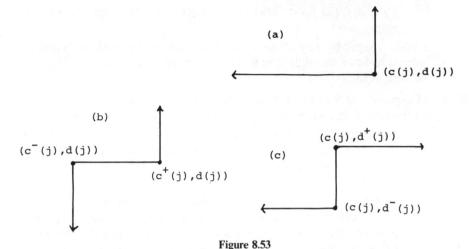

Figure 8.53

(*Remark.*   This leads to an analog method for solving any piecewise quadratic optimal distribution problem. According to the network equilibrium theorem, it suffices to solve the corresponding equilibrium problem involving polygonal relations. An electrical network can be constructed in which each of the latter is realized as the characteristic curve of a certain "black box." When the resulting currents and voltages are measured, they give the solutions to the optimal distribution problem and its dual.)

8.61.   (*Black Boxes*). Show that any optimal distribution problem with piecewise quadratic costs can be reformulated so that every cost function f_j is either of the elementary linear type in Section 7I or of the form $f_j(\xi)$ $= \frac{1}{2}\alpha\xi^2$ for all $\xi \in R$, where $\alpha > 0$.

8.62.   (*Sensitivity Analysis*). In an optimal circulation problem, consider an arc j_0 whose cost function has the simple form $f_{j_0}(\xi) = d_0\xi$ for all $\xi \in R$, and let $h(d_0)$ denote the infimum in the problem as a function of d_0 as a parameter. (*Note.*   Questions of sensitivity analysis of general optimal distribution problems and optimal differential problems with respect to a parameter $d^+(j)$ or $d^-(j)$ can typically be reduced to this special case.) Show that in passing from Figure 8.54*a* to the black box indicated in Figure 8.54*b*, one obtains the representation

$$h(d_0) = \inf_\xi \{ f(\xi) + d_0\xi \} = -g(-d_0),$$

where f and g are the functions associated with the black box as in 8N(4) (5). Draw conclusions about h and its derivatives from the black box theorem.

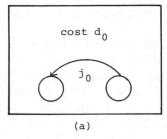
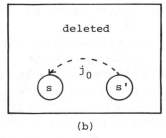

(a)   (b)

Figure 8.54

(*Remark.* In the case of piecewise linear problems the procedure explained in Section 9E will trace the characteristic curve Γ corresponding to *f* and *g* and thus will compute all the values of *h* and its derivatives.)

8Q.* COMMENTS AND REFERENCES

The theory of convex functions is a highly developed subject with numerous applications, especially to problems of optimization and to the analysis of equations and other conditions that correspond to variational principles. Here we have had to rush through just one branch of the theory, concerned with functions of a single variable, for needed background. A more thorough treatment can be found in Rockafellar [1970, 1974]. The notion of convex functions conjugate to each other has a long history in special cases, such as symmetric functions of one variable (used in the norms of L^p spaces and Orlicz spaces) and smooth strictly convex functions on domains in R^n (the Legendre transform). But the simplicity and generality of the correspondence was not fully captured until the work of Fenchel [1947]. Even with that it was not until the 1960s that general conjugate functions began to find their way into many applications.

The classical theory of electrical and hydraulic networks starts from the characteristic curves Γ_j and the equilibrium problem in Section 8H (see examples in Section 8I). The so-called *linear* case is thus the one where every Γ_j is a line, possibly horizontal or vertical. The corresponding functions f_j and g_j are then all quadratic, linear, or represent the assignment of fixed values to certain variables. From an optimization point of view this case therefore corresponds to minimizing a quadratic function subject to a system of linear equations. In hydraulic and some electrical applications it is necessary to consider nonlinear characteristic curves for certain arcs; typically Γ_j^{-1} is defined as the graph of a certain "conductance function." This is often referred to as the case of a *nonlinear* network, although in optimization it corresponds to minimizing a *nonquadratic* convex function subject still to linear equations. Not until economic applications began to be made did the possibility of a much more general equivalence between optimization problems and equilibrium conditions emerge.

In more recent times the trend has come full circle with the increasing use of optimization techniques in the solution of equilibrium problems even in electrical and hydraulic networks. For example, see Bertsekas [1976, 1978] and Collins, Cooper, Helgason, Kennington, and LeBlanc [1978]. A general introduction to hydraulic networks can be found in Hall [1976] and to mechanical networks in Hall [1972]. The articles collected in the book of Willson [1974] present the electrical point of view and serve as a reminder that characteristic curves relating flow and tension are not the only things of interest in networks. Arc elements representing inductors and capacitors, not to mention multi-terminal elements like transistors, are very important in electrical theory but have yet to find their natural generalization, if any, to other settings such as economic models. Furthermore monotonicity of characteristic curves is sometimes too restrictive a condition.

Duffin [1947] extended the quadratic variational principle for "linear" networks to "nonlinear" networks in terms of integrals of conductance and resistance functions, and this was carried farther by Birkhoff and Diaz [1956]. Characteristic curves in the form of more general monotone relations were explored by Millar [1951], but not in great mathematical depth. J. B. Dennis [1959] described connections between electrical variational principles, possibly involving nonlinear resistors and diodes, and how they might be used in an analog method of solving optimization problems (see the remark in Exercise 8.60).

Minty [1960] pushed the theory much farther forward in allowing Γ_j to be an arbitrary maximal monotone relation for every arc and defining the functions f_j and g_j by a type of integration over the intervals $\tilde{C}_j$ and $\tilde{D}_j$, the projections of Γ_j. He proved the existence theorem for network equilibrium in Section 8L (the concepts of unbalanced circuit in Section 8J and unbalanced cut in Section 8K are due to him), as well as a variational principle closely related to the one in Section 8H. He also established the result in Section 8N about the characteristic curve of a "black box" being again maximal monotone, if nonempty.

Minty took $b = 0$ in his formulation, but this is no real restriction, since one can always pass to the augmented network to express things in terms of circulations. More important in judging Minty's contribution is the fact that he dealt only with f_j on $\tilde{C}_j$ and g_j on $\tilde{D}_j$. He did not extend f_j and g_j by taking limits at the endpoints in order to get the possibly larger intervals C_j and D_j, and he did not give direct formulas of conjugacy between f_j and g_j like those in 8E(3) (4). In his variational principle the constraints $x(j) \in C(j)$ and $v(j) \in D(j)$ are replaced by $x(j) \in \tilde{C}(j)$ and $v(j) \in \tilde{D}(j)$. Thus the optimization is relative to the sets of *regularly* feasible solutions, rather than feasible solutions as we have defined them. With this restriction the [inf] = [sup] relation in the duality theorem in Section 8H is true only under the additional hypothesis that both problems have regularly feasible solutions, or that one of them has an optimal solution that is regularly feasible.

The distinction between feasibility and regular feasibility was overlooked by several authors after Minty, who got the impression that his results were true for C_j and D_j. For instance, Berge and Ghouila-Houri [1962] stated the duality theorem falsely, as did others. Iri [1969] avoided the pitfall by treating only the case where $\tilde{C}(j)$ and $\tilde{D}(j)$ are closed (hence equal to $C(j)$ and $D(j)$), which as we have seen in Section 8E places an awkward restriction of "nonasymptotic behavior" on f_j, g_j, and Γ_j.

The complete forms of the duality and existence theorems presented in Sections 8H and 8L were developed by Rockafellar [1963], [1967]. This holds also for the black box theorem in Section 8N. The descent theorems in Section 8J and 8K and boundedness theorems in Section 8M have not previously been stated explicitly, although they underly Rockafellar's proofs of existence. The uniqueness results in Exercises 8.55, 8.56, and 8.57 extend those of Minty [1960] and earlier writers.

The network equilibrium problem undergoes a far-reaching extension in applications where the flux and tension in an arc are not fixed numbers but functions of time that are interrelated in a certain way. Many such applications can be covered by allowing $x(j)$ and $v(j)$ to be elements not just of R but some Hilbert space H. Then Γ_j is not a "curve" but the graph of a so-called maximal monotone operator from H to H, possibly multivalued. See Dolezal [1979].

9

ALGORITHMS FOR CONVEX COSTS

The optimal distribution algorithm and optimal differential algorithm described in Chapter 7 are special cases of methods of descent that can be applied to solve any dual pair of network optimization problems with convex costs. In particular, piecewise linear problems can be solved without first reformulating them as linear problems. Since the dual of a nonelementary linear problem may only be piecewise linear, this more general framework is essential in understanding how even a linear problem can always be solved by a dual approach. It leads also to versions of the thrifty adjustment alogrithm and out-of-kilter algorithm that can be applied directly to any piecewise linear problem.

The general descent algorithms in their purest form may be hard to implement for problems whose cost functions are not piecewise linear, and their convergence to an optimum in that case is uncertain. A simple modification (the "fortified" algorithms) takes care of the convergence question and thereby provides a constructive proof of the part of the duality theorem in Section 8H that has been left hanging. However, this modification does not help in implementing the methods in the general case. For that purpose discretized versions of the algorithms are developed. These methods require the boundedness property of both problems in order to work to the fullest extent. But they use only function values, not derivatives, and they are capable of calculating "approximately" optimal solutions in terms of practical estimates. In effect the cost functions are replaced by piecewise linear functions with a finer and finer grid, but these approximations need not be represented explicitly, just generated locally as calculation proceeds.

For piecewise linear problems there is yet another method, the general out-of-kilter algorithm. This starts with an arbitrary flow and potential and by a sequence of iterations of the painted network algorithm adjusts them until the equilibrium conditions are satisfied. A closely related procedure traces the maximal monotone relation obtained by synthesizing a subnetwork, as in

Section 8N, in the piecewise linear case. It too can be used to solve certain optimization problems.

The notation throughout this chapter is the same as in Chapter 8: there is a connected network with supplies $b(i)$ satisfying $b(N) = 0$, and each arc j has a conjugate pair of closed proper convex functions f_j and g_j with associated maximal monotone relation Γ_j.

9A.   OPTIMAL DISTRIBUTION ALGORITHM

The first method that will be described generalizes the one in Section 7C to the setting of an arbitrary optimal distribution problem and its dual. It is based on the results in Section 8J on improvement of flows.

Given any regularly feasible solution x to the optimal distribution problem, consider the feasible differential problem with respect to the span intervals

$$[d_x^-(j), d_x^+(j)] = [f_{j-}'(x(j)), f_{j+}'(x(j))]$$
$$= \{\eta \mid (x(j), \eta) \in \Gamma_j\}. \tag{1}$$

(Note that these would not meet the requirements for span intervals if x were not *regularly* feasible.)

Invoke any method that constructs a potential u with $d_x^-(j) \le v(j) \le d_x^+(j)$ for all $j \in A$, or if none exists, produces an elementary circuit P with $d_x^+(P) < 0$. (For instance, the feasible differential algorithm and the tension rectification algorithm accomplish this. In the piecewise linear case the first leads to refinements discussed in Section 9B, and the second to the method in Section 9F.)

If a potential u is obtained, terminate; x solves the optimal distribution problem, and u solves the corresponding optimal differential problem.

If a circuit P is obtained, the function $f(t) = \Phi(x + te_P)$ has $f_+'(0) = d_x^+(P) < 0$. Let

$$\alpha = \min\{t > 0 \mid f_+'(t) \ge 0\}. \tag{2}$$

Then $\alpha > 0$. If $\alpha = \infty$, that is, if $f_+'(t) < 0$ for all $t \in R$, terminate; P is unbalanced, the optimal distribution problem has no optimal solution, and the optimal differential problem has no regularly feasible solution. If $\alpha < \infty$, then α is the lowest of the values of t at which f attains its minimum. Let $x' = x + \alpha e_P$. Then x' is another regularly feasible solution, and $\Phi(x') < \Phi(x)$. Iterate with x'.

Justification

If a potential u is obtained, then for all arcs j one has $f_{j-}'(x(j)) \le v(j) \le f_{j+}'(x(j))$, which means that $(x(j), v(j)) \in \Gamma_j$. Of course div $x = b$ also, because x is a feasible solution. The network equilibrium theorem in Section 8H then asserts that x and u are optimal for the respective problems.

If an elementary circuit P is obtained instead, one has

$$0 > \sum_{j \in P^+} d_x^+(j) - \sum_{j \in P^-} d_x^-(j) = \sum_{j \in P^+} f_{j+}'(x(j)) - \sum_{j \in P^-} f_{j-}'(x(j)) = f_+'(0)$$

(3)

(see 8J(3)). Since f is a closed proper convex function (see Exercise 8.15), f_+' is nondecreasing. If $f_+'(t) < 0$ for all $t > 0$, it follows that f itself is a decreasing function (see Exercise 8.10(b)). Then P is unbalanced, according to the descent theorem for flows in Section 8J, and one may conclude from the existence theorem in Section 8L that the optimal distribution problem has no optimal solution, and the optimal differential problem has no regularly feasible solution.

If it is not true that $f_+'(t) < 0$ for all $t > 0$, then $0 < \alpha < \infty$ in (2) by (3) and the right continuity of f_+'. Indeed, since f_+' and f_-' obey the limit laws in 8A(8), α is actually the least of the values t such that $f'(t) \leq 0 \leq f_+'(t)$, and these are the values which minimize f (see Exercise 8.11). In particular $f(\alpha) < f(0)$, because (3) precludes the minimum being attained at 0. This translates into $\Phi(x') < \Phi(x)$ for $x' = x + \alpha e_P$. Moreover $f_+'(\alpha) > -\infty$ and $f_-'(\alpha) < \infty$ (because $f_-'(\alpha) \leq 0 \leq f_+'(\alpha)$), so x' is regularly feasible (see Exercise 8.18).

Calculation of the Stepsize α

Formula (2) for α corresponds to minimizing $f(t)$ exactly and may therefore be difficult to implement for general nonlinear problems, which in any event are probably better handled by the methods in Section 9I and 9J. However, the formula poses no obstacle when the functions f_j are all piecewise quadratic, as in Section 8C. Then f too is piecewise quadratic, and its breakpoints are the values of t such that either $x(j) + t$ is a breakpoint of f_j for some $j \in P^+$, or $x(j) - t$ is a breakpoint of f_j for some $j \in P^-$.

To be specific, suppose that $t_1 < t_2 < \cdots < t_m$ are the positive breakpoints of f, and let $t_0 = 0$, $t_{m+1} = \infty$. Suppose the constant value of f'' for $t_{k-1} < t < t_k$ is $a_k \geq 0$. Then (by Exercise 9.2)

$$f_+'(t) = d_x^+(P) + \sum_{0 < k < r} \left[f_+'(t_k) - f_-'(t_k) + (a_k - a_{k+1})t_k \right] + a_r t$$

$$\text{for } t_{r-1} \leq t < t_r \text{ and } r = 1, 2, \dots \quad (4)$$

(stop with $r = m$ if $f(t) = \infty$ for $t > t_m$). From this it is easy to determine the α given by (2). Check the breakpoints one by one to find the lowest index r such that $f_+'(t_r) \geq 0$; if none exists, take $r = m + 1$. If $a_r = 0$, then $\alpha = t_r$. If

$a_r > 0$, then α is the value of t solving the linear equation

$$0 = d_x^+(P) + \sum_{0<k<r} \left[f_+'(t_k) - f_-'(t_k) + (a_k - a_{k+1})t_k \right] + a_r t$$

(all terms but the last being constants).

The calculation is even simpler for piecewise linear problems, since then $a_1 = a_2 = \cdots = 0$. It is just a matter of adding the jump terms $f_+'(t_k) - f_-'(t_k)$ one by one to $d_x^+(P)$ until the sum turns non-negative; then α is that breakpoint value t_r. (If all the terms have been exhausted and the sum is still negative, then $\alpha = \infty$.)

This is all valid of course even if some of the values t_k are not truly breakpoints of f, as may after all be a nuisance sometimes to check. It is important merely that all possible breakpoints be *included* among the t_k's.

Termination

Although the algorithm in this form does produce a sequence of flows that satisfy all the constraints and exhibit lower and lower costs, there is no general guarantee that the sequence converges to a solution or even that the sequence of costs approaches the minimum. Something more is needed. This is clear from the fact that the algorithm reduces to the one in Section 7C when applied to linear problems (see Example 1 in Section 9B), and it therefore includes the max flow algorithm as a special case (see Exercise 7.8). The max flow algorithm, without refinements, is known to generate nonminimizing sequences in certain situations where the commensurability condition is not fulfilled (see Exercise 3.12). A nonlinear example is sketched in Exercise 9.27.

For piecewise linear problems it is easy to make sure that the algorithm will terminate with optimal solutions to both problems, and this is discussed in Section 9B. For more general problems, even piecewise quadratic, the convergence properties are poorly understood. Modifications described in Sections 9H and 9J enable one to determine ε-optimal solutions for any $\varepsilon > 0$. The real difficulty, however, is that there may be a great many circuits P with $d_x^+(P) < 0$ at any stage, and the speed of descent may depend heavily on which of these happens to be selected by the subroutine that is invoked. Subroutines that, like the feasible differential algorithm, embody min path ideas are probably better in this respect, because they favor circuits with lower values of $d_x^+(P)$ than might be detected less systematically.

9B.   APPLICATION TO PIECEWISE LINEAR PROBLEMS

When the algorithm in Section 9A is used on a linear optimal distribution problem, it reduces to the algorithm of the same name in Section 7C, as shown in Example 1 in this section. But in contrast to the latter, it can be applied to solve any piecewise linear optimal distribution problem directly, without reformulation. Moreover it will furnish at the same time a solution to the corresponding piecewise linear optimal differential problem. The dual approach to the optimal differential problem, hampered under the earlier form of the algorithm by the need for recasting the problem first as an elementary one of the type in Section 7I, is made much simpler and more effective. Example 2 illustrates this in a particular case. After the two examples, some general features of the piecewise linear case are described.

EXAMPLE 1. (*Linear Case*)

For a linear optimal distribution problem, the cost functions have the form

$$f_j(x(j)) = \begin{cases} d(j)x(j) + p(j) & \text{if } c^-(j) \le x(j) \le c^+(j), \\ \infty & \text{otherwise,} \end{cases}$$

so that the spans associated with a feasible solution x are

$$d_x^+(j) = f_{j+}'(x(j)) = \begin{cases} \infty & \text{if } x(j) = c^+(j), \\ d(j) & \text{if } x(j) < c^+(j), \end{cases}$$

$$d_x^-(j) = f_{j-}'(x(j)) = \begin{cases} d(j) & \text{if } x(j) > c^-(j), \\ -\infty & \text{if } x(j) = c^-(j). \end{cases}$$

Also the function $f(t) = \Phi(x + te_p)$ is linear over the interval where it is finite, so that minimizing $f(t)$ in t (when $f_+'(0) < 0$) is equivalent to finding the largest value of t such that $x + te_p$ is feasible. The algorithm therefore reduces to the one in Section 7C.

EXAMPLE 2. (*Dual Linear Case*)

A linear optimal differential problem can be solved by applying the algorithm of the preceding section to the piecewise linear optimal distribution problem which is its dual. How does this work out? The cost functions all have the form

$$f_j(x(j)) = \begin{cases} d^+(j)[x(j) - c(j)] - q(j) & \text{if } x(j) \ge c(j), \\ d^-(j)[x(j) - c(j)] - q(j) & \text{if } x(j) \le c(j) \end{cases}$$

(see Exercises 8.31 and 7.20, and Example 2 in Section 8F). Assume for simplicity that the spans $d^+(j)$, $d^-(j)$, are all finite. Then any flow x satisfying $\operatorname{div} x = b$ is a feasible solution, and the corresponding intervals in the algorithm are

$$[d_x^-(j), d_x^+(j)] = \begin{cases} [d^+(j), d^+(j)] & \text{if } x(j) > c(j), \\ [d^-(j), d^+(j)] & \text{if } x(j) = c(j), \\ [d^-(j), d^-(j)] & \text{if } x(j) < c(j). \end{cases}$$

The breakpoints of $f(t) = \Phi(x + te_P)$ are values of t such that either $x(j) + t = c(j)$ for some $j \in P^+$, or $x(j) - t = c(j)$ for some $j \in P^-$, and at such a point the jump in the slope of f is (assuming for simplicity that j is unique):

$$f_+'(t) - f_-'(t) = f_{j+}'(c(j)) - f_{j-}'(c(j)) = d^+(j) - d^-(j).$$

Thus let t_1 be the lowest of the positive numbers of the form $[c(j) - x(j)]e_P(j)$, t_2 the next lowest, and so on, and let j_r be the arc of P corresponding to t_r. The said procedure for calculating α then reduces to the following. Add the terms $d^+(j_r) - d^-(j_r)$ to $d_x^+(P)$ one by one (in order) until the sum turns non-negative; α is the t_r at that stage. (If all the terms have been exhausted, and the sum is still negative, then $\alpha = -\infty$.)

This description of the algorithm turns out to be valid verbatim even if the spans $d^+(j)$, $d^-(j)$, are allowed to be infinite (Exercise 9.3).

Finite Termination

In the general context of piecewise linear problems, termination of the optimal distribution algorithm can be obtained through a commensurability condition or through certain subroutines (see the refinements that follow).

The commensurability condition is related to the method explained in Section 9A for calculating the stepsize α. The formula for α in the piecewise linear case says α will be a value of t such that, for some arc j, either $x(j) + t$ or $x(j) - t$ is a breakpoint of f_j. From this the following fact can be derived. *If all the functions f_j are piecewise linear and their breakpoints and the initial flow values are commensurable, and if the infimum in the optimal distribution problem is finite, then the algorithm is sure to terminate in finitely many iterations with optimal solutions to the problem and its dual.* In fact the breakpoint analysis shows that the flows stay in the same commensurability class throughout the computation, and hence the values of α that are generated are bounded below by a certain $\delta > 0$. There are only finitely many possible negative slope values that can occur as $f_+'(0)$ in the various iterations [only certain sums and differences of slope values for the functions f_j; see the derivative formula 8J(3)]; hence these are bounded above by a certain $-\varepsilon < 0$. The decrease from $\Phi(x)$ to $\Phi(x')$ must therefore be at least $\varepsilon\delta$, so an infinite number of iterations is incompatible with the infimum in the problem being finite.

Of course every feasible solution to a piecewise linear problem is regularly feasible and hence eligible to be used as the starting flow. One that meets the commensurability requirement (when the given breakpoints are all commensurable) can be found by solving the underlying feasible distribution problem by either of the algorithms in Chapter 3 (see the commensurability arguments in Sections 3I and 3K used in establishing that these algorithms terminate). The finiteness of the infimum in the problem is equivalent to the nonexistence of an elementary circuit P with $d^+(P) < 0$ (see Exercise 8.47), and this too can be tested algorithmically.

Refinements Using Feasible Differential Subroutine

The feasible differential algorithm with restart provision (see Section 6H) can be used as the main subroutine in the optimal distribution algorithm. The initial potential for the subroutine is in this case taken to be the potential with which the subroutine terminated on the previous round. When the problem is piecewise linear, this approach has an important feature generalizing what was observed for the linear case at the end of Section 7C: at no stage is the distance of $(x(j), v(j))$ from Γ_j ever increased for any arc j (Exercise 9.5).

The feasible differential subroutine can in turn be executed in multipath form, and by the argument at the end of Section 7L, this will guarantee finite termination for the optimal distribution algorithm when applied to any piecewise linear problem. Not only is the resulting procedure highly efficient therefore, but it does not depend on commensurability for termination.

The same refinements can be used of course in solving problems that are not piecewise linear, but it can no longer be expected that the distance of $(x(j), v(j))$ from Γ_j will never be increased, or that the algorithm will terminate finitely.

9C. OPTIMAL DIFFERENTIAL ALGORITHM

The corresponding procedure for solving a dual pair of problems in terms of an improving sequence of potentials has a closely parallel form.

Given any regularly feasible solution u to the optimal differential problem, consider the feasible distribution problem with respect to the supplies $b(i)$ and capacity intervals

$$[c_u^-(j), c_u^+(j)] = [g_{j^-}'(v(j)), g_{j^+}'(v(j))]$$

$$= \{\xi \mid (\xi, v(j)) \in \Gamma_j\}. \tag{1}$$

(Again regular feasibility is needed for these to be nonempty.) Invoke any method that constructs a flow x with $c_u^-(j) \le x(j) \le c_u^+(j)$ for all $j \in A$ or, if none exists, produces a cut $Q = [S, N \setminus S]$ with $c_u^+(Q) < b(S)$. [The feasible distribution algorithm (Section 3I) and flow rectification algorithm (Section 3K) do this; see refinements later in this section.]

If a flow x is obtained, terminate; x solves the optimal distribution problem, and u solves the optimal differential problem.

If a cut Q is obtained, the function $f(t) = -\Psi(u + te_{N\setminus S})$ has $f'_+(0) < 0$. Let

$$\alpha = \min\{t > 0 | f'_+(t) \geq 0\}. \tag{2}$$

Then $\alpha > 0$. If $\alpha = \infty$, that is, if $f'_+(t) < 0$ for all $t \in R$, terminate; Q is unbalanced, the optimal differential problem has no optimal solution, and the optimal distribution problem has no regularly feasible solution. If $\alpha < \infty$, then α is the lowest of the values of t at which f attains its minimum. Let $u' = u + \alpha e_{N\setminus S}$. Then u' is another regularly feasible solution and $\Psi(u') > \Psi(u)$. Iterate with u'.

Justification

This is practically the same as the justification in Section 9A, except for notation, and is left as Exercise 9.6.

Calculation of α

The remarks in Section 9A about how α can be determined in the piecewise linear or quadratic case carry over completely, the only change being that the term $c_u^+(Q) - b(S)$ appears in place of $d_x^+(P)$.

Termination for Piecewise Linear Problems

In the piecewise linear case the breakpoints for g_j are the various slope values of its conjugate f_j (see Section 8E). The result can therefore be stated in terms of the optimal distribution problem as follows. *If all the functions f_j are piecewise linear and their slope values and the initial potential values are commensurable, and if the optimal distribution problem has a feasible solution, then the algorithm is sure to terminate in finitely many iterations with optimal solutions to the problem and its dual.* The existence of a feasible solution to the optimal distribution is equivalent (by the duality theorem) to the finiteness of the supremum in the optimal differential problem, since the latter does have feasible solutions (the potentials handled by the algorithm). The proof of termination follows the pattern for the proof for the optimal distribution algorithm in Section 9B (Exercise 9.7).

Refinements

In the piecewise linear case use of the feasible distribution algorithm as the subroutine leads to a number of shortcuts. The end result is the version of the thrifty adjustment algorithm in Section 9D. The flow rectification algorithm as subroutine leads, on the other hand, to the general out-of-kilter algorithm in Section 9F. For problems that are not piecewise linear, there do not seem to be special advantages to attach to either subroutine. Both can of course be executed in multipath form.

EXAMPLE 3. (*Linear Case*)

The linear optimal differential problem has costs

$$g_j(v(j)) = \begin{cases} c(j)v(j) + q(j) & \text{if } d^-(j) \le v(j) \le d^+(j), \\ \infty & \text{otherwise.} \end{cases}$$

Then for any feasible solution u one has

$$c_u^+(j) = g_{j+}'(v(j)) = \begin{cases} \infty & \text{if } v(j) = d^+(j), \\ c(j) & \text{if } v(j) < d^+(j), \end{cases}$$

$$c_u^-(j) = g_{j-}'(v(j)) = \begin{cases} c(j) & \text{if } v(j) > d^-(j), \\ -\infty & \text{if } v(j) = d^-(j). \end{cases}$$

In this manner everything reduces to the linear optimal differential algorithm in Section 7G.

EXAMPLE 4. (*Dual Linear Case*)

The dual approach to solving a linear optimal distribution problem with costs

$$f_j(x(j)) = \begin{cases} d(j)x(j) + p(j) & \text{if } c^-(j) \le x(j) \le c^+(j), \\ \infty & \text{otherwise,} \end{cases}$$

is to apply the optimal differential algorithm to the dual problem. The conjugate functions are

$$g_j(v(j)) = \begin{cases} c^+(j)[v(j) - d(j)] - p(j) & \text{if } v(j) \ge d(j), \\ c^-(j)[v(j) - d(j)] - p(j) & \text{if } v(j) \le d(j), \end{cases}$$

so that

$$[c_u^-(j), c_u^+(j)] = \begin{cases} [c^+(j), c^+(j)] & \text{if } v(j) > d(j), \\ [c^-(j), c^+(j)] & \text{if } v(j) = d(j), \\ [c^-(j), c^-(j)] & \text{if } v(j) < d(j). \end{cases}$$

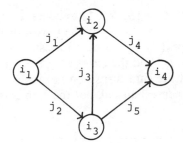

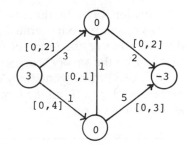

Figure 9.1

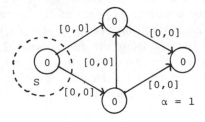

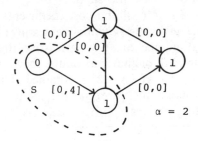

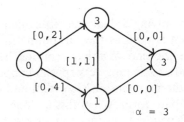

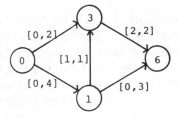

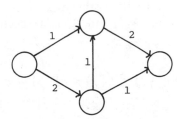

Figure 9.2

The calculation of α in the case of a cut $Q = [S, N \setminus S]$ reduces to the following procedure (to be verified as Exercise 9.8). Let t_1 be the lowest of the positive numbers of the form $[d(j) - v(j)]e_Q(j)$, t_2 the next lowest, and so on, and let j_r be the arc of Q corresponding to t_r. Add the terms $c^+(j_r) - c^-(j_r)$ to $c_u^+(Q) - b(S)$ one by one (in order) until the sum turns non-negative; α is the corresponding t_r. (If all the terms have been exhausted, but the sum is still negative, then $\alpha = \infty$.)

EXAMPLE 5

The algorithm in the special case of Example 4 works as follows when applied to the linear optimal distribution problem in Example 7 in Section 7C. The data in the problem are given again in Figure 9.1 for convenient reference. The nodes show the supplies $b(i)$, whereas the arcs show the capacity intervals $[c^-(j), c^+(j)]$ and cost coefficients $d(j)$; the constants $p(j)$ are all 0. Figure 9.2 gives the potential values $u(i)$ (at the nodes) and intervals $[c_u^-(j), c_u^+(j)]$ (along the arcs) in successive iterations, starting from the zero potential. Breakthrough occurs on the fourth round: the flow $x(j_1) = x(j_3) = x(j_5) = 1$, $x(j_2) = x(j_4) = 2$, is obtained. This solves the optimal distribution problem, whereas the potential $u(i_1) = 0$, $u(i_2) = 3$, $u(i_3) = 1$, $u(i_4) = 6$, solves the dual problem.

9D. THRIFTY ADJUSTMENT ALGORITHM (PIECEWISE LINEAR)

In Chapter 7 it was observed that when, in solving a dual pair of elementary problems, the feasible distribution algorithm is used as the subroutine in the optimal differential algorithm (the flow at the end of the subroutine being taken as the initial flow the next time around), something very special happens. Whole blocks of iterations mimic those in the max tension algorithm and can therefore be realized very efficiently in min path form. When this is done, one obtains an overall procedure with an independent justification of its own, namely, the thrifty adjustment algorithm.

As a matter of fact all this carries over to piecewise linear problems with hardly any changes. The characteristic curves Γ_j are all of staircase form of course, and one has to deal with both types of intervals $[d_x^-(j), d_x^+(j)]$ and $[c_u^-(j), c_u^+(j)]$, as defined in 9A(1) and 9C(1), respectively. According to these definitions,

$$(x(j), v(j)) \in \Gamma_j \Leftrightarrow c_u^-(j) \le x(j) \le c_u^+(j) \Leftrightarrow d_x^-(j) \le v(j) \le d_x^+(j).$$

$$(1)$$

Statement of the Algorithm

To begin with, x and u are any flow and potential satisfying the equivalent conditions in (1) for all $j \in A$. (In particular then, u is a feasible solution to the optimal differential problem.) For $y = \operatorname{div} x$, let

$$N^+ = \{i | b(i) > y(i)\} \quad \text{and} \quad N^- = \{i | b(i) < y(i)\}.$$

If $N^+ = \varnothing = N^-$, terminate; x and u are optimal solutions to the respective problems (see Section 8H). Otherwise $N^+ \neq \varnothing \neq N^-$ (because $b(N) = 0$ by blanket assumption).

Apply the min path algorithm to N^+, N^-, and the span intervals $[d_x^-(j), d_x^+(j)]$, with starting potential $u_0 = u$ (even though this u_0 might not be constant on N^+ and on N^-). If this subroutine ends with a cut $Q: N^+ \downarrow N^-$ of infinite span with respect to the intervals $[d_x^-(j), d_x^+(j)]$, terminate. Then Q is an unbalanced cut, the optimal distribution problem has no feasible solutions, and the optimal differential algorithm has $[\sup] = +\infty$ (see Sections 8K and 8L).

Otherwise the min path subroutine ends with a certain potential $u_0 + w$ and a path $P: N^+ \to N^-$ having $d_x^+(P)$ finite and as low as possible. Replace $u = u_0$ by $u = u_0 + w$, and for the new intervals $[c_u^-(j), c_u^+(j)]$ calculate

$$\alpha = \min \begin{cases} c_u^+(j) - x(j) & \text{for } j \in P^+, \\ x(j) - c_u^-(j) & \text{for } j \in P^-, \\ b(i) - y(i) & \text{for the initial node } i \text{ of } P, \\ y(i') - b(i') & \text{for the terminal node } i' \text{ of } P. \end{cases} \tag{2}$$

Then $0 < \alpha < \infty$. Replace x by $x + \alpha e_P$, and return to the start of the procedure with this new x and u, which will still satisfy the conditions in (1) for all $j \in A$.

Justification and Interpretation

The formal justification can be patterned on the special case in Section 7J, and this is left to the reader as Exercise 9.10. The economic interpretation for the elementary case, outlined in Section 7K, also works for the general setting (Exercise 9.12). In particular, $d_x^+(P)$ is the marginal cost of sending additional flow along P from a surplus node ($i \in N^+$) to a shortage node ($i' \in N^-$), and the min path subroutine therefore determines the cheapest way to do this. One has $d_x^+(P) < \infty$ if and only if the additional flow can be sent along P without violating the underlying capacity constraints in the optimal distribution problem. The quantity α gives the most that can be sent along P before an increase in the marginal cost takes place and forces a reinspection of the possibilities. The marginal costs of the paths used in successive iterations form a nondecreasing sequence.

Initiation

At the start of the algorithm, u can be taken to be any potential that satisfies the constraints $v(j) \in [d^-(j), d^+(j)]$ in the optimal differential problem (where, it will be recalled, $[d^-(j), d^+(j)]$ is the projection of Γ_j on the vertical axis). Then $x(j)$ can be chosen arbitrarily in the corresponding interval $[c_u^-(j), c_u^+(j)]$ for each $j \in A$ so as to have the conditions in (1) fulfilled.

Note that in the important case where the underlying capacity intervals $[c^-(j), c^+(j)]$ in the optimal distribution problem (the projections of the curves Γ_j on the horizontal axis) are all bounded, one has $[d^-(j), d^+(j)] = [-\infty, \infty]$ for all $j \in A$. Then *any* potential u will serve initially. In particular, one can take $u \equiv 0$ in this case, and then the interval $[c_u^-(j), c_u^+(j)]$ from which $x(j)$ is selected will consist of the values in $[c^-(j), c^+(j)]$ where the cost function f_j is at its minimum.

Termination and Refinements

The commensurability condition given in Section 9C for the piecewise linear problems likewise ensures finite termination for the thrifty adjustment algorithm, which is not surprising in view of the remarks at the beginning of this section. Finite termination is also guaranteed when the min path subroutine is executed in multipath form along the lines in Section 7L. This approach also heightens efficiency by tending to achieve several iterations of the thrifty adjustment algorithm at one time.

EXAMPLE 6.   (*Approximation by Double Stochastic Matrices*)

Given an $m \times n$ matrix $[a(k, l)]$, we seek among the class of all $m \times n$ matrices $[x(k, l)]$ satisfying the constraints

$$x(k, l) \geq 0 \quad \text{for } k = 1, \ldots, m \text{ and } l = 1, \ldots, n,$$

$$\sum_{l=1}^{n} x(k, l) = 1,$$

$$\sum_{k=1}^{m} x(k, l) = 1,$$

one that minimizes the norm

$$\sum_{k=1}^{m} \sum_{l=1}^{n} |x(k, l) - a(k, l)|.$$

This is a piecewise linear optimal distribution problem in the complete bipartite network shown in Figure 9.3. Assuming for simplicity that $a(k, l) > 0$,

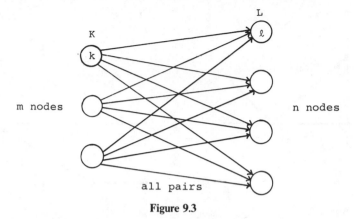

Figure 9.3

we have all the cost functions and characteristic curves as illustrated in Figure 9.4. The supplies are $b(k) = 1$ for $k \in K$ and $b(l) = -1$ for $l \in L$.

The thrifty adjustment algorithm may be executed with $x(k, l) = a(k, l)$ and $u(k) = u(l) = 0$ initially. In the general iteration one has

$$N^+ = \left\{ k \Big| \sum_l x(k, l) < 1 \right\} \cup \left\{ l \Big| \sum_k x(k, l) > -1 \right\},$$

$$N^- = \left\{ k \Big| \sum_l x(k, l) > 1 \right\} \cup \left\{ l \Big| \sum_k x(k, l) < -1 \right\}.$$

The spans used in the min path subroutine are

$$[d_x^-(k, l), d_x^+(k, l)] = \begin{cases} [1, 1] & \text{if } x(k, l) > a(k, l), \\ [-1, 1] & \text{if } x(k, l) = a(k, l), \\ [-1, -1] & \text{if } 0 < x(k, l) < a(k, l), \\ (-\infty, -1] & \text{if } 0 = x(k, l). \end{cases}$$

The calculation of α uses

$$[c_u^-(k, l), c_u^+(k, l)] = \begin{cases} [a(k, l), \infty) & \text{if } u(l) - u(k) = 1, \\ [a(k, l), a(k, l)] & \text{if } -1 < u(l) - u(k) < 1, \\ [0, a(k, l)] & \text{if } u(l) - u(k) = -1, \\ (-\infty, 0] & \text{if } u(l) - u(k) < -1. \end{cases}$$

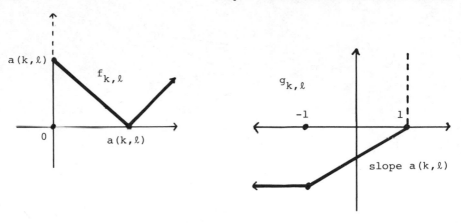

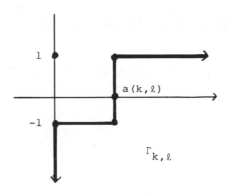

Figure 9.4

If the entries $a(k, l)$ are all integers, the number of iterations cannot exceed

$$r = \sum_k \left| \sum_l a(k, l) - 1 \right| + \sum_l \left| \sum_k a(k, l) - 1 \right|.$$

It is interesting that in this case the solution matrix must be integral, hence actually a permutation matrix.

9E.* APPLICATION TO BLACK BOXES

The thrifty adjustment algorithm can very easily be adapted as a tool for answering questions about the "black boxes" in Section 8N in the piecewise linear case.

The basic situation is this: we have a piecewise linear optimal distribution problem with special supply constraints

$$y(s) = -y(s') = \xi, \qquad y(i) = 0 \quad \text{for } i \neq s, s' \tag{1}$$

(where s and s' are two distinguished nodes), and we want to compute the function $f(\xi)$ giving the minimum cost in this problem as a function of ξ, along with its one-sided derivatives. We know from Section 8N that (if $f(\xi)$ is finite for at least one value of ξ) the function f is closed, proper, and convex. Furthermore a pair (ξ, η) lies on the corresponding curve Γ if and only if there is a potential u with

$$u(s') - u(s) = \eta \tag{2}$$

which, together with a flow x satisfying (1), has $(x(j), v(j)) \in \Gamma_j$ for all j. The latter conditions mean that x and u solve the pair of problems corresponding to the supply function specified by (1). In particular then, we can determine for a given ξ the value $f(\xi)$ and a point (ξ, η) on Γ by solving the optimal distribution problem for the supply constraints (1) and taking, for any solution u to the corresponding optimal differential problem, the number η given by (2). Note that the function g conjugate to f, which also is of importance in the black box context, has $g(\eta) = \xi\eta - f(\xi)$.

As soon as a single point (ξ, η) on Γ has been determined, along with corresponding x and u, the entire curve Γ can be traced by means of the thrifty adjustment algorithm, first in one direction from (ξ, η) and then in the other. The complete functions f and g can be constructed at the same time by integrating Γ and its inverse from the known initial values $f(\xi)$ and $g(\eta)$.

Tracing the Characteristic Curve

The trick for tracing Γ "northeasterly" from (ξ, η) is simply to apply the thrifty adjustment algorithm as if

$$b(s) = \infty, b(s') = -\infty, \text{ and } b(i) = 0 \quad \text{for all } i \neq s, s'. \tag{3}$$

In other words, the last two terms in 9D(2) are dropped, and one has $N^+ = \{s\}$ and $N^- = \{s'\}$ at all times.

Since the conditions $(x(j), v(j)) \in \Gamma_j$ are maintained throughout the procedure, each time x or u is modified a new point on Γ is generated. The modification of u occurs at the end of the min path subroutine and involves increasing $u(s')$ by an amount $\beta \geq 0$ while $u(s)$ and the values $x(j)$ stay fixed. This amounts to passing from (ξ, η) to a point (ξ, η') with $\eta' \geq \eta$, or in other words, to tracing a vertical segment of Γ (except in iterations where u is not actually changed at all, i.e., where $\beta = 0$). If the min path subroutine has ended in an unbalanced cut, this means that the infinite vertical segment proceeding upward from (ξ, η) belongs to Γ, and there is nothing more to learn about Γ in the "northeasterly" direction.

The modification of x, on the other hand, increases $y(s)$ (decreases $y(s')$) by an amount $\alpha > 0$ while leaving u fixed. This then corresponds to passing from (ξ, η') to a point (ξ', η') with $\xi' > \xi$, or in other words, to tracing a *horizontal* segment of Γ. The case $\alpha = \infty$ (which now could occur because of the deletion of the last two terms in 9D(2)) means that Γ terminates with the infinite horizontal segment proceeding rightward from (ξ, η').

This continues step by step until an infinite segment, vertical or horizontal, is encountered. The tracing of Γ in this direction is then finished.

To trace Γ "southwesterly" from (ξ, η), the only change necessary is to replace (3) by

$$b(s) = -\infty, b(s') = \infty, \text{ and } b(i) = 0 \quad \text{for all } i \neq s, s'. \tag{4}$$

Again a sequence of vertical and horizontal segments will be generated, but these will be downward and leftward.

Termination

Since the procedure is based on the thrifty adjustment algorithm, the results about termination of the latter are applicable. The execution of the min path subroutine in multipath form (see Section 7L) is enough to guarantee that the entire characteristic curve Γ (and with it all of the functions f, g, and their one-sided derivatives) can be determined in a finite number of steps. The commensurability condition in Section 9C would suffice for determining the portion of Γ corresponding to any bounded ξ-interval.

Incidentally, these observations furnish a constructive proof of the following theorem: *if the characteristic curves Γ_j of all the components of a black box are of staircase form (f_j and g_j piecewise linear), then the characteristic curve Γ of the black box itself is of staircase form (f and g piecewise linear).*

EXAMPLE 7

A black box is displayed in Figure 9.5; the arcs have the indicated capacity intervals $[c^-(j), c^+(j)]$ and cost coefficients $d(j)$. This is a piecewise linear example (actually linear) for which the characteristic curve Γ can be determined by the foregoing procedure.

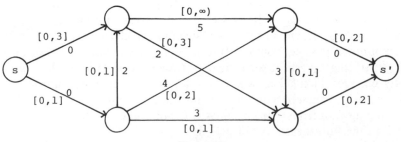

Figure 9.5

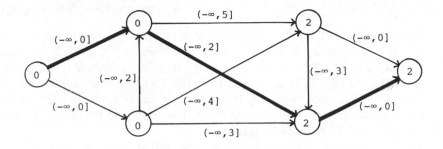

Iteration 1: α = 2, β = 2

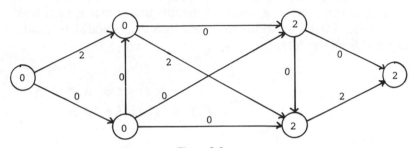

Figure 9.6

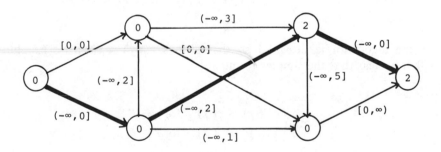

Iteration 2: α = 1, β = 2

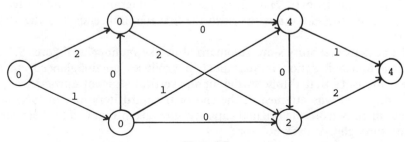

Figure 9.7

The curves Γ_j all fit the pattern in Figure 7.6 and determine intervals $[d_x^-(j), d_x^+(j)]$ by formula 7C(3). Taking $x \equiv 0$ and $u \equiv 0$ initially, we have $v(j)$ in $[d_x^-(j), d_x^+(j)]$ for all $j \in A$ and may therefore conclude that the point $(\xi, \eta) = (0, 0)$ lies on Γ. It is obvious from the data that there can be no flow through the black box with $\xi < 0$, so Γ must contain the infinite vertical segment proceeding downward from $(0, 0)$. Thus only the portion of Γ in the first quadrant of R^2 needs to be traced.

The first diagram in Figure 9.6 shows the shifted span intervals $[d_x^-(j) - v(j), d_x^+(j) - v(j)]$ used in the min path subroutine in the first iteration with $x \equiv 0$, $u \equiv 0$ (actually the same as the intervals $[d_x^-(j), d_x^+(j)]$ in this case, since $v \equiv 0$). It also shows an optimal path $P : s \to s'$ with respect to these spans and, at the nodes, the values of the function w that is calculated. Here $\beta = w(s') = 2$. First w is added to u, and then αe_P is added to x, where α is determined from the formulas

$$
\alpha = \min \begin{cases} c_u^+(j) - x(j) & \text{for } j \in P^+, \\ x(j) - c_u^-(j) & \text{for } j \in P^-, \end{cases}
$$

$$
[c_u^-(j), c_u^+(j)] = \begin{cases} [c^+(j), c^+(j)] & \text{if } v(j) > d(j), \\ [c^-(j), c^+(j)] & \text{if } v(j) = d(j), \\ [c^-(j), c^-(j)] & \text{if } v(j) < d(j). \end{cases}
$$

Since the algorithm is designed so that always $c_u^-(j) \leq x(j) \leq c_u^+(j)$ but $\alpha > 0$, it is clear that the simpler formula

$$
\alpha = \min \begin{cases} c^+(j) - x(j) & \text{for } j \in P^+, \\ x(j) - c^-(j) & \text{for } j \in P^-, \end{cases} \tag{5}
$$

serves for the application at hand. In particular, this yields $\alpha = 2$ in the first iteration. This iteration therefore traces Γ vertically from $(0, 0)$ to $(0, 2)$ and then horizontally from $(0, 2)$ to $(2, 2)$. The second diagram in Figure 9.6 records the flow x and potential u with which the first iteration ends and the second begins.

The next two iterations are summarized in like manner in Figures 9.7 and 9.8. If a fourth iteration is attempted, it produces an unbalanced cut and thereby indicates that Γ follows the infinite vertical segment upward from the point $(4, 5)$ that was attained at the end of the third iteration. In fact this is obvious directly from the original capacity intervals in Figure 9.5: there can be no flow through the black box for $\xi > 4$.

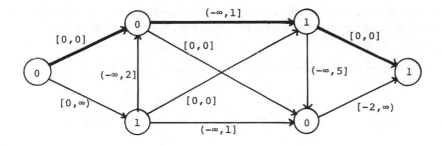

Iteration 3:   α = 1, β = 1.

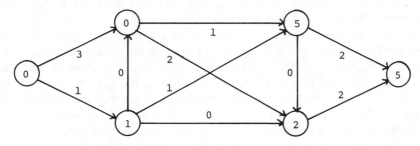

Figure 9.8

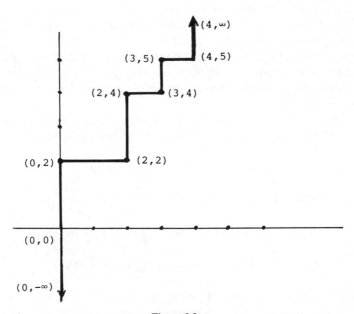

Figure 9.9

407

Figure 9.9 shows the whole shape of Γ. The functions f and g can be calculated from this and the values $f(0) = 0$, $g(0) = 0$, observed as corresponding to $x \equiv 0$, $u \equiv 0$.

9F.   OUT-OF-KILTER ALGORITHM (PIECEWISE LINEAR)

The various methods discussed so far all concentrate their attack on just one of the problems in a dual pair, although they furnish solutions to both on termination. The method that will be described now appears to be of quite a different character: the two problems are treated with complete symmetry, and no feasible solution to either is needed initially.

The method was explained in Section 7M for a pair of elementary problems in duality, but we now look at the general piecewise linear case, where each arc j has a characteristic curve consisting of a finite number of horizontal and vertical segments. An interesting fact is that despite the apparent novelty, the procedure could be obtained straightforwardly from either of the basic descent algorithms. When working with flows that are feasible, it is essentially equivalent to the optimal distribution algorithm with the tension rectification algorithm as the main subroutine. When working with potentials that are feasible, it is equivalent to the optimal differential algorithm with the flow rectification algorithm as the main subroutine.

The fundamental symmetry in the method is reflected by the fact that each iteration invokes the painted network algorithm. Improvements are made with a path or a cut, whichever is available. However, this pleasing simplicity and symmetry disappears from sight when attempts are made to speed things up by incorporating min path techniques. The more efficient algorithm that can be engineered in such a fashion resembles surprisingly closely the ones already viewed in Sections 9A and 9D. See the end of Section 9G for the discussion of such refinements.

An arc j is said to be *out of kilter* with respect to a flow x and potential u if $(x(j), v(j)) \notin \Gamma_j$. Thus j is *in kilter* if and only if the equivalent conditions

$$c_u^-(j) \le x(j) \le c_u^+(j) \quad \text{or} \quad d_x^-(j) \le v(j) \le d_x^+(j),$$

are satisfied, where as usual

$$c_u^-(j) = g'_{j-}(v(j)) \quad \text{and} \quad c_u^+(j) = g'_{j+}(v(j)),$$

$$d_x^-(j) = f'_{j-}(x(j)) \quad \text{and} \quad d_x^+(j) = f'_{j+}(x(j)).$$

The algorithm relies on a way of modifying either x or u so as to reduce "overall out-of-kilterness" and thereby move closer to optimality. This involves painting the arcs of the network as follows, according to the location of $(x(j), v(j))$ relative to Γ_j (see Figure 9.10):

green if $c_u^-(j) < x(j) < c_u^+(j)$
white if $x(j) < c_u^+(j)$ and $v(j) > d_x^-(j)$
black if $x(j) > c_u^-(j)$ and $v(j) < d_x^+(j)$
red   if $d_x^-(j) < v(j) < d_x^+(j)$

(The reader should verify that these four cases do exhaust all the possibilities.)

Note that every out-of-kilter arc is white or black, although some white or black arcs may be in kilter. For white arcs the point $(x(j), v(j))$ is eligible to be moved some distance "east or south." For black arcs it is "west or north;" for green arcs, "east or west;" for red arcs, "north or south."

Statement of the Algorithm

The procedure is initiated with any flow x satisfying $\text{div } x = b$ and any potential u, not necessarily feasible solutions to the respective problems. (Even the constraint $\text{div } x = b$ causes no real pain, for if necessary, the problem could be reformulated as one for circulations, so that $b = 0$; then $x \equiv 0$ could be used at the start.)

If there are no out-of-kilter arcs, terminate; x and u are optimal. Otherwise select any out-of-kilter arc $\bar{j}$, paint the network as before, and apply Minty's lemma (along with the corresponding version of the painted network algo-

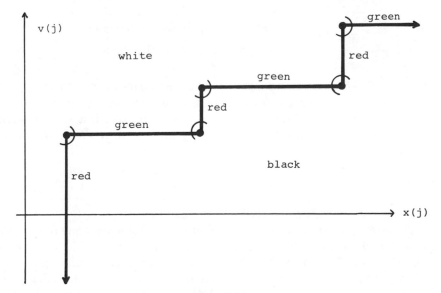

Figure 9.10

rithm; see Section 2H). This yields either a compatible elementary circuit P containing $\bar{j}$ or a compatible cut Q containing $\bar{j}$.

In the case of the circuit define

$$\alpha = \min\begin{cases} c_u^+(j) - x(j) & \text{for } j \in P^+, \\ x(j) - c_u^-(j) & \text{for } j \in P^-, \\ \bar{\alpha}, \end{cases}$$

where (in order to avoid "overshooting")

$$\bar{\alpha} = \begin{cases} c_u^-(\bar{j}) - x(\bar{j}) & \text{if } \bar{j} \text{ is white (hence in } P^+), \\ x(\bar{j}) - c_u^+(\bar{j}) & \text{if } \bar{j} \text{ is black (hence in } P^-). \end{cases}$$

Then $\alpha > 0$. If $\alpha = \infty$, terminate; P is an unbalanced circuit, so the optimal differential problem has no feasible solution, whereas the optimal distribution problem either has no feasible solution or has $[\inf] = -\infty$. If $\alpha < \infty$, replace x by $x' = x + \alpha e_P$, and return to the beginning with x' and the same u.

In the case of the cut define

$$\alpha = \min\begin{cases} d_x^+(j) - v(j) & \text{for } j \in Q^+. \\ v(j) - d_x^-(j) & \text{for } j \in Q^-, \\ \bar{\alpha}, \end{cases}$$

where

$$\bar{\alpha} = \begin{cases} d_x^-(\bar{j}) - v(\bar{j}) & \text{if } \bar{j} \text{ is black (hence in } Q^+) \\ v(\bar{j}) - d_x^+(j) & \text{if } \bar{j} \text{ is white (hence in } Q^-). \end{cases}$$

Then $\alpha > 0$. If $\alpha = \infty$, terminate; Q is an unbalanced cut, so the optimal distribution problem has no feasible solution, whereas the optimal differential problem either has no feasible solution or has $[\sup] = \infty$. If $\alpha < \infty$, replace u by $u' = u + \alpha e_{N \setminus S}$ (and v by $v' = v + \alpha e_Q$), and return to the beginning with u' and the same x.

Justification

In the case of a circuit P the choice of the painting ensures $c_u^+(j) - x(j) > 0$ for all $j \in P^+$ and $x(j) - c_u^-(j) > 0$ for all $j \in P^-$. Moreover any out-of-kilter arc in P (and there is at least one, i.e., $\bar{j}$) has $c_u^-(j) - x(j) > 0$ if it is in P^+ or $x(j) - c_u^+(j) > 0$ if it is in P^-. Therefore $\bar{\alpha} > 0$, and consequently $\alpha > 0$.

If $\alpha < \infty$, the flow $x' = x + \alpha e_P$ has $\operatorname{div} x' = \operatorname{div} x = b$. It yields points $(x'(j), v(j))$ which lie α units "east" of the corresponding points $(x(j), v(j))$

for arcs $j \in P^+$ and "west" for $j \in P^-$. (The new point coincides with the old point for $j \notin P$.) The choice of α guarantees, among other things, that

$$x(j) < x'(j) \le c_u^+(j) \quad \text{for } j \in P^+,$$
$$x(j) > x'(j) \ge c_u^-(j) \quad \text{for } j \in P^-. \tag{1}$$

Monotonicity of the derivatives of f_j then implies

$$d_x^+(j) \le d_{x'}^-(j) \le v(j) \quad \text{for } j \in P^+,$$
$$d_x^-(j) \ge d_{x'}^+(j) \ge v(j) \quad \text{for } j \in P^-. \tag{2}$$

Let us take as a measure of "out-of-kilterness" the quantities

$$D_j^1(x(j), v(j)) = \max\{0, c_u^-(j) - x(j), x(j) - c_u^+(j)\},$$
$$D_j^2(x(j), v(j)) = \max\{0, d_x^-(j) - v(j), v(j) - d_x^+(j)\}. \tag{3}$$

(These give the distance of $x(j)$ from $[c_u^-(j), c_u^+(j)]$ and of $v(j)$ from $[d_x^-(j), d_x^+(j)]$, respectively, when these intervals are nonempty.) Clearly j is in kilter if and only if $D_j^1(x(j), v(j)) = 0$ and $D_j^2(x(j), v(j)) = 0$. From (1) and (2) one sees that for all $j \in A$

$$D_j^1(x(j), v(j)) \ge D_j^1(x'(j), v(j)) \ge 0, \tag{4}$$

$$D_j^2(x(j), v(j)) \ge D_j^2(x'(j), v(j)) \ge 0. \tag{5}$$

In this sense the new pair x', v, is an improvement over x, v. All arcs that were in kilter for x, v, are still in kilter for x', v. Furthermore the decrease in (4) is strict (by the amount α) for any out-of-kilter arc $j \in P$ such that $D_j^1(x(j), v(j)) < \infty$ (i.e., such that $j \in P^+$ with $v(j) \le d^+(j)$, or $j \in P^-$ with $v(j) \ge d^-(j)$).

Note incidentally that $\bar{\alpha}$ is the largest of the values $D_j^1(x(j), v(j))$ for $j \in P$, so that there is no interest in taking $x' = x + te_P$ for $t > \bar{\alpha}$. This is why $\bar{\alpha}$ was introduced in the formula for α. If it turns out that $\alpha = \bar{\alpha}$ in this formula, none of the arcs in P will be left out of kilter.

If $\alpha = \infty$, one has $c_u^+(j) = \infty$ for all $j \in P^+$ and $c_u^-(j) = -\infty$ for all $j \in P^-$. Furthermore $\bar{\alpha} = \infty$ in this case, so either $c_u^-(j) = \infty$ for some $j \in P^+$ or $c_u^+(j) = -\infty$ for some $j \in P^-$. Since $[d^-(j), d^+(j)] = D(j) = \tilde{D}(j)$ by piecewise linearity (see Section 8E), these conditions can be translated into the fact that $v(j) \ge d^+(j)$ for all $j \in P^+$ and $v(j) \le d^-(j)$ for all $j \in P^-$, at least one of the inequalities being strict. Then

$$d^+(P) = \sum_{j \in P^+} d^+(j) - \sum_{j \in P^-} d^-(j) < \sum_{j \in P^+} v(j) - \sum_{j \in P^-} v(j) = v \cdot e_P = 0$$

(since P is a circuit and v a differential; see Exercise 2.21), so P is unbalanced. The assertions about the two problems in this case are consequences of the existence theorems in Section 8L; see Exercise 8.47.

The argument for the case where Minty's lemma produces a cut is analogous and is left as Exercise 9.16.

9G.* TERMINATION AND REFINEMENTS

Finite termination of the out-of-kilter algorithm, regardless of the feasibility of the initial flow and potential, will now be established in terms of a commensurability condition. Refinements in terms of min path techniques will then be described. These too ensure finite termination.

At each iteration there is a set of non-negative values $D_j^1(x(j), v(j))$ and $D_j^2(x(j), v(j))$, $j \in A$, which measure the "out-of-kilterness" of the flow x and potential u, as defined in 9F(3). Circumstances under which the algorithm must terminate can be derived from the fact that each of these values is either decreased or left the same in passing to the flow and potential for the next iteration. It has been seen moreover that in iterations involving a circuit P the number α is subtracted from $D_j^1(x(j), v(j))$ for each out-of-kilter arc $j \in P$ (in particular, $\bar{j}$). Analogously in iterations involving a cut Q, α is subtracted from $D_j^2(x(j), v(j))$ for each out-of-kilter arc $j \in Q$ (in particular, $\bar{j}$). However, the subtraction has no effect if $D_j^1(x(j), v(j))$ or $D_j^2(x(j), v(j))$ is $+\infty$, and this is a troublesome point that must be taken into account.

A *strong* iteration is one that produces a decrease in $D_j^1(x(j), v(j))$ or $D_j^2(x(j), v(j))$ for at least one $j \in A$. A *weak* iteration is one that leaves these values the same for every $j \in A$. A weak iteration involving a circuit P must in particular have $D_j^1(x(j), v(j)) = \infty$ for all out-of-kilter arcs $j \in P$, which means that $v(j) > d^+(j)$ for all out-of-kilter arcs $j \in P^+$ and $v(j) < d^-(j)$ for all out-of-kilter arcs $j \in P^-$. Similarly, a weak iteration involving a cut Q must have $D_j^2(x(j), v(j)) = \infty$ for all out-of-kilter arcs $j \in Q$, which means that $x(j) > c^+(j)$ for all out-of-kilter arcs $j \in Q^+$ and $x(j) < c^-(j)$ for all out-of-kilter arcs $j \in Q^-$.

The following fact will be proved. *The out-of-kilter algorithm must terminate in a finite number of iterations if both of the following conditions hold*:

1. *The breakpoint and slope values for all the functions f_j are commensurable along with all the initial flux and tension values $x(j)$ and $v(j)$.*

2. *It is impossible for the algorithm to encounter an infinite sequence of iterations which are all weak.*

Moreover Condition 1 implies Condition 2 if the optimal distribution problem and the optimal differential problem both have feasible solutions.

Alternate forms of the last condition, about the existence of feasible solutions, are furnished by Exercise 8.47. In Condition 1 it should be recalled that the slope values for f_j are the breakpoint values for g_j, and vice versa.

Sometimes Condition 2 can be verified directly. This is true, for instance, in the special case that gives the flow rectification algorithm (see Exercise 9.14). Another case is apparent from the preceding description of weak iterations: there cannot be a weak iteration involving a circuit if u is feasible, nor one involving a cut if x is feasible. Thus if the algorithm is initiated with x and u both feasible, it is immediate that Condition 2 holds.

To establish the first claim, suppose for simplicity that all the numbers in Condition 1 are integral (as could be arranged by changing scales if necessary). Then no matter what form the iteration takes, α will be integral, so that integrality of the flow and potential will be maintained and all the values $D_j^1(x(j), v(j))$ and $D_j^2(x(j), v(j))$ will be decreased by an integral amount if changed at all. These values are all bounded below by zero, and under Condition 2 they cannot all stay constant for an infinite sequence of iterations. Hence only a finite number of iterations can occur before they are all zero and the algorithm terminates (if it did not terminate earlier with an unbalanced circuit or cut).

The proof of the second claim, that Condition 2 holds under Condition 1 if feasible solutions to both problems merely exist (even if the algorithm is not initiated with such) is much harder. As a start, observe that whenever $x(j)$ is changed, the quantity

$$\text{dist}[x(j), C(j)] = \max\{0, c^-(j) - x(j), x(j) - c^+(j)\}$$

is decreased by an integral amount, if it is not already zero. Thus if an arc j were affected by infinitely many iterations, it would have to have either $\text{dist}[x(j), C(j)] = 0$ or $\text{dist}[v(j), D(j)] = 0$ after a certain stage. But from the description of weak iterations it is clear that once one has $\text{dist}[x(j), C(j)] = 0$, it is impossible thereafter for the arc j to be affected by a weak iteration involving a cut, and similarly, once one has $\text{dist}[v(j), D(j)] = 0$, it is impossible thereafter for j to be affected by a weak iteration involving a circuit.

Thus if an arc j were affected by infinitely many iterations but stayed out of kilter, it would eventually have to be in one of the four states PW, PB, QW, or QB illustrated in Figures 9.11 and 9.12. In state PW one has $\text{dist}[x(j), C(j)]$

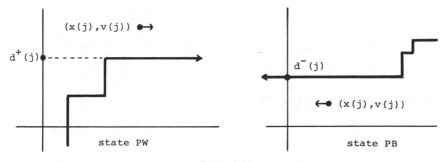

Figure 9.11

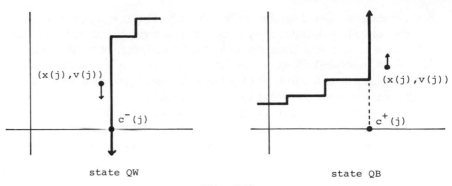

state QW   state QB

Figure 9.12

$= 0$ and

$$f_{j}'^{-}(x(j)) = f_{j}'^{+}(x(j)) = d^{+}(j) < v(j),$$

where $v(j)$ is henceforth constant, but $x(j)$ increases infinitely often by integral amountsl. (Thus j is forever white and affected only by iterations involving circuits P, hence the designation "PW.") The other states have analogous descriptions, obvious from the figures.

The argument will aim at a contradiction. Suppose there is an infinite sequence of iterations, all weak. Disregarding finitely many iterations at the beginning if necessary, one can reduce the situation to the case where every out-of-kilter arc affected at all by the sequence is in one of the four illustrated states. It must be shown that this is incompatible with the existence of feasible solutions $\bar{x}$ and $\bar{u}$ to the two problems.

It will be convenient to denote the sets of arcs in the four states in question by $A_{PW}, A_{PB}, A_{QW}, A_{QB}$, and to let A_K be the set of arcs in kilter and A_U the set of arcs unaffected by the sequence of iterations. Then

$$f_{j}(x(j)) + g_{j}(v(j)) = x(j)v(j) \quad \text{for all } j \in A_K,$$

$$f_{j}(x(j)) + g_{j}(d^{+}(j)) = x(j)d^{+}(j) \quad \text{for all } j \in A_{PW},$$

$$f_{j}(x(j)) + g_{j}(d^{-}(j)) = x(j)d^{-}(j) \quad \text{for all } j \in A_{PB},$$

$$f_{j}(c^{-}(j)) + g_{j}(v(j)) = c^{-}(j)v(j) \quad \text{for all } j \in A_{QW},$$

$$f_{j}(c^{+}(j)) + g_{j}(v(j)) = c^{+}(j)v(j) \quad \text{for all } j \in A_{QB}.$$

Consider now the (finite) quantity

$$\gamma = \sum_{j \in A_K} f_{j}(x(j)) + \sum_{j \in A_{PW}} f_{j}(x(j)) + \sum_{j \in A_{PB}} f_{j}(x(j))$$

$$+ \sum_{j \in A_K} g_{j}(v(j)) + \sum_{j \in A_{QW}} g_{j}(v(j)) + \sum_{j \in A_{QB}} g_{j}(v(j)) + b \cdot u, \quad (1)$$

where of course $b \cdot u = -x \cdot v$. The equations just given imply that

$$\gamma = \sum_{j \in A_{PW}} \left[x(j)[d^+(j) - v(j)] - g_j(d^+(j)) \right]$$

$$+ \sum_{j \in A_{PB}} \left[x(j)[d^-(j) - v(j)] - g_j(d^-(j)) \right]$$

$$+ \sum_{j \in A_{QW}} \left[v(j)[c^-(j) - x(j)] - f_j(c^-(j)) \right]$$

$$+ \sum_{j \in A_{QB}} \left[v(j)[c^+(j) - x(j)] - f_j(c^+(j)) \right]$$

$$- \sum_{j \in A_U} x(j)v(j).$$

Every iteration in the sequence being considered involves at least one out-of-kilter arc, hence an arc in one of the states PW, PB, QW, or QB. This implies γ is reduced each time by a positive, integral amount. Hence $\gamma \to -\infty$.

On the other hand, for feasible solutions $\bar{x}$ and $\bar{u}$ (and $\bar{v} = \Delta\bar{u}$) one has $x \cdot \bar{v} = -b \cdot \bar{u}$, $\bar{x} \cdot v = -b \cdot u$, and by the conjugacy of f_j and g_j both

$$f_j(x(j)) \geq x(j)\bar{v}(j) - g_j(\bar{v}(j)) \quad \text{and} \quad g_j(v(j)) \geq \bar{x}(j)v(j) - f_j(\bar{x}(j)).$$

It follows from the definition (1) of γ that

$$\gamma \geq \sum_{j \in A_K \cup A_{PW} \cup A_{PB}} \left[x(j)v(j) - g_j(v(j)) \right]$$

$$+ \sum_{j \in A_K \cup A_{QW} \cup A_{QB}} \left[\bar{x}(j)v(j) - f_j(\bar{x}(j)) \right] - \bar{x} \cdot v$$

$$= b \cdot \bar{u} - \sum_{j \in A_U \cup A_{QW} \cup Q_{AB}} x(j)\bar{v}(j) - \sum_{j \in A_K \cup A_{PW} \cup A_{PB}} g_j(\bar{v}(j))$$

$$- \sum_{j \in A_U \cup A_{PW} \cup A_{PB}} \bar{x}(j)v(j) - \sum_{j \in A_K \cup A_{QW} \cup A_{QB}} f_j(\bar{x}(j)).$$

Thus γ is bounded below by an expression that is constant from iteration to iteration (since $x(j)$ is never changed for j in A_U, A_{QW} or A_{QB}, whereas $v(j)$ is never changed for j in A_U, A_{PW}, or A_{PB}). This contradicts the earlier conclusion that $\gamma \to -\infty$ and finishes the proof.

Refinements

Greater efficiency can be achieved by min path techniques. The guiding principle is this: when x is feasible, the out-of-kilter algorithm can be identified with the version of the feasible distribution algorithm that uses the tension rectification algorithm as the main subroutine—except that instead of descending to the minimum of $f(t) = \Phi(x + te_P)$ in the case that an improvement circuit is detected, one may descend only to a lower breakpoint of f.

Indeed, the tension rectification algorithm, when applied to the intervals $[d_x^-(j), d_x^+(j)] \neq \varnothing$ and the potential u, calls the same arcs out of kilter and paints the arcs of the network exactly as in Figure 9.10. Its iterations produce a series of cuts that improve u by moving $v(j)$ closer to $[d_x^-(j), d_x^+(j)]$ for at least one out-of-kilter arc without moving it further away for any other arc. These cuts and improvements fit the prescriptions of the out-of-kilter algorithm. When the tension rectification algorithm terminates with a circuit P, its compatibility with the painting implies it too fits the prescriptions of the out-of-kilter algorithm, and (as seen in Section 9A) it satisfies $d_x^+(P) < 0$. Replacing x by $x + \alpha e_P$ is then a descent step, although the value of α may differ from the one specified for the optimal distribution algorithm. (For more on this aspect of descent, see Exercise 9.18.)

It is a simple matter, in view of this, to incorporate into the out-of-kilter algorithm the refinements suggested in Section 6I for the tension rectification algorithm. These involve a min path subroutine. The next step is to execute the subroutine in multipath form in the manner outlined for the thrifty adjustment algorithm in Section 7L (see the remarks concerning the elementary case of the out-of-kilter algorithm in Section 7M).

Of course these routines can be invoked even in the case where x is not feasible. Then what one gets corresponds more or less to the flow rectification algorithm implemented with the Dinic-Karzanov method (see Section 3K).

One consequence of this approach is that finite termination is guaranteed without any commensurability condition. The end result, interestingly enough, is not very different in effect from the refined version of the optimal distribution algorithm based on the feasible distribution algorithm as a subroutine (see Section 9B).

9H. FORTIFIED ALGORITHMS AND THE DUALITY THEOREM

The simplicity of the algorithms in Sections 9A and 9C is overshadowed in the general case of problems that are not piecewise linear by the fact that convergence is not assured. Slightly strengthened versions will now be developed that are guaranteed to produce "ε-optimal solutions" for arbitrary $\varepsilon > 0$. These are used to complete the proof of the duality theorem in Section 8H.

An *ε-optimal solution* to the optimal distribution problem is a feasible solution x such that $\Phi(x)$ differs from the infimum in the problem by at most ε. For the optimal differential problem ε-optimal solutions are defined similarly.

Fortified Optimal Distribution Algorithm

Fix $\varepsilon > 0$, and let $\delta = \varepsilon/|A|$. Given any feasible solution x (*not* necessarily regularly feasible), try to solve the feasible differential problem with respect to span the intervals

$$[d_x^-(j), d_x^+(j)] = [f_j^{\delta}(x(j)), f_j^{\delta}(x(j)],$$ (1)

where

$$f_{j+}^{\delta}(\xi) = \inf_{\zeta > \xi} \frac{f_j(\zeta) - f_j(\xi) + \delta}{\zeta - \xi} \geq f_{j+}'(\xi),$$

$$f_{j-}^{\delta}(\xi) = \sup_{\zeta < \xi} \frac{f_j(\zeta) - f_j(\xi) + \delta}{\zeta - \xi} \leq f_{j-}'(\xi).$$ (2)

(These values give the slopes of the two tangent lines indicated in Figure 9.13. Implementable expressions for them can be furnished in some cases, such as quadratic costs; see Exercise 9.19.)

If a potential u is obtained, terminate; x and u are ε-optimal solutions to the optimal distribution problem and its dual, in fact with $\Phi(x) - \Psi(u) < \varepsilon$. If an elementary circuit P with $d_x^+(P) < 0$ is obtained instead, then the function $f(t) = \Phi(x + te_P)$ has inf $f < f(0) - \delta$. Choose any α such that $f(\alpha) \leq f(0) - \delta$, and let $x' = x + \alpha e_P$. (In particular, one could choose α to minimize f, if possible.) Then x' is another feasible solution to the optimal distribution problem, and $\Phi(x') \leq \Phi(x) - \delta$. Iterate with x'.

Justification

A crucial property of the values (2) is that

$$f_{j-}^{\delta}(\xi) \leq \eta \leq f_{j+}^{\delta}(\xi) \quad \Leftrightarrow \quad f_j(\xi) + g_j(\eta) - \xi\eta \leq \delta$$ (3)

(Exercise 9.20). Since $\inf_\eta\{g_j(\eta) - \xi\eta\} = -f_j(\xi)$, this implies in particular

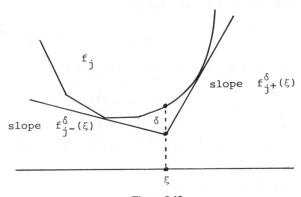

Figure 9.13

that the real interval (1) is nonempty when $f_j(x(j)) < \infty$, even if x is not *regularly* feasible.

If an iteration produces a potential u, one has by (3) that

$$f_j(x(j)) + g_j(v(j)) - x(j)v(j) \le \delta \quad \text{for all} \quad j \in A.$$

Summing this over all arcs and using the formula $-x \cdot v = (\text{div } x) \cdot u = b \cdot u$, one obtains

$$\sum_{j \in A} f_j(x(j)) + \sum_{j \in A} g_j(v(j)) + \sum_{i \in N} b(i)u(i) \le \delta |A|,$$

or in other words, $\Phi(x) - \Psi(j) \le \varepsilon$. In particular then, $\Psi(u) > -\infty$, so u is feasible as well as x. But as demonstrated in Section 8H, the inequality

$$\Phi(x) \ge \begin{bmatrix} \text{inf in optimal} \\ \text{distribution problem} \end{bmatrix} \ge \begin{bmatrix} \text{sup in optimal} \\ \text{differential problem} \end{bmatrix} \ge \Psi(u) \quad (4)$$

holds for arbitrary feasible solutions to the two problems. From the fact that $\Phi(x) - \Psi(u) \le \varepsilon$, (4) implies both $\Psi(x) - [\inf] \le \varepsilon$ and $[\sup] - \Psi(u) \ge \varepsilon$. Thus x and u are ε-optimal.

If a circuit P is obtained instead, one has

$$0 > d_x^+(P) = \sum_{j \in P^+} f_{j+}^{\delta}(x(j)) - \sum_{j \in P^-} f_{j-}^{\delta}(x(j))$$

$$= \sum_{j \in P} \inf_{t_j > 0} \left[f_j(x(j) + t_j e_P(j)) - f_j(x(j)) + \delta \right] / t_j. \quad (5)$$

For a moment consider arbitrary fixed values $t_j > 0$, and let

$$s = 1 \Big/ \sum_{j \in P} (1/t_j) > 0.$$

For each $j \in P$, one has $t_j > s$ and consequently

$$\left[f_j(x(j) + t_j e_P(j)) - f_j(x(j)) \right] / t_j \ge \left[f_j(x(j) + t e_P(j)) - f_j(x(j)) \right] / s$$

by the monotonicity of difference quotients of convex functions. Therefore

$$\sum_{j \in P} \left[f_j(x(j) + t_j e_P) - f_j(x(j)) + \delta \right] / t_j$$

$$\ge (\delta/s) + \sum_{j \in P} \left[f_j(x(j) + s e_P(j)) - f_j(x(j)) \right] / s$$

$$= \left[f(s) - f(0) + \delta \right] / s.$$

Applying this estimate to (5), one sees that

$$0 > \inf_{t>0} \left[f(t) - f(0) + \delta \right]/t.$$

Thus there does exist α with $f(\alpha) \leq f(0) - \delta$, as asserted.

Termination

Since the cost is decreased by at least δ with each descent, an infinite number of iterations is impossible, unless the infimum in the optimal distribution problem is $-\infty$ (and hence by virtue of (4), the supremum in the optimal differential problem is also $-\infty$).

Fortified Optimal Differential Algorithm

Fix $\varepsilon > 0$ and let $\delta = \varepsilon/|A|$. Given any feasible solution u, try to solve the feasible distribution problem with supplies $b(i)$ and capacity intervals

$$\left[c_u^-(j), c_u^+(j) \right] = \left[g_j'^\delta(v(j)), g_j'^\delta(v(j)) \right], \tag{6}$$

where

$$g_j'^\delta(\eta) = \inf_{\zeta > \eta} \frac{g_j(\zeta) - g_j(\eta) + \delta}{\zeta - \eta} \geq g_{j+}'(\eta),$$

$$g_j'^\delta(\eta) = \sup_{\zeta < \eta} \frac{g_j(\zeta) - g_j(\eta) + \delta}{\zeta - \eta} \leq g_{j-}'(\eta). \tag{7}$$

If a flow x is obtained, terminate; x and u are ε-optimal solutions to the optimal distribution problem and its dual, in fact with $\Phi(x) - \Psi(u) \leq \varepsilon$. If a cut $Q = [S, N \setminus S]$ is obtained instead, then the function $f(t) = -\Psi(u + te_{N\setminus S})$ has $\inf f < f(0) - \delta$. Choose any α such that $f(\alpha) \leq f(0) - \delta$, and let $u' = u + \alpha e_{N\setminus S}$. Then u' is another feasible solution to the optimal differential problem, and $\Psi(u') \geq \Psi(u) + \delta$. Iterate with u'.

Justification

See Exercise 9.22.

Termination

Again, since the value of Ψ is increased by at least δ in each nonterminal iteration, there cannot be infinitely many iterations unless the supremum in the optimal differential problem is $+\infty$ (in which case the infimum in the optimal distribution problem is likewise $+\infty$ by (4)).

Remaining Proof of the Duality Theorem

Since (4) was proved in Section 8F, the only thing left to show is that the finiteness of either [inf] or [sup] implies [inf] $\leq$ [sup]. If [inf] is finite, the fortified optimal distribution algorithm can be applied with arbitrary $\varepsilon > 0$ and must terminate with feasible solutions x and u satisfying $\Phi(x) - \Psi(u) \leq \varepsilon$. Then [inf] $\leq$ [sup] $+ \varepsilon$ by (4). Similarly, if [sup] is finite the fortified optimal differential algorithm can be applied with arbitrary $\varepsilon > 0$ and must terminate with feasible solutions x and u satisfying $\Phi(x) - \Psi(u) \leq \varepsilon$. Then again [inf] $\leq$ [sup] $+ \varepsilon$ by (4). Since this is true for all $\varepsilon > 0$, one has [inf] $\leq$ [sup] as claimed.

9I. DISCRETIZED DESCENT ALGORITHMS

For general nonlinear problems such that the altered derivative expressions used in the fortified algorithms are not convenient for computation, one can follow much the same procedure using difference quotients instead. Whole multiples of a fundamental step size δ are admitted only, so that in effect the given cost functions are represented merely by their values at a discrete set of points with spacing δ and by linear interpolation in between. The drawback in this case is that the algorithm only produces "quasi-optimal" solutions. These are also ε-optimal for a certain ε, which, however, is generally not known until after the algorithm terminates.

In order to get ε-optimal solutions for an ε specified in advance, the procedure may be repeated with a decreasing sequence of δ values. In some cases an estimate based on bounds for the second derivatives of the cost functions can be used instead. These refinements will be discussed in Section 9J.

For $\xi \in C(j)$ and $\delta > 0$, define

$$\Delta^{\delta}_{+}f_{j}(\xi) = \left[f_{j}(\xi + \delta) - f_{j}(\xi) \right]/\delta,$$

$$\Delta^{\delta}_{-}f_{j}(\xi) = \left[f_{j}(\xi) - f_{j}(\xi - \delta) \right]/\delta. \tag{1}$$

Observe that

$$\Delta^{\delta}_{+}f_{j}(\xi) \downarrow f_{j}'^{+}(\xi) \quad \text{as } \delta \downarrow 0,$$

$$\Delta^{\delta}_{-}f_{j}(\xi) \uparrow f_{j}'^{-}(\xi) \quad \text{as } \delta \downarrow 0, \tag{2}$$

and in particular,

$$\left[\Delta^\delta_- f_j(\xi), \Delta^\delta_+ f_j(\xi) \right] \supset \left[f'_{j-}(\xi), f'_{j+}(\xi) \right]. \tag{3}$$

An important role will be played by the expression

$$w_j(\xi, \delta) = \delta \max \left\{ \begin{matrix} \Delta^\delta_+ f_j(\xi) - f'_{j+}(\xi) \\ f'_{j-}(\xi) - \Delta^\delta_- f_j(\xi) \end{matrix} \right\} \geq 0, \tag{4}$$

where the first difference in the maximum is counted as 0 if $\Delta^\delta_+ f_j(\xi) = f'_{j+}(\xi) = +\infty$, and the second is counted as 0 if $\Delta^\delta_- f_j(\xi) = f'_{j-}(\xi) = -\infty$. (The two intervals in (3) coincide if and only if $w_j(\xi, \delta) = 0$.) Estimates of optimality will be in terms of the quantity

$$w(x, \delta) = \sum_{j \in A} w_j(x(j), \delta) \leq \infty, \tag{5}$$

which is well defined for any feasible flow x.

Discretized Optimal Distribution Algorithm

Fix $\delta > 0$. Given any feasible solution x to the optimal distribution problem, try to solve the feasible differential problem with span intervals

$$\left[d_x^-(j), d_x^+(j) \right] = \left[\Delta^\delta_- f_j(x(j)), \Delta^\delta_+ f_j(x(j)) \right]. \tag{6}$$

If a potential u is obtained (such that $d_x^-(j) \leq v(j) \leq d_x^+(j)$ for all $j \in A$, where $v = \Delta u$), terminate. For the value $\varepsilon = w(x, \delta)$ if finite, x is an ε-optimal solution to the optimal distribution problem, and u is an ε-optimal solution to the dual problem, in fact with $\Phi(x) - \Psi(u) \leq \varepsilon$. Furthermore x is δ-*quasi-optimal* in the sense that $\Phi(x + te_P) \geq \Phi(x)$ for all elementary circuits P and all $t \geq \delta$.

If an elementary circuit P is obtained such that $d_x^+(P) < 0$, then $\Phi(x + \delta e_P) < \Phi(x)$. If there is no integer $n > 0$ such that

$$\Phi(x + (n + 1)\delta e_P) \geq \Phi(x + n\delta e_P), \tag{7}$$

terminate. In this event P is an unbalanced circuit, so the optimal distribution problem has no optimal solution, and the dual problem has no regularly feasible solution; in particular, the optimal distribution problem does not have the boundedness property in Section 8M. Otherwise determine the smallest integer n such that (7) holds, and let $x' = x + n\delta e_P$. Then x' is a feasible solution with $\Phi(x') < \Phi(x)$. Iterate with x'.

Justification

From the definition of the span intervals, one has

$$d_x^+(P) = \sum_{j \in P^+} \frac{f_j(x(j) + \delta) - f_j(x(j))}{\delta} - \sum_{j \in P^-} \frac{f_j(x(j)) - f_j(x(j) - \delta)}{\delta}$$

$$= \sum_{j \in A} \frac{f_j(x(j) + \delta e_P(j)) - f_j(x(j))}{\delta}$$

$$= \frac{\Phi(x + \delta e_P) - \Phi(x)}{\delta}.$$

Thus x fails to be δ-quasi-optimal if and only if there is an elementary circuit P with $d_x^+(P) < 0$. Suppose P is such a circuit, and let $f(t) = \Phi(x + te_P)$. If there is no integer $n > 0$ satisfying (7), then

$$0 > [f((n + 1)\delta) - f(n\delta)]/\delta \geq f_+'(n\delta)$$

for all integers $n > 0$, and consequently $f_+'(t) < 0$ for all $t \in R$ (since f_+' is nondecreasing). This implies f is a decreasing function (see Exercise 8.10(b)). Then P is an unbalanced cut by the descent theorem for flows in Section 8J; the assertions about the nonexistence of optimal solutions and regularly feasible solutions follow in this case from the existence theorem in Section 8L. If there is an integer $n > 0$ satisfying (7), then $n\delta$ minimizes $f(t)$ over all the values of t that are multiples of δ. In particular, $f(n\delta) \leq f(\delta) < f(0)$.

The only thing left to show is that when a potential u is obtained one has

$$\Phi(x) - \Psi(u) \leq w(x, \delta) = \varepsilon. \tag{8}$$

Indeed, the inequality $\Phi(x) - \Psi(u) \leq \varepsilon < \infty$ implies by the duality theorem that x and u are ε-optimal. The proof of (8) is based on the identity

$$\Phi(x) - \Psi(u) = \sum_{j \in A} f_j(x(j)) + \sum_{i \in N} b(i)u(i) + \sum_{j \in A} g_j(v(j))$$

$$= \sum_{j \in A} \left[f_j(x(j)) + g_j(v(j)) - x(j)v(j) \right].$$

In view of definitions (5) and (6), it suffices to establish that

$$f_j(x(j)) + g_j(v(j)) - x(j)v(j) \leq w_j(x(j), \delta)$$

$$\text{when} \quad \Delta_-^\delta f_j(x(j)) \leq v(j) \leq \Delta_+^\delta f_j(x(j)),$$

or in simpler notation with $f_j(\xi) = f_j(x(j)) < \infty$, that

$$f_j(\xi) + g_j(\eta) - \xi\eta \le w_j(\xi, \delta) \quad \text{when} \quad \Delta_-^\delta f_j(\xi) \le \eta \le \Delta_+^\delta f_j(\xi). \tag{9}$$

If actually $f_j'(\xi) \le \eta \le f_{j+}'(\xi)$, then

$$f_j(\xi) + g_j(\eta) - \xi\eta = 0 \tag{10}$$

(see 8E(7)(12)), so the conclusion in (9) is trivially valid.

Consider next the case where $f_{j+}'(\xi) < \eta \le \Delta_+^\delta f_j(\xi)$, and suppose first that $f_{j+}'(\xi) > -\infty$. Since f_j and g_j are conjugate to each other, one has

$$f_j(\xi) + g_j(\eta) - \xi\eta = -\inf_\zeta r(\zeta), \tag{11}$$

where the function

$$r(\zeta) = f_j(\zeta) - f_j(\xi) + (\xi - \zeta)\eta \tag{12}$$

is closed proper convex with $r(\xi) = 0$,

$$r(\xi + \delta) = \delta[\Delta_+^\delta f_j(\xi) - \eta] \ge 0,$$

$$r_+'(\xi) = f_{j+}'(\xi) - \eta < 0.$$

These properties of r imply that $\inf_\zeta r(\zeta)$ is attained somewhere in the interval $(\xi, \xi + \delta)$. Since

$$[f(\xi) - f(\xi)]/(\zeta - \xi) \ge f_{j+}'(\xi) \quad \text{for all } \zeta > \xi,$$

one has by (12) that

$$r(\zeta) \ge (\zeta - \xi)[f_{j+}'(\xi) - \eta] \quad \text{for all} \quad \zeta > \xi,$$

and consequently

$$\inf_{\zeta \in R} r(\zeta) = \inf_{\zeta \in (\xi, \xi+\delta)} r(\zeta) \ge \inf_{\zeta \in (\xi, \xi+\delta)} (\zeta - \xi)[f_{j+}'(\xi) - \eta]$$

$$\ge \delta[f_{j+}'(\xi) - \eta] \ge \delta[f_{j+}'(\xi) - \Delta_+^\delta f_j(\xi)].$$

Applied to (11), this says that

$$f_j(\xi) + g_j(\eta) - \xi\eta \le \delta[\Delta_+^\delta f_j(\xi) - f_{j+}'(\xi)]. \tag{13}$$

Of course (13) holds also if $f_{j+}'(\xi) = -\infty$, since the right side is $+\infty$ in that event.

A parallel argument demonstrates in the case where $\Delta^\delta_- f_j(\xi) \leq \eta < f'_j(\xi)$ that

$$f_j(\xi) + g_j(\eta) - \xi\eta \leq \delta\left[f'_j(\xi) - \Delta^\delta_- f_j(\xi)\right]. \tag{14}$$

Thus no matter where η lies in the interval $[\Delta^\delta_- f_j(\xi), \Delta^\delta_+ f_j(\xi)]$, either (10) or (13) or (14) holds. Hence (9) is true as claimed.

Termination

Suppose the optimal distribution problem has the boundedness property (defined in Section 8M). Then there are no unbalanced circuits, so an integer $n > 0$ satisfying (7) does exist at each iteration. Moreover the algorithm must terminate after a finite number of iterations with a flow x and a potential u having the optimality properties described.

This can be seen from the fact that the flow at the end of iteration k has the form

$$x^k = x^0 + n_1 \delta e_{P_1} + \cdots + n_k \delta e_{P_k},$$

where x^0 is the initial flow. Its flux in each arc j differs from $x^0(j)$ merely by a multiple of δ. There are only finitely many flows of this form in any bounded set of flows. Since one has $\Phi(x^k) < \cdots < \Phi(x^1) < \Phi(x^0)$, and since the set of feasible solutions x satisfying $\Phi(x) \leq \Phi(x^0)$ is bounded by assumption, the iterations cannot continue indefinitely.

Dual Version

The analogous version of the optimal differential algorithm in terms of the difference quotients

$$\Delta^\delta_+ g_j(\eta) = \left[g_j(\eta + \delta) - g_j(\eta)\right]/\delta,$$

$$\Delta^\delta_- g_j(\eta) = \left[g_j(\eta + \delta) - g_j(\eta)\right]/\delta, \tag{15}$$

involves the expressions

$$z_j(\eta, \delta) = \delta \max\left\{ \begin{matrix} \Delta^\delta_+ g_j(\eta) - g'_{j+}(\eta) \\ g'_{j-}(\eta) - \Delta^\delta_- g_j(\eta) \end{matrix} \right\} \geq 0, \tag{16}$$

where the first difference in the maximum is counted as 0 if $g'_{j+}(\eta) = \Delta^\delta_+ g_j(\eta) = +\infty$, and the second if $g'_{j-}(\eta) = \Delta^\delta_- g_j(\eta) = -\infty$. Estimates of optimality are in terms of the quantity

$$z(u, \delta) = \sum_{j \in A} z_j(v(j), \delta) \leq \infty, \tag{17}$$

where u is a feasible solution to the optimal differential problem and $v = \Delta u$.

Discretized Optimal Differential Algorithm

Fix $\delta > 0$. Given any feasible solution u to the optimal differential problem, try to solve the feasible distribution problem with supplies $b(i)$ and capacity intervals

$$\left[c_u^-(j), c_u^+(j) \right] = \left[\Delta_-^\delta g_j(v(j)), \Delta_+^\delta g_j(v(j)) \right]. \tag{18}$$

If a flow x is obtained such that $c_u^-(j) \le x(j) \le c_u^+(j)$ for all $j \in A$ and $\operatorname{div} x = b$, terminate. For the value $\varepsilon = z(u, \delta)$ if finite, u is an ε-optimal solution to the optimal differential problem, and x is an ε-optimal solution to the dual problem, in fact with $\Phi(x) - \Psi(u) \le \varepsilon$. Furthermore u is δ-*quasi-optimal* in the sense that $\Psi(u + te_{N\setminus S}) \le \Psi(u)$ for all $S \subset N$ and $t \ge \delta$.

If a cut $Q = [S, N\setminus S]$ is obtained such that $c_u^+(Q) < b(S)$, then $\Psi(u + \delta e_{N\setminus S}) > \Psi(u)$. If there is no integer $n > 0$ such that

$$\Psi\left(u + (n+1)\delta e_{N\setminus S}\right) \le \Psi\left(u + n\delta e_{N\setminus S}\right), \tag{19}$$

terminate. In this event Q is an unbalanced cut, so the optimal differential problem has no optimal solution, and the dual problem has no regularly feasible solution; in particular, the optimal differential problem does not have the boundedness property in Section 8M. Otherwise determine the smallest integer n such that (19) holds, and let $u' = u + n\delta e_{N\setminus S}$. Then u' is a feasible solution with $\Psi(u') > \Psi(u)$. Iterate with u'.

Termination

If the optimal differential problem has the boundedness property, this procedure must terminate in a finite number of iterations in a flow x and a potential u having the optimality properties described. The proof of this and the other assertions in the statement of the algorithm is parallel to that for the discretized optimal distribution algorithm and is consigned to Exercise 9.23.

9J.   CALCULATING ε-OPTIMAL SOLUTIONS

The fortified algorithms are capable of calculating ε-optimal solutions to any dual pair of problems for any preassigned $\varepsilon > 0$, but they may be hard to implement in the general case. In order to accomplish the same thing with the discretized algorithms, either there must be some way of determining an appropriate step size δ in advance from the structure of the problems, or the procedure must be iterated with a decreasing sequence of δ values. The latter two approaches will now be examined in detail. Because of the transparent duality, only the results for the flow algorithm will be stated.

Choosing the Right Step Size in Advance

In rather general circumstances knowledge of the functions f_j leads to an upper bound for $w(x, \delta)$ in terms of δ alone. Then for any $\varepsilon > 0$ it is possible to choose a corresponding $\delta > 0$ such that the discretized optimal distribution algorithm, when it terminates with a flow x and a potential u, will have $w(x, \delta) \leq \varepsilon$. This ensures that x and u will be ε-solutions for the *given* ε.

The following result covers piecewise quadratic problems, in particular. Suppose that:

1. For each arc j there is a known constant $\sigma_j \geq 0$ such that, except for at most a finite set of points in $[c^-(j), c^+(j)]$ (which together with $c^-(j)$ and $c^+(j)$, if finite, will be called "breakpoints"), the second derivative $f_j''(\xi)$ exists and satisfies $f_j''(\xi) \leq \sigma_j$.
2. The breakpoints and initial flow values are all multiples of a certain $\delta_0 > 0$.

Then if the step size is of the form $\delta = \delta_0/m$ with m a positive integer, every flow x generated by the algorithm, and in particular, the terminal flow, will satisfy

$$w(x, \delta) \leq \sigma\delta^2, \quad \text{where } \sigma = \sum_{j \in A} \sigma_j \geq 0. \tag{1}$$

Termination with ε-solutions for a preassigned $\varepsilon > 0$ is thus assured by selecting m large enough that $\sigma\delta^2 \leq \varepsilon$, or in other words, by choosing

$$m \geq \delta_0\sqrt{\sigma/\varepsilon}.$$

Indeed, Case 2 implies that the breakpoints will all be "grid points" for the step size δ. Then $f_j''(\xi) \leq \sigma_j$ by Case 7 for all in $(x(j), x(j) + \delta)$ if $x(j) < c^+(j)$, and for all ξ in $(x(j) - \delta, x(j))$ if $x(j) > c^-(j)$. In the first case a double application of the mean value theorem yields the estimate

$$\Delta^\delta_+ f_j(x(j)) - f_{j+}'(x(j)) = f_j'(\xi') - f_{j+}'(x(j))$$

$$= (\xi' - x(j))f_j''(\xi'') \leq \sigma_j\delta,$$

where ξ' and ξ'' are certain numbers satisfying $x(j) < \xi'' < \xi' < x(j) + \delta$. In the second case one obtains similarly the estimate

$$f_{j-}'(x(j)) - \Delta^\delta_- f_j(x(j)) \leq \sigma_j\delta.$$

Thus $w_j(x(j), \delta) \leq \sigma_j\delta^2$, so that (1) follows from the definition 9I(5) of $w(x, \delta)$.

Iterated Algorithm

The other approach, given any $\varepsilon > 0$, is to apply the discretized optimal distribution algorithm with arbitrary initial feasible solution x_0 and step size $\delta_0 > 0$ to get (either an unbalanced circuit or) x_1 and u_1 satisfying

$$\Phi(x_1) - \Psi(u_1) \leq w(x_1, \delta_0), \qquad \Phi(x_1) < \Phi(x_2).$$

If it turns out that $w(x_1, \delta_0) \leq \varepsilon$, then x_1 and u_1 are ε-optimal and the goal has been achieved. If not, let $\delta_1 = \delta_0/p$ (where $p > 1$ is a fixed integer selected in advance), and apply the algorithm again with initial feasible solution x_1 and step size δ_1. This produces (either an unbalanced circuit or) x_2 and u_2, satisfying

$$\Phi(x_2) - \Psi(x_2) \leq w(x_2, \delta_1), \qquad \Phi(x_2) < \Phi(x_1).$$

If $w(x_2, \delta_1) \leq \varepsilon$, stop; x_2 and u_2 are ε-optimal. Otherwise start the algorithm again with x_2 and $\delta_2 = \delta_1/p$, and so forth.

Termination

If the following three conditions are satisfied, one will eventually have $w(x_k, \delta_{k-1}) \leq \varepsilon$, so that the iterated algorithm will terminate with x_k and u_k as ε-optimal solutions. The conditions are:

1. The optimal distribution problem has the boundedness property in Section 8M.
2. The optimal differential problem has the boundedness property in 8M.
3. For some integer $m > 0$, the values $c_0(j), c^-(j)$ and $c^+(j)$ for each arc differ only by multiples of δ_0/p^m.

Note that Condition 3 is certainly true if $x_0(j)$ and (if finite) $c^-(j)$ and $c^+(j)$ are finite decimals, $\delta_0 = 10^{-r}$ for some integer $r > 0$, and $p = 10$. The purpose of Condition 3 is to ensure that $c^-(j)$ and $c^+(j)$ (to the extent that they are finite) are "grid points" relative to δ_k once $k \geq m$. Of course Conditions 1 and 2 rule out the presence of unbalanced circuits or cuts, in particular.

This termination result will be demonstrated in the next section as a corollary of a still stronger property of the iterated algorithm.

9K.* OPTIMIZING SEQUENCES AND PIECEWISE LINEARIZATION

The fact that the optimal distribution algorithm works especially well for piecewise linear problems suggests the following approach to the general case. Approximate the given cost functions f_j in some way by piecewise linear functions and solve the corresponding "approximate" problem. Repeat with finer and finer approximations, using as the initial flow on each round the solution to the approximate problem in the preceding round. Hopefully, if this is carried out in the right way, the flows that are generated will form a minimizing sequence for the optimal distribution problem whereas the potentials that solve the duals of the approximate problems will form a maximizing sequence for the optimal differential problem.

As a matter of fact the discretized optimal distribution algorithm with step size δ_k and initial flow x_k is nothing more than the optimal distribution algorithm applied to a certain approximate problem, namely, the one in which each f_j is replaced by f_j^k as indicated in Figure 9.14. Specifically, f_j^k is the function defined by interpolating linearly between the values of f_j at the grid points for δ_k, that is, the points of the form $\xi = x_k(j) + n\delta_k$, n integral. At such points ξ one has $f_j^k(\xi) = f_j(\xi)$ and

$$f_j^{k\prime}(\xi) = \Delta_+^{\delta_k} f_j(\xi), \qquad f_j^{k\prime}(\xi) = \Delta_-^{\delta_k} f_j(\xi).$$

Hence if $f^k(t) = \Phi^k(x + te_P)$, where Φ^k is the cost function in the approximate problem with each f_j replaced by f_j^k and x is a flow whose value $x(j)$ are all grid points for δ_k, one has $f^k(t) = f(t) \, (= \Phi(x + te_P))$ for all t of the form $n\delta_k$. Moreover for such t one has

$$f_+^{k\prime}(t) = \sum_{j \in P^+} \Delta_+^{\delta_k} f_j\big(x(j) + t\delta_k\big) - \sum_{j \in P^-} \Delta_-^{\delta_k} f_j\big(x(j) - t\delta_k\big)$$

$$= \Delta_+^{\delta_k} f(t).$$

Each stage of the discretized algorithm with step size δ_k thus tests a certain "grid point" flow x to see if there is a circuit P such that $f_+^{k\prime}(0) < 0$. If there is

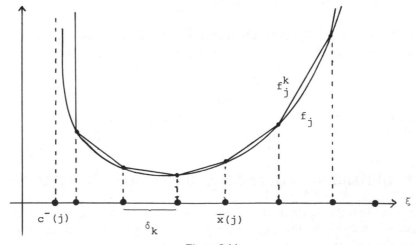

Figure 9.14

such a P (and it is not unbalanced), then $x' = x + \alpha e_P$, where α is the lowest value of t minimizing $f^k(t)$. If there is not, then the algorithm terminates. In the latter case x is optimal for the *approximate* problem, and this therefore is the true meaning of the "quasi-optimality" notion introduced for the given problem.

The functions f_j^k may not quite fit the definition of "piecewise linear" given in Section 8C because their graphs may consist of an *infinite* number of segments. When the optimal distribution problem has the boundedness property, however, there is an overall upper bound to the number of grid points for δ_k that can actually come into play in the course of computation. Thus there is no real difference in principle between the case described here and one in which the functions f_j^k are truncated into truly "piecewise linear" approximations.

From this discussion it is clear that the approach of "successive approximation" just outlined can be realized by applying the discretized algorithm repeatedly with $\delta_k = \delta_0/p^k$, $k = 0, 1, \ldots$, as in Section 9J. The rest of this section is devoted to proving the following result along these lines, which immediately implies the result stated at the end of Section 9J.

Suppose the iterated form of the discretized optimal distribution algorithm is executed without the ε stopping condition, and with Conditions 1, 2, and 3 satisfied (see the end of Section 9J), thereby generating infinite sequences $\{x_k\}$ and $\{u_k\}$. Then there is a constant $\beta \geq 0$ such that

$$0 \leq w(x_k, \delta_{k-1}) \leq \beta\delta_0/p^{k-1} \text{ for all } k \text{ sufficiently large.} \tag{1}$$

In view of the properties of the discretized algorithm that were established in Section 9I, this implies of course that x_k and u_k are ε_k-optimal, where $\varepsilon_k = w(x_k, \delta_{k-1}) \to 0$. Hence $\{x_k\}$ is a minimizing sequence for the optimal distribution problem whereas $\{u_k\}$ is a maximizing sequence for the optimal differential problem. It follows from Conditions 1 and 2 that the sequences $\{x_k\}$ and $\{v_k\}$ (where $v_k = \Delta u_k$) are bounded and hence have cluster points. These cluster points furnish optimal solutions to the two problems, as explained in Section 8M.

Proof of (1). It is clear from the observations about the discretized algorithm that $\Phi(x_k) = \Phi^k(x_k) = \min \Phi^k$, where "min" in what follows is to be understood as taken subject to $\text{div } x = b$. The first stage of the proof of (1) consists in showing that

$$\min \Phi^k \to \min \Phi \quad \text{as} \quad k \to \infty, \tag{2}$$

from which one can conclude at least that $\Phi(x^k) \to \min \Phi$.

The definition of f_j^k implies that $f_j^k(\xi) \geq f_j^{k+1}(\xi) \geq \cdots \geq f_j(\xi)$ for all ξ. Equality holds throughout the chain if ξ is a grid point for δ_k, or also if $\xi < c^-(j)$ or $\xi > c^+(j)$ (all terms then being $+\infty$). Furthermore one has

$$f_j^k(\xi) \downarrow f_j(\xi) \quad \text{as} \quad k \to \infty \quad \text{if} \quad c^-(j) < \xi < c^+(j), \tag{3}$$

as can be seen by the following argument. For numbers ξ_0 and ξ_1 satisfying

$$c^-(j) < \xi_0 < \xi < \xi_1 < c^+(j),$$

the values $f_j'^+(\xi_0)$ and $f_j'^+(\xi_1)$ are finite. Once k is large enough ($\delta_k = \delta_0/p^k$ small enough), there will be a grid point $\bar{\xi}$ for δ_k with

$$\xi_0 < \bar{\xi} < \xi \le \bar{\xi} + \xi_k < \xi_1.$$

Then $f_j^k(\bar{\xi}) = f_j(\bar{\xi})$, so that

$$f_j^k(\xi) - f_j(\xi) = \int_{\bar{\xi}}^{\xi} \left[f_{j+}^{k\prime}(t) - f_{j+}'(t) \right] dt$$

$$\le \int_{\bar{\xi}}^{\xi} \left[f_{j-}'(\xi_1) - f_{j+}'(\xi_0) \right] dt \le \delta_k \left[f_{j-}'(\xi_1) - f_{j+}'(\xi_0) \right].$$

This inequality implies (3).

In order to derive (2) from (3), consider any $\mu > 0$ and any optimal solution $\tilde{x}$ to the optimal distribution problem. (At least one optimal solution exists because of Assumption 1 of the boundedness theorem see Section 8M.) The function $f(t) = \Phi(\tilde{x} + t(x^0 - \tilde{x}))$ is closed proper convex (where x^0 is the feasible solution given at the start of the algorithm), so $f(t) \to f(0)$ as $t \downarrow 0$. Thus there is a value $t \in (0, 1)$ such that $f(t) < f(0) + \varepsilon$. The flow

$$x^* = \tilde{x} + t(x^0 - \tilde{x}) = (1 - t)\tilde{x} + tx^0$$

is then a feasible solution that satisfies

$$\Phi(x^*) < \Phi(\tilde{x}) + \mu = [\min \Phi] + \mu \tag{4}$$

and has for every arc j either $x^*(j) = \tilde{x}(j) = x^0(j)$ or $c^-(j) < x^*(j) < c^+(j)$, or both. For arcs of the first category, $x^*(j)$ is a grid point for δ_k when $k \ge m$ (see Assumption 3), so that $f_j^k(x^*(j)) = f_j(x^*(j))$ when $k \ge m$. For arcs of the second category, one has $f_j^k(x^*(j)) \to f_j(x^*(j))$ as $k \to \infty$ by (2). Therefore

$$\Phi^k(x^*) \to \Phi(x^*) \quad \text{as } k \to \infty. \tag{5}$$

It follows from (4), (5), and the inequality $\Phi^k \ge \Phi$ that, for all k sufficiently large,

$$[\min \Phi] + \mu \ge \Phi^k(x^*) \ge \min \Phi^k \ge \min \Phi.$$

Since μ was arbitrary, (2) holds as claimed. Thus $\{x_k\}$ is a minimizing sequence for the optimal distribution problem.

Assumption 2 has not yet been used, but it enters in showing that actually $w(x_k, \delta_{k-1}) \to 0$, specifically through its implication that all optimal solutions to the optimal distribution problem are regularly feasible (see Exercise 8.17 and the existence theorem for potentials in Section 8M). It is impossible under such circumstances for the flux values $x^k(j)$ in any arc to cluster at a finite endpoint $c^-(j)$ with $f'_{j+}(c^-(j)) = -\infty$, or $c^+(j)$ with $f'_{j-}(c^+(j)) = -\infty$. For if this were the case, a subsequence of the bounded sequence $\{x^k\}$ would converge to a flow taking on such a flux value $c^-(j)$ or $c^+(j)$, and this flow would be an optimal solution (see Exercise 8.52) which is not regularly feasible. Hence there exist finite values $\xi_0(j)$ and $\xi_1(j)$ for each arc j such that

$$c^-(j) \le \xi_0(j) \le x^k(j) \le \xi_1(j) \le c^+(j) \tag{6}$$

for all k sufficiently large and

$$\xi_0(j) > c^-(j) \quad \text{for each } j \text{ with} \quad c^-(j) > -\infty \quad \text{and} \quad f'_{j+}(c^-(j)) = -\infty,$$

$$\xi_1(j) < c^+(j) \quad \text{for each } j \text{ with} \quad c^+(j) < +\infty \quad \text{and} \quad f'_{j-}(c^+(j)) = +\infty.$$

$$\tag{7}$$

Note that these properties imply

$$-\infty < f'_{j+}(\xi_0(j)) \le f'_{j-}(\xi_1(j)) < +\infty \quad \text{if } c^-(j) \ne c^+(j). \tag{8}$$

In view of Assumption 3, moreover, $\xi_0(j)$ and $\xi_1(j)$ can be chosen so as to be grid points for δ_k when k is sufficiently large. Decreasing $\xi_0(j)$ and increasing $\xi_1(j)$ slightly if necessary, one can arrange further that for arcs j with $\xi_0(j) > c^-(j)$ one has

$$x_k(j) - \delta_{k-1} \ge \xi_0(j) \quad \text{for all } k \text{ sufficiently large,}$$

whereas for arcs j with $\xi_1(j) < c^+(j)$ one has

$$x_k(j) + \delta_{k-1} \le \xi_1(j) \quad \text{for all } k \text{ sufficiently large.}$$

Now for k large enough for all these properties to hold, consider the quantities $w_j(x_k(j), \delta_{k-1})$ (defined in 9I(4)). There are only four cases to look at, due to the way things have been set up:

1. $\xi_0(j) \le x_k(j) - \delta_{k-1}$ and $x_k(j) + \delta_{k-1} \le \xi_1(j)$.
2. $c^-(j) + \xi_0(j) = x_k(j)$ and $x_k(j) + \delta_{k-1} \le \xi_1(j)$.
3. $\xi_0(j) \le x_k(j) - \delta_{k-1}$ and $x_k(j) = \xi_1(j) = c^+(j)$.
4. $c^-(j) = \xi_0(j) = x_k(j) = \xi_1(j) = c^+(j)$.

In Case 1 one has

$$w_j\big(x_k(j), \delta_{k-1}\big) \leq \delta_{k-1} \max \left\{ \begin{matrix} f_j'^-(\xi_1(j)) - f_j'^+(x_k(j)) \\ f_j'^-(x_k(j)) - f_j'^+(\xi_0(j)) \end{matrix} \right\}$$

$$\leq \delta_{k-1}\big[f_j'^-(\xi_1(j)) - f_j'^+(\xi_0(j)) \big].$$

In Case 2 one has

$$w_j\big(x_k(j), \delta_{k-1}\big) \leq \delta_{k-1} \max\big\{ f_j'^-(\xi_1(j)) - f_j'^+(x_k(j)), 0 \big\}$$

$$\leq \delta_{k-1}\big[f_j'^-(\xi_1(j)) - f_j'^+(\xi_0(j)) \big].$$

Similarly, in Case 3

$$w_j\big(x_k(j), \delta_{k-1}\big) \leq \delta_{k-1} \max\big\{ 0, f_j'^-(x_k(j)) - f_j'^+(\xi_0(j)) \big\}$$

$$\leq \delta_{k-1}\big[f_j'^-(\xi_1(j)) - f_j'^+(\xi_0(j)) \big],$$

whereas in Case 4

$$w_j\big(x_k(j), \delta_{k-1}\big) = \max\{0, 0\} = 0.$$

It follows that for $A_0 = \{ j \mid c^-(j) \neq c^+(j) \}$ and

$$\beta = \sum_{j \in A_0} \big[f_j'^-(\xi_1(j)) - f_j'^+(\xi_0(j)) \big],$$

one has (1) as claimed, where $\delta_0 / p^{k-1} = \delta_{k-1}$.

9L.* CONVEX SIMPLEX METHOD

The simplex method for flows, presented in Section 7D for the case of a linear optimal distribution problem, can be generalized to piecewise linear and more general convex problems. It is no longer possible in such a context to work only with flows that are extreme solutions to the underlying feasibility problem, since optimality may not be attained at an extreme solution. The possibility of breakpoints of f_j interior to the interval $[c^-(j), c^+(j)]$ causes difficulty even with the formulation of the method, unless a condition of nondegeneracy is met. On the other hand, in the nondegenerate case the method can be regarded as a special realization of the optimal distribution algorithm in Section 9A, where a subroutine for finding an improvement circuit in each iteration is introduced in terms of spanning trees. The robustness of the method, in comparison to the others that have been discussed, therefore depends roughly on whether spanning trees and their associated flows and potentials can be manipulated efficiently enough to make up for the apparent advantage of subroutines that incorporate min path techniques instead.

Without referring to any particular representation of the functions f_j in terms of subintervals of $[c^-(j), c^+(j)]$ as in the piecewise linear of piecewise quadratic cases, let us define the *first-order breakpoints* of f_j to be in general the points $\xi \in [c^-(j), c^+(j)]$, where $f'_j(\xi) < f'_{j+}(\xi)$. Thus $c^-(j)$ and $c^+(j)$ are first-order breakpoints, when finite. A point ξ strictly between $c^-(j)$ and $c^+(j)$ fails to be a first-order breakpoint if and only if f is differentiable at ξ (with $f'_j(\xi) = f'_{j+}(\xi) = f'_{j-}(\xi)$, finite). The number of first-order breakpoints is finite if and only if f_j is *piecewise differentiable*.

In what follows, we suppose as in Section 7D that the network is connected and that a node s has been singled out. A differential v is then uniquely determined by its values for the arcs in a spanning tree, and a corresponding potential u with $u(s) = 0$ can be constructed by integration along the paths of the routing (with base s) determined by the tree.

Statement of the Algorithm

Let x be a feasible solution to the optimal distribution problem that is *nondegenerate* in the sense that there is a spanning tree F with

$$F \subset \left\{ j \in A \mid x(j) \text{ is not a first-order breakpoint for } f_j \right\}. \qquad (1)$$

For such an F, let u be the unique potential with $u(s) = 0$ such that $v = \Delta u$ satisfies $v(j) = f'_j(x(j))$ for all $j \in F$. Check whether

$$d_x^-(k) \leq v(k) \leq d_x^+(k) \quad \text{for all arcs } k \in A \setminus F \qquad (2)$$

(where $[d_x^-(k), d_x^+(k)]$ is defined by 9A(1), as usual). If yes, terminate; x and u solve the optimal distribution problem and its dual.

If no, let $\bar{k}$ be any arc for which (2) is violated. There is a unique elementary circuit $P_{\bar{k}}$ that uses $\bar{k}$ and otherwise only arcs belonging to F (see Section 4H). Let P be $P_{\bar{k}}$ if $v(\bar{k}) > d_x^+(\bar{k})$, but the reverse of $P_{\bar{k}}$ if $v(\bar{k}) < d_x^-(\bar{k})$. Then $d_x^+(P) < 0$, so that the function $f(t) = \Phi(x + te_P)$ has $f'_+(0) < 0$.

Let $\alpha = \min\{t \geq 0 \mid f'_+(t) \geq 0\}$. Then $\alpha > 0$. If $\alpha = \infty$, terminate; P is unbalanced, the optimal distribution problem has no optimal solution, and its dual has no regularly feasible solution. If $\alpha < \infty$, replace x by $x + \alpha e_P$. If for the new x there is an arc $\bar{j} \in F$ such that $x(\bar{j})$ is a first-order breakpoint for f_j, replace F by $[F \setminus \{\bar{j}\}] \cup \{\bar{k}\}$, but otherwise leave F as it is. Then return to the beginning with x and F.

Justification

Note that (1) will again be satisfied if there is no more than one $\bar{j}$ as described, and none when $x(\bar{k})$ is a breakpoint for $f_{\bar{k}}$. Provided this is true in every iteration (*nondegeneracy assumption*), the method clearly reduces to a form of the optimal distribution algorithm, since

$$\left[d_x^-(j), d_x^+(j)\right] = \left[f_j'(x(j)), f_j'(x(j))\right] \quad \text{for all } j \in F$$

by (1). When F is altered, it remains a spanning tree because $x(\bar{j})$ could not have shifted from a nonbreakpoint to a breakpoint of f_j unless $\bar{j}$ were in the circuit $P_{\bar{k}}$.

Effectiveness

If the choice of $\bar{k}$ is left completely open, all the arcs that violate (2) being equally eligible, the algorithm might not generate a minimizing sequence (see Exercise 9.27). A simple but good rule is to select $\bar{k}$ for which the violation of (2) in terms of distance of $v(k)$ from $[d_x^-(k), d_x^+(k)]$, is greatest.

Extension to Degenerate Cases

When $x(j)$ is a first-order breakpoint of f_j for some $j \in F$, the rule for associating a potential u with x and F falls apart. If $c^-(j) < x(j) < c^+(j)$, the one-sided derivatives $f_j'^+(x(j))$ and $f_j'^-(x(j))$ are both finite but different, and no simple substitute for the formula $v(j) = f_j'(x(j))$ is available. However, in the cases where $c^-(j) = x(j) < c^+(j)$ or $c^-(j) < x(j) = c^+(j)$, in other words, where $x(j)$ is a breakpoint that is not internal, one can naturally take $v(j) = f_j'^+(x(j))$ and $v(j) = f_j'^-(x(j))$, respectively.

This provision allows the algorithm to be executed with "breakpoint" replaced by "internal breakpoint" in (1). As long as preference in the choice of $\bar{j}$ is given to the case where $x(\bar{j})$ is an internal breakpoint of $f_{\bar{j}}$, and this kind of breakpoint is never encountered for more than one $\bar{j}$ (none when $x(\bar{k})$ is an internal breakpoint for $f_{\bar{k}}$), the modified form of (1) will be maintained. Of course $\alpha = 0$ may be possible under this extension, so the circuit P may have $d_x^+(P) = 0$. Iterations can therefore take place where F is altered but the flow x remains the same. No longer will one have a true descent algorithm, and something like Bland's priority rule in Section 7D will be needed to prevent the possibility of an infinite degenerate sequence.

Note that any problem with every f_j piecewise differentiable can be reduced to one where there are no first-order *internal* breakpoints by black box representations like the ones used for linear problems (Exercise 9.30). In fact that convex simplex method can be formulated still more broadly, so that it makes sense, at least, without any breakpoint condition like (1) at all. This is shown in Section 11J in the general case of monotropic programming. Again iterations with step-size $\alpha = 0$ can occur with such a formulation, so that degeneracy does remain an underlying difficulty.

Dual Version

The convex simplex method can also be developed in terms of potentials rather than flows, so as to get an extension of the simplex method in Section 7H that corresponds to the general optimal differential algorithm (exercise 9.31).

9M.* EXERCISES

9.1. (*Optimal Distribution Algorithm*). Prove that a feasible solution x to the optimal distribution problem is optimal if and only if for every elementary circuit P the function $f(t) = \Phi(x + te_P)$ has $f'_+(0) \geq 0$.

(*Note.* This sharpens the result in Exercise 8.19.)

9.2. (*Optimal Distribution Algorithm*). Verify the formula for $f'_+(t)$ given in 9A(4) for the piecewise quadratic case.

9.3. (*Optimal Distribution Algorithm*). Verify the statement in the last sentence of Section 9A.

9.4. (*Optimal Distribution Algorithm*). Let $\bar{x}$ be an arbitrary flow, and consider the problem of approximating $\bar{x}$ in the sense of finding a circulation x that minimizes

$$\Phi(x) = \sum_{j \in A} |x(j) - \bar{x}(j)| = \|x - \bar{x}\|_1.$$

(This is a piecewise linear optimal distribution problem with $b(i) = 0$ for all nodes i and $f_j(\xi) = |\xi - \bar{x}(j)|$ for all arcs j.) What does the optimal distribution algorithm reduce to in this case, and what is needed to ensure termination?

9.5. (*Optimal Distribution Algorithm*). Prove that when the feasible differential algorithm is used as the subroutine in the optimal distribution algorithm, as indicated in Section 9B, the distance of $(x(j), v(j))$ from Γ_j is at no stage increased for any arc j.

(*Hint.* Generalize the argument at the end of Section 7C.)

9.6. (*Optimal Differential Algorithm*). Fill in the justification of the optimal differential algorithm in Section 9C, following the lines for the optimal distribution algorithm. (State without proof the analog of Exercise 8.18 that needs to be invoked in one place.)

9.7. (*Optimal Differential Algorithm*). Prove the assertions in Section 9C about the termination of the optimal differential algorithm.

9.8. (*Optimal Differential Algorithm*). Verify the assertions about how α can be calculated in Example 4 in Section 9C, and show they are valid even if the capacity bounds $c^+(j)$, $c^-(j)$, are allowed to be infinite.

9.9. (*Thrifty Adjustment Algorithm*). Justify the algorithm in Section 9D by extending the argument for the special case of it in Section 7J.

9.10. (*Thrifty Adjustment Algorithm*). Verify the claims at the beginning of Section 9D about how the thrifty adjustment algorithm can be derived (see Exercise 7.34 for the case of elementary problems).

9.11. (*Thrifty Adjustment Algorithm*). Show that the economic interpretation in Section 7K can be extended to the more general algorithm in Section 9D.

9.12. (*Thrifty Adjustment Algorithm*). Demonstrate that in the case of a linear optimal distribution problem the thrifty adjustment algorithm reduces to the version in Exercise 7.36. Similarly for the linear optimal differential problem and Exercise 7.41.

9.13. (*Black Box Application*). Use the procedure in Section 9E to determine the entire characteristic curve Γ for the network in Figure 9.15, where the numbers specify $[d^-(j), d^+(j)]$ and $c(j)$, each arc having a characteristic curve of the form in Figure 9.16.

9.14. (*Flow Rectification Algorithm*). Demonstrate that the flow rectification algorithm in Section 3K is essentially equivalent to the general out-of-kilter algorithm as applied to the piecewise linear optimal distribution problem with costs

$$
f_j(x(j)) = \begin{cases} 0 & \text{if } x(j) \in [c^-(j), c^+(j)], \\ \infty & \text{if } x(j) \notin [c^-(j), c^+(j)], \end{cases}
$$

and with initial potential $u = 0$.

9.15. (*Tension Rectification Algorithm*). Following the lines of the preceding exercise, show that the tension rectification algorithm in Exercise 6.6

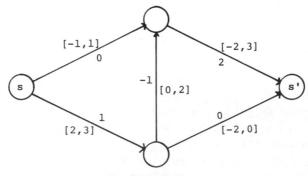

Figure 9.15

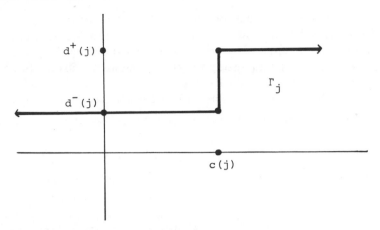

Figure 9.16

can likewise be identified with a special case of the out-of-kilter algorithm.

9.16. (*Out-of-Kilter Algorithm*). Furnish the justification for the part of the out-of-kilter algorithm where Minty's lemma produces a cut.

9.17. (*Out-of-Kilter Algorithm*). Use the general out-of-kilter algorithm to solve the linear optimal distribution problem in Figure 9.1. Here each cost function is of the form $f_j(x(j)) = d(j)x(j)$ on a capacity interval of the form $[0, c(j)]$. These intervals and the costs $d(j)$ are indicated along the arcs, whereas the supplies $b(i)$ are the numbers at the nodes. (*Note.* Different choices of the starting flow x and potential u, not necessarily feasible but with $\operatorname{div} x = b$, will demonstrate the various possibilities inherent in the algorithm.)

9.18. (*Out-of-Kilter versus Descent*). Suppose the general out-of-kilter algorithm is initialized with a flow x and a potential u that are feasible solutions to the respective problems. Prove that in each iteration none of the quantities

$$\Delta_j(x(j), v(j)) = f_j(x(j)) + g_j(v(j)) - x(j)v(j) \geq 0$$

is increased, and at least one is reduced; thus the sum

$$\sum_{j \in A} \Delta_j(x(j), v(j)) = \Phi(x) - \Psi(u)$$

is always reduced. (Optimality is achieved when this quantity reaches zero.) Show in fact that in iterations involving a circuit P, one has $f'_+(0) < 0$ for $f(t) = \Phi(x + te_P)$, whereas in iterations involving a cut $Q = [S, N \setminus S]$, one has $f'_+(0) < 0$ for $f(t) = -\Psi(u + te_{N \setminus S})$. (*Remark.* In the case where the initial x and u are feasible, the out-of-kilter algorithm is therefore like a joint version of the optimal distribution and optimal differential algorithms, descending sometimes with one and sometimes with the other. The rules for selecting α could just as well be replaced by the ones for the latter algorithms: let α be the lowest value of t minimizing the $f(t)$ in question. This might speed up convergence, but it would also require dealing explicitly with both f_j and g_j, as well as Γ_j, so a lot more structure would have to be programmed.)

9.19. (*Fortified Derivatives*). Suppose f_j is of the general quadratic form

$$
f_j(x(j)) = \begin{cases} a(j)x(j)^2 + d(j)x(j) + p(j) \\ \qquad \text{if } x(j) \in [c^-(j), c^+(j)], \\ +\infty \quad \text{if } x(j) \notin [c^-(j), c^+(j)], \end{cases}
$$

where $a(j) \geq 0$. Verify that the definitions 9H(2) for arbitrary $\delta > 0$ then yield, at least when $c^-(j)$ and $c^+(j)$ are finite, the expressions

$$
f_{j+}'^{\delta}(x(j)) = \begin{cases} -\infty \quad \text{if } x(j) < c^-(j), \\ 2a(j)x(j) + d(j) + 2\sqrt{\delta a(j)} \\ \qquad \text{if } c_\delta^-(j) \leq x(j) < c_\delta^-(j), \\ 2a(j)x(j) + d(j) + a(j)c^+(j) \\ \qquad + \delta/[c^+(j) - x(j)] \\ \qquad \text{if } c_\delta^+(j) \leq x(j) < c^+(j), \\ +\infty \quad \text{if } x(j) > c^+(j), \end{cases}
$$

where

$$
c_\delta^+(j) = \max\{ c^-(j), c^+(j) - \sqrt{\delta/a(j)} \},
$$

and

$$
f_{j-}'^{\delta}(x(j)) = \begin{cases} -\infty \quad \text{if } x(j) \leq c^-(j), \\ 2a(j)x(j) + d(j) + a(j)c^-(j) \\ \qquad + \delta/[c^-(j) - x(j)] \\ \qquad \text{if } c^-(j) < x(j) \leq c_\delta^-(j), \\ 2a(j)x(j) + d(j) - 2\sqrt{\delta a(j)} \\ \qquad \text{if } c_\delta^-(j) < x(j) \leq c^+(j), \\ +\infty \quad \text{if } x(j) > c^+(j), \end{cases}
$$

where

$$
c_\delta^-(j) = \min\{ c^+(j), c^-(j) + \sqrt{\delta/a(j)} \}.
$$

Here $\sqrt{\delta/a(j)}$ is to be interpreted as $+\infty$ if $a(j) = 0$. Show that under certain other conventions of a similar nature these formulas remain valid even though $c^+(j)$ and $c^-(j)$ might be infinite.

(*Remark.*   This demonstrates that the fortified algorithms in Section 9H are readily implementable for all quadratic problems. A generalization could be made to piecewise quadratic problems, but the formulas become rather complicated. It would be better probably to reformulate such problems as quadratic problems, by means of the tricks explained for the piecewise linear case at the ends of Sections 7A and 7E; (see also Exercise 8.61.)

9.20.   (*Fortified Derivatives*). Prove for arbitrary $\delta > 0$ that

$$f_j'^{\delta}(\xi) \leq \eta \leq g_{j+}'^{\delta}(\xi) \quad \Leftrightarrow \quad f_j(\xi) + g_j(\eta) - \xi\eta \leq \delta,$$

from which it follows also by symmetry that

$$g_j'^{\delta}(\eta) \leq \xi \leq g_{j+}'^{\delta}(\eta) \quad \Leftrightarrow \quad f_j(\xi) + g_j(\eta) - \xi\eta \leq \delta.$$

(Use definitions 9H(2) and the conjugacy relations 8E(3)(4).)

9.21.   (*Fortified Algorithms*). Prove that the following specific choice of α in the fortified optimal distribution algorithm is sure to yield $f(\alpha) < f(0) - \delta$:

$$\alpha = 1 \Big/ \sum_{j \in P} (1/\alpha_j)$$

where $\alpha_j > 0$ is for each $j \in P$ the largest of the values t_j for which

$$\inf_{t_j > 0} \big[f_j(x(j) + t_j e_P(j)) - f_j(x(j)) + \delta \big]/t_j$$

is attained. (If there is no such t_j, let $\alpha_j = \infty$, or in other words, interpret $1/\alpha_j$ as 0. If the latter happens for all $j \in P$, let $\alpha = \infty$, and interpret this as saying that $f(t) < f(0) - \delta$ for all t sufficiently high.)

(*Hint.*   Refine the argument given to justify the algorithm.)

(*Remark.*   A similar result holds for the fortified optimal differential algorithm.)

9.22.   (*Fortified Algorithms*). Provide the justification for the fortified optimal differential algorithm.

9.23. (*Discretized Algorithms*). Provide the justification for the discretized optimal differential algorithm and the assertions in Section 9I about its termination.

9.24. (*Discretized Algorithms*). Formulate and prove the results for the discretized optimal differential algorithm that correspond to those in Section 9J. (These results lead to a parallel version of the theory in Section 9K.)

9.25. (*Piecewise Linear Approximation*). Show that when f_j is replaced by the approximation f_j^k described in Section 9I, this amounts to replacing g_j by the function g_j^k which is the pointwise maximum of all the linear (i.e., affine) functions l such that $l \leq g$ and the slope of l is of the form $x_k(j) + n\delta$, n integral. (Demonstrate that this function g_j^k is the conjugate of f_j^k.)

(*Remark.* f_j^k is an "inner" approximation of f_j, whereas g_j^k is an "outer" approximation to g_j. These two kinds of approximation are therefore dual to each other. To solve the optimal distribution problem by successive "outer" approximation, one would use the iterated form of the discretized optimal differential algorithm.)

9.26. (*Convex Simplex Method*). Suppose in the method of Section 9L that no f_j has any breakpoints at all, that is, that f_j is differentiable on $[c^-(j), c^+(j)] = (-\infty, \infty)$ for all $j \in A$. Show that any spanning tree F can be chosen at the start and need never be altered and that the objective $\Phi(x)$ can be regarded as a differentiable function ϕ of the variables $\xi_k = x(k)$ for $k \in A \setminus F$ (without any constraints on these variables entering the minimization). Moreover the method reduces in this case to the following. Given a vector $\xi = (\ldots, \xi_k, \ldots)$; if $(\partial\phi/\partial\xi_k)(\xi) = 0$ for all $k \in A \setminus F$, terminate with ξ optimal. Otherwise select *any* $\bar{k} \in A \setminus F$ with $(\partial\phi/\partial\xi_{\bar{k}})(\xi) \neq 0$, and minimize ϕ with respect to $\xi_{\bar{k}}$ (all other coordinates being held fixed) to get a new point ξ'.

9.27. (*Convex Simplex Method*). Use the special case in Exercise 9.26 to demonstrate that unless some further rule is introduced for the choice of the arc $\bar{k} \in A \setminus F$, the convex simplex method could generate a nonminimizing sequence of flows, that is, one whose objective function values, though decreasing, do not converge to the minimum.

(*Hint.* Consider a case where one keeps choosing either $\bar{k} = k_1$ or $\bar{k} = k_2$ without ever inspecting other arcs $k \in A \setminus F$.)

(*Remark.* Since this exercise concerns a sequence of feasible flows that could have been generated by the optimal distribution algorithm, as pointed out in Section 9L, it illustrates again the need for a good rule for choosing the particular improvement circuit that is used. Arbitrary choices may lead to convergence troubles.)

9.28. (*Convex Simplex Method*). Show that the simplex method in Section 9L (extended to the degenerate case as described) reduces, when applied to a linear optimal distribution problem, to a slightly generalized version of the simplex method for flows in Section 7D.

9.29. (*Convex Simplex Method*). Prove that in the degenerate case, provided no internal breakpoints are encountered, the extended version of the convex simplex method in Section 9L will not generate an infinite sequence of nondescent iterations ($\alpha = 0$) as long as Bland's priority rule is employed in the selection of the arc $\bar{k}$. (The argument in Section 7D is directly applicable.)

9.30. (*Piecewise Differentiable Problems*). Explain how a piecewise differentiable optimal distribution problem (each f_j having only finitely many first-order breakpoints) can be reformulated as one having no *internal* first-order breakpoints for any arc.

(*Hint.*   Make use of black box representations like those in Sections 7A, 7E, and 7I.)

9.31. (*Dual Convex Simplex Method*). Formulate a dual version of the method Section 9L that works instead with a feasible solution to the optimal differential problem, rather than one to the optimal distribution problem, at each iteration (see the algorithm in Section 7H for the linear case).

9N.* COMMENTS AND REFERENCES

The optimization problems in this chapter are in particular convex programming problems with linear constraints and could be solved by some of the general methods for this class. The texts of Avriel [1976] and Luenberger [1973], for example, treat such methods and explain in detail how to conduct line searches, where a function $f(t)$ must be minimized to determine a step length α. Line searches are commonly used in all kinds of numerical optimization. For a line search algorithm especially well suited to the efficient minimization of convex functions whose derivatives may have jumps, see Mifflin [1983].

The aim here has not been to survey the general algorithms of convex programming in a special framework but to concentrate on ideas peculiar to networks, or at least separable programming. Highly developed techniques for handling paths, cuts, and constraints on flows and potentials motivate the search for ways of generating directions and step lengths with relatively little effort and manipulation of data. This may make it possible to tackle nonlinear problems of very large size, even though the individual descent directions may not be as well chosen as they could be if full use were made of the information available in each iteration.

For example, the fundamental descent procedures in convex programming (steepest descent and quasi-Newton methods with gradient reduction, etc.), as applied to an optimal distribution problem, would typically require line searches in terms of

$$f(t) = \Phi(x + tz) = \sum_{j \in A} f_j(x(j) + tz(j)),$$

where z is some circulation having $z(j) \neq 0$ for a large number of arcs j. Each j with $z(j) \neq 0$ would introduce a nontrivial term in the sum and make the work harder. In networks with thousands of arcs, this could be a serious difficulty.

In contrast, direction vectors of the form $z = e_P$ (or $z = e_Q$ when dealing with potentials) bring only a small proportion of the available arcs into any line search. The general optimal distribution and optimal differential algorithms, presented for the first time here, generate such direction vectors exclusively. They are the natural heirs to some of the well-known methods for various types of linear problems (see the remarks in Section 7P), but no algorithms previously described for nonlinear problems have had this property. The fortified versions of these algorithms in Section 9H are also new of course, as well as the constructive proof they provide of the duality theorem. (The original proof of Rockafellar [1967] was an argument by induction.) The newness carries over to the discretized optimal distribution and optimal differential algorithms in Sections 9I, 9J, and 9K. In all these methods a circuit or cut giving a direction of descent is obtained by solving a feasible differential or feasible distribution problem, a task for which the algorithms in Chapters 3 and 6 can be put to good use.

Another respect in which the algorithms in this chapter differ significantly in nature and intent from standard ones in convex programming is that they do not insist on f_j being finite and differentiable everywhere on $[c^-(j), c^+(j)]$. Thus $f_j(x(j))$ may tend to $+\infty$ as $x(j) \downarrow c^-(j)$ or as $x(j) \uparrow c^+(j)$, and even when this is not true, f_j is not required to have a finite one-sided derivative at either endpoint. More important, jumps in the slope of f are also allowed: there may be points $x(j)$ between $c^-(j)$ and $c^+(j)$ where $f_j'^-(x(j)) < f_j'^+(x(j))$.

This degree of generality is absolutely essential in obtaining a theory that can be applied equally to the functions f_j in a flow problem and their conjugates g_j, the cost functions in the corresponding problem for potentials (see Section 8E). Such generality is indispensible therefore to methods for solving a problem by way of its dual. The generality is also valuable in a direct sense. Piecewise linear or quadratic cost functions can be treated without introducing black box representations that might greatly add to the number of arcs in the network and the dimensionality of the problem.

The especially prominent role of duality in the development of the algorithms in this chapter is not surprising. Although a comprehensive theory of duality does exist for general optimization problems having enough convexity properties (see Rockafellar [1970], [1974] and [1978], for example), it does not come near the one expounded in Chapter 8 in the sharpness of its results and the ease with which the problem dual to a given problem can be written down in explicit detail. Only linear programming duality reaches the same level. Linear programming duality is widely familiar, and many applications of it have been found, not only in linear programming itself but also in terms of approximations and direction-finding mechanisms in nonlinear programming. One of the purposes of this book is to demonstrate the possibility of comparable advances using separable programming duality.

Methods based on piecewise linearization are rarely practical in general convex programming. Cutting plane algorithms and their duals fall into this category, but they are mainly used when other methods are not applicable. Nevertheless, piecewise linearization has long been recognized as an attractive approach in separable programming, in particular for networks with nonlinear costs; see Charnes and Cooper [1958] and Miller [1963]. This idea has been applied to hydraulic networks by Collins et al. [1978], who used a least-squares subroutine to determine a "best" piecewise linear approximation to each nonlinear f_j. They reported good results with this method (better than the other things they tried, i.e., the convex simplex algorithm, the Frank-Wolfe method (described later in this section), and a Newton-Raphson approach to solving the equilibrium conditions), even though the approximation phase turned out to be as time-consuming as the optimization. In effect they replaced each arc and its piecewise linearized cost function by a "black box," so as to obtain a *linear* optimal distribution problem to which the simplex method for flows (Section 7D) could be applied. They approximated only once and did not address the question of how much error this might have caused.

Finer and finer piecewise linearizations, if coupled to the simplex method and its demand for reformulation in terms of a linear problem, do run into trouble with higher and higher dimensionality. Perhaps the trouble could be avoided by a network form of the piecewise linear simplex method proposed by Müller-Merbach [1970] as a generalization of Dantzig's upper bounding technique. An alternative would be to use the optimal distribution and optimal differential algorithms developed here. These do not require reformulation in terms of a linear problem and hold the possibility of highly efficient implementation (see Sections 9B and 9D). In particular, the thrifty adjustment algorithm and the out-of-kilter algorithm are available as versions of the basic descent algorithms specifically designed for piecewise linear problems.

The thrifty adjustment algorithm is new in the form in Section 9D, although in the case of linear problems it overlaps in some ways with the earlier min cost flow algorithm described in Section 7P. Its main inspiration comes from the black box application in Section 9E and Minty's ideas for calculating the characteristic curve of a network with every Γ_j of staircase form. (Minty [1960] worked with circuits and cuts as in the out-of-kilter method rather than a min path subroutine.) The piecewise linear form of the out-of-kilter algorithm in Section 9F was sketched by Lawler [1976], but the proof of finite termination in Section 9G is new. Lawler discussed convergence only in the case where *every* potential u is a feasible solution to the optimal differential problem (which is true for piecewise linear problems whenever all the intervals

$[c^-(j), c^+(j)]$ are bounded), and he did not come up with refinements comparable to those mentioned at the end of Section 9G.

A different idea for circumventing the escalating dimensionality associated with repeated piecewise linearizations has been explored by Meyer [1979] (see also Meyer and Smith [1978], Kao and Meyer [1979]). This consists of two-segment representations in the pattern of Figure 9.17. The flow x_{k+1} is obtained by solving the subproblem specified by the functions f_j^k; these in turn are obtained from the flow x_k in the preceding iteration by interpolation relative to certain truncated intervals $[c_k^-(j), c_k^+(j)]$. Since only two segments are used at a time, there is no escalation in dimensionality even if the simplex method is chosen for solving the subproblem, which is what Meyer had in mind (although it is apparent now that other methods for piecewise linear problems could be used instead). He develops a test for ε-optimality that goes beyond earlier results of Geoffrion [1977] on upper and lower bounds associated with cost function approximation.

The reader of Section 9K will recognize a close connection between Meyer's approach and our discretized descent algorithms as applied iteratively in the manner of Section 9J. The discretized algorithms keep a fixed step size δ_k until the approximate problem corresponding to complete piecewise linearization of each f_j with spacing δ (see Figure 9.14) is solved. However, the piecewise linearized function never has to be represented explicitly; only the portion corresponding to the subinterval $[x_k(j) - \delta, x_k(j) + \delta]$ (possibly truncated by intersection with $[c^-(j), c^+(j)]$) needs to be generated at any one time. This is used to find a descent direction (not necessarily the "best"), and then line search is carried out in discretized form to get x_{k+1}. Meyer instead concentrates on a problem of local optimality relative to a neighborhood of x_k defined by intervals $[x_k(j) - \delta_k^-(j), x_k(j) + \delta_k^+(j)]$, where the numbers $\delta_k^-(j)$

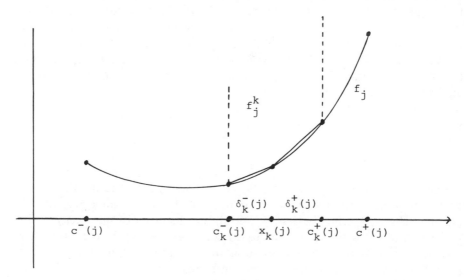

Figure 9.17

and $\delta_k^+(j)$ can be tailored to the shape of f_j and the experience gained in earlier iterations. Thus x_{k+1} is chosen with greater care but also greater effort. Of course this may also mean that the difference $z = x_{k+1} - x_k$ furnishes a particularly good direction of descent. A possible modification of Meyer's method would be to do a line search in the direction of z before setting up the next two-segment subproblem.

Another method besides piecewise linearization that has been tried on nonlinear network problems is the algorithm of Frank and Wolfe [1956]. In the experience of Collins et al. [1978], who implemented it in the more advanced PARTAN form (see the text of Luenberger [1973]), it is not as effective as piecewise linearization. There are theoretical reasons too for doubting the effectiveness of this method, which in any case cannot handle jumps in derivatives and accordingly seems to fall short of requirements.

The idea of the Frank-Wolfe algorithm is the following. Given any feasible solution x to the optimal distribution problem, define

$$d(j) = f_{j^+}'(x(j)) = f_{j^-}'(x(j)) \quad \text{if } c^-(j) < x(j) < c^+(j),$$

$$d(j) = f_{j^+}'(x(j)) \quad \text{if } x(j) = c^-(j),$$

$$d(j) = f_{j^-}'(x(j)) \quad \text{if } x(j) = c^+(j).$$

(The assumption must be made that these values are not only well defined but finite. In particular, f_j must be differentiable at $x(j)$ if $c^-(j) < x(j) < c^+(j)$.) Next solve the *linearized* problem with cost coefficients $d(j)$ and the same capacity intervals $[c^-(j), c^+(j)]$ (see Figure 9.18) to get $\tilde{x}$. Finally, do a line search in the direction of $z = x - \tilde{x}$.

The chief advantage presumably is that a direction of descent is found by means of a *linear* subproblem to which the simplex method can be applied

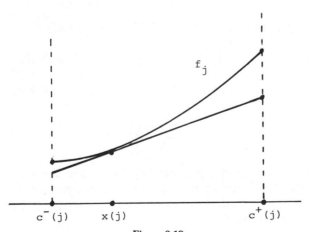

Figure 9.18

without trouble over escalating dimensionality. It is clear, though, that the direction obtained may depend more on distant aspects of the constraints than on the nature of the objective function. Surely a more local subproblem like the one of Meyer, which is no harder really to solve, would be more reliable in providing a vector z for a line search.

The convex simplex method is due to Zangwill [1967], [1969]. It was specialized to networks by Collins et al. [1978] with mixed success, but Helgason and Kennington [1977] (two of the coauthors) have pursued the matter further and have improved the implementation. Interest in this method is closely tied to the success in recent years of the simplex method for flows (see Section 7P for references to the software). Although an extension to the case of cost functions with jumps in derivatives has been provided in Section 9L (and still more broadly in Section 11J), difficulties must be expected in practice with degeneracy. The fact that the method tends to become ill formulated when jumps are allowed and degeneracy is encountered, suggests something basically lacking about the whole approach. As noted in Section 7P, degeneracy in some classes of network optimization problems is the norm rather than the exception. An extension of the "strong" spanning tree idea of Cunningham [1976] explained in the linear case in Section 7D might help in coping with this degeneracy. Anyway, existing proofs of convergence of the convex simplex method (Zangwill [1967], [1969]), being based on nondegeneracy assumptions and other conditions not entirely appropriate to network applications as developed here, indicate the need for more work on this subject.

Dembo and Klincewicz [1981] have modified the convex simplex method to take second-derivative information into account without disrupting the network structure, in the hope of speeding up convergence near an optimal solution. Their procedure is roughly to choose not just one arc $\bar{k}$ for the improvement process but a whole set K consisting of arcs that would have been eligible. For each $k \in K$ they determine the value $\lambda(k)$ that minimizes the quadratic expansion

$$q_k(t) = f_k(x(k)) + f_k'(x(k))t + \tfrac{1}{2}f_k''(x(k))t^2$$

relative to the interval $[c^-(k) - x(k), c^+(k) - x(k)]$ (with the obvious interpretation when $x(k) = c^-(k)$ or $x(k) = c^+(k)$; note that f_k cannot have internal breakpoints of either first or second order). Then line search is performed in the direction of

$$z = \sum_{k \in K} \lambda(k) e_{P_k}.$$

Depending on what happens, the spanning tree F and the arc set $K \subset A \setminus F$ may then be adjusted. Degeneracy is still a source of trouble of course, as is the need for managing K and F in such a way as to make best use of the second-order information.

Although much remains to be done in establishing the effectiveness of this approach, there is no doubt that the development of algorithms that use second derivatives and yet take good advantage of network structure is an important project. Especially valuable would be methods that allow for break-points, since they could be applied directly to piecewise quadratic costs, for example.

At present, besides the method of Dembo and Klincewicz, there are very special algorithms of Bachem and Korte [1978] and Cottle [1983] for minimiz-ing a quadratic convex function over the set of all doubly stochastic matrices or the like. (These are designed for a general approximation problem similar to the one in Example 6 in Section 9D.) More promising for application to general problems are the gradient projection ideas that have been pursued by Bertsekas [1979] in a multicommodity model. The augmented Lagrangian methods of nonlinear programming have also been applied to a network problem by Bertsekas [1976].

10

LINEAR SYSTEMS OF
VARIABLES

All the problems we have treated so far fit into the framework of the two linear systems of variables associated with a network: the x, y system and the u, v system, both expressed by a node-arc incidence matrix E. Even the more combinatorial problems in the preceding chapters can be viewed as centering on special properties of these systems, because elementary paths P and cuts Q can be identified with certain flows e_P and tensions e_Q. Indeed, everything about a network can be recovered from its incidence matrix and therefore from knowledge of the relationships among the variables $x(j)$ and $y(i)$, or $v(j)$ and $u(i)$.

A remarkable fact is that very much of the theory we have been building up can be extended to *general* linear systems of variables, where E is an *arbitrary* real matrix. Associated with such systems are combinatorial structures, called *oriented real matroids*, which capture the properties of paths and cuts and make it possible to develop results quite analogous to those in the network case. Such results belong to *monotropic programming*, a broad subject that includes general linear programming, quadratic (convex) programming with linear constraints, and even piecewise linear or piecewise quadratic programming.

A monotropic programming problem is an optimization problem that can be expressed in the form of either the optimal distribution problem or the optimal differential problem in Chapter 8, but with variables interrelated by an arbitrary real matrix E, not just an incidence matrix. Questions about the optimal solutions to such problems and the methods of determining them will be taken up in Chapter 11. The present chapter is devoted to underlying facts about feasible solutions and more broadly the combinatorial aspects of general linear systems of variables and the constraints that may be imposed on them.

The first task is that of describing the primal and dual systems associated with a matrix E and the special kinds of index sets P and Q, called "elementary supports," that can be made to play the roles of circuits and cuts (Sections 10A and 10B). Although matroid theory is really involved, the approach taken

448

is not axiomatic but focuses rather on constructive features of the Tucker representations for the linear systems in question (Section 10C). This leads naturally to the development of "pivoting" algorithms for various purposes, in particular for the resolution of the alternatives of the painted index theorem in Section 10G (which is a generalization of Minty's lemma) and for the feasibility theorems in Section 10I.

   Although problems in monotropic programming need not have anything to do with networks, there are many optimization models that are based on networks but require the more general setting of monotropic programming. Examples are "networks with gains" and flow problems where there is more than one "commodity" or kind of traffic, or where the flux in one arc can interact with the flux in another arc. The linear systems in these examples are discussed in Sections 10D, 10E, and 10F.

10A.   PRIMAL AND DUAL VARIABLES

Let E be an arbitrary real matrix, but to maintain notational flexibility as in network programming, write the general entry of E as $e(i, j)$, where i and j range over abstract index sets I and J, both nonempty and finite. For $x \in R^J$ define $y \in R^I$ by

$$y(i) = \sum_{j \in J} e(i, j)x(j) \quad \text{for } i \in I, \tag{1}$$

whereas for $u \in R^I$ define $v \in R^J$ by

$$v(j) = - \sum_{i \in I} u(i)e(i, j) \quad \text{for } j \in J. \tag{2}$$

The variables $x(j)$ and $y(i)$ make up the *primal* linear system associated with E, whereas the variables $u(i)$ and $v(j)$ make up the *dual* linear system. These homogeneous systems are represented jointly by the tableau in Figure 10.1.

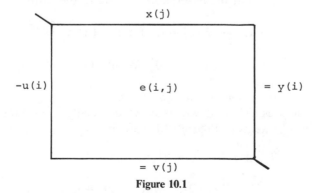

Figure 10.1

Equations (1) and (2) imply

$$\sum_{j \in J} v(j)x(j) = - \sum_{i \in I, j \in J} u(i)e(i, j)x(j) = - \sum_{i \in I} u(i)y(i).$$

Thus $v \cdot x = -u \cdot y$ always. In fact this formula characterizes the duality between the two systems. For $y \in R^I$ and $v \in R^J$ to satisfy (2), it is not just necessary but sufficient to have $u \cdot y + v \cdot x = 0$ for all $x \in R^J$ and $y \in R^I$ satisfying (1). Indeed, under (1) one has

$$u \cdot y + v \cdot x = \sum_{j \in J} \left[v(j) + \sum_{i \in I} u(i)e(i, j) \right] x(j),$$

and this holds for all $x \in R^J$ if and only if u and v satisfy (2). Dually, for $x \in R^J$ and $y \in R^I$ to satisfy (1), it is sufficient, as well as necessary, to have $u \cdot y + v \cdot x = 0$ for all $u \in R^I$ and $v \in R^J$ satisfying (2). This is seen from using (2) to write

$$u \cdot y + v \cdot x = \sum_{i \in I} u(i) \left[y(i) - \sum_{j \in J} e(i, j)x(j) \right].$$

The primal linear system can be identified with a certain subset of the vector space $R^{I \cup J}$, namely,

$$\mathscr{C} = \{ (y, x) | y \in R^I, x \in R^J, \text{ and (1) holds} \}.$$

The dual linear system can be identified with

$$\mathscr{D} \{ (u, v) | u \in R^I, v \in R^J, \text{ and (2) holds} \}.$$

Note that $\mathscr{C}$ and $\mathscr{D}$ are closed under addition and scalar multiplication; they are linear subspaces of $R^{I \cup J}$. Thinking of $u \cdot y + v \cdot x$ as the inner product $(u, v) \cdot (y, x)$, one can put the observations above in the form

$$\mathscr{D} = \{ (u, v) | (u, v) \cdot (y, x) = 0 \quad \text{for all} \quad (y, x) \in \mathscr{C} \} = \mathscr{C}^{\perp},$$

$$\mathscr{C} = \{ (y, x) | (u, v) \cdot (y, x) = 0 \quad \text{for all} \quad (u, v) \in \mathscr{D} \} = \mathscr{D}^{\perp}.$$

Thus $\mathscr{C}$ and $\mathscr{D}$ are *orthogonally complementary*.

Two subspaces of R^J itself will play an especially important role in the discussion of the systems in Figure 10.1, namely,

$$\mathscr{C} = \{ x | Ex = 0 \},$$

$$\mathscr{D} = \{ v | \exists u \text{ with } -uE = v \}. \tag{3}$$

Clearly $\mathscr{C}$ consists of the vector $x \in R^J$ such that $(0, x) \in \bar{\mathscr{C}}$, whereas $\mathscr{D}$ is the projection of $\bar{\mathscr{D}}$ on R^J. Another description is this: $\mathscr{D}$ is the subspace of R^J generated by the rows of E, whereas $\mathscr{C}$ is the subspace consisting of all vectors orthogonal to the rows of E. It follows that

$$\mathscr{C} = \{ x \in R^J | v \cdot x = 0 \quad \text{for all } v \in \mathscr{D} \} = \mathscr{D}^\perp,$$

$$\mathscr{D} = \{ v \in R^J | v \cdot x = 0 \quad \text{for all } x \in \mathscr{C} \} = \mathscr{C}^\perp. \tag{4}$$

In other words, $\mathscr{C}$ and $\mathscr{C}$ too are orthogonally complementary.

Obviously, there is nothing special about the nature of the subspaces $\mathscr{C}$ and $\mathscr{D}$ that arise in this way from matrices E, aside from orthogonal complementarity. *Any* pair of subspaces $\mathscr{C}$ and $\mathscr{D}$ satisfying (4) can be expressed (nonuniquely) in terms of some matrix E as in (3): take any finite set of vectors generating $\mathscr{D}$, and let these vectors be the rows of E. In a larger picture the spaces $\bar{\mathscr{C}}$ and $\bar{\mathscr{D}}$ can be viewed as associated with the matrix $\bar{E} = [-I, E]$, say, in the same sense that $\mathscr{C}$ and $\mathscr{D}$ are associated with E.

Tucker representations are often useful in connection with $\mathscr{C}$ and $\mathscr{D}$. To get such a representation of $\mathscr{C}$, solve the homogeneous systems

$$\sum_{j \in J} e(i, j) x(j) = 0 \quad \text{for } i \in I$$

for a maximal set of variables $x(j)$, so as to obtain an equivalent system of the form

$$\sum_{k \in J \setminus F} a(j, k) x(k) = x(j) \quad \text{for } j \in F, \tag{5}$$

where the variables $x(k)$ can take on arbitrary values. The vectors $x \in \mathscr{C}$ are then precisely the ones satisfying (5). It follows by way of (4) that the vectors $v \in \mathscr{D}$ are the ones satisfying

$$- \sum_{j \in F} v(j) a(j, k) = v(k) \quad \text{for } k \in J \setminus F. \tag{6}$$

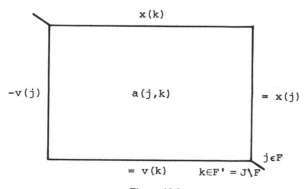

Figure 10.2

This is a Tucker representation for $\mathcal{D}$. Both representations are expressed by the tableau in Figure 10.2.

It is possible to pass from any such tableau for $\mathcal{C}$ and $\mathcal{D}$ to any other by a sequence of *pivoting* transformations, where a single index $j \in F$ is exchanged for an index $\bar{k} \notin F$, with a $(\bar{j}, \bar{k}) \neq 0$. The formula for such a transformation, as derived in Section 4I, is perfectly valid in general; see Figure 10.3. The pair $(\bar{j}, \bar{k})$ is the *pivot*.

Observe that the original tableau in Figure 10.1 gives in effect a pair of Tucker representations for the spaces $\mathcal{C}$ and $\mathcal{D}$. On the other hand, one could start out with arbitrary systems given in terms of coefficients $a(j, k)$ as in

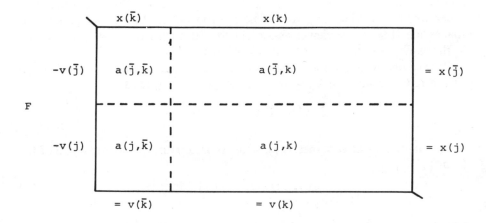

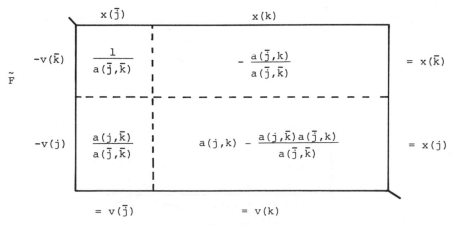

Figure 10.3

Figure 10.2, regard them as defining two orthogonally complementary sub-spaces $\mathscr{C}$ and $\mathscr{D}$ in R^J, and then represent these spaces by a matrix E as explained here. There is no difference then between the tableaus in Figures 10.1 and 10.2, except for notation and interpretation.

The theory of primal and dual linear systems thus coincides in large measure with the theory of subspaces orthogonally complementary to each other. It has the added feature, however, that the subspaces are not treated abstractly but in terms of "variables" whose relationships can be summarized in a Tucker tableau, only a finite number of such tableaus being possible in any given case. The combinatorial aspect sharply distinguishes the theory from anything in classical linear algebra.

10B. ELEMENTARY VECTORS AND SUPPORTS

The combinatorial side of the theory of linear systems of variables is concerned with certain signed subsets of the index set J that are induced by the subspaces $\mathscr{C}$ and $\mathscr{D}$. Recall that a *signed set* is simply a set S together with a partition of S into subsets S^+ and S^-. The *reverse* of S is obtained by reversing the designation of S^+ and S^-.

A signed subset P of J is called a *support of $\mathscr{C}$*, or a *primal support*, if there is a vector $x \in \mathscr{C}$ such that

$$P^+\{j \in J|x(j) > 0\}, \qquad P^- = \{j \in J|x(j) < 0\}. \tag{1}$$

Then P is said to be the *support of x*; note that only the zero vector has empty support. A primal support P is *elementary* if it is nonempty but does not properly include any other primal support (even with signs disregarded), or in other words, if there is no other nonempty primal support P_0 with $P_0^+ \cup P_0^- \subset P^+ \cup P^-$ but $P_0^+ \cup P_0^- \neq P^+ \cup P^-$.

An *elementary vector of $\mathscr{C}$* is defined to be a vector $x \in \mathscr{C}$ whose support P as in (1) is elementary in this sense.

The elementary primal supports make up a finite collection of signed subsets of J that will be seen to have many interesting properties. In the ordinary network case, where J is the arc set and $\mathscr{C}$ the circulation space, a signed set P is an elementary primal support if and only if it comprises an elementary circuit (see Exercise 4.6). The elementary vectors of $\mathscr{C}$ are then the elementary circulations, i.e. the flows of the form αe_P, where e_P is the incidence vector associated with such a circuit and $\alpha > 0$.

In general, *any two elementary vectors x and x' of $\mathscr{C}$ that have the same support must be positive multiples of each other*. Indeed, if

$$\{j|x(j) > 0\} = \{j|x'(j) > 0\} = P^+,$$

$$\{j|x(j) < 0\} = \{j|x'(j) < 0\} = P^-,$$

the number

$$\alpha = \min\{ x'(j)/x(j) | j \in P \} > 0$$

has the property that the vector $x'' = x' - \alpha x \in \mathscr{C}$ has its support properly contained in P; since P is elementary this implies $x'' = 0$, hence $x' = \alpha x$.

Each elementary primal support P of $\mathscr{C}$ thus corresponds to a vector x of $\mathscr{C}$ that is uniquely determined up to a positive multiple. It is convenient to normalize by choosing the multiple so that

$$\sum_{j \in P} |x(j)| = |P| \qquad \text{(the number of elements in } P\text{)}. \tag{2}$$

This choice is guided by the network case, where the unique circulation that is supported by a given elementary circuit P and satisfies (2) is $x = e_P$. We therefore define in general, for any elementary support P of $\mathscr{C}$:

$$e_P = \left[\begin{array}{c} \text{the unique elementary } x \in \mathscr{C} \text{ having} \\ P \text{ as its support and satisfying (2)} \end{array} \right]. \tag{3}$$

Every other elementary vector of $\mathscr{C}$ corresponding to P can then be expressed in the form αe_P with $\alpha > 0$. Note, however, that although this notation agrees with what was introduced for networks, e_P in the general case will *not* be merely an "incidence" vector. The components of e_P might *not* just be $+1$, -1, or 0, but certain other real numbers that only have to satisfy $e_P(j) > 0$ for $j \in P^+$, $e_P(j) < 0$ for $j \in P^-$, $e_P(j) = 0$ for $j \notin P$, and

$$\sum_{j \in P^+} e_P(j) - \sum_{j \in P^-} e_P(j) = |P| = |P^+| + |P^-|.$$

Elementary vectors and supports of the subspace $\mathscr{D}$ are defined similarly. A *support of $\mathscr{D}$*, or a *dual support*, is a signed subset Q of J such that there is a vector $v \in \mathscr{D}$ with

$$Q^+ = \{ j \in J | v(j) > 0 \} = 0, \qquad Q^- = \{ j \in J | v(j) < 0 \}, \tag{4}$$

in which event Q is said to be the *support of v*. If Q is a nonempty dual support that does not properly include any other nonempty dual support (even with signs disregarded), it is *elementary*, and any $v \in \mathscr{D}$ having Q as its support is termed an *elementary vector* of $\mathscr{D}$. Again there is only one such vector up to a positive scalar multiple, and to fix this multiple we impose the equation

$$\sum_{j \in J} |v(j)| = |Q|. \tag{5}$$

Thus we define for any elementary support Q of $\mathcal{D}$:

$$e_Q = \left[\begin{array}{l} \textit{the unique elementary } v \in \mathcal{D} \textit{ having} \\ Q \textit{ as its support and satisfying (5)} \end{array} \right]. \tag{6}$$

Every other elementary vector of $\mathcal{D}$ with Q as its support is of the form αe_Q with $\alpha > 0$.

In the network case an elementary dual support Q is an elementary cut (see Exercise 4.19), and e_Q is the incidence vector for Q. But in general, e_Q might *not* be simply an incidence vector with only $+1$, -1, or 0 as components.

The nature of the elementary primal and dual supports can be pinned down precisely in some cases besides that of an ordinary network when the structure of the linear systems that are involved is simple enough. The example of networks with gains is discussed in Sections 10D and 10E.

Conformal Realization

Every nonempty primal support P as in (1) not only includes some elementary primal support, which would be true from the definition of "elementary," but includes one in the *conformal* sense: there exists an elementary primal support P_1 such that $P_1^+ \subset P^+$ and $P_1^- \subset P^-$. This will be seen from the "painted index theorem" in Section 10G (see Exercise 10.13). As a consequence of this fact it can be shown that every nonzero vector $x \in \mathcal{C}$ with support P can be expressed in the form

$$x = \alpha_1 e_{P_1} + \cdots + \alpha_r e_{P_r}, \qquad \alpha_k \geq 0,$$

where each P_k is an elementary primal support such that $P_k^+ \subset P^+$ and $P_k^- \subset P^-$ (see Exercise 10.14). This generalizes the conformal realization theorem of Section 4B in the case of circulations in a network.

Similarly, every nonzero vector $v \in \mathcal{D}$ with support Q as in (4) can be expressed in the form

$$v = \alpha_1 e_{Q_1} + \cdots + \alpha_r e_{Q_r}, \qquad \alpha_k \geq 0,$$

where each Q_k is an elementary dual support with $Q_k^+ \subset Q^+$ and $Q_k^- \subset Q^-$.

10C. BASES

An index set $F \subset J$ that yields a Tucker tableau for the spaces $\mathcal{C}$ and $\mathcal{D}$ as in Figure 10.2 will be called a *basis* in J. There are only finitely many of these, and in the case of an ordinary network they correspond to maximal forests (Sections 4F and 4G). Thus bases, like primal and dual supports, make up a collection of special "configurations" of elements of J whose properties are worthy of study.

The basis theorem in Section 4H, relating maximal forests to elementary circuits and cuts, generalizes to a result that serves as one of the best tools for determining that the elementary supports actually are in some situations. It also lays a foundation for computing with primal and dual supports by way of Tucker representations and pivoting. The general theorem is developed in this section.

Let e^j denote the jth column of E. The equation $Ex = 0$ which defines $\mathscr{C}$ then takes the form

$$\sum_{j \in J} x(j)e^j = 0. \tag{1}$$

For any subset F of J, this can be written as

$$\sum_{j \in F} x(j)e^j = - \sum_{k \in F'} x(k)e^k, \quad \text{where } F' = J \setminus F. \tag{2}$$

To have F furnish a Tucker representation for $\mathscr{C}$, each (arbitrary) assignment of values to the variables $x(k)$ for $k \in F'$ must yield unique values for the variables $x(j)$ for $j \in F$ through (2). Therefore F is a basis in J if and only if the vectors e^j for $j \in F$ form a basis (in the usual sense of linear algebra) for

$$\mathscr{B} = [\text{subspace spanned by all the columns of } E]$$

$$= \{ y \in R^I \mid \exists x \in R^J \text{ with } Ex = y \}. \tag{3}$$

Let us carry terminology further by calling F an *independent set* in J if F does not include any nonempty [elementary] primal support P, and a *spanning set* in J if $J \setminus F$ does not include any nonempty [elementary] dual support Q. Then in fact F *is an independent set in J if and only if the vectors e^j for $j \in F$ are linearly independent.* Indeed, linear dependence of $\{ e^j \mid j \in F \}$ is tantamount to the existence of coefficients $x(j)$, satisfying (1), that vanish for all $j \notin F$ but not for all $j \in F$; such coefficients define an $x \in \mathscr{C}$ whose support P is included in F. (Recall too that every nonempty primal support, if not itself elementary, must by definition include some elementary primal support.) Likewise, F *is a spanning set in J if and only if the vectors e^j for $j \in F$ span the subspace β in (3).* Namely, since $\mathscr{C}$ and $\mathscr{D}$ are orthogonally complementary, the existence of a nonempty dual support Q in F means that there are coefficients $v(k)$ for $k \in F' = J \setminus F$, not all zero, with

$$\sum_{k \in F'} x(k)v(k) = 0 \quad \text{for all } x \in \mathscr{C}.$$

This is equivalent to saying that in (2) the values of the variables $x(k)$ for $k \in F'$ cannot be chosen freely, that is, not every vector expressible as a linear combination of columns of E can be expressed in terms of $\{ e^j \mid j \in F \}$ alone.

It follows from linear algebra that a basis F in J can be characterized variously as a maximal independent set of indexes, a minimal spanning set, or as a spanning set of indexes that happens also to be independent. These characterizations lead to a description of the coefficients that occur in Tucker tableaus in terms of elementary vectors of $\mathscr{C}$ and $\mathscr{D}$.

General Basis Theorem.   *Let F be a basis in J, and let the numbers $a(j, k)$ for $j \in F, k \in F' = J \setminus F$, be the coefficients in the corresponding Tucker tableau for $\mathscr{C}$ and $\mathscr{D}$:*

1. *(Column Supports). For each $k \in F'$ there is a unique elementry primal support P_k that contains k in its positive part and otherwise uses only elements of F. The corresponding elementary vectors e_{P_k} form an algebraic basis for $\mathscr{C}$ and are determined by*

$$\lambda(k)e_{P_k}(j) = \begin{cases} \alpha(j, k) & \text{for all } j \in F \\ 0 & \text{for all } j \in F' \text{ except } j = k, \\ 1 & \text{for } j = k, \end{cases} \tag{4}$$

where

$$\lambda(k) = \left(1 + \sum_{j \in F} |a(j, k)|\right) \Big/ \left(1 + |\{ j \in F : a(j, k) \neq 0\}|\right) > 0.$$

2. *(Row Supports). For each $j \in F$, there is a unique elementary dual support Q_j that contains j in its positive part and otherwise uses only elements of F'. The corresponding elementary vectors e_{Q_j} form an algebraic basis for $\mathscr{D}$ and are given by*

$$\mu(j)e_{Q_j}(k) = \begin{cases} -a(j, k) & \text{for all } k \in F' \\ 0 & \text{for all } k \in F \text{ except } k = j, \\ 1 & \text{for } k = j, \end{cases} \tag{5}$$

where

$$\mu(j) = \left(1 + \sum_{k \in F'} |a(j, k)|\right) \Big/ \left(1 + |\{ k \in F' : a(j, k) \neq 0\}|\right) > 0.$$

The proof of the theorem is easy in light of the said characterizations of bases, and it is left as Exercise 10.1.

The positive scale factors $\lambda(k)$ and $\mu(j)$ arise from the normalizing equations imposed in the definitions of e_P and e_Q in Section 10B. They can be neglected in the many situations where only the direction of an elementary vector is of interest, rather than its magnitude, as for instance in optimization algorithms involving line search (see the end of Section 10L). In any event $\lambda(k)$ and $\mu(j)$ reduce to 1 whenever the coefficients $a(j, k)$ are all $+1$, -1, or 0, as with networks. (For this "unimodular" case, see Exercises 10.2 and 10.3.)

The vectors e_{P_k} in Part 1 of the theorem form the *fundamental basis for $\mathscr{C}$ associated with F*, and similarly the vectors e_{Q_j} in Part 2 for $\mathscr{D}$. Not every basis for $\mathscr{C}$ (or $\mathscr{D}$) that is comprised entirely of elementary vectors is a fundamental basis or can be made into one by rescaling (see Exercise 10.4). However, each elementary vector does have some multiple occurring in at least one fundamental basis. More specifically, *for every elementary primal support P and every $k \in P^+$, there is a basis F in J such that $k \notin F$ and $P = P_k$; likewise, for every elementary dual support Q and every $j \in Q^+$, there is a basis F in J such that $j \in F$ and $Q = Q_j$.*

This is easy to see because $P \setminus \{k\}$, being a proper subset of an elementary primal support, cannot include any other nonempty primal support and therefore constitutes an independent set. As such it must be contained in some maximal independent set, that is, a basis F in J, and this F cannot meet j without including all of P and therefore violating the definition of independence. The proof of the assertion about dual supports is parallel (Exercise 10.5).

The significance of this remark is that to identify all the elementary primal and dual supports for a given pair of subspaces $\mathscr{C}$ and $\mathscr{D}$, it suffices first to characterize all the bases F in J and then to determine the special supports P_k and Q_j that correspond to them as in the theorem.

10D.* NETWORKS WITH GAINS

An excellent illustration of the concepts introduced so far in this chapter, because of its proximity to the ordinary network case and its numerous applications, is that of a network G each of whose arcs $j \in A$ has assigned to it a number $\gamma(j) \in (0, \infty)$, the *gain* of j. No longer must it be true that what enters an arc equals what comes out: $x(j)$ is interpreted as the flux in j at the *initial* node; the flux at the other end of j is $\gamma(j)x(j)$. Thus if $\gamma(j) > 1$, some sort of amplification is imagined as taking place in j, whereas if $0 < \gamma(j) < 1$, there is attenuation. The latter may be natural, for instance, in transportation models where losses or spoilage can occur en route. (Note that amplification of a negative flux is equivalent to attenuation of a positive quantity flowing in the opposite direction. Accordingly, the reversal of an arc j replaces $\gamma(j)$ by $1/\gamma(j)$.)

In this situation the divergence at a node i, still thought of as the difference between what departs from i and what arrives there, is

$$y(i) = \sum_{j \in [i, N \setminus i]^+} x(j) - \sum_{j \in [i, N \setminus i]^-} \gamma(j)x(j). \tag{1}$$

The flux and divergence variables therefore constitute the primal linear system

10A(1) for the coefficients

$$e(i, j) = \begin{cases} 1 & \text{if } i \text{ is the initial node of } j, \\ -\gamma(j) & \text{if } i \text{ is the terminal node of } j, \\ 0 & \text{otherwise.} \end{cases} \tag{2}$$

Thus we are interested in a matrix E that resembles the incidence matrix for G, except that the -1 in each column has been replaced by an arbitrary negative real number $-\gamma(j)$.

The corresponding dual linear system is

$$v(j) = \gamma(j)u(i') - u(i) \quad \text{for } j \sim (i, i'), \tag{3}$$

where $u(i)$ is still called the potential at i and $v(j)$ the tension in j.

The *gain* $\gamma(P)$ of a path P in G is defined of course as the product of the factors $\gamma(j)$ for each arc traversed positively and $1/\gamma(j)$ for each arc traversed negatively. A circuit P is said to be *passive* if $\gamma(P) = 1$ and *active* if $\gamma(P) \neq 1$: a *generating* circuit if $\gamma(P) > 1$ and an *absorbing* circuit if $\gamma(P) < 1$. An active circuit can serve as a source or sink, so the presence of such circuits clearly leads to phenomena of a different character than previously encountered.

There is a trivial case where the gains in G can be regarded as artifacts coming from the use of different scales of measurement of flux at different nodes. Suppose $x(j)$ is replaced by $\sigma(j)\xi(j)$, and $y(i)$ by $\rho(i)\eta(i)$, where $\sigma(j)$ and $\rho(i)$ are positive scale factors. Is it possible to choose these factors in such a fashion that, for all nodes i, (1) reduces to

$$\eta(i) = \sum_{j \in [i, N \setminus i]^+} \xi(j) - \sum_{j \in [i, N \setminus i]^-} \xi(j), \tag{4}$$

or in other words, to the divergence formula for an ordinary network without gains? If so, there is nothing really new to be investigated, and we shall say that the gains in G are *artificial*.

Passivity Theorem.   *For a connected network G with gains the following are equivalent*:

1. *The gains in G are artificial.*
2. *There is a positive function ρ on N such that $\gamma(j) = \rho(i')/\rho(i)$ when $j \sim (i, i')$.*
3. *All circuits in G are passive.*
4. *The rank of the matrix E defined by* (2) *is less than $|N|$ (the number of rows of E).*

Proof.   The reduction of (1) to (4), as referred to in Condition 1, occurs if and only if $\sigma(j) = \rho(i)$ when $j \in [i, N \setminus i]^+$, and $\gamma(j)\sigma(j) = \rho(i')$ when $j \in [i', N \setminus i']^-$. These formulas say that $\sigma(j) = \rho(i)$ and $\gamma(j)\rho(i) = \rho(i')$ when $j \sim (i, i')$. Thus Condition 1 is equivalent to Condition 2.

The equivalence of Conditions 2 and 3 is seen by considering $v(j) = \log \gamma(j)$ and $u(i) = \log \rho(i)$. In terms of these values Condition 2 is the assertion that $v(j) = u(i') - u(i)$ when $j \sim (i, i')$. A potential u with this property exists if and only if the tension of v along every circuit P is zero (see Section 6A). But when $v(j) = \log \gamma(j)$, the tension of v along P is $\log \gamma(P)$, and this is zero if and only if $\gamma(P) = 1$. The condition therefore translates into Condition 3.

The rank property in Condition 4 means the existence of numbers $\rho(i)$, not all zero, such that

$$\sum_{i \in N} \rho(i) e(i, j) = 0 \quad \text{for all} \quad j \in A,$$

or by (2), $\rho(i') = \rho(i)\gamma(j)$ when $j \sim (i, i')$. This generalizes to

$$\rho(i') = \rho(i)\gamma(P) \quad \text{for any path} \quad P: i \to i'.$$

Since G is connected and $\gamma(P) > 0$, we conclude that the numbers $\rho(i)$, if they do exist, must *all* be nonzero and of the same sign, and they can just as well be taken then as positive. Hence Condition 4 too is equivalent to Condition 2, and the theorem has been proved.

In view of the passivity theorem, the case really of interest in the sense of offering something not already encompassed by earlier theory, is the one where G is a connected network with gains that has *at least one active circuit*. Then from Condition 4 one has

$$|N| = \text{rank } E = \dim \mathcal{D} = |A| - \dim \mathcal{C} = \dim \mathcal{B} \tag{5}$$

(where $\mathcal{B}$ is the space in 10C(3)). Thus $\mathcal{B}$ is all of R^N, and for each $v \in \mathcal{D}$ there is a *unique* u satisfying (3).

The bases and elementary supports in this case will be characterized in Section 10E.

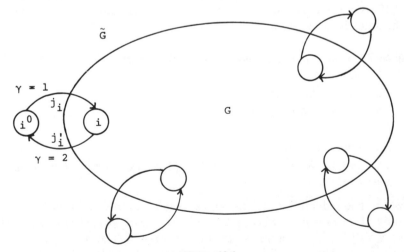

Figure 10.4

Enlarged Network

We have frequently found it useful, in dealing with ordinary networks, to identify $\mathscr{C}$ and $\bar{\mathscr{D}}$ with the circulation and differential spaces of the augmented network $\bar{G}$ formed by adding a single "distribution node" to G (see Section 1G). This trick fails when G is a network with gains that has at least one active circuit. Nevertheless, one can accomplish much the same thing in this case by adjoining an active two-arc circuit to each node i as in Figure 10.4. Here the variable $y(i)$ for G is identified with the flux in the arc j_i in the enlarged network $\tilde{G}$. The gain in the comparison arc $j_{i'}$ ensures that when conservation is imposed at all nodes of $\tilde{G}$, the amount that is generated by the added circuit and that must be transmitted away from i on the adjacent arcs of G is the same quantity $y(i)$. Thus the elements of $\mathscr{C}$ correspond one to one with the circulations in $\tilde{G}$.

10E.*   A GENERALIZATION OF CIRCUITS AND CUTS

The nature of the bases and elementary supports in the nontrivial case of a network with gains will now be elucidated. Some new concepts will be needed.

A *sprout* in G is a connected subnetwork that contains, up to reversal, exactly one elementary circuit, and this circuit is active; see Figure 10.5. A union of one or more disjoint sprouts is a *field*, and a *spanning field* in G is a field that uses every node; see Figure 10.6.

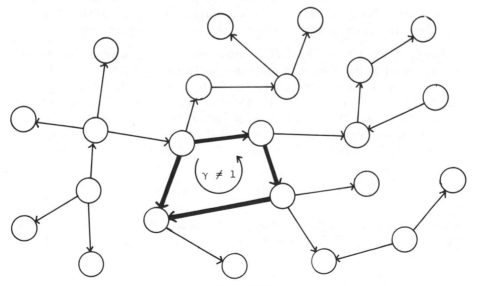

Figure 10.5

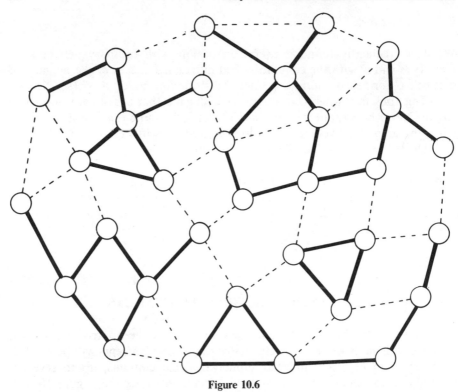

Figure 10.6

Proposition 1. *Let G be a connected network with gains that has at least one active circuit. Then F is a basis for the associated linear systems* 10D(1)(3) *if and only if the subnetwork formed by the arcs of F is a spanning field for G.*

    Proof. Consider any arc set F, and let $F_1, \ldots, F_p$ denote the subsets corresponding to the components of the subnetwork of G formed by F. Let N_k be the set of nodes incident to F_k. For F to be a basis, it is necessary and sufficient that $|F| = \text{rank } E$ and that there be no $v \in \mathcal{D}$ vanishing on F except $v = 0$. Recall from 10D(5) that for a connected network with gains that has at least one active circuit, rank $E = |N|$, and there is for each $v \in \mathcal{D}$ a *unique u* satisfying (3). The condition then is that $|F| = |N|$ and that the equations

$$0 = \gamma(j)u(i') - u(i) \quad \text{for all} \quad (i, i') \sim j \in F \tag{1}$$

should imply $u \equiv 0$ on N. Certainly this would entail every node in n being incident to some arc in F, so that $N = N_1 \cup \cdots \cup N_p$. Thus F is a basis if and only if

$$|F_1| + \cdots + |F_p| = |N_1| + \cdots + |N_p| \tag{2}$$

and, for each k, the subsystem of (1) for arcs $j \in F_k$ implies that $u(i) = 0$ for all $i \in N_k$. The second property holds if and only if the connected subnetwork formed by F_k contains an active circuit; this is clear from the passivity theorem in Section 10D, as applied to the subnetwork. In this event $|F_k| \geq |N_k|$ for all k because any spanning tree for the subnetwork formed by F_k has $|N_k| - 1$ arcs; therefore if (2) holds one must actually have $|F_k| = |N_k|$ for all k. The condition for F to be a basis reduces then to the following: the subnetwork formed by each F_k must consist of a tree to which one additional arc has been joined to form an active elementary circuit, and the union of these subnetworks must include every node of G. This finishes the proof of Proposition 1.

The way is now open to determining all the elementary primal and dual supports for the linear systems at hand. We have noted in Section 10C that each elementary primal support can be obtained by taking a basis F, an arc $k \notin F$, and determining an $x \in \mathscr{D}$ (unique up to scalar multiples) that vanishes outside F, except that $x(k) \neq 0$. The characterization in Proposition 1 makes it easy to size up all the possibilities.

A *goggle* in G is a signed set of arcs forming one of the following configurations (see Figure 10.7):

TYPE 1.   Two disjoint elementary circuits, one generating and one absorbing, together with an elementary path having only its initial node in the generating circuit and only its terminal node in the absorbing circuit.

TYPE 2.   Two elementary circuits, one generating and one absorbing, that have a (single) joint portion.

TYPE 3.   Two elementary circuits, one generating and one absorbing, that meet in exactly one node.

Type 3 can be thought of as a degenerate form of the other two types.

Proposition 2.   *Let G be a connected network with gains that has at least one active circuit. Then the elementary primal supports P are the passive elementary circuits in G and the goggles of Types 1, 2, and 3.*

The details of proof are left as Exercise 10.8. Note that the normalized elementary vectors e_P corresponding to the configurations in question can easily be computed when the exact gains of the arcs are specified.

The elementary dual supports can be determined similarly by considering an arbitrary basis F and an arc $j \in F$. There is, up to scalar multiples, a unique $v \in \mathscr{D}$ whose support includes j but no other arcs of F. According to Proposition 1, F forms a union of disjoint sprouts. The removal of j from F would convert one of these into a tree, or the union of a sprout and a tree (or single node), depending on whether j belonged to the active circuit in the sprout. The desired v must vanish on all the remaining arcs of F. What does this tell us

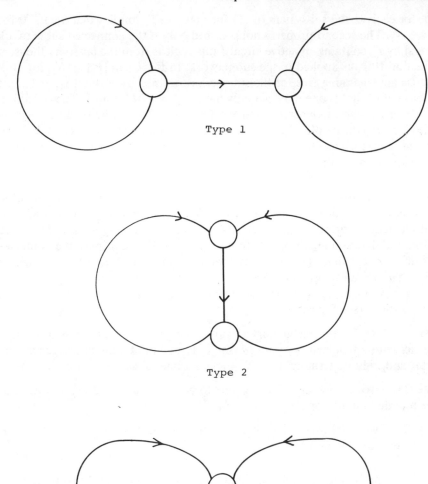

Type 1

Type 2

Type 3

Figure 10.7

about the associated u? We have observed in the proof of Proposition 1 that if $v(j) = 0$ for all arcs of a sprout, then $u(i) = 0$ for all nodes of the sprout. Thus u must vanish at all nodes of G except those of the tree (or isolated node) that was created when j was removed from F. On the tree itself, the relations (1) allow u to be determined up to a multiple by integration, with values all positive or all negative. Obviously then, the support of v will include all the arcs that join the tree to nodes not in it, as well as certain of the arcs that form elementary circuits when added to the tree. Some reflection on this situation leads to an abstract description of the supports in question.

Define a *passive subnetwork* H of G to be a subnetwork (with at least one node) that has no active circuits, and say that it is *full* if there does not exist another passive subnetwork having the same node set but an arc set that properly includes the arc set of H.

If H is a full passive subnetwork and j is an arc that is not in H but has both endpoints in H, the definition of "fullness" implies there is an active circuit P that traverses j in the positive direction and otherwise uses only arcs that belong to H; see Figure 10.8. Although P may not be unique, any two such circuits must have *the same gain* (different from 1), due to the passivity of H (Exercise 10.9). Call j a *generator* for H if $\gamma(P) > 1$ and an *absorber* if $\gamma(P) < 1$.

Associate now with each full passive subnetwork H the signed set Q defined as follows: Q^+ consists of the arcs $j \in A$ that are generators for H or have terminal node, but not initial node, in H; Q^- consists of the arcs that are absorbers for H or have initial node, but not terminal node in H. (Positive flux in an arc $j \in Q^+$ would add material to H, whereas for an arc $j \in Q^-$, it would remove it.) We shall call Q the *activator* of H and the reverse of Q the *deactivator*.

The notion of the deactivator of H generalizes, in the context of a network with gains, the cut associated with the node set of H.

An *elementary* activator or deactivator Q is one that is associated with a full passive subnetwork H that is connected and such that, if H and all the arcs incident to it were deleted from G, each component of the remaining network (if any) would contain at least one active circuit.

Proposition 3. *Let G be a connected network with gains that has at least one active circuit. Then the elementary dual supports Q are the elementary activators and deactivators in G.*

Again the details of the proof are left to the reader, as Exercise 10.10.

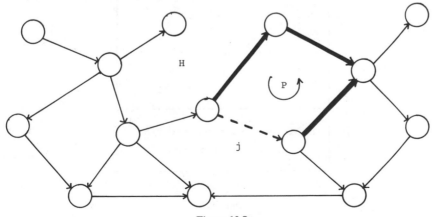

Figure 10.8

10F.*   MULTICOMMODITY SYSTEMS AND FACTORIZATION

Linear systems that are closely related to the ordinary network case, and therefore are amenable to special treatment and representation, arise not only when gains are introduced but also when flow interactions can take place among various arcs. In some models based on a network G, for instance, there might be a requirement that $x(j_1)$ be a certain multiple of $x(j_2)$, or that the arcs belonging to a certain family all have the same flux, even though the arcs in question might not have nodes in common. It might be necessary to work with additional variables which, like the divergences $y(i)$, are linear functions of the flow variables $x(j)$, such as quantities representing resources consumed by the overall flow.

Abstractly we can think of the linear systems in this class as having the form in Figure 10.9. The extra row indexes h correspond not to nodes in G but "interactions." The associated variables $y(h)$ might, in various applications, be required to have specific values or to lie in certain intervals. There could be cost terms connected with them.

A particularly instructive example is furnished by *multicommodity transportation problems*, or *traffic problems*, where an arc must handle several different kinds of flow at the same time. To keep things simple, let us imagine r different "commodities" all being transported around the same connected network G_*. Denote by x_k the flow corresponding to the kth commodity: $x_k(j)$ is the flux of this commodity in the arc j. Besides constraints and costs on the separate vectors x_k and their divergences y_k, there may also be constraints and costs on the total flow $z = x_1 + x_2 + \cdots + x_r$, and this is the essence of the situation.

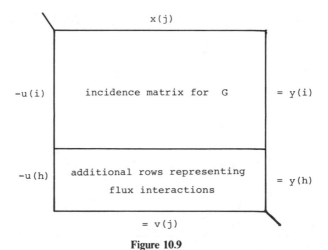

Figure 10.9

For instance, in an arc j we may be concerned that $z(j)$ should not exceed a certain total capacity bound.

Figure 10.10 expresses the corresponding linear system, with E_* the incidence matrix for G_*. Note that the staircase portion of the tableau amounts to the incidence matrix for the disconnected network G consisting of r disjoint copies of G_*. We can therefore regard this tableau as a special case of the one in Figure 10.9.

Models of vehicular traffic lead to the same kind of linear system. We need only interpret $x_k(j)$ as the number of cars in highway link j that originated in location s_k and are headed for location s'_k, where s_k and s'_k are certain nodes of G_*. We would then want $y_k(i) = 0$ for $i \neq s_k, s'_k$, so that $y_k(s_k) = -y_k(s'_k)$. The latter quantity could be required to equal a known amount or be subjected to other conditions.

Having the same network G_* serve for every kind of flow x_k in Figure 10.10 enhances symmetry without causing any real loss in generality. In some applications it may happen that certain arcs j are forbidden to certain kinds of flow; thus x_k may be limited in effect to a subnetwork G_*^k of G_* for $k = 1, \ldots, r$. However, this can be taken into account in the system of Figure 10.10 by imposing constraints of the form $x_k(j) \in [0, 0]$ at the stage when capacity intervals are introduced (see Section 10I).

It would be easy to generalize to the case where sums $z(j) = \sum_{k=1}^{r} x_k(j)$ are replaced by weighted combinations $z(j) = \sum_{k=1}^{r} \lambda_k(j) x_k(j)$. In highway traffic,

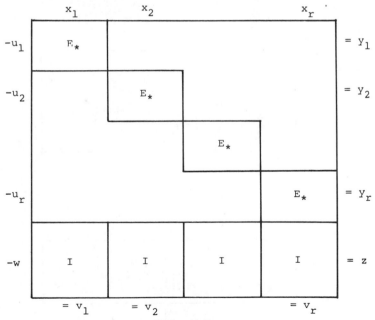

Figure 10.10

for example, this would facilitate distinctions between different kinds of
vehicles such as private cars, trucks, and buses. The transit time for a highway
link j could depend on such a weighted combination.

   Needless to say, traffic problems with time-dependent costs and requirements can be modeled in the preceding manner but in a dynamic network as
defined in Section 1H.

   The question we want now to address is whether the special structure of the
linear systems in Figures 10.9 and 10.10 leads to any simplification in setting

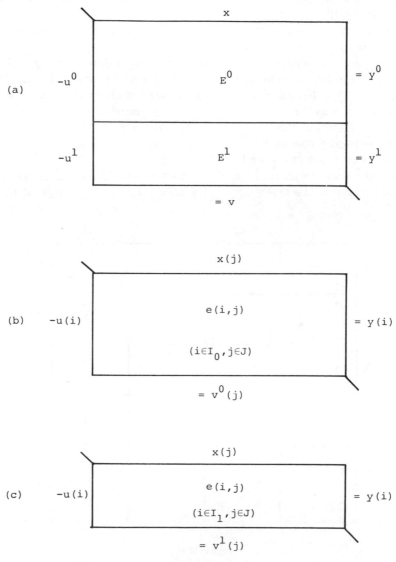

Figure 10.11

up Tucker representations and determining primal and dual elementary supports for the corresponding spaces $\mathscr{C}$ and $\mathscr{D}$, or $\overline{\mathscr{C}}$ and $\overline{\mathscr{D}}$. For flexibility in formulating the answer, as well as applications to other situations not necessarily involving networks, we shall focus on a linear system of the general type in Figure 10.11a, which we regard as a juxtaposition of the ones in Figures 10.11b and c, with $v = v^0 + v^1$. We assume that the subsystem of Figure 10.11b is relatively easy to work with (perhaps being based on network structure), and we look for ways of taking advantage of this. In other words, we suppose that Tucker representations of the spaces

$$\mathscr{C}^0 = \left\{ x \in R^J | E^0 x = 0 \right\}, \qquad \mathscr{D}^0 = \left\{ v^0 \in R^J | \ \exists u^0 \ \text{with} \ -u^0 D^0 = v^0 \right\}$$

$$(1)$$

are readily available and ask whether this leads to shortcuts in manipulating Tucker representations of the spaces

$$\mathscr{C} = \left\{ x \in R^J | \ E^0 x = 0 \ \text{and} \ E^1 x = 0 \right\}$$

$$\mathscr{D} = \left\{ v \in R^J | \ \exists u^0, u^1, \ \text{with} \ -u^0 E^0 - u^1 E^1 = v \right\} \qquad (2)$$

Later it will be observed that results along such lines carry over to the extended spaces $\overline{\mathscr{C}}$ and $\overline{\mathscr{D}}$ associated with Figure 10.11a, as well as certain intermediate spaces.

Factorization Theorem. *Let $F \subset J$ be any index set that is a basis relative to the full matrix E (with upper and lower portions E^0 and E^1) and therefore yields a Tucker tableau for $\mathscr{C}$ and $\mathscr{D}$ with coefficients $a(j, k)$ for $j \in F$, $k \in J \setminus F$. Then F includes an index set F_0 that is a basis relative to the submatrix E^0 and therefore yields a Tucker tableau for $\mathscr{C}^0$ and $\mathscr{D}^0$ with coefficients $a_0(j, k)$ for $j \in J_0, k \in F \setminus J_0$. Moreover one has*

$$a(j, k) = a_0(j, k) + \sum_{h \in F \setminus F_0} a_0(j, h) a(h, k) \quad \text{for all } j \in F_0, k \in J \setminus F.$$

$$(3)$$

Proof. Let $\mathscr{B}$ and $\mathscr{B}^0$ denote the column spaces of E and E^0,

$$\mathscr{B} = \left\{ (y^0, y^1) | \ \exists x \ \text{with} \ E^0 x = y^0, E^1 x = y^1 \right\},$$

$$\mathscr{B}^0 = \left\{ y^0 | \ \exists x \ \text{with} \ E^0 x = y^0 \right\},$$

and observe that

$$\mathscr{B}^0 = \left\{ y^0 | \ \exists y^1 \ \text{with} \ (y^0, y^1) \in \mathscr{B} \right\}.$$

Therefore any set of columns of E that spans $\mathscr{B}$ gives at the same time a set of columns of E^0 that spans $\mathscr{B}^0$. In other words, any spanning set of indexes relative to E is also such relative to E^0 and therefore includes some basis relative to E^0. Applying this to F, we get the first assertion of the theorem. The second assertion is just a matter of substituting

$$x(h) = \sum_{k \in J \setminus F} a(h, k)x(k) \quad \text{for } h \in F \setminus F_0$$

into the equation

$$x(j) = \sum_{k \in J \setminus F} a_0(j, k)x(k) + \sum_{h \subset F \setminus F_0} a_0(j, h)x(h) \quad \text{for } j \in F_0$$

and combining terms.

Consequences for Pivoting

The relationships in the factorization theorem are summarized in Figure 10.12. The first tableau there gives Tucker representations for $\mathscr{C}$ and $\mathscr{D}$ associated with F, and since $F \subset F_0$, the indexes $h \in F \setminus F_0$ correspond to rows. The second

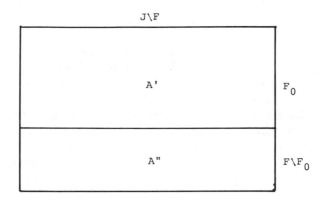

$$A' = A'_0 + A''_0 A''$$

Figure 10.12

tableau gives Tucker representations for $\mathscr{C}^0$ and $\mathscr{D}^0$ associated with F_0. The indexes $h \in F \setminus F_0$ correspond then to columns. With the indicated partitioning, (3) says that $A' = A'_0 + A''_0 A''$.

The conclusion to be drawn is that to know the first tableau in Figure 10.12, we merely need to know the coefficients in the A'' portion along with the ("readily available") coefficients in the second tableau. Any entry in the A' portion can be generated by (3) upon demand.

Storage requirements in the manipulation of Tucker representations can greatly be reduced by this observation. Indeed, pivoting transformations can be carried out without ever having to calculate the matrix A'.

Suppose, for instance, that we want to pivot in the first tableau in Figure 10.12 on an entry $a(\bar{h}, \bar{k})$ in the A'' portion, as in Figure 10.13a. This amounts to trading an index $\bar{k} \notin F$ for an index $\bar{h} \in F$ such that $\bar{h} \notin F_0$. The general pivoting formulas express the new coefficients in the A'' portion entirely in terms of the old ones without any of the coefficients in the A' portion being involved, and of course the second tableau in Figure 10.12 remains unchanged because F_0 is the same. This case reduces therefore to an ordinary pivoting transformation on the subtableau corresponding to the A'' portion alone.

If, on the other hand, we want to pivot on an entry $a(\bar{j}, \bar{k})$ in the A' portion of the first tableau, as in Figure 10.13b, we will be trading an index $\bar{k} \notin F$ for an index $\bar{j} \in F$ such that actually $\bar{j} \in F_0$. To get the new A'' coefficients from the general pivoting formulas, it will be necessary to calculate the A' coefficients in the $\bar{j}$ row using (3), but no other rows of A' will have to be generated. Afterwards we must pivot on the entry $a_0(\bar{j}, \bar{k})$ in the second tableau of Figure 10.12, since F_0 has been changed by substituting $\bar{k}$ for $\bar{j}$. This second

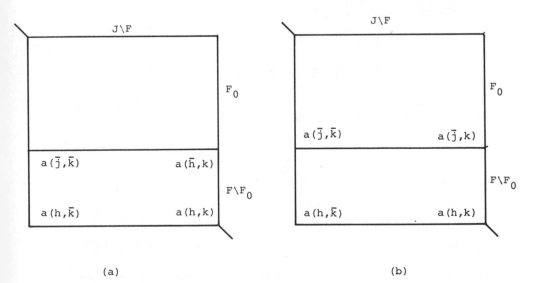

(a)   (b)

Figure 10.13

transformation is, by assumption, relatively easy. In a network setting it might well be executed in a special combinatorial mode in terms of spanning trees and their associated fundamental circuits (see Sections 4H and 4I).

What has been said about the manipulation of Tucker tableaus could of course be translated into the language of primal and dual elementary supports by way of the basis theorem in Section 10C.

Representations of Extended Spaces

If our interest lies not in $\mathscr{C}$ and $\mathscr{D}$ but the extended spaces

$$\bar{\mathscr{C}} = \{(y^0, y^1, x) | y^0 = E^0 x \quad \text{and} \quad y^1 = E^1 x\},$$

$$\bar{\mathscr{D}}\{(u^0, u^1, x) | v = -u^0 E^0 - u^1 E^1\},$$

this can be satisfied by applying the technique just described to the matrix

$$\bar{E} =$$

	I_0	I_1	J	
$-I$		0	E^0	I_0'
	0	$-I$	E^1	I_1'

in terms of its natural partitioning into an upper portion $\bar{E}^0$ and lower portion $\bar{E}^1$. (Here I_0' and I_1' are second copies of the index sets I_0 and I_1.) The tableaus in Figure 10.11a and b then furnish initial Tucker representations which correspond to the two tableaus in Figure 10.12 for this case.

Similar considerations apply to the spaces

$$\tilde{\mathscr{C}} = \{(y^1, x) | \quad E^0 x = 0 \quad \text{and} \quad E^1 x = y^1\},$$

$$\tilde{\mathscr{D}} = \{(u^1, v) | \quad \exists u^0 \quad \text{with} \quad v = -u^0 E^0 - u^1 E^1\},$$

which in many applications are the main ones to consider. These can be interpreted as corresponding to the matrix

$$\tilde{E} =$$

	I_1	J	
0		E^0	I_0'
$-I$		E^1	I_1'

Here we work with the index set $J_1 = I_1 \cup J$. For any basis $F_0 \subset J$ relative to E^0, the set $F = I_1 \cup F_0$ is a basis relative to E. Corresponding to these bases we have an initial pair of tableaus, as shown in Figure 10.14, which are ready to play for $\tilde{\mathscr{C}}$ and $\tilde{\mathscr{D}}$ the role that the tableaus in Figure 10.12 did for $\mathscr{C}$ and $\mathscr{D}$.

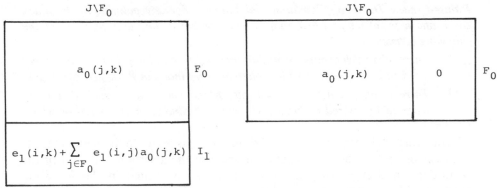

Figure 10.14

Application to Traffic Systems

In the case of Figure 10.10 the staircase portion E^0 of the tableau is the incidence matrix of a network G consisting of r disjoint copies of a certain connected network G_*. A basis F_0 relative to E^0 corresponds to a maximal forest in G and is obtained by choosing a spanning tree F_0^k from each copy of G_*. The associated Tucker tableau for G decomposes into the r tableaus for G_* associated with these spanning trees.

Then in Figure 10.12, where J is the union of r disjoint copies of the arc set for G_*, every coefficient in the second tableau can be generated combinatorially from the fundamental circuits in G_* associated with the F_0^k's. The same is true when the space $\bar{\mathscr{C}}$ and $\bar{\mathscr{D}}$, or $\hat{\mathscr{C}}$ and $\hat{\mathscr{D}}$, are treated in place of $\mathscr{C}$ and $\mathscr{D}$ as described earlier. Thus in all cases the factorization theorem provides a substantially reduced format for Tucker representations and pivoting.

Linear systems of this kind will be taken up again in the optimization context in Sections 11F and 11G.

10G.   PAINTED INDEX THEOREM AND ALGORITHM

Returning now to the context of a general pair of linear systems of variables, we look at a fundamental and rather surprising fact: Minty's lemma about the circuits and cuts in a network (Section 2H) is just a special case of a deeper result about primal and dual supports.

In what follows, we shall mean by a *painting* of the index set J of course a partition of J into four subsets (some possibly empty) whose elements will be called "green," "white," "black," and "red," respectively. Recall that a *primal support* is a support of $\mathscr{C}$, and a *dual support* is a support of $\mathscr{D}$, where $\mathscr{C}$ and $\mathscr{D}$ are the general subspaces in Section 10A.

Painted Index Theorem (Combinatorial Form).   *For any painting of the index set J and any particular $l \in J$ that is white or black, one and only one of the following is true*:

1. *There is an [elementary] primal support P containing l, such that every index in P^+ is green or white, whereas every index in P^- is green or black.*
2. *There is an [elementary] dual support Q containing l, such that every index in Q^+ is red or black, whereas every index in Q^- is red or white.*

The word "elementary" is placed in brackets here to indicate that its presence or absence does not affect the truth of the assertions. In view of the definitions in Section 10B, the result can also be stated in the following manner, without reference to supports.

Painted Index Theorem (Vector Form).   *For any painting of the index set J and any particular $l \in J$ that is white or black, one and only one of the following is true*:

1. *There is an [elementary] vector x of $\mathscr{C}$ with $x(l) \neq 0$, such that $x(j) \geq 0$ for every white index, $x(j) \leq 0$ for every black index, and $x(j) = 0$ for every red index.*
2. *There is an [elementary] vector v of $\mathscr{D}$ with $v(l) \neq 0$, such that $v(j) \geq 0$ for every black index, $v(j) \leq 0$ for every white index, and $v(j) = 0$ for every green index.*

In this form the theorem amounts to a statement about the complementary feasibility of two systems of homogeneous linear equations and inequalities. This is especially clear when the relationships among the variables $x(j)$ and $v(j)$ are expressed by a Tucker tableau, as in Figure 10.2. In the tableau format, where it will be recalled that the matrix entries could well be arbitrary real numbers (and $\mathscr{C}$ and $\mathscr{D}$ could then be defined accordingly), such a result is known classically as a *transposition theorem*. The combinatorial version of the painted index theorem in terms of supports, on the other hand, has the advantage of emphasizing the discrete nature of the result (since there are only finitely many primal and dual supports), as well as the powerful analogy with networks.

A general, algorithmic proof will be provided later in this section. This involves starting with any Tucker tableau for $\mathscr{C}$ and $\mathscr{D}$ and pivoting through a sequence of other such tableaus until one is reached that indicates the fulfillment of either Alternative 1 or 2. The two alternatives are mutually exclusive, even with the word "elementary" omitted, as is readily seen in the vector setting. If there were both an $x \in \mathscr{C}$ meeting conditions in Alternative 1 and a $v \in \mathscr{D}$ meeting the conditions in Alternative 2, then

$$x \cdot v = x(l)v(l) = \underset{\substack{\text{white} \\ \neq l}}{\sum} x(j)v(j) + \underset{\substack{\text{black} \\ \neq l}}{\sum} x(j)v(j) < 0,$$

and yet at the same time $x \cdot v = 0$ by the orthogonality of $\mathscr{C}$ and $\mathscr{D}$. Thus a procedure which always ends in Alternative 1 or 2 does suffice to establish the theorem.

To prepare for the pivoting algorithm, we need to translate the theorem into a statement about the patterns of signs that can be found in various Tucker representations. Let us speak of a *compatible column* in a tableau as a column that meets the requirements in Figure 10.15a for signs versus colors. Here "arb" denotes an entry that can have an arbitrary value. Thus, for instance, a column whose index is white must have a nonpositive number in every row whose index is black, among other things. The "inc" under "red" signifies that any column whose index is red is deemed incompatible. The requirements for a *compatible row* are given analogously in Figure 10.15b.

A compatible column yields an elementary primal support P such that every index in P^+ is green or white, and every index in P^- is green or black. Indeed, if k is the index of the column and k is white, one can take P to be the support P_k described in the general basis theorem in Section 10C, whereas if k is black, one can take P to be the reverse of P_k; if k is green, both choices would work (all indexes in P actually being green according to the specifications in Figure 10.15a).

Likewise, *a compatible row yields an elementary dual support Q such that every index in Q^+ is red or black, and every index in Q^- is red or white.* If j is the index of the row and j is black, one can take Q to be the support Q_j in Section 10C, whereas if j is white, one can take the reverse of Q_j; if k is red, either will do (the indexes in Q all being red).

Let us say further that a column *uses* a particular l if l is either the index of the column or the index of some row whose entry in that column is nonzero, that is, if l belongs to the elementary support determined by the column; similarly for rows.

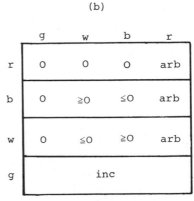

(a)

	g	w	b	r
r	O	O	O	inc
b	O	≤O	≥O	inc
w	O	≥O	≤O	inc
g	arb	arb	arb	inc

column compatibility

(b)

	g	w	b	r
r	O	O	O	arb
b	O	≥O	≤O	arb
w	O	≤O	≥O	arb
g	inc	inc	inc	inc

row compatibility

Figure 10.15

The upshot then is that *if we can come up with a Tucker tableau having a compatible column that uses l, we have determined constructively that Alternative 1 holds in the theorem, whereas in the case of a compatible row that uses l, we have determined Alternative* 2. A procedure for achieving this will now be formulated in terms of pivoting. Supplementary rules that ensure termination of the procedure, and thereby complete the proof of the painted index theorem, will be explained in the next section.

Painted Index Algorithm

Start with any Tucker tableau. The given white or black index l may correspond to either a row or a column (it is called the *lever* index); see Figure 10.16.

If l corresponds to a row, check whether this row is compatible. If it is, terminate with Alternative 2 of the theorem. If not, there is an entry in the l row that fails the compatibility test. Let $\bar{k}$ be the index of any column containing such an entry, and check whether this column is compatible. If it is, terminate with Alternative 1 of the theorem. If not, there is an entry in the $\bar{k}$ column that fails the compatibility test. Let $\bar{j}$ be the index of any row containing such an entry. Pivot on $(\bar{j}, \bar{k})$ and return to the beginning of the procedure. (Here $l \neq \bar{j}$ if $\bar{k}$ is not green; $\bar{k}$ cannot be red.)

Dually, if l corresponds to a column, check whether this column is compatible. If it is, terminate with Alternative 1 of the theorem. If not, there is an entry in the l column that fails the compatibility test. Let $\bar{j}$ be the index of any row containing such an entry, and check whether this row is compatible. If it is, terminate with Alternative 2 of the theorem. If not, there is an entry in the $\bar{j}$ row that fails the compatibility test. Let $\bar{k}$ be the index of any column containing such an entry. Pivot on $(\bar{j}, \bar{k})$ and return to the beginning of the procedure. (Here $l \neq \bar{k}$ if $\bar{j}$ is not red; $\bar{j}$ cannot be green.)

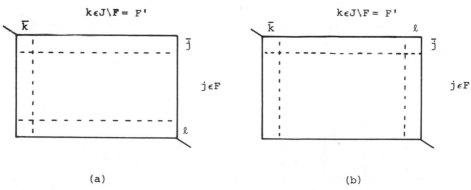

(a)   (b)

Figure 10.16

Remark

A particularly simple case, which may help the reader's understanding, is the one where all indexes are white. Then l must stay fixed as a row index or a column index. A compatible column is one whose entries are all non-negative, whereas a compatible row is one whose entries are all nonpositive. Thus if l corresponds to a row, say, we look for a positive entry in this row, and if we find one, we look for a negative entry in that column. Success in both searches determines a pivot entry and results in a new tableau having l still in the same position but a certain row index interchanged with a certain column index.

Network Cases

The painted index algorithm reduces essentially to the painted network algorithm when applied in a network setting in a certain way (see Exercises 10.17 and 10.18). Reduction to a direct "geometric" procedure also happens for networks with gains (see Exercise 10.19).

Clearly we are dealing here with a conceptual algorithm of great generality. It is capable of much refinement in implementation, particularly when the underlying combinatorial structure lends itself to direct manipulation.

10H.*   TERMINATION AND PREPROCESSING

Two general refinements of the painted index algorithm will now be described.

Without anything extra the painted index algorithm might continue indefinitely without terminating in Alternatives 1 or 2. One way to prevent this from happening is to invoke *Bland's priority rule*: number the elements of J arbitrarily ("priority"), and whenever there is more than one index that could be selected as $\bar{j}$ or as $\bar{k}$, take the one whose priority is highest.

Proof of Termination under Bland's Rule

Consider what the situation would be if the algorithm did go on for infinitely many iterations. Since there are only finitely many possible Tucker tableaus, some cycle of tableaus would have to repeat itself indefinitely. Denote by J_0 the set of indexes whose position changes at some time during the cycle, and by J_r and J_c the sets of indexes that remain in row position and in column position, respectively. Each index in J_0 must be chosen at least once as $\bar{j}$ during the cycle, and also least once as $\bar{k}$. It is clear therefore that none of the indexes in J_0 is green or red, and that $l \notin J_0$. Suppose $l \notin J_r$ for simplicity (the case where $l \in J_c$ could be argued in parallel fashion), and let l' denote the index of lowest priority in J_0.

In an iteration of the cycle when l' is a column index and is chosen as $\bar{k}$, l' must be the *only* index in J_0 corresponding to a column whose entry in the l row spoils the compatibility of that row, for otherwise Bland's rule would have been violated. Much in the same way, in an iteration of the cycle when l' is a row index and l' is chosen as $\bar{j}$, a certain $\bar{k}$ column must just have been selected, and l' must be the *only* index in J_0 corresponding to a row whose entry in the $\bar{k}$ column spoils the compatibility of that column. Consider now the modified painting in which the color of l' is reversed between white and black, the indexes in J_c are all made red, and those in J_r, except for l itself, are all made green. With respect to this, the l row is indeed compatible in the situation where l' is chosen as $\bar{k}$, and so is the $\bar{k}$ column in the situation where l' is chosen as $\bar{j}$. Both the compatible row and the compatible column in question use l'. Thus for a certain painting and a certain index l', both alternatives of the painted index theorem hold. But the two alternatives were seen earlier to be mutually exclusive. A contradiction has therefore been laid bare that demonstrates that an infinite sequence of iterations cannot occur.

Red-Green Preprocessing Rule

For both practical and theoretical purposes the following supplement to the painted index algorithm is of importance. *Before executing the algorithm itself, perform as many consecutive pivot steps as possible in which the pivot* $(\bar{j}, \bar{k})$ *satisfies these conditions*:

1. $\bar{j}$ *is red or $\bar{k}$ is green (or both).*
2. $\bar{j}$ *is not green, and $\bar{k}$ is not red.*

Select such pivots arbitrarily, except for giving preference to ones with $\bar{j} \neq l$ and $\bar{k} \neq l$, when available. In simpler terms: pivot red row indexes into column position and green column indexes into row position until no longer possible without merely exchanging red for red or green for green, but in accomplishing this try to avoid moving the lever index l. Clearly the number of pivot steps taken under the rule can never exceed the number of red indexes originally in row position plus the number of green indexes originally in column position. We shall refer to this preliminary procedure as *red-green preprocessing*. Its advantages are as follows.

Proposition 4. *Suppose the painted index algorithm is applied to a Tucker tableau after red-green preprocessing.*

   If there is a primal support P containing l and having all of its other indexes green, or if there is a dual support Q containing l and having all of its other indexes red (special cases of Alternatives 1 and 2 of the painted index theorem, mutually exclusive), then the algorithm will terminate immediately, either with l as the index of a compatible column giving such a P, or with l as the index of a compatible row giving such a Q.

Otherwise the algorithm will proceed without l or any red or green index ever being a candidate for selection as $\bar{j}$ or $\bar{k}$.

Thus red-green preprocessing takes care of the red and green indexes once and for all. Thereafter l is sure to stay in fixed position, and all inspection for compatibility is concentrated in the reduced tableau formed by the rows and columns with black or white indexes.

To prove the proposition, we appeal to the tableau structure in Figure 10.17, which we claim must be present after red-green preprocessing. Indeed, rows with red indexes must have 0's in all columns except those with red indexes, or another pivot $(\bar{j}, \bar{k})$ meeting Conditions 1 and 2 could be chosen. Likewise, columns with green indexes must have 0's in all rows except those with green indexes, if indeed there are any such rows. The striped area of Figure 10.17 is a reminder that no row with a green index is eligible for j, and no column with a red index is eligible for $\bar{k}$. Entries in this area can be ignored in testing for compatibility.

Consider first the case where l happens to be a column index (black or white) in a tableau with such structure. The elementary primal support P corresponding to the l column does not involve any red indexes, so there cannot exist a dual support Q containing l and having all other indexes red (for that would contradict the exclusivity asserted by the painted index theorem when applied with all nonred indexes other than l temporarily repainted green). If the l column has only zero entries in the rows with black or white indexes, it is a compatible column, and the algorithm immediately terminates.

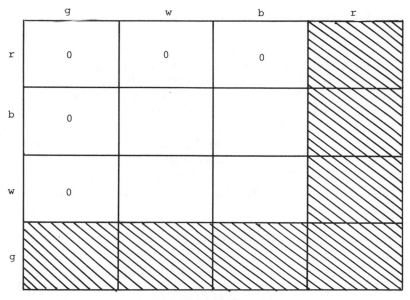

Figure 10.17

Then P contains, besides l, only green indexes. If the l column has a nonzero entry in a row with a black or white index j, then such a row yields an elementary dual support Q that contains l but no green indexes. In this event there cannot exist a primal support P' that contains l and has all of its other indexes green (for that would contradict the exclusivity asserted by the painted index theorem when applied with all nongreen indexes other than l temporarily repainted red).

Everything is similar in the case where l happens to be a row index (black or white) in a tableau of the kind in Figure 10.17. The elementary dual support Q corresponding to the l row does not involve any green indexes, so there cannot exist a primal support P containing l and having all other indexes green (for that would contradict the exclusivity asserted by the painted index theorem when applied with all nongreen indexes other than l temporarily repainted red). If the l row has only zero entries in the columns with black or white indexes, it is a compatible row, and the algorithm immediately terminates. Then Q contains, besides l, only red indexes. If the l row has a nonzero entry in a column with a black or white index k, then such a column yields an elementary primal support P that contains l but no red indexes. In this event there cannot exist a dual support Q' that contains l and has all of its other indexes red (for that would contradict the exclusivity asserted by the painted index theorem when applied with all the nonred indexes other than l temporarily green).

Lever Row or Lever Column?

The fact that the lever index l stays put as a row or column index after red-green processing means that the painted index algorithm in this case reverts to one of two entirely separate modes, depending on the position of l. Actually, if the algorithm does not immediately terminate in the manner described in the proposition, the row or column corresponding to l in the subtableau formed by the rows and columns with black or white indexes must have a nonzero entry. Thus a pivot step could be taken in which l is exchanged for some other black or white index. In other words, in executing the painted index algorithm, *one can freely choose between having l stay fixed in row position or having l stay fixed in column position*. We can speak of these two cases as the *primal mode* and the *dual mode* of implementation. They are certainly different, but there is no general reason for preferring one to the other. Particular applications may dictate a preference, however, as will be seen later.

10I. CONSTRAINTS AND FEASIBILITY

We turn now to the investigation of constraints on a linear system of variables that are more general than those represented in the vector form of the painted index theorem. Associate with each index $j \in J$ a nonempty real interval $C(j)$, not necessarily closed. For a fixed vector $b \in R^I$ and matrix $E \in R^{I \times J}$, we look for solutions to the following.

Primal Feasibility Problem.   *Find $x \in R^J$ such that*

$$\sum_{j \in J} e(i, j)x(j) = b(i) \quad \text{for all} \quad i \in I, \tag{1}$$

$$x(j) \in C(j) \quad \text{for all} \quad j \in J. \tag{2}$$

The existence of solutions x to this problem will be characterized in terms of a condition on the elementary supports Q for the subspace $\mathscr{D}$ corresponding to E as in Section 10A. Note that when $b = 0$, (1) reduces to the requirement that x belong to the subspace $\mathscr{C}$ corresponding to E.

Some reflections on the nature of Equation (1) will enable us to state the main result in notation similar to that of the feasible distribution theorem in Section 3H. Recall that there exists an x satisfying (1) if and only if the vector b belongs to the subspace $\mathscr{B}$ of R^I defined in 10C(3), the column space of E. There cannot be a *unique* such x, however, unless $\mathscr{C}$, which is the subspace of R^J consisting of all vectors that satisfy the corresponding homogeneous system of equations, is zero-dimensional, a case not of real interest here.

The existence of at least one x satisfying (1) implies a property of elementary dual supports Q, namely (from the relation $-u \cdot y = v \cdot x$ connecting primal and dual variables),

$$-\sum_{i \in I} u(i)b(i) = \sum_{j \in J} e_Q(j)x(j) \tag{3}$$

for *every* u corresponding to the vector $v = e_Q \in \mathscr{D}$. Thus the quantity on the left side of (3) is in fact independent of the *particular* u used to represent e_Q and may be denoted simply by b_Q, so that we have

$$b_Q := -\sum_{i \in I} u(i)b(i) \quad \text{for arbitrary } u \text{ with } -uE = e_Q, \tag{4}$$

$$\sum_{j \in J} e_Q(j)x(j) = b_Q \quad \text{for every } x \text{ satisfying (1)}. \tag{5}$$

In the network case where Q is a cut $[S, N \setminus S]$, b_Q is the divergence $b(S)$ (see Section 3H).

We introduce next for each elementary dual support Q the nonempty real interval

$$C(Q) := \sum_{j \in J} e_Q(j)C(j). \tag{6}$$

This consists of all the values taken on by

$$e_Q \cdot x = \sum_{j \in J} e_Q(j) x(j)$$

as each variable $x(j)$ ranges arbitrarily over $C(j)$. If every $C(j)$ is a *closed* interval $[c^-(j), c^+(j)]$ (possibly with $c^+(j)$ or $c^-(j)$ infinite), one has $C(Q) = [c^-(Q), c^+(Q)]$ where

$$c^+(Q) = \sum_{j \in Q^+} e_Q(j) c^+(j) + \sum_{j \in Q^-} e_Q(j) c^-(j),$$

$$c^-(Q) = \sum_{j \in Q^+} e_Q(j) c^-(j) + \sum_{j \in Q^-} e_Q(j) c^+(j). \tag{7}$$

Our generalization of the feasible distribution theorem in Section 3H can now be formulated.

Primal Feasibility Theorem (Combinatorial Form). *Suppose that there is at least one $x \in R^J$ satisfying (1) (i.e., that $b \in \mathscr{B}$). Then one and only one of the following holds*:

1. *There is an x satisfying both (1) and (2).*
2. *There is an elementary dual support Q such that $C(Q) \subset (-\infty, b_Q)$ (or in the case of closed intervals $C(j)$, $c^+(Q) < b_Q$).*

This result can also be stated as follows.

Primal Feasibility Theorem (Vector Form). *Suppose that there is at least one $x \in R^J$ satisfying (1) (i.e., that $b \in \mathscr{B}$). Then one and only one of the following holds*:

1. *There is an x satisfying (1) and (2).*
2. *There is a vector pair u, v, with $v = -uE$ such that $v \cdot x < -u \cdot b$ for all vectors x satisfying (2); in this case v can be taken to be an elementary vector of $\mathscr{D}$, and u can be any vector with $-uE = v$.*

Obviously, when v is an elementry vector of $\mathscr{D}$, we have $v = \lambda e_Q$ for some elementary dual support Q, $\lambda > 0$, and the vector form of Alternative 2 reduces to the combinatorial form. Alternative 2, even in the general vector form with v not necessarily elementary, precludes Alternative 1 because any x satisfying (1) would have to have $v \cdot x = -u \cdot b$. Thus the two alternatives of the theorem are mutually exclusive. The fact that the two cannot *both* fail to hold will follow constructively from the termination argument for the "primal rectification algorithm" in Section 10J. This algorithm produces either an x as in Alternative 1 or a Q as in Alternative 2, depending on the given data, but it only applies to the case of closed intervals $C(j) = [c^-(j), c^+(j)]$. The extension to general nonempty intervals $C(j)$, as needed to complete the proof of the theorem in its full statement, is outlined in Exercise 10.20.

A dual support Q as in Alternative 2 of the theorem corresponds to an "irreducibly infeasible subsystem" of the constraints in the primal feasibility problem; see Exercise 10.21.

Corollary (Case of b = 0).   *One and only one of the following holds:*
1.   *There is a vector $x \in \mathscr{C}$ satisfying (2).*
2.   *There is an elementary dual support Q such that $C(Q) \subset (-\infty, 0)$ (or in the case of closed intervals $C(j)$, $c^+(Q) < 0$); equivalently, there is an [elementary] vector $v \in \mathscr{D}$ such that $v \cdot x < 0$ for every $x \in R^J$ satisfying (2).*

The corollary can be applied to an arbitrary subspace $\mathscr{C}$ of R^J with $\mathscr{D} = \mathscr{C}^\perp$. The particular representation of $\mathscr{C}$ in terms of a matrix E makes no difference. In this manner it is possible to analyze extended feasibility problems with (1) replaced by

$$y(i) \in C(i) \quad \text{for all } i \in I, \qquad y = Ex, \tag{8}$$

where $C(i)$ is a nonempty interval for each i: think of such a problem as one for the subspace $\bar{\mathscr{C}}$ of $R^{I \cup J}$. (Elementary supports $\bar{Q} \subset I \cup J$ for the complementary space $\bar{\mathscr{D}} = \bar{\mathscr{C}}^\perp$ would enter in.) The primal feasibility problem as standardized here corresponds of course to the case of (8) where $C(i) = [b(i), b(i)]$ for all $i \in I$. The chain of reasoning shows that this problem, and for that matter even its special case in the corollary, can actually cover just as wide a range of situations as the seemingly extended problem with (8) in place of (1).

A result that generalizes the feasible differential theorem will be formulated next. For each $j \in J$, let $D(j)$ be a nonempty real interval, not necessarily closed.

Dual Feasibility Problem.   *Find a $v \in R^J$ such that*

$$\exists u \in R^I \quad \text{with} \quad -uE = v, \qquad (i.e., v \in \mathscr{D}), \tag{9}$$

$$v(j) \in D(j) \quad \text{for all } u \in J. \tag{10}$$

For each elementary primal support P, let

$$D(P) = \sum_{j \in P} e_P(j) D(j). \tag{11}$$

Thus $D(P)$ is a nonempty interval consisting of all the values taken on by $e_P \cdot v$ as the variables $v(j)$ range freely over the intervals $D(j)$ (without regard to whether $v \in \mathcal{D}$ or not). If each $D(j)$ happens to be a *closed* interval $[d^-(j), d^+(j)]$, one has $D(P) = [d^-(P), d^+(P)]$, where

$$d^+(P) = \sum_{j \in P^+} e_P(j)d^+(j) + \sum_{j \in P^-} e_P(j)d^-(j),$$

$$d^-(P) = \sum_{j \in P^+} e_P(j)d^-(j) + \sum_{j \in P^-} e_P(j)d^+(j). \qquad (12)$$

Dual Feasibility Theorem.   *One and only one of the following holds*:
1.   *There is a vector v satisfying both (9) and (10).*
2.   *There is an elementary primal support P such that $D(P) \subset (-\infty, 0)$ (or in the case of closed intervals, $d^+(P) < 0$); equivalently, there is an [elementary] $x \in \mathcal{C}$ such that $v \cdot x < 0$ for every $v \in R^J$ satisfying (10).*

The word "elementary" is placed in brackets in the last part of Alternative 2 to indicate that without it the theorem would still be true.

Again we recognize that by appropriate choice of the representing matrix E in (9), we could apply this result to an arbitrary subspace $\mathcal{D}$ of R^J (with $\mathcal{C} := \mathcal{D}^\perp$). In particular, we could use this approach in connection with the subspace $\bar{\mathcal{D}} \subset R^{I \cup J}$ to analyze extended problems in which (9) is replaced by the more restrictive condition

$$\exists u \in R^I \quad \text{with} \quad -uE = v \quad \text{and} \quad u(i) \in D(i) \quad \text{for all} \quad i \in I, \quad (13)$$

where each $D(i)$ is a nonempty interval. (The earlier case corresponds to $D(i) = (-\infty, \infty)$.)

Logically speaking, the dual feasibility theorem is merely a restatement of the corollary to the primal feasibility theorem in notation appropriate to $\mathcal{D}$ rather than $\mathcal{C}$. Thus it would not need a separate proof. Nevertheless, we shall furnish in the next section a direct algorithm for determining which of Alternatives 1 or 2 holds in a given case. This is convenient because of the dual role of the spaces $\mathcal{C}$ and $\mathcal{D}$ in various optimization algorithms in Chapter 11 and our desire to make this duality easy to exploit without a lot of reformulation.

The primal feasibility theorem can be viewed as a specialized kind of *separation theorem* (in the sense of convex analysis): either the (nonempty, convex) sets

$$K_1 = \{ x \in R^J | \, (1) \text{ holds} \},$$

$$K_2 = \{ x \in R^J | \, (2) \text{ holds} \},$$

have a point in common (Alternative 1), or there is a $v \in R^J$ such that

$$v \cdot x_1 < v \cdot x_2 \quad \text{for all} \quad x_1 \in K_1, x_2 \in K_2$$

(an equivalent statement of the general vector form of Alternative 2). A broader form of the same result, where K_1 is any nonempty "partial convex polyhedron" and likewise for K_2, can be easily derived; see Exercise 10.27. The well-known "lemma of Farkas," often referred to in linear programming theory, is also a special case (see Exercise 10.26).

10J. RECTIFICATION ALGORITHMS

The primal and dual feasibility problems in Section 10I in the case of *closed* intervals can be solved by methods parallel to the flow rectification algorithm in Section 3K and the tension rectification algorithm in Section 6I but which rely on the alternatives in the painted index theorem rather than Minty's lemma. If the problems happen to be unsolvable, this is discovered through detection of an elementary support P or Q as in Alternatives 2 of the two feasibility theorems.

The basic conceptual algorithms will be presented in this section and some of their refinements in Sections 10K and 10L.

Primal Rectification Algorithm

The intervals $C(j)$ in the primal feasibility problem are assumed to be closed:

$$C(j) = [c^-(j), c^+(j)] \quad \text{for all} \quad j \in J.$$

A vector $x \in R^J$ is on hand with the property that

$$\sum_{j \in J} e(i, j)x(j) = b(i) \quad \text{for all} \quad i \in I \tag{1}$$

but *not* necessarily

$$c^-(j) \leqq x(j) \leqq c^+(j) \quad \text{for all} \quad j \in J. \tag{2}$$

Define index sets

$$J^+ = \{ j \,|\, x(j) > c^+(j)\}, \quad J^- = \{ j \,|\, x(j) < c^-(j)\}.$$

The indexes in $J^+ \cup J^-$ are said to be *out of kilter*. If $J^+ = \varnothing = J^-$, x does satisfy (2) and is therefore a solution to the primal feasibility problem; terminate. Otherwise paint the indexes $j \in J$ as follows:

$$\begin{aligned}
&\text{green} && \text{if } c^-(j) < x(j) < c^+(j) \\
&\text{white} && \text{if } x(j) \leqq c^-(j) \text{ and } x(j) < c^+(j) \\
&\text{black} && \text{if } x(j) > c^-(j) \text{ and } x(j) \geqq c^+(j) \\
&\text{red} && \text{if } c^-(j) = x(j) = c^+(j)
\end{aligned}$$

Select as l any index in either J^+ or J^- (the indexes in J^+ are necessarily black, whereas those in J^- are necessarily white). Apply the painted index algorithm (or any other procedure for resolving the alternatives of the painted index theorem).

If this produces an elementary dual support Q as in Alternative 2 of the painted index theorem, terminate: Q actually satisfies $c^+(Q) < b_Q$ as in Alternative 2 of the primal feasibility theorem, and the primal feasibility problem is therefore unsolvable.

If it produces instead an elementary primal support P as in Alternative 1 of the painted index theorem, calculate

$$\alpha = \min \begin{cases} [c^+(j) - x(j)]/e_P(j) & \text{for } j \in P^+, \\ [c^-(j) - x(j)]/e_P(j) & \text{for } j \in P^-, \\ \bar{\alpha}, \end{cases} \tag{3}$$

where

$$\bar{\alpha} = \begin{cases} [c^-(l) - x(l)]/e_P(l) & \text{if } l \in P^+ \text{ (i.e., } l \in J^-), \\ [c^+(l) - x(l)]/e_P(l) & \text{if } l \in P^- \text{ (i.e., } l \in J^+). \end{cases} \tag{4}$$

Then $0 < \alpha < \infty$. Let $x' = x + \alpha e_P$; this will be closer to feasibility than x, in the sense that (1) also holds for x', and in addition

$$\text{dist}(x'(j), C(j)) \leq \text{dist}(x(j), C(j)) \quad \text{for all } j \in J,$$

$$\text{dist}(x'(l), C(l)) < \text{dist}(x(j), C(l)). \tag{5}$$

Return to the beginning with x' in place of x (and a correspondingly modified painting and possibly smaller sets J^+ and J^-).

Justification

If the painted index subroutine produces an elementary dual support Q with every $j \in Q^+$ red or black, and every $j \in Q^-$ red or white, we have by the choice of the painting that

$$x(j) \geq c^+(j) \quad \text{for all} \quad j \in Q^+ \quad \text{and} \quad x(j) \leq c^-(j) \quad \text{for all} \quad j \in Q^-.$$

Moreover at least one of these inequalities is strict, namely, for $j = l$, because $l \in J^+ \cup J^-$. Therefore

$$e_Q \cdot x > \sum_{j \in Q^+} e_Q(j)c^+(j) + \sum_{j \in Q^-} e_Q(j)c^-(j),$$

or in other words, $b_Q > c^+(Q)$. Thus Alternative 2 of the primal feasibility theorem holds, and as seen in the preceding section, this precludes the existence of any solution to the primal feasibility problem.

If the painted index subroutine produces an elementary primal support P with every $j \in P^+$ green or white, and every $j \in P^-$ green or black, we have

$$x(j) < c^+(j) \quad \text{for all} \quad j \in P^+ \quad \text{and} \quad x(j) > c^-(j) \quad \text{for all} \quad j \in P^-.$$

Also l is one of the white or black indexes in P, so

$$\text{either} \quad l \in P^+ \quad \text{with} \quad x(l) < c^-(l), \quad \text{or} \quad l \in P^+ \quad \text{with} \quad x(l) > c^+(l).$$

The numbers $\bar{\alpha}$ and α are therefore finite and positive as claimed. From the definition of α we obtain $x(j) < x'(j) \leq c^+(j)$ for all $j \in P^+$, $x(j) > x'(j) \geq c^-(j)$ for all $j \in P^-$, $x(j) = x'(j)$ for all $j \notin P$, and furthermore

$$\text{either} \quad x(l) < x'(l) \leq c^-(l) \quad \text{or} \quad x(l) > x'(l) \geq c^+(l), \tag{6}$$

depending on whether $l \in P^+$ or $l \in P^-$. The distance relations (5) are therefore correct.

The purpose of the $\bar{\alpha}$ term in the formula for α is of course to prevent "overshooting": if $\alpha = \bar{\alpha}$, l will be removed from $J^+ \cup J^-$.

Termination

The primal rectification algorithm contains the flow rectification algorithm of Section 3K as a special case, so it is clear that without some extra precaution a nonterminating sequence of iterations might be possible. On the other hand, we do not in general have anything like the integrality properties of flows at our disposal, so a termination argument based on commensurability is not available.

Properties of the painted index algorithm, when used as the subroutine in a pivoting implementation of the primal rectification algorithm, come to our rescue. That algorithm requires precautions of its own to ensure termination. We have seen in Section 10H that Bland's priority rule suffices, and that "red-green preprocessing" enhances the algorithm further by quickly narrowing the focus of activity down to the subtableau labeled by the row and column indexes that are white or black. In Bland's rule the indexes in J are numbered arbitrarily, once and for all, and indexes with lower numbers are said to have higher "priority." In the selection of a pivot row index j or pivot column index $\bar{k}$, preference is given to the index with highest priority when more than one index would fit the conditions. Red-green preprocessing transforms the Tucker tableau before such selection (based on compatibility tests) even takes place, simply by pivoting red indexes into column position and green indexes into row position until this is no longer possible without just trading red for red or green for green. Thereafter rows and columns labeled by red or green indexes play no role in the pivot selection process (see Section 10H).

It turns out that a minor sharpening of these supplementary rules for the painted index subroutine will keep the primal rectification algorithm from floundering. We shall say that the painted index subroutine is executed *with OK priorities* if:

1. Bland's rule is used with out-of-kilter ("OK") indexes always regarded as having higher priority than in-kilter indexes.
2. The out-of-kilter index of highest priority is the one chosen to be the lever index l (see remark later in the section).

This can be accomplished as follows with a fixed numbering of all the indexes $j \in J$. When checking a row or column for an incompatible entry, first look at the entries corresponding to out-of-kilter indexes, starting with the lowest label and working upward, and then look at the entries corresponding to in-kilter entries, also starting with the lowest label and working upward. (Here red and green indexes can be skipped over, in the aftermath of red-green preprocessing.) Let the first incompatible entry detected in this way determine the index selected as $\bar{j}$ or $\bar{k}$, whichever the case may be.

We shall prove that *if the primal rectification algorithm is implemented with the painted index algorithm as subroutine (the Tucker tableau left over from one round of the latter being used to initiate the next), and if this subroutine is executed with red-green preprocessing and OK priorities as just defined, then termination of the primal rectification algorithm after finitely many pivot steps is assured.* The proof will be by contradiction. We suppose the procedure in the form described leads in some situation to an infinite sequence of pivot steps, and we show that this yields an absurd conclusion.

Let us speak of a "configuration" as a Tucker tableau (determined by a choice of basis F) together with some painting of the indexes $j \in J$ and a designation of which of these are out of kilter. There are only finitely many possible configurations, so we may suppose, by deleting an initial portion of our sequence if necessary, that the algorithm generates a cycle of configurations which repeats over and over. Since out-of-kilter indexes can become in-kilter but not vice versa, and since our second priority rule causes us to choose the same index as l as long as it stays out of kilter, we can suppose that throughout this cycle the same set of indexes is always out of kilter, and the same index is kept as l.

We do know that the painted index subroutine takes only a finite number of iterations on each round (red-green preprocessing included), inasmuch as a version of Bland's priority rule is being followed. It cannot finish with a dual support Q, for that would terminate the primal rectification algorithm, so it must finish each time with a compatible column yielding a primal support P that leads to an improvement in x, specifically a change in the value of $x(l)$.

This change is always upward if l is white and always downward if l is black. Thus $x(l)$ must change at least once each time we go through our repeating cycle, and the change is always in the same direction.

We argue now that in the course of this cycle, whenever the painted index subroutine (after red-green preprocessing) chooses a pivot column $\bar{k}$, the index $\bar{k}$ must be out of kilter. Indeed, $\bar{k}$ is obtained by inspecting a certain row; it is the column index of highest priority corresponding to an entry that fails the compatibility test for that row. The row in question is the l row if l happens to be in row position as in Figure 10.16a, whereas if l is a column index, it is a certain $\bar{j}$ row as in Figure 10.16b, where $\bar{j}$ itself was chosen because it corresponded to an entry that failed the compatibility test for the l column (and must therefore have been black or white, since red-green preprocessing ensures that entries in rows or columns having red or green indexes cannot fail any compatibility test). Either way we are speaking of a row that uses l but no green indexes. If $\bar{k}$ were in kilter, that would mean by our first priority rule that the row contained no incompatible entries corresponding to out-of-kilter column indexes. The row would then yield an elementary dual support Q containing l which meets all requirements for compatibility with respect to the painting at hand, except possibly for certain of its indexes that are in kilter and not green. The relation $e_Q \cdot x = b_Q$ corresponding to this Q (see 10I(5)) could then be written as

$$\sum_{\substack{j \in Q^+ \\ \text{white} \\ \text{out of kilter}}} e_Q(j)x(j) + \sum_{\substack{j \in Q^- \\ \text{black} \\ \text{out of kilter}}} e_Q(j)x(j) = b_Q - \sum_{\substack{j \in Q \\ \text{nongreen} \\ \text{in kilter}}} e_Q(j)x(j),$$

(7)

where the index l appears in one of the sums on the left. The left side of this equation expresses a quantity that can only decrease and in fact must have a lower value each time we go through our endlessly repeating cycle and return to the same configuration with an improved x. The right side, on the other hand, only has finitely many possible values, because

$$x(j) = c^+(j) \quad \text{or} \quad x(j) = c^-(j) \quad \text{when } j \text{ is in kilter, not green.} \quad (8)$$

These two properties of (7) are inconsistent, and this validates our claim about $\bar{k}$ always being an out-of-kilter index. In particular, any column yielding a P that improves x is labeled by such a $\bar{k}$.

We observe next that in the course of the cycle no red-green preprocessing will actually come into play; there will never be a pivot $(\bar{j}, \bar{k})$ involving a red or green index. In fact, since red indexes can only be moved from row position to column position and green indexes in the opposite direction, such a pivot is impossible unless the changes in x which modify the painting are able to turn some row index red or column index green. But the changes in x occur only for indexes used by a certain column yielding a compatible primal support P. As we have just seen, the index of this column must be some out-of-kilter $\bar{k}$ which, since it is not brought into kilter by the improvement in x, does not change color, nor can any of the out-of-kilter row indexes used by P change color. Thus no column index can turn green, and the only row indexes that might conceivably turn red are in-kilter row indexes used by P. But the nature of the painting and the improvement process are such that no change in x can ever make an in-kilter index turn red.

We claim further that *after any improvement in x, the very next pivoting step in our cycle must use a row whose index $\bar{j}$ is in kilter*. We have just seen that only in-kilter row indexes used by the P column used to make the improvement can have switched colors and affected the column's compatibility. As a matter of fact the compatibility of the column must have been destroyed in this way, for otherwise an immediate further improvement would be available using the same P, contrary to the choice of the step-size α in (3) as the largest possible; this is discussed in more detail in Section 10K. (Any index j yielding the minimum in (3) will change color and have this effect. Note that $\alpha < \bar{\alpha}$, because l is not brought in kilter by the improvement.) Thus if the column is the next one to be inspected for compatibility, in the process of determining a pivot row, it will fail the test but only because of some entry corresponding to an in-kilter row index which will then be chosen as $\bar{j}$. To see that the former P column truly is the next to be inspected, we need to look at Figure 10.16 again. If l happens to be a column index, the painted index subroutine must have finished by producing P from column l, so certainly we do reinspect this column immediately. If l is a row index (perpetually out of kilter), nothing has changed in the l row, so that when we start up the painted index algorithm again we choose the same column, namely, the one we had terminated with the last time because it was compatible which is the column that yielded P.

The argument is nearing its end. We have seen that in the supposed cycle of configurations that repeats itself over and over, the pivots (j, k) can use only black or white indexes; $\bar{k}$ must always be out of kilter, but at least once in the cycle $\bar{j}$ must be in kilter. But this implies that each time we go through the cycle the number of out-of-kilter indexes in column position is increased by at least one, an irreversible process that shows we *cannot* keep returning to the same configurations. The contradiction completes the proof that the primal rectification algorithm must eventually terminate if it is implemented in the manner described.

Remark

Part 2 of the OK priority rule is more restrictive than necessary, as the preceding proof indicates. All one really needs is the assurance that once an out-of-kilter index is chosen as l, it is kept as l until brought in kilter.

Note that the OK priority rule becomes superfluous when there is only one index out of kilter. That is indeed the case in the aggregated version of the algorithm which will be presented in Section 10L.

We turn now to the corresponding procedure for the dual flexibility problem.

Dual Rectification Algorithm

The intervals $D(j)$ in the dual feasibility problem are assumed to be closed:

$$D(j) = [d^-(j), d^+(j)] \quad \text{for all } j \in J.$$

A vector $v \in R^J$ is on hand which for some $u \in R^I$ satisfies

$$v(j) = -\sum_{i \in I} u(i)e(i, j) \quad \text{for all } j \in J \tag{9}$$

but *not* necessarily

$$d^-(j) \leqq v(j) \leqq d^+(j) \quad \text{for all } j \in J. \tag{10}$$

Define the index sets

$$J^+ = \{ j | v(j) > d^+(j)\}, \qquad J^- = \{ j | v(j) < d^-(j)\}.$$

The indexes in $J^+ \cup J^-$ are said to be *out of kilter*. (All others are *in kilter*.) If $J^+ = \emptyset = J^-$, v does satisfy (8) and is therefore a solution to the dual feasibility problem; terminate. Otherwise paint the indexes $j \in J$ as follows:

$$
\begin{array}{ll}
\text{red} & \text{if } d^-(j) < v(j) < d^+(j) \\
\text{black} & \text{if } v(j) \leqq d^-(j) \text{ and } v(j) < d^+(j) \\
\text{white} & \text{if } v(j) > d^-(j) \text{ and } v(j) \geqq d^+(j) \\
\text{green} & \text{if } d^-(j) = v(j) = d^+(j)
\end{array}
$$

Select as l any index in either J^+ or J^- (the indexes in J^+ are necessarily white, whereas those in J^- are necessarily black). Apply the painted index algorithm (or any other procedure for resolving the alternatives of the painted index theorem).

If this produces an elementary primal support P as in Alternative 1 of the painted index theorem, terminate: P actually satisfies $d^+(P) < 0$ as in Alternative 2 of the dual feasibility theorem, and the dual feasibility problem is therefore unsolvable.

If it produces instead an elementary dual support Q as in Alternative 2 of the painted index theorem, calculate

$$\alpha = \min \begin{cases} [d^+(j) - v(j)]/e_Q(j) & \text{for } j \in Q^+, \\ [d^-(j) - v(j)]/e_Q(j) & \text{for } j \in Q^-, \\ \bar{\alpha}, \end{cases} \tag{11}$$

where

$$\bar{\alpha} = \begin{cases} [d^-(l) - v(l)]/e_Q(l) & \text{if } l \in Q^+ \text{ (i.e., } l \in J^-), \\ [d^+(l) - v(l)]/e_Q(l) & \text{if } l \in Q^- \text{ (i.e., } l \in J^+). \end{cases} \tag{12}$$

Then $0 < \alpha < \infty$. Let $v' = v + \alpha e_Q$. Then v' again satisfies (9) for some $u' \in R^I$ (i.e., $u' = u + \alpha \bar{u}$ for any $\bar{u}$ such that $-\bar{u}E = e_Q$), and v' is closer to feasibility than v in the sense that

$$\text{dist}(v'(j), D(j)) \leqq \text{dist}(v(j), D(j)) \quad \text{for all } j \in J,$$

$$\text{dist}(v'(l), D(l)) < \text{dist}(v(l), D(l)).$$

Return to the beginning with v' in place of v (and a correspondingly modified painting and possibly smaller sets J^+ and J^-).

Justification and Termination Property

The arguments for this procedure are closely parallel to the ones for the preceding algorithm and are left to the reader. In particular: *if the dual rectification algorithm is implemented with the painted index algorithm as subroutine (the Tucker tableau left over from one round of the latter being used to initiate the next), and if this subroutine is executed with red-green preprocessing and OK priorities as defined in connection with the primal algorithm, then termination is assured.*

10K.* SHORTCUTS IN PIVOTING IMPLEMENTATION

The rectification algorithms in the preceding section can sometimes be implemented directly in terms of underlying combinatorial structure. This is true in the network setting in Sections 3K and 6I as well as in networks with gains, for example. The statement of the algorithms is designed to emphasize such a possibility. In general, however, a pivoting implementation in terms of the painted index algorithm may be necessary. We now explore some of the details in how this may be carried out.

Improvement Phase of the Primal Algorithm

Each time the primal rectification algorithm replaces x by an improved vector $x' = x + \alpha e_P$, it is necessary to reapply the painted index subroutine to a possibly different painting and choice of l from a possibly smaller set $J^+ \cup J^-$. Clearly the Tucker tableau left over from the previous application of the painted index subroutine serves for the restart. If one insists on red-green preprocessing as in Section 10H, whose value has been demonstrated in the termination arguments in Section 10J, certain pivot steps will automatically be taken in this tableau before the search begins anew for a compatible column leading to a further improvement. Let us look more closely at this phase of the calculations.

To begin with, a shortcut is available in passing from x to x'. The formula $x' = x - \alpha e_P$ with α given by 10J(3) suggests that the elementary vector e_P must be determined in full, by applying the basis theorem in Section 10C to a certain compatible column of the tableau, but this is not actually necessary.

The situation is indicated in Figure 10.18a. The painted index subroutine has terminated in a compatible column with index $\bar{k}$ which (if red-green preprocessing has been used) must be either white or black; the lever index l is either $\bar{k}$ or one of the white or black indexes $j \in F$. Corresponding to this column is an elementary vector e_{P_k} described in Part 1 of the basis theorem, and the desired e_P is either this vector or its negative, depending on whether $\bar{k}$ is white or black.

(a)

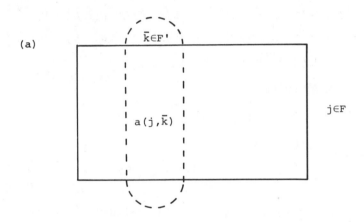

(b)

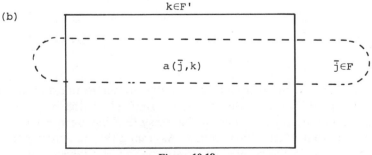

Figure 10.18

The components of e_P, as calculated in this manner, involve a normalization factor $\lambda(\bar{k})$ which has a reciprocal scaling effect on the value of α in 10J(3). In forming $x' = x + \alpha e_P$, this normalization really makes no difference. The net rule for updating x thus works out simply to the following: *if $\bar{k}$ is white*, one has

$$x'(j) = x(j) - \tau a(j, k) \quad \text{for } j \in F,$$

$$x'(\bar{k}) = x(\bar{k}) + \tau,$$

$$x'(k) = x(k) \quad \text{for } k \in F', k \neq \bar{k}, \tag{1}$$

where

$$\tau = \min \begin{cases} [c^+(j) - x(j)]/a(j, \bar{k}) & \text{for } j \in F \text{ with } a(j, \bar{k}) > 0, \\ [c^-(j) - x(j)]/a(j, \bar{k}) & \text{for } j \in F \text{ with } a(j, \bar{k}) < 0, \\ c^+(\bar{k}) - x(\bar{k}), \\ \bar{\tau}, \end{cases} \tag{2}$$

$$\bar{\tau} = \begin{cases} [c^-(l) - x(l)]/a(l, \bar{k}) & \text{if } l \in F \text{ with } a(l, \bar{k}) > 0, \\ [c^+(l) - x(l)]/a(l, \bar{k}) & \text{if } l \in F \text{ with } a(l, \bar{k}) < 0, \\ c^-(l) - x(l) & \text{if } l = \bar{k}, \end{cases} \tag{3}$$

whereas *if $\bar{k}$ is black*, one has

$$x'(j) = x(j) - \tau a(j, \bar{k}) \quad \text{for } j \in F,$$

$$x'(\bar{k}) = x(\bar{k}) - \tau,$$

$$x'(k) = x(k) \quad \text{for } k \in F', k \neq \bar{k} \tag{1'}$$

where

$$\tau = \min \begin{cases} [x(j) - c^-(j)]/a(j, \bar{k}) & \text{for } j \in F \text{ with } a(j, \bar{k}) > 0, \\ [x(j) - c^+(j)]/a(j, \bar{k}) & \text{for } j \in F \text{ with } a(j, \bar{k}) < 0, \\ x(\bar{k}) - c^-(\bar{k}), \\ \bar{\tau} \end{cases} \tag{2'}$$

$$\bar{\tau} = \begin{cases} [x(l) - c^+(l)]/a(l, \bar{k}) & \text{if } l \in F \text{ with } a(l, \bar{k}) > 0, \\ [x(l) - c^-(l)]/a(l, \bar{k}) & \text{if } l \in F \text{ with } a(l, \bar{k}) < 0, \\ x(l) - c^+(l) & \text{if } l = \bar{k}. \end{cases} \tag{3'}$$

What happens to the painting used in the primal rectification algorithm, when x is replaced by x' according to these formulas? Obviously, colors remain the same except possibly for $\bar{k}$ and the indexes $j \in F$ having $a(j, \bar{k}) \neq 0$, since it is only there that x' can differ from x. Assuming that red-green preprocessing was employed in the painted index subroutine, the tableau must exhibit, before color modification, the structure shown in Figure 10.17 (aside from permutations of rows and columns), but it is possible that $\bar{k}$ may now switch to green, or some of the indexes $j \in F$ with $a(j, \bar{k}) \neq 0$ may switch to red. If so, red-green preprocessing in the next round of the painted index subroutine can come into play.

Specifically, if $\bar{k}$ switches to green it will be necessary to pivot on $(\bar{j}, \bar{k})$ for some $\bar{j} \in F$ with $a(\bar{j}, \bar{k}) \neq 0$ such that the color of $\bar{j}$, after modifications, is not green. That such a $\bar{j}$ exists is easy to see. In fact $\bar{j}$ *can be taken to be any index yielding the minimum in the formula* (2) (or (2')) *for* τ. (Count the case where $\tau = \bar{\tau}$ as one in which the index l gives the minimum.) Such an index $\bar{j}$ will have either $x'(\bar{j}) = c^+(\bar{j})$ or $x'(\bar{j}) = c^-(\bar{j})$, so its color, after modifications, will not be green.

It is less likely that an index $j \in F$ with $a(j, \bar{k}) \neq 0$ may switch to red, but this happens if j is one of the indexes in $J^+ \cup J^-$ and at same time a $\bar{j}$ yielding the minimum in the formula for τ, as just described. Unless there is more than one such index, it is apparent therefore that red-green preprocessing in the restart of the painted index subroutine will not require more than a single $(\bar{j}, \bar{k})$ pivot transformation.

Improvement Phase of the Dual Algorithm

These considerations all carry over to the dual rectification algorithm. On termination of the painted index subroutine, one has a compatible row corresponding to an index $\bar{j}$ which is either white or black (at least if red-green preprocessing has been used). The situation is indicated in Figure 10.18b. The shortcut rule for calculating $v' = v + \alpha e_Q$ turns out to be the following: *if $\bar{j}$ is white*, one has

$$v'(k) = v(k) - \tau a(\bar{j}, k) \quad \text{for } k \in F'$$

$$v'(\bar{j}) = v(\bar{j}) + \tau$$

$$v'(j) = v(j) \quad \text{for } j \in F, j \neq \bar{j}, \tag{4}$$

where

$$\tau = \min \begin{cases} [v(k) - d^+(k)]/a(\bar{j}, k) & \text{for } k \in F' \text{ with } a(\bar{j}, k) < 0, \\ [v(k) - d^-(k)]/a(\bar{j}, k) & \text{for } k \in F' \text{ with } a(\bar{j}, k) > 0, \\ d^+(\bar{j}) - v(\bar{j}), \\ \bar{\tau}, \end{cases} \tag{5}$$

$$\bar{\tau} = \begin{cases} [v(l) - d^-(l)]/a(\bar{j}, l) & \text{if } l \in F' \text{ with } a(\bar{j}, l) < 0, \\ [v(l) - d^+(l)]/a(\bar{j}, l) & \text{if } l \in F' \text{ with } a(\bar{j}, l) > 0, \\ d^-(l) - v(l) & \text{if } l = \bar{j}, \end{cases} \tag{6}$$

whereas *if $\bar{j}$ is black*, one has

$$v'(k) = v(k) + \tau a(\bar{j}, k) \quad \text{for } k \in F',$$

$$v'(\bar{j}) = v(\bar{j}) - \tau,$$

$$v'(j) = v(j) \quad \text{for } j \in F, j \neq \bar{j}, \tag{4'}$$

where

$$\tau = \min \begin{cases} [d^-(k) - v(k)]/a(\bar{j}, k) & \text{for } k \in F' \text{ with } a(\bar{j}, k) < 0 \\ [d^+(k) - v(k)]/a(\bar{j}, k) & \text{for } k \in F' \text{ with } a(\bar{j}, k) > 0, \\ v(\bar{j}) - d^-(\bar{j}), \\ \bar{\tau}, \end{cases} \tag{5'}$$

$$\bar{\tau} = \begin{cases} [d^+(l) - v(l)]/a(\bar{j}, l) & \text{if } l \in F' \text{ with } a(\bar{j}, l) < 0, \\ [d^-(l) - v(l)]/a(\bar{j}, l) & \text{if } l \in F' \text{ with } a(\bar{j}, l) > 0, \\ v(l) - d^+(l) & \text{if } l = \bar{j}. \end{cases} \tag{6'}$$

The painting associated with v' can differ from the one associated with v only in $\bar{j}$ and the indices $k \in F'$ having $a(\bar{j}, k) \neq 0$. It is possible that the color of $\bar{j}$ may switch to red, and in that event red-green processing at the restart of the painted index algorithm will involve pivoting on $(\bar{j}, \bar{k})$, where $\bar{k}$ *can be selected as any index yielding the minimum in formula* (5) (or (5')) *for* τ. It is also possible that an index $k \in F'$ with $a(\bar{j}, k) \neq 0$ may switch to red, but this happens only if k is an index in $J^+ \cup J^-$ that is also a $\bar{k}$ yielding the minimum in the formula for τ. Unless there is more than such index, the red-green preprocessing can be accomplished with at most one $(\bar{j}, \bar{k})$ pivot transformation.

Lever Position

In executing the painted index subroutine with red-green preprocessing, one has the choice of insisting that the lever index l be either in row position or column position (the primal and dual modes discussed at the end of Section 10H). Although both rectification algorithms will work whichever position l is in, a rule of thumb that appears to facilitate performance is to place l in row position in the primal rectification algorithm and in column position in the dual rectification algorithm.

10L.*   AUGMENTED AND AGGREGATED FORMATS

In applications of the primal rectification algorithm which will be made in Chapter 11, namely, as a subroutine in general procedures of optimization, the typical outcome is not the construction of an x solving the primal feasibility problem but the detection of a dual elementary support Q such that $c^+(Q) < b_Q$. The vector $v = e_Q$ may then be used for a "direction of descent," for instance. Often in these cases it is not enough just to have v: one also needs a u with $-uE = v$. On the other hand, it may be possible to get by with a v that is not necessarily an elementary vector e_Q but merely satisfies the weaker vectorial condition in Alternative 2 of the primal feasibility theorem in Section 10I. When the intervals $C(j)$ are closed, the latter condition on v takes the form: there exists u such that $-uE = v$ and

$$\sum_{i \in I} u(i)b(i) + \sum_{j: v(j)>0} v(j)c^+(j) + \sum_{j: v(j)<0} v(j)c^-(j) < 0. \quad (1)$$

This section is devoted to a discussion of modes of implementation that may be advantageous in this setting.

Augmented Format

Let us recall first that the primal feasibility problem with closed intervals $[c^-(j), c^+(j)]$ and given constants $b(i)$ can always be translated into a similar problem in the variables $x(j)$ and $y(j)$ with corresponding constants all zero. Thus we look instead for a pair (y, x) in the subspace

$$\mathscr{C} = \{(y, x) \in R^I \times R^J | y = Ex\}$$

such that

$$x(j) \in [c^-(j), c^+(j)] \text{ for } j \in J \text{ and } y(i) \in [b(i), b(i)] \text{ for } i \in I. \quad (2)$$

Applying the corollary of the primal feasibility theorem in Section 10I, we observe that such a pair (y, x) fails to exist if and only if there is a pair (u, v) in the complementary subspace

$$\mathscr{D} = \{(u, v) \in R^I \times R^J | v = -uE\} = \mathscr{C}^\perp$$

such that (1) holds. Indeed, (1) is equivalent to the statement that $v \cdot x + u \cdot y < 0$ for all (y, x) satisfying (2). Of course the corollary in question asserts that (u, v) can actually be taken to be an *elementary* vector of the subspace $\overline{\mathscr{D}}$. Note, however, that the pairs (u, v) which are elementary for $\overline{\mathscr{D}}$ do not correspond in any simple way to the ones such that v is elementary for $\mathscr{D}$, as in the primal feasibility theorem itself.

The feasibility problem in this reformulation is said to be in the *augmented format*. It can be solved by the primal rectification algorithm with E itself taken as the coefficient matrix for the initial Tucker representation of $\mathscr{C}$ and $\overline{\mathscr{D}}$; see Figure 10.1. Thus a Tucker representation of $\mathscr{C}$ and $\mathscr{D}$ does not have to be determined first, nor is it necessary to have an x such that $Ex = b$. Arbitrary

$x(j)$ values can be chosen at the start and combined with the $y(i)$ values given by $y = Ex$. The painting of the indexes $j \in J$, already specified in the statement of the primal rectification algorithm in Section 10J, then automatically undergoes a natural extension to the indices $i \in I$: such an i is:

$$
\begin{array}{ll}
\text{white} & \text{if } y(i) < b(i) \\
\text{black} & \text{if } y(i) > b(i) \\
\text{red} & \text{if } y(i) = b(i)
\end{array}
$$

The algorithm proceeds just as before, but in terms of the index set $I \cup J$. It terminates either with a pair $(y, x) \in \mathscr{C}$ satisfying (2) or an *elementary* pair $(u, v) \in \overline{\mathscr{D}}$ satisfying (1). Again finite termination is assured if red-green preprocessing is added to the painted index subroutine (and the latter is executed with the OK priority rule, say). Although an elementary vector (u, v) of $\overline{\mathscr{D}}$ merely yields a $v \in \mathscr{D}$, not necessarily an *elementary* vector of $\mathscr{D}$, this v could be forced, if necessary, to yield an elementary vector of $\mathscr{D}$ by a process of conformal realization (see Exercise 10.29).

In one important case when the algorithm in this format terminates with $(u, v) \in \overline{\mathscr{D}}$ satisfying (1), v is indeed an elementary vector of $\mathscr{D}$ (so that $v = \lambda e_Q$ for some $\lambda > 0$ and elementary dual support Q satisfying $c^+(Q) < b_Q$). *This is true, namely, when the terminal vector x does at least have $Ex = b$* (i.e., $y(i) = b(i)$ for every $i \in I$), *hence in particular when the algorithm is initiated with such an x* (since the distance between $y(i)$ and $b(i)$, like the distance between $x(j)$ and $[c^-(j), c^+(j)]$ never increases in the course of the calculations). In asserting this, we assume that red-green preprocessing has been followed in applying the painted index subroutine. Then, since every $i \in I$ is red, the tableau on termination must have the general structure shown in Figure 10.19, except for permutations of rows and columns (see Section 10H). As many indexes i as possible will have been brought into column position, so that each index $h \in I$ still in row position, if any, can have nonzero entries

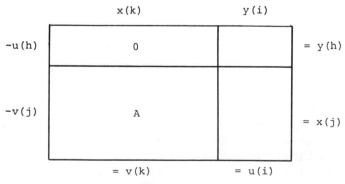

Figure 10.19

only in columns with indexes $i \in I$. The portion of the tableau designated in Figure 10.19 by the matrix A then gives a Tucker representation of the spaces $\mathscr{C}$ and $\mathscr{D}$ associated with some basis $F \subset J$. The pair (u, v), having been obtained from a row of the extended tableau with index $j \in F$, has v corresponding similarly to a row of the subtableau given by A. Therefore v is actually an elementary vector of $\mathscr{D}$.

Aggregated Infeasibility

Another interesting case to record is the one where the primal rectification algorithm in augmented format as just described is initiated with an x that satisfies $c^-(j) \le x(j) \le c^+(j)$ for all $j \in J$. Certainly this is always easy to do: choose arbitrary values $x(j)$ in the intervals $[c^-(j), c^+(j)]$, and let the corresponding values $y(i)$ (determined from $y = Ex$) fall where they may. Then the only indexes that the algorithm can ever single out as levers l for the process of adjustment are indexes $i \in I$ with $y(i) \ne b(i)$.

A simple trick in this situation serves to aggregate all the infeasibility in a single "artificial" index l corresponding to a new row of the augmented tableau, as indicated in Figure 10.20: select any weighting vector $w \in R^I$ that, for the initial $y = Ex$, satisfies

$$w(i) > 0 \quad \text{for all } i \in I \text{ with } y(i) > b(i),$$

$$w(i) < 0 \quad \text{for all } i \in I \text{ with } y(i) < b(i), \tag{3}$$

and take

$$e(l, j) = \sum_{i \in I} w(i)e(i, j) \quad \text{for all } j \in J, \tag{4}$$

$$b(l) = \sum_{i \in I} w(i)b(i). \tag{5}$$

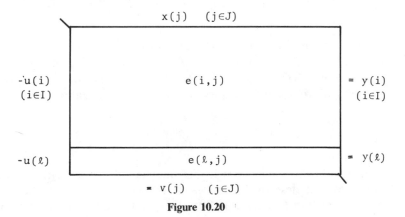

Figure 10.20

The variable $y(l)$ associated with this new row obviously is given by

$$y(l) = \sum_{i \in I} w(i) y(i),$$

so that $y(l) > b(l)$ unless $y(i) = b(i)$ for every $i \in I$. Thus we need only concentrate on bringing the value of $y(l)$ down to $b(l)$. All the $y(i)$ values will come into line accordingly.

The primal rectification algorithm can therefore be applied to the tableau in Figure 10.20 *with the artificial index l as the lever in every iteration*. Improvements in y will not upset the validity of (3), so the weights $w(i)$ can be kept fixed. The procedure will terminate either in values of the x, y system in Figure 10.20 which satisfy (2) and $y(l) \in [b(l), b(l)]$, thus yielding a solution x to the primal feasibility problem, or it will terminate in values of the u, v system in Figure 10.20 which satisfy an extended form of (1):

$$u(l)b(l) + \sum_{i \in I} u(i)b(i) + \sum_{j \,:\, v(j)>0} v(j)c^+(j) + \sum_{j \,:\, v(j)<0} v(j)c^-(j) < 0,$$

$$(6)$$

where

$$v(j) = - \sum_{i \in I} u(i)e(i, j) - u(l)e(l, j). \qquad (7)$$

In the latter case we observe from (4) and (5) that by setting

$$\tilde{u}(i) = u(i) + u(l)w(i) \quad \text{for all } i \in I \qquad (8)$$

we have $v = -\tilde{u}E$ and

$$\sum_{i \in I} \tilde{u}(i)b(i) + \sum_{j \,:\, v(j)>0} v(j)c^+(j) + \sum_{j \,:\, v(j)<0} v(j)c^-(j) < 0. \qquad (9)$$

Thus v is a vector in $\mathscr{D}$ which satisfies (1) with respect to $\tilde{u}$ rather than u. As mentioned earlier, such a v often serves just as well as an elementary vector e_Q of $\mathscr{D}$ that satisfies $c^+(Q) < b_Q$, and anyway, an e_Q with the latter property could be derived from v if necessary by conformal realization (see Exercise 10.29).

Remarks

The primal rectification algorithm in the augmented format with aggregated infeasibility, as just described, reduces in the network case to the feasible distribution algorithm (Exercise 10.30). This underscores the interest in the procedure and also illuminates the difficulties connected with the fact that it produces a $v \in \mathcal{D}$ that satisfies (1) but is not necessarily an *elementary* vector of $\mathcal{D}$. Indeed, the feasible distribution algorithm, when it terminates with a cut Q, does not necessarily have Q elementary.

10M.* EXTREME SOLUTIONS

If in the primal feasibility problem in Section 10I the constraint intervals are all closed,

$$C(j) = [c^-(j), c^+(j)] \quad \text{for all } j \in J,$$

it is possible to distinguish certain solutions of a special nature and use them to represent all other solutions. A solution x to the problem is said to be *extreme* if the index set

$$F_x = \{ j \in J | c^-(j) < x(j) < c^+(j) \}$$

is independent in the sense of Section 10C, or in other words, if there exists a basis F such that $F_x \subset F$, that is,

$$\text{for all } j \notin F, \text{ either } \quad x(j) = c^+(j) \quad \text{or} \quad x(j) = c^-(j). \tag{1}$$

This property of F_x can easily be tested by pivoting (Exercise 10.31).

Extreme solutions can be identified geometrically with the vertexes of the convex polyhedron in R^J consisting of all solutions to the problem; see Exercise 10.34.

To state the representation theorem, we need some additional terminology. An elementary primal support P will be said to be of *unlimited feasibility* if $c^+(j) = +\infty$ for all $j \in P^+$ and $c^-(j) = -\infty$ for all $j \in P^-$, and of *doubly unlimited feasibility* if both P and its reverse are of unlimited feasibility, that is, if $C(j) = (-\infty, \infty)$ for every $j \in P$. The first condition implies that for any solution x to the primal feasibility problem, the vector $x + t e_P$ will also be a solution, for arbitrary $t \geq 0$. The second condition says the same thing for arbitrary $t \in R$.

An elementary primal support P of doubly unlimited feasibility exists if and only if the index set

$$\{ j \in J | c^+(j) = +\infty \text{ and } c^-(j) = -\infty \}$$

fails to be independent. Then no set of form F_x can be independent either, so there could be no extreme solutions at all.

Extremal Representation Theorem (Primal Case). *Assume that the primal feasibility problem with closed intervals $C(j)$ has at least one solution, but there are no elementary primal supports of doubly unlimited feasibility. Then there are* *only finitely many extreme solutions. Moreover a vector x solves the problem if and only if it can be represented as*

$$x = \sum_{k=1}^{r} \lambda_k x_k + \sum_{l=1}^{q} \mu_l e_{P_l}, \tag{2}$$

where each x_k is an extreme solution, $\lambda_k \geq 0$, $\sum_{k=1}^{n} \lambda_k = 1$, each P_l is an elementary primal support of (singly) unlimited feasibility, and $\mu_l \geq 0$.

Proof. If x can be represented as in (2), we have $Ex_k = b$ for $k = 1, \ldots, r$ and $Ee_{P_l} = 0$ for $l = 1, \ldots, q$ (since $e_{P_l} \in \mathcal{C}$), so

$$Ex = \sum_{k=1}^{r} \lambda_k Ex_k + 0 = \left(\sum_{k=1}^{r} \lambda_k \right) b = b.$$

Furthermore we have $c^-(j) \leq x_k(j) \leq c^+(j)$ for all $j \in J$, so that

$$c^-(j) \leq \min_{k=1,\ldots,r} x_k(j) \leq \sum_{k=1}^{r} \lambda_k x_k(j) \leq \max_{k=1,\ldots,r} x_k(j) \leq c^+(j).$$

Since the term $\mu_l e_{P_l}(j)$ is positive only if $c^+(j) = +\infty$ and negative only if $c^-(j) = -\infty$, it is evident from this that $c^-(j) \leq x(j) \leq c^+(j)$ for all $j \in J$. Thus x is a solution to the problem.

The remainder of the proof consists of demonstrating constructively that any solution x to the problem can indeed be represented in the form (2). This will establish in particular that at least one extreme solution exists. Of course there can be only finitely many extreme solutions, because a vector x satisfying $Ex = b$ is uniquely determined by the specification of a basis F and the values $x(j)$ for $j \notin F$, and in the case of an extreme solution only finitely many such specifications are possible; see (1).

Extremal Representation Algorithm

Given any solution x, form the index set F_x and test whether or not it is independent (see Exercise 10.31). If so, x is extreme. If not, one obtains an elementary primal support P such that

$$c^-(j) < x(j) < c^+(j) \quad \text{for all} \quad j \in P.$$

Since P cannot be of doubly unlimited feasibility (according to the hypothesis of the theorem), it can be supposed, passing to the reverse support if necessary, that the number

$$\mu' = \max\{\mu \in R | c^-(j) \leq x(j) - \mu e_P(j) \leq c^+(j), \forall j \in J\} \in (0, \infty)$$

exists. Let $x' = x - \mu' e_P$. Then $Ex' = Ex = b$ (because $e_P \in \mathscr{C}$), so x' is another solution to the problem and

$$x = x' + \mu' e_P \quad \text{with } F_{x'} \subset F_x, F_{x'} \neq F_x. \tag{3}$$

If P is of (singly) unlimited feasibility, the procedure is next repeated for x'. If not, the number

$$\mu'' = \max\{\mu \in R \mid c^-(j) \leqq x(j) + \mu e_P(j) \leqq c^+(j), \forall j \in J\} \in (0, \infty)$$

also exists and yields a solution

$$x'' = x + \mu'' e_P \quad \text{with } F_{x''} \subset F_x, F_{x''} \neq F_x.$$

Then for $\lambda' = \mu''/(\mu' + \mu'')$ and $\lambda'' = \mu'/(\mu' + \mu'')$ one has

$$x = \lambda' x' + \lambda'' x'', \qquad \lambda' > 0, \qquad \lambda'' > 0, \qquad \lambda' + \lambda'' = 1 \tag{4}$$

In this case the procedure is applied to x'', as well as x'.

The procedure thus decomposes a given solution x either into form (3) with P of unlimited feasibility or form (4); the solutions x' and x'' in these decompositions have associated index sets $F_{x'}$ and $F_{x''}$ strictly smaller than F_x. Finitely many applications of the procedure yield, when the decompositions are nested back together, a representation of the desired type (2).

Determining an Extreme Solution

A shortcut is possible in the extremal representation algorithm if the objective is not to obtain a complete representation (2) but just to determine *some* extreme solution to the primal feasibility problem. Then there is no need to look for decompositions (4). Having obtained (3), reapply the procedure immediately to x', regardless of whether P is of unlimited feasibility or not.

Note that when the primal rectification algorithm is employed to calculate a solution to the primal feasibility problem, this solution might not be extreme. For instance, if the initial vector x happens to be a solution that is not extreme, the algorithm will terminate immediately with this x. However, if the algorithm is initiated with an x whose associated index set F_x (defined in the same way as before, even though x may not be feasible) happens to be independent, then this property will be maintained from iteration to iteration, provided that red-green preprocessing is used (Exercise 10.35). In that case one will also have F_x independent on termination, so that x will be an extreme solution.

Dual Situation

Solutions to the dual feasibility problem in Section 10I, with closed intervals

$$D(j) = [d^-(j), d^+(j)] \quad \text{for all } j \in J,$$

can be represented similarly. An *extreme solution* to that problem is a solution v such that the *complement* of

$$F_v' = \{\, j \in J \mid d^-(j) < v(j) < d^+(j) \,\}$$

is a spanning set of indexes in the sense of Section 10C. An elementary dual support Q is of *unlimited feasibility* if $d^+(j) = +\infty$ for all $j \in Q^+$ and $d^-(j) = -\infty$ for all $j \in Q^-$; it is of *doubly* unlimited feasibility if in fact $D(j) = (-\infty, \infty)$ for all $j \in Q$.

Extremal Representation Theorem (Dual Case).   *Assume that the dual feasibility problem with closed intervals $D(j)$ has at least one solution, but there are no elementary dual supports of doubly unlimited feasibility. Then there are only finitely many extreme solutions. Moreover a vector v solves the problem if and only if it can be represented as*

$$v = \sum_{k=1}^{r} \lambda_k v_k + \sum_{l=1}^{q} \mu_l e_{Q_l},$$

where each v_k is an extreme solution, $\lambda_k \geq 0$, $\sum_{k=1}^{r}\lambda_k = 1$, Q_l is an elementary dual support of (singly) unlimited feasibility, and $\mu_l \geq 0$.

The proof of this result, which can be given in algorithmic form, is really no different from the proof in the primal case and is left to the reader as Exercise 10.38.

Integrality

In the network setting, extreme solutions to problems with integral data are integral. This holds more generally when certain unimodularity assumptions are satisfied; see Exercise 10.36.

10N.*   EXERCISES

10.1. (*Bases*). Prove the general basis theorem in Section 10C. (Generalize, as needed, the argument for the network version of the theorem in Section 4H.)

10.2. (*Unimodularity*). A subspace $\mathscr{C}$ of R^J is said to be *unimodular* if for every elementary support P, the vector e_P has all its components equal to $+1$, -1, or 0 (so that e_P is just the *incidence vector* for P). Show that this is true if and only if in every Tucker representation for $\mathscr{C}$, the matrix coefficients $a(j, k)$ are all $+1$, -1 or 0.

(*Note.*   This means that the matrix in any Tucker tableau is *totally unimodular*; see the discussion in Section 4M).

10.3. (*Unimodularity*). Show that a subspace $\mathscr{C}$ of R^J is unimodular if and only if its orthogonal complement is unimodular. (Use the preceding exercise.)

10.4. (*Bases*). Show that in the network of Figure 10.21, the vectors $e_{P_1}, e_{P_2}, e_{P_3}$ corresponding to the elementary circuits $P_1 : i_0 \rightarrow i_1 \rightarrow i_2 \rightarrow i_3 \leftarrow i_0$, $P_2 : i_0 \rightarrow i_2 \rightarrow i_3 \leftarrow i_0$, $P_3 : i_0 \rightarrow i_2 \rightarrow i_3 \rightarrow i_4 \leftarrow i_0$, form a basis for the circulation space $\mathscr{C}$, but not a fundamental basis (i.e., one that corresponds to a spanning tree F).

10.5. (*Bases*). Prove that for every elementary dual support Q and every $j \in Q^+$, there is a basis F in J such that $j \in F$ and $Q = Q_j$, the support corresponding to j in the general basis theorem. (See Section 10C for the argument in the case of primal supports P.)

10.6. (*Elementary Vectors*). For every elementary support P of $\mathscr{C}$, one has $|P| \leq |J| - \dim \mathscr{C} + 1$; likewise, for every elementary vector Q of $\mathscr{D}$, one has $|Q| \leq |J| - \dim \mathscr{D} + 1$. Prove this from the remarks near the end of Section 10C.

(*Note.* The number of columns in any Tucker representation equals $\dim \mathscr{C}$, whereas the number of rows is $\dim \mathscr{D}$.)

10.7. (*Networks with Gains*). Write down in tableau notation as in Figure 10.1 the linear systems corresponding to the enlarged network with gains in Figure 10.4 in Section 10D. Explain the sense in which the elements of the associated space $\hat{\mathscr{D}}$ correspond to elements of the $\overline{\mathscr{D}}$ space for G.

10.8. (*Networks with Gains*). Supply a detailed proof for Proposition 2 in Section 10E, using Proposition 1 and the remarks preceding the statement of Proposition 2.

10.9. (*Networks with Gains*). Let H be a passive subnetwork of G as defined in Section 10E, and let j be an arc that is not in H but has both endpoints in H. Prove that if P_1 and P_2 are circuits that traverse j in the positive direction and, aside from j, use only arcs belonging to H, then $\gamma(P_1) = \gamma(P_2)$.

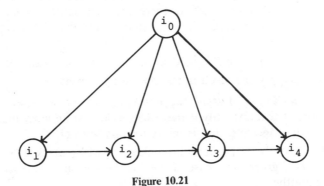

Figure 10.21

10.10. (*Networks with Gains*). Do the same as in Exercise 10.8, but for Proposition 3 in Section 10E.

10.11. (*Networks with Gains*). Let G be a connected network with gains having at least one active circuit. Prove that no signed set Q can be both an elementary activator (for some H) and an elementary deactivator (for some H').

10.12. (*Painted Index Theorem*). Prove Tucker's complementarity theorem: there exist $\bar{x} \in \mathscr{C}$ and $\bar{v} \in \mathscr{D}$ such that $\bar{x}(j) \geq 0$, $\bar{v}(j) \geq 0$, and $\bar{x}(j) + \bar{v}(j) > 0$ for all $j \in J$.

(*Hint.* Start by applying the painted index theorem with every $j \in J$ painted white.)

10.13. (*Conformal Realization*). A signed set P in J is said to conform to a vector x if $x(j) > 0$ for all $j \in P^+$ and $x(j) < 0$ for all $j \in P^-$. Prove that if $x \in \mathscr{C}$, then for any j with $x(j) \neq 0$ there is an elementary support P of $\mathscr{C}$ that uses j and conforms to x.

(*Hint.* Apply the painted index theorem to a certain painting; see Section 4B. Remember that $x \perp e_Q$ for every elementary support Q of $\mathscr{D}$.)

10.14. (*Conformal Realization*). Building on the fact in the preceding exercise, prove that every nonzero $x \in \mathscr{C}$ can be expressed in the form

$$x = \alpha_1 e_{P_1} + \cdots + \alpha_r e_{P_r},$$

where $P_1, \ldots, P_r$ are elementary supports of $\mathscr{C}$ that conform to x and the coefficients $\alpha_1, \ldots, \alpha_r$ are positive.

(*Hint.* Look at what the algorithm in Section 4C would reduce to in the case of a circulation x, and generalize to a non-network setting.)

(*Remark.* the analogous result also holds of course for $\mathscr{D}$.)

10.15. (*Characterization of Supports*). Recall from Section 4C that two signed sets are *in conformity* if no element belongs to the positive part of one and the negative part of the other. Prove that a signed set P is a support of $\mathscr{C}$ if and only if there exist elementary signed supports $P_1, \ldots, P_r$ of $\mathscr{C}$, pairwise in conformity, such that $P^+ = \cup_{k=1}^r P_k^+$ and $P^- = \cup_{k=1}^r P_k^-$.

(*Note.* In this case P is said to be the *conformal union* of $P_1, \ldots, P_r$.)

(*Hint.* Apply the result in the preceding exercise.)

10.16. (*Characterization of Dual Supports*). Prove that a signed set Q is a support of $\mathscr{D}$ if and only if there does not exist an elementary support P of $\mathscr{C}$ that meets Q and is in conformity with Q.

(*Hint.* Apply the painted index theorem with j black if $j \in Q^+$, white if $j \in Q^-$, green otherwise. Take l to be various elements of Q, one after another.)

(*Remark.* Exercises 10.15 and 10.16 demonstrate that all the supports of $\mathscr{C}$ and $\mathscr{D}$ can be constructed simply from knowledge of all the elementary supports of either $\mathscr{C}$ or $\mathscr{D}$. This is interesting especially because it is an assertion just about "sign patterns." Even though the elementary vectors e_P and e_Q may have general real components, these are not needed in the characterization of the supports, merely their signs.)

10.17. (*Painted Index Algorithm*). Demonstrate that the painted index algorithm reduces to the painted network algorithm in the following situation. Given the network G, two different nodes s and s' and a painting, extend the painting to the augmented network $\overline{G}$ as in Figure 10.22. Regard the tableau representing the x, y and u, v systems for G (see Figure 1.12) as the Tucker tableau for $\overline{G}$ corresponding to the spanning tree comprised of all the distribution arcs. Apply the painted index algorithm with $l = j_s$ (a row index).

10.18. (*Painted Index Algorithm*). Show that the painted index algorithm reduces in the following situation to the version of the painted network algorithm corresponding to Minty's lemma in Section 2H (except that where the latter branches out from only one node of the arc l until the other node is reached, it is permitted instead to branch out from both ends, "forward" and "backward.") Regard the tableau representing

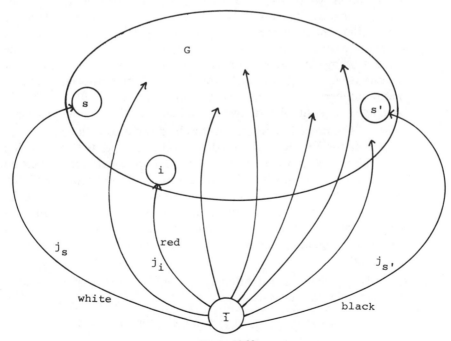

Figure 10.22

the x, y and u, v systems for G (see Figure 1.12) as the Tucker tableau for the augmented network $\overline{G}$ that corresponds to the spanning tree for $\overline{G}$ comprised of the distribution arcs j_i. (The given arc l of G is associated with a column of this tableau.) Paint all the distribution arcs red.

10.19. (*Painted Index Algorithm*). What does the painted index algorithm reduce to in the case of a connected network with gains having at least one active circuit? Specifically, in imitation of the preceding exercise, given a painting and an arc l of G, pass to the enlarged network $\tilde{G}$ in Figure 10.4 (see the end of Section 10D) and paint all the added arcs red. These arcs make up an initial field F for $\tilde{G}$. Start with the corresponding Tucker tableau (see Exercise 10.7), for which l will be a column index, and applying the painted index algorithm. How can the resulting procedure be described directly in terms of the network G? (You will need to use the characterizations in Section 10E.)

10.20. (*Feasibility*). Derive the primal feasibility theorem in Section 10I from the special case of it where the intervals are all closed; likewise for the dual feasibility theorem.

(*Hint.* Any nonclosed interval is the union of a nested sequence of closed intervals. Using this idea, represent the given feasibility problem by a sequence of "approximate" problems with closed intervals, and apply the theorem in its special case to each of these in turn. Argue that if Alternative 1 holds for one of the approximate problems, it holds for the original problem. Making use of the fact that there are only finitely many elementary supports altogether, argue that if each time Alternative 2 is obtained for one of the approximate problems the corresponding elementary support is tested to see whether it also yields Alternative 2 for the original problem, then after inspecting only a finite number of problems in the sequence, one will be able to conclude that either Alternative 1 or 2 holds for the original problem.)

(*Note.* The procedure in the hint can be dovetailed with any effective algorithm for the case of closed intervals, such as the rectification algorithms of Section 10J in the pivoting implementation of Section 10K, to get a finitely terminating algorithm for general feasibility problems where the intervals are not necessarily closed. This could be used to test for the existence of unbalanced supports as in Chapter 11, and so forth, at least in principle, but the robustness of such a procedure is open to question. Finite termination is no guarantee of practicality.)

10.21. (*Feasibility*). Show that if Alternative 2 holds in the primal feasibility theorem, then the subsystem

$$Ex = b \quad \text{and} \quad x(j) \in C(j) \quad \text{for all} \quad j \in Q$$

is itself infeasible, but it would become feasible if the condition

$x(j) \in C(j)$ were omitted for any single $j \in Q$.

(*Hint.* Temporarily regard $C(j)$ as $(-\infty, \infty)$ for all $j \in J \setminus Q$, and apply the theorem.)

(*Remark.* This says that Q describes an "irreducibly infeasible subsystem" of the original constraint system. Notice from Exercise 10.6 that necessarily $|Q| \le p$, where $p = |J| - \text{rank } E + 1$. Thus the given constraint system is feasible if and only if every subsystem involving p of the conditions $x(j) \in C(j)$ is feasible.)

10.22. (*Feasibility*). Demonstrate that the painted index theorem, vector form, can be viewed as a special case of the corollary (for $b = 0$) of the primal feasibility theorem in Section 10I.

(*Hint.* Introduce $C(j)$ to correspond to the desired condition on $x(j)$. Analyze carefully the meaning then of $C(Q) < 0$.)

10.23. (*Feasibility*). Prove Stiemke's transposition theorem: for any matrix E, the equation $Ex = 0$ has a solution x with all components positive, if and only if there does not exist u such that the vector uE is nonzero and has all components non-negative; dually, the equation $Ex = 0$ has a nonzero solution x with all components non-negative, if and only if there does not exist u such that the vector uE has all components positive.

(*Hint.* Apply the corollary of the primal feasibility theorem in the first case and the dual feasibility theorem in the second.)

10.24. (*Feasibility*). Prove that for any skew symmetric matrix K, there is a vector u such that uK and u have all components non-negative and $uK + u$ has all components positive.

(*Hint.* Apply the dual feasibility theorem to the case of the partitioned matrix $E = [K, I, K + I]$. Use the fact that $z \cdot Kz = 0$ for all z.)

10.25. (*Feasibility*). Let $\mathscr{C}$ and $\mathscr{D}$ be any pair of orthogonally complementary subspaces in R^J, and let $c \in R^J$. Prove there exists a vector $x \in \mathscr{C}$ satisfying $x(j) \le c(j)$ for all $j \in J$ if and only if there does not exist a vector $v \in \mathscr{D}$ such that $c \cdot v < 0$.

10.26. (*Lemma of Farkas*). Prove the following result about an arbitrary finite collection $\{e^j | j \in J\}$ of vectors in R^I and a vector $b \in R^I$. Either there exist coefficients $x_j \ge 0$ such that $\sum_{j \in J} x_j e^j = b$, or there is a vector $u \in R^I$ such that $u \cdot e^j \ge 0$ for all $j \in J$, but $u \cdot b < 0$.

(*Hint.* Regard the vectors e^j as the columns of a matrix E, and apply the primal feasibility theorem in Section 10I with $C(j) = [0, \infty)$ for all j.)

(*Remark.* The primal rectification algorithm in Section 10K, implemented as in Section 10L, is capable, according to the hint, of deciding constructively between the alternatives.)

10.27. (*Separation Theorem*). A set $K \subset R^I$ is called a *partial convex polyhedron* if it can be represented in the form

$$K = \left\{ u \mid -u \cdot e^j \in D(j) \text{ for all } j \in J \right\}$$

for some finite collection of vectors $e^j \in R^I$ and nonempty real intervals $D(j)$. (It is a *convex polyhedron* if this can be done with the intervals all closed.) Suppose K_1 and K_2 are both of this type, corresponding to (disjoint) index sets J_1 and J_2. Prove that $K_1 \cap K_2 = \varnothing$ if and only if there is a vector $y \in R^I$ satisfying

$$u_1 y > u_2 \cdot y \quad \text{for all } u_1 \in K_1, u_2 \in K_2.$$

(*Hint.* For the necessity, treat the vectors e^j for $j \in J = J_1 \cup J_2$ as the columns of a matrix E and apply the dual feasibility theorem. From the equation $0 = v \cdot e_P = -uEe_P$ for all $u \in R^I$, conclude that

$$\sum_{j \in J_1} e_P(j) e^j = - \sum_{j \in J_2} e_P(j) e^j,$$

and let y be the latter vector.)

(*Remark.* The suggested proof of this result is constructive: the dual rectification algorithm in Section 10J will produce either an element of $K_1 \cap K_2$ or a vector y with the indicated properties, provided the intervals are closed. The nonclosed case could be covered by extending the procedure as in Exercise 10.20, at least in principle.)

10.28. (*Conformal Realization*). Suppose x is a vector which satisfies Alternative 2 of the dual feasibility theorem but is not elementary. Consider any conformal realization of x in the sense of Exercise 10.14. Prove that at least one of the P_k's has $D(P_k) \subset (-\infty, 0)$.

10.29. (*Conformal Realization*). Suppose v is a vector which satisfies Alternative 2 of the primal feasibility theorem (vector form) but is not necessarily elementary. Consider any conformal realization of v in the sense of Exercise 10.14 (but relative to the space $\mathcal{D}$):

$$v = \alpha_1 e_{Q_1} + \cdots + \alpha_r e_{Q_r}, \qquad \alpha_k > 0,$$

where each Q_k is an elementary dual support which conforms to v, that is, has

$$Q_k^+ \subset \{ j \mid v(j) > 0 \} \quad \text{and} \quad Q_k^- \subset \{ j \mid v(j) < 0 \}.$$

Prove that at least one of the Q_k's has $C(Q_k) \subset (-\infty, b_{Q_k})$.
(*Hint.* Introducing arbitrary vectors $u_k \in R^I$ such that $-u_k E = e_{Q_k}$, translate the condition on v in Alternative 2 into the statement that

$$\alpha_1 (t_1 - b_{Q_1}) + \cdots + \alpha_r (t_r - b_{Q_r}) < 0$$

for all choices of $t_1 \in C(Q_1), \ldots, t_r \in C(Q_r)$.)

10.30. (*Aggregated Infeasibility*). Show when the primal rectification algorithm in the augmented format with aggregated infeasibility (Section 10L) is applied in the network case (where E is an incidence matrix), it reduces essentially to the feasible distribution algorithm of Chapter 3.

10.31. (*Independence Test*). The following procedure tests whether a given index set $F_0 \subset J$ is independent in the sense of Section 10C; verify the claims. Starting from an arbitrary Tucker representation, perform an arbitrary sequence of pivoting transformations in which the pivot $(\bar{j}, \bar{k})$ has $\bar{k} \in F_0$ but $\bar{j} \notin F_0$; terminate when this is no longer possible, as must be true after a finite number of transformations (no more than the number of elements of F_0 not in the basis corresponding to the initial Tucker representation). If at that stage one has $F_0 \subset F$, where F is the basis corresponding to the terminal representation (i.e., all the indexes in F_0 have been brought into row position), then F_0 is an independent set. If not, then F_0 fails to be independent. Indeed, in that case there must be an index $k \in F_0 \setminus F$ (i.e., an index of F_0 still in column position), and for any such k the corresponding elementary primal support P_k derived from the k column of the terminal tableau (as in the basis theorem in Section 10C) is included in F_0.

(*Remark*.   Instead of starting from a Tucker representation for the spaces $\mathscr{C}, \mathscr{D}$, one could start from the Tucker representation for the spaces $\bar{\mathscr{C}}, \bar{\mathscr{D}}$ having E as its coefficient matrix, as in Figure 10.1. The procedure and its properties would be the same.)

10.32. (*Spanning Test*). The following procedure tests whether a given index set $F_0 \subset J$ is a spanning set in the sense of Section 10C; verify the claims. Pivot as in the preceding exercise. If on termination one has $F_0 \supset F$ (i.e., the indexes not in F_0 have all been brought into row position), then F_0 is a spanning set. If not, then F_0 fails to be a spanning set. Indeed, in that case there must be an index $j \in F \setminus F_0$ (i.e., an index not in F_0 which is still in row position), and for any such j the corresponding elementary dual support Q_j derived from the j row of the terminal tableau (as in the basis theorem in Section 10C) is included in $J \setminus F_0$.

(*Remark*.   Again one could proceed instead from the tableau with coefficient matrix E; see the remark about the preceding exercise. Combining the independence test and spanning test, one gets of course a *basis test* for an index set F_0.)

10.33. (*Extreme Solutions*). Let x be a solution to the primal feasibility problem in the case of closed intervals $C(j)$. Prove that x is an extreme solution if and only if (there are no primal elementary supports of doubly unlimited feasibility and) the set F_x (as defined in Section 10M) is *minimal*, in the sense that there is no solution x' different from x and having $F_x \subset F_{x'}$.

10.34. (*Extreme Solutions as Extreme Points*). Let x be a solution to the primal feasibility problem in the case of closed intervals $C(j)$. Prove that x is an extreme solution (as defined in Section 10M) if and only if it cannot be expressed as $(1 - \lambda)x' + \lambda x''$, where x' and x'' are also solutions (not both identical to x) and $0 < \lambda < 1$.

(*Note.* The latter property means by definition that x is an *extreme point*, or *vertex*, of the convex polyhedron formed by all the solutions to the problem.)

10.35. (*Calculating Extreme Solutions*). Suppose the primal feasibility problem is solved by means of the primal rectification algorithm, starting from a vector x such that the index set F_x is independent in J. Verify that if red-green preprocessing is used, the solution obtained will necessarily be extreme.

10.36. (*Integrality*). Suppose the subspace $\mathscr{C}$ is unimodular in the sense of Exercise 10.2, and the numbers $b(i)$, $c^+(j)$ and $c^-(j)$ in the primal feasibility problem are all integers. Prove that in this case every extreme solution to the problem is integral, that is, has all integral components.

(*Hint.* For an extreme solution x, consider a basis $F \supset F_x$.)

10.37. (*Dual Extreme Solutions*). Suppose the dual feasibility problem does have at least one solution. Show that it has an extreme solution if and only if there is no elementary dual support Q of doubly unlimited feasibility.

10.38. (*Dual Extreme Solutions*). Develop an algorithmic proof of the dual case of the extremal representation theorem in Section 10M that is parallel to the proof in the primal case.

10P.*   COMMENTS AND REFERENCES

It was A. W. Tucker [1956, 1960, 1963] who focused attention on the combinatorial properties of linear systems of variables and their role in the theory of inequalities. Tucker's research was inspired by the simplex method of linear programming and sought to make pivoting operations a fundamental tool of constructive linear algebra. An interesting feature of Tucker's results was that they chiefly concerned the sign patterns that are possible in certain situations, rather than the actual magnitudes of the numbers having these signs.

Rockafellar [1969] showed how such results could be presented in a still more combinatorial mode in terms of elementary vectors and supports. He pointed out that the subject really amounted to a theory of "oriented matroids," which he did not develop axiomatically, however. Among the results he gave were the painted index theorem, the conformal realization theorem (Exercises 10.14 and 10.15), the characterization of dual supports (Exercise 10.16), and an abstract version of the complementarity theorem of Tucker [1956] (Exercise 10.12).

The notion of a *matroid*, invented by Whitney [1935], has become an important part of combinatorial mathematics. Texts of Tutte [1971] and Welsh [1976] provide general expositions. There are applications to combinatorial optimization (e.g., see Lawler [1976]) and to electrical engineering (see Bruno and Weinberg [1976]) among other areas, and the subject is currently quite active. Matroid theory is roughly the study of collections of subsets of a finite set J which have properties like those of the collection of all elementary supports of a subspace of R^J. Until the mid 1970s, however, "orientation" had not been incorporated into the framework. Thus, although there were plenty of results pertaining to elementary primal and dual supports P and Q, there were almost none (except those given by Rockafellar [1969]) that took P^+ and P^-, Q^+ and Q^-, into account. The only attempt at introducing orientation abstractly was that of Minty [1966], which, however, was limited to the very special "unimodular" case (corresponding to Exercises 10.2 and 10.3).

The axiomatic theory of oriented matroids is now well underway, thanks to contributions by Bland [1974, 1977, 1979], Bland and Las Vergnas [1978, 1979], Folkman and Lawrence [1978], Lawrence [1982], and Lawrence and Weinberg [1980, 1981]. It is able to treat objects somewhat more general than collections of signed supports of elementary vectors, but we have kept to the latter here because the computational possibilities of the more general objects are still rather unexplored. Our aim has been to establish the power of matroidal concepts in extending procedures and results from network programming into related areas of linear and convex programming, and signed supports serve that aim well. Nevertheless, the axiomatic theory has influenced our treatment in important ways, in particular through the priority pivoting rule of Bland [1977, 1978].

Bland hit upon this rule in his efforts to translate the simplex method of linear programming into the context of oriented matroids, and he formulated it in such terms. We have employed the idea in Section 10G to obtain a simple, algorithmic proof of the painted index theorem. Previous proofs of this result (Rockafellar [1969]) or its manifestations in various classical transposition theorems (see Tucker [1956] and the results he attributes to Motzkin [1933], Stiemke [1915], Farkas [1902] and Gordon [1873]) have all resorted to induction arguments.

The painted index algorithm is certainly a convenient and versatile tool in theory. How robust is it, though, in computational practice? Little can be said a priori on this issue, since so much depends on the way the algorithm is implemented, and this in turn may depend on the special nature of the linear systems being manipulated. What we are dealing with here is a prototype pivoting method which is designed for use as a subroutine in other algorithms, and it is hard to evaluate its performance out of context. In many of the applications, such as in various versions of the rectification algorithms in this chapter and the simplex method and out-of-kilter algorithm of Chapter 11, we can expect it to terminate frequently with red-green preprocessing alone. Pivoting iterations beyond that are really aimed at coping with degeneracy of a sort long familiar in linear programming (see Dantzig [1963]). It is in that respect that Bland's rule might have to come into play as a guarantee of termination. Whether such an additional refinement is actually worth the bother in implementation is debatable. Linear programming experience suggests that the answer is no.

Instead of simply using Bland's rule in the painted index algorithm, it would be possible to adapt the procedure of Balinski and Gomory [1963] to the subtableau formed by the white and black indexes after red-green preprocessing. That procedure, a variant of the simplex method, provides pivoting rules that in finitely many steps can bring about the desired sign patterns. The choice of pivot row and column is more complicated, however, and involves a ratio test and in general the inspection of more than just one row and column for compatibility.

One should remember too in connection with the painted index algorithm that the basic formulas for modifying a Tucker tableau, once a pivot has been chosen, do not have to be taken literally. The numerical transformation can be carried out fully or partially, as needed, in a variety of ways. Much thought has gone into this matter in refining the simplex method of linear programming, and the techniques are equally applicable here. Certain factorizations and basis representations can be very helpful. See Gill, Murray, Saunders, and Wright [1982] for a discussion of such numerical linear algebra. Techniques of Hellerman and Rarick [1971], Tomlin [1972], and Forrest and Tomlin [1972] have in particular proved valuable. Obviously, the performance of the painted index algorithm may heavily be influenced by the incorporation of such pivoting techniques.

In some situations beyond those in ordinary network programming, everything is immediately much simpler because all the coefficients in all the Tucker representations are just $+1$, -1, or 0. This is the *unimodular* case mentioned in Exercises 10.2 and 10.3, where $e_P(j)$ and $e_Q(j)$ likewise can only be $+1$, -1, or 0. Then, at the very least, the pivoting formulas can be executed in a sort of "Boolean" manner.

The unimodular case corresponds in matroidal terms to the early work of Minty [1966] on orientation. It has recently been developed in great detail by Hamacher [1980, 1981], Burkard and Hamacher [1981], as a domain in which network programming ideas and algorithms receive an especially complete extension. On the practical side, however, the linear systems that arise in applications and have the unimodularity property typically do come from networks themselves, even if sometimes in disguise. Seymour [1980] has shown in fact that every unimodular system can be built in a certain manner from the primal and dual systems associated with various networks.

Networks with gains offer a different domain for a close generalization of ordinary network programming, and one that does have many noteworthy applications. Although unimodularity is not present for networks with gains, the pivoting transformations in the painted index algorithm and elsewhere can nonetheless be executed in a purely combinatorial mode. The important pioneer paper on this subject is by Jewell [1962]; a more complete presentation can be found in the book by Helgason and Kennington [1980]. These authors describe some of the applications as well as give more detail on implementation. They also consider *negative* gains, which we have excluded for the sake of simplicity. See also Glover, Klingman, and Stutz [1973] and Glover, Elam, and Klingman [1979] for developments on implementation, and Glover and Klingman [1973] for a result akin to the passivity theorem in Section 10C. Although absorbing circuits, special basis structure, and so forth, play a fundamental role in these works, the full characterization of the primal and dual elementary supports which we have provided in Section 10D is new, as is our approach to networks with gains in terms of the generalized circulation space $\mathscr{C}$ and its complement $\mathscr{D}$.

Networks with gains have also been treated by Dantzig [1963], Frank and Frisch [1971], Maurres [1972], Grinold [1973], and Deo [1974].

A linear system of variables can sometimes be reduced to a system that corresponds to an ordinary network. The passivity theorem in Section 10C gives a constructive criterion for systems arising from networks with gains, but an algorithm for accomplishing the task more generally, when possible, has been furnished by Bixby and Cunningham [1980].

Besides the unimodular case and networks with gains, there are many other kinds of linear systems whose structure can be exploited in pivoting transformations and other procedures. The structures in Section 10F, which concern multicommodity flows or more broadly the addition of complicating side constraints or interactive variables to a system that might otherwise be relatively easy to deal with, are among the most interesting. There are numerous models in applied optimization that involve multicommodity systems. We are not able to treat such problems with much individual attention in this book (the main results of a special nature are given in Sections 11F and 11G), but the surveys of Assad [1978] and Kennington [1978] can connect the reader with a large literature. The book of Helgason and Kennington [1980] is a good reference more generally for network-related structures of linear systems and how to handle them.

The feasibility theorems in Section 10I, when stated in vector form with the word "elementary" omitted, are classical results well known in linear programming theory, although not usually expressed in terms of intervals as here. For closed intervals they correspond to separation theorems for convex polyhedra and have traditionally been derived from the lemma of Farkas [1902] (stated in Exercise 10.26). Extensions to nonclosed intervals have long been known too. These include some of the transportation theorems mentioned earlier in connection with the painted index theorem; more generally see Kuhn [1956].

The versions of the feasibility theorems in terms of *elementary* vectors and supports have a shorter history. They stem from Rockafellar [1969]; see Camion [1968] for an independent proof given at the same time in an abstract algebraic setting. Other proofs have been given by Rockafellar [1970, Sec. 22] and Minty [1974]. These results generalize the theorems of Minty [1960] for the network case as well as earlier theorems on irreducibly inconsistent subsystems of a system of linear inequalities (see Fan [1956] and Exercise 10.21).

All the previous proofs of the general feasibility theorems rely on induction arguments and are nonconstructive. The difference here is that we have provided constructive proof in terms of the rectification algorithms in Section 10J and moreover have turned the results into a sort of algorithmic module that can be plugged into numerous situations. For example, we obtain as a by-product constructive proofs of the lemma of Farkas (Exercise 10.26) and the separation theorem for partial convex polyhedra (Exercise 10.27).

The feasibility theorems reduce in the case of networks to the combinatorial results presented in Chapters 3 and 6, but they can be specialized similarly also to other linear systems for which the primal and dual elementary supports can be characterized directly, such as in networks with gains. See Hassin [1981] for some results that correspond to this kind of specialization, along with references to their applications.

Like the painted index algorithm, the primal and dual rectification algorithms are prototype methods that can be implemented in a multitude of ways. They can even be applied to the same problem in several different formats, as explained in Section 10L. They require—as a subroutine—the painted index algorithm or some other procedure capable of resolving the alternatives in the painted index theorem (as is certainly available not only for ordinary networks but also networks with gains), and they are therefore heir to all the computational tricks and troubles associated with such a subroutine, some of which have already been discussed. Furthermore the rectification algorithms are directed toward application as subroutines themselves in optimization algorithms like the ones that will be described in Chapter 11. Thus it is impossible to pass judgement on their practical efficiency without pinning the discussion down to some particular variant, implementation, and context. From a general point of view one can say that these algorithms are closely related to linear programming methods for ascertaining feasibility (see Dantzig [1963] and Exercises 11.67 and 11.68) but exhibit more flexibility and versatility than the latter in their formulation and to this extent can lead to better performance when applied sensitively in various settings. As a practical matter the OK priority rule can probably be ignored in most cases without noticeable adverse effect.

The rectification algorithms will be seen in Section 11K to be special instances of the out-of-kilter algorithm for general piecewise linear programming. Like the latter they are essentially new as extensions beyond the framework of network programming.

11

MONOTROPIC PROGRAMMING

The preceding chapter has laid the groundwork for a broad generalization of the network optimization theory in Chapters 7, 8, and 9. Analogues of elementary circuits and cuts have been developed for arbitrary linear systems of variables and used to prove theorems on the feasibility of constraints. They have also been put to use in algorithms for the solution of feasibility problems.

The task ahead is the application of these ideas to the monotropic programming problems that are obtained when the incidence matrix E in the optimal distribution and optimal differential problems is replaced by an arbitrary real matrix. The existence and duality theory in Chapter 8 can be extended with hardly more than changes in notation and terminology. This provides a powerful tool for the treatment of all piecewise linear or quadratic programming problems, in particular.

Perhaps the most interesting feature of the extended theory, however, is the development of optimization algorithms in which descent takes place in "elementary directions" only. These correspond to network methods in which flows or potentials are improved in just one circuit or cut at a time. In some situations beyond those in ordinary networks, such as networks with gains or traffic problems, "elementary directions" of descent can be generated by special techniques which rely on combinatorial substructure as discussed in Sections 10E and 10F. The rectification algorithms and their relatives in Sections 10J, 10K, and 10L serve for the general case in various implementations based on pivoting in Tucker tableaus.

Besides the basic descent algorithms that correspond to the optimal distribution and optimal differential algorithms in Chapter 9, "fortified" and "discretized" methods are described. Primal and dual simplex methods are formulated for arbitrary monotropic programming problems with particular attention to the piecewise linear case. For the latter there is also a generalized form of the out-of-kilter algorithm that handles primal and dual variables symmetrically.

11A.   OPTIMIZATION AND EQUILIBRIUM

Throughout this chapter we shall be dealing with the primal and dual linear systems that correspond to an arbitrary real matrix E expressed in terms of index sets I and J as in Section 10A:

$$y(i) = \sum_{j \in J} e(i, j)x(j) \quad \text{for } i \in I, \tag{1}$$

$$v(j) = - \sum_{i \in I} u(i)e(i, j) \quad \text{for } j \in J. \tag{2}$$

For each $j \in J$, let f_j and g_j be closed proper convex functions on R conjugate to each other, and let Γ_j be the corresponding characteristic curve (see Sections 8A, 8B, and 8E). Let $C(j)$ and $D(j)$ be the intervals where f_j and g_j are finite, and let b be a vector in the column space $\mathcal{B}$ of the matrix E (see 10C(3)).

Monotropic programming deals with optimization problems that can be cast in either of the following basic forms.

Primal Optimization Problem.   *Minimize*

$$\Phi(x) = \sum_{j \in J} f_j(x(j))$$

subject to the constraints

$$x(j) \in C(j) \quad \text{for all } j \in J, \tag{3}$$

$$y(i) = b(i) \quad \text{for all } i \in I, \tag{4}$$

where $y(i)$ is given by (1).

Dual Optimization Problem.   *Maximize*

$$\Psi(u) = - \sum_{i \in I} b(i)u(i) - \sum_{j \in J} g_j(v(j))$$

subject to the constraints

$$v(j) \in D(j) \quad \text{for all } j \in J, \tag{5}$$

where $v(j)$ is given by (2).

These are the same as the optimal distribution and optimal differential problems in Sections 8D and 8G, except that E need no longer be an incidence matrix. A corresponding generalization of the network equilibrium problem in Section 8H plays a significant role in monotropic programming too.

Equilibrium Problem.   *Find* $x \in R^J$ *and* $u \in R^I$ *(and corresponding* $y \in R^I$ *and* $v \in R^J$ *given by* (1) *and* (2)*) such that* (4) *holds and*

$$(x(j), v(j)) \in \Gamma_j \quad \text{for all } j \in J. \tag{6}$$

Each of these three problems generates the other two by way of the fundamental relations between f_j, g_j and Γ_j (see Sections 8B and 8E). It will be seen in the theorems of Section 11D that, as in the network case, solving any one of them is tantamount to solving them all. Certainly it can be noted right away that

$\Phi(x) \geq \Psi(u)$ *for every* x *satisfying* (3) *and every* u;

moreover equality holds if and only if (6) *is satisfied*   (7)

This is obvious from the equation

$$-\sum_{i \in I} u(i)b(i) = \sum_{j \in J} v(j)x(j) \quad \text{when } y(i) = b(i)$$

and the fact (see Section 8E) that the inequality

$$f_j(x(j)) \geq v(j)x(j) - g_j(v(j))$$

always holds, and holds with equality if and only if $(x(j), v(j)) \in \Gamma_j$.

Notation has been set up so that $f_j(x(j)) = \infty$ when $x(j) \notin C(j)$ and $g_j(v(j)) = \infty$ when $v(j) \notin D(j)$. The constraints (3) correspond therefore to $\Phi(x) < \infty$ and the constraints (5) to $\Psi(u) > -\infty$. This saves us from having always to list these constraints explicitly. In minimizing Φ and maximizing Ψ, regions where $\Phi = \infty$ or $\Psi = -\infty$ are avoided automatically.

A vector x satisfying (3) and (4) is said to be a *feasible solution* to the primal problem, and a vector u satisfying (5) a feasible solution to the dual problem. We speak of *regularly* feasible solutions when (3) and (5) can be strengthened respectively to

$$x(j) \in \tilde{C}(j) \quad \text{for all } j \in J, \tag{8}$$

$$v(j) \in \tilde{D}(j) \quad \text{for all } j \in J. \tag{9}$$

where $\tilde{C}(j)$ and $\tilde{D}(j)$ are the projections of Γ_j on the horizontal and vertical axes of R^2. Recall that $\tilde{C}(j)$ and $\tilde{D}(j)$ are intervals contained in $C(j)$ and $D(j)$ and having the same closures $[c^-(j), c^+(j)]$ and $[d^-(j), d^+(j)]$ as the latter (see Sections 8B and 8E for an explanation of the possible discrepancies between these intervals). An *optimal solution* to the primal problem is a feasible solution that yields the minimum in question; analogously for the dual problem.

The distinction between feasibility and regular feasibility will be crucial in stating general results about the existence of optimal solutions (see Section 11D) and in formulating certain algorithms (e.g., in Section 11H), but of course it makes no difference in the many cases where $\tilde{C}(j) = C(j)$ and $\tilde{D}(j) = D(j)$.

The existence of feasible solutions is addressed by the theorems in Section 10I, and the same results with $C(j)$ and $D(j)$ replaced by $\tilde{C}(j)$ and $\tilde{D}(j)$ cover the existence of regularly feasible solutions. When the intervals in question are closed, such solutions can be determined by the rectification algorithms in Section 10J. Elementary primal and dual supports P and Q have an important role in that context and will be equally prominent in connection with optimal solutions (see Section 11C). This is one of the prime characteristics of monotropic programming and, practically speaking, distinguishes it as a natural domain for solution procedures based on pivoting transformations of linear systems of variables.

Alternative Formulations

In the case where $b = 0$, the primal problem becomes

$$\text{minimize} \quad \sum_{j \in J} f_j(x(j)) \quad \text{over all} \quad x \in \mathscr{C} \quad \text{satisfying (3)}, \qquad (10)$$

while the dual problem becomes

$$\text{maximize} \quad -\sum_{j \in J} g_j(v(j)) \quad \text{over all} \quad v \in \mathscr{D} \quad \text{satisfying (5)}. \qquad (11)$$

Here $\mathscr{C}$ and $\mathscr{D}$ are the complementary subspaces defined in Section 10A. The equilibrium problem reduces then to

$$\text{find} \quad x \in \mathscr{C} \quad \text{and} \quad v \in \mathscr{D} \quad \text{satisfying (8)}. \qquad (12)$$

This formulation can in turn be brought to bear on the augmented subspaces $\overline{\mathscr{C}}$ and $\overline{\mathscr{D}}$ in $R^I \times R^J$ (see Section 10A) in order to get problems that are more general in appearance, but appearance only, than the ones we started with, specifically the "full" problems:

$$\text{minimize} \quad \overline{\Phi}(x) = \sum_{i \in I} f_i(y(i)) + \sum_{j \in J} f_j(x(j)) \quad \text{subject to}$$
$$y(i) \in C(i) \quad \text{and} \quad x(j) \in C(j) \quad \text{for all } i \in I, j \in J; \qquad (13)$$

$$\text{maximize} \quad \Psi(u) = - \sum_{i \in I} g_i(u(i)) - \sum_{j \in J} g_j(v(j)) \quad \text{subject to}$$

$$u(i) \in D(i) \quad \text{and} \quad v(j) \in D(j) \quad \text{for all } i \in I, j \in J; \tag{14}$$

$$\text{find} \quad x \in R^J \quad \text{and} \quad u \in R^I \quad \text{satisfying}$$

$$(y(i), u(i)) \in \Gamma_i \quad \text{and} \quad (x(j), v(j)) \in \Gamma_j \quad \text{for } i \in I, j \in J; \tag{15}$$

where f_i, g_i, Γ_i, $C(i)$, $D(i)$, are elements associated with the indexes $i \in I$ just like those associated with $j \in J$.

For theoretical purposes, the initial trio of problems could be replaced by (10), (11), and (12) or by (13), (14), and (15): each formulation could serve equally well as the basic model. Any problem that fits one of the patterns can be translated into the others. Our choice of which trio to focus on as primary has been dictated entirely by convenience for our broad purposes in the rest of this chapter. Although problems (10), (11), and (12) are the simplest and are nicely symmetric, the explicit incorporation of a possibly nonzero vector b is helpful in many applications and has little drawback. Specialization to the case where $b = 0$ is always immediate. Translating results for (10), (11), and (12) into the framework of the primal, dual, and equilibrium problems as introduced initially here takes more effort in comparison.

By the way, any optimization problem that could be set up in the primal form could also be set up in the dual form. This is clear from the equivalence with (10) and (11), which in turn differ from each other only in notation: any closed proper convex function g_j on R is the conjugate of some other such function f_j, and any subspace $\mathscr{D}$ of R^J is the complement of some other subspace $\mathscr{C}$. Thus the terms "primal" and "dual" refer only to a scheme of pairing and not to any categorization of problems according to their intrinsic nature.

Our blanket assumption that $b \in \mathscr{B}$, that is, that (4) can be satisfied at least, ensures that the first sum in the formula for $\Psi(u)$, namely, $u \cdot b$, does not actually depend on any particular u corresponding to v but only on v itself (see the beginning of Section 10C). In this way the dual problem can be viewed as one over the space $\mathscr{D}$, even when $b \neq 0$. Of course this is a moot point when rank $E = |I|$ because then there is a *unique* u for each $v \in \mathscr{D}$.

11B. EXAMPLES OF MONOTROPIC PROGRAMMING

The optimal distribution problem in Section 8D is an obvious special case of the primal optimization problem in the preceding section. The optimal differential problem in Section 8G is a special case of the dual problem. These are not the only network problems that fall under monotropic programming, however.

Networks with gains, as described in Sections 10D and 10E, provide further examples. The primal and dual problems in that case are similar in interpretation to the optimal distribution and optimal differential problems in an ordinary network, but the matrix E differs from an ordinary incidence matrix in having general coefficients $-\gamma(j) \in (-\infty, 0)$ in place of the -1's. Networks with flow interactions, in particular, multicommodity and traffic networks, likewise involve linear systems more complicated than those of an ordinary network but still based largely on incidence matrices; see Section 10F. Optimization problems in such networks require the broader approach of monotropic programming.

The examples mentioned so far are concerned with special kinds of linear systems of variables, but other cases of monotropic programming problems can be obtained by specializing the elements f_j, g_j and Γ_j.

EXAMPLE 1.   (*Linear and Piecewise Linear Programming*)

A very important case is the one where the functions f_j and g_j are piecewise linear, and correspondingly the relations Γ_j are of staircase form (see Sections 8C and 8F), so in particular

$$\tilde{C}(j) = C(j) = [c^-(j), c^+(j)] \quad \text{and} \quad \tilde{D}(j) = D(j) = [d^-(j), d^+(j)].$$

$$(1)$$

We speak of this case as *piecewise linear programming*. Note that the dual of a piecewise linear problem is piecewise linear in this scheme.

If every f_j is actually linear relative to $C(j)$, that is, has no breakpoints between $c^-(j)$ and $c^+(j)$, the primal problem is a *linear programming* problem in the general sense. Likewise, if every g_j is linear relative to $D(j)$, the dual problem is linear. However, linearity of the primal does *not* imply linearity of the dual, or vice versa, but merely piecewise linearity. This is clear from the conjugacy relations in Section 8E. Full duality is not manifested within the confines of the class of linear programming problems in the general sense.

Pairs of *linear* problems that *are* dual to each other can be obtained only through further restrictions on f_j or g_j. Situations where nontrivial upper *and* lower bounds are placed on the variables cannot be admitted, as it turns out. One cannot have either $-\infty < c^-(j) < c^+(j) < \infty$ or $-\infty < d^-(j) < d^+(j) < \infty$.

An important case where both the primal and dual monotropic programming problems are linear is the one where f_j, g_j, and Γ_j are *elementary* in the sense of Section 7I (see Figure 8.33 and Figure 7.16):

$$f_j(x(j)) = d(j)x(j) \quad \text{on} \quad C(j) = [c(j), \infty), \tag{2}$$

$$g_j(v(j)) = c(j)v(j) \quad \text{on} \quad D(j) = (-\infty, d(j)], \tag{3}$$

$$(x(j), v(j)) \in \Gamma_j \Leftrightarrow \begin{cases} x(j) \geq c(j), \quad v(j) \leq d(j), \\ [x(j) - c(j)][v(j) - d(j)] = 0 \end{cases} \tag{4}$$

(where $c(j)$ and $d(j)$ are finite). The primal and dual problems then have the form

$$\text{minimize} \quad d \cdot x \quad \text{subject to} \quad x \geq c, \qquad Ex = b, \tag{5}$$

$$\text{maximize} \quad -u \cdot (b - Ec) \quad \text{subject to} \quad -uE \leq d, \tag{6}$$

and the equilibrium problem is

$$\text{find} \quad x \in R^J \quad \text{and} \quad u \in R^I \quad \text{such that} \tag{7}$$
$$x - c \geq 0, \qquad Ex = b, \qquad uE + d \geq 0, \qquad [x - c][uE + d] = 0.$$

We may speak in this case of *elementary* monotropic programming. For $c = 0$, (5) and (6) are the so-called *standard* problems of general linear programming, and (7) expresses the so-called *complementary slackness conditions* associated with them.

Any piecewise linear problem can be reduced to a linear problem (Exercise 11.9), and any linear problem to an elementary one with $c = 0$ (Exercise 11.10). Such reformulations are classical, but they require wholesale changes in the linear system of variables and make duality relations correspondingly difficult to interpret.

EXAMPLE 2.   (*Quadratic and Piecewise Quadratic Programming*)

Another important case of monotropic programming, likewise with property (1), occurs when f_j and g_j are piecewise quadratic and the relations Γ_j polygonal (see Section 8C and 8F). We speak of this as *piecewise quadratic programming*. It encompasses general quadratic (convex) programming (with linear constraints), despite the somewhat special looking separable form of the objective functions Φ and Ψ. To be convinced, consider any problem of the type

$$\text{minimize} \quad d \cdot z + \tfrac{1}{2}z \cdot Dz \quad \text{subject to} \quad z \geq 0, \qquad Az = a, \tag{8}$$

where D is a positive semidefinite matrix. Express D as $M^T M$ for some matrix

M (where M^T is a transpose of M); this device of classical linear algebra entails no serious computational effort. Problem (8) then becomes

$$\text{minimize} \quad d \cdot z + \tfrac{1}{2} w \cdot w \quad \text{subject to} \quad z \geq 0 \quad \text{and}$$

$$\left[-\frac{A \mid 0}{-M \mid I} \right] \cdot \left[\begin{array}{c} z \\ w \end{array} \right] = \left[\begin{array}{c} a \\ 0 \end{array} \right]. \tag{9}$$

Setting $x = [z, w]$, one obtains a monotropic programming problem in primal form with

$$f_j(\xi) = \begin{cases} d(j)\xi & \text{if } \xi \geq 0, \\ \infty & \text{if } \xi < 0 \end{cases} \tag{10}$$

for indexes j corresponding to components of z, and

$$f_j(\xi) = \tfrac{1}{2}\xi^2 \tag{11}$$

for indexes j corresponding to components w.

The dual of this problem is obtained by calculating the conjugate functions. For (1) the conjugate is

$$g_j(\eta) = \begin{cases} 0 & \text{if } \eta \leq d(j), \\ \infty & \text{if } \eta > d(j), \end{cases}$$

whereas the function (11) is self-conjugate:

$$g_j(\eta) = \tfrac{1}{2}\eta^2.$$

Splitting u into $[r, s]$ in accordance with the row blocks in the partitioned matrix in (9), one gets as the dual:

$$\text{maximize} \quad -r \cdot a - \tfrac{1}{2}|s|^2 \quad \text{subject to} \quad rA - sM \leq d.$$

EXAMPLE 3.   (*Abstract Monotropic Programming*)

For a better understanding of the nature of monotropic programming, let us look at the following kind of problem:

$$\text{minimize} \quad F(z) = \sum_{j \in J} f_j(l_j(z_j)) \quad \text{over all} \quad z \in R^n \quad \text{satisfying}$$

$$\text{the constraints} \quad l_j(z) \in C(j) \quad \text{for all} \quad j \in J, \tag{12}$$

where $\{ l_j | j \in J \}$ is a finite collection of linear functions on R^n (some perhaps giving components of z), and each f_j is a closed proper convex function on R with $C(j)$ as its interval of finiteness. (For certain indexes, f_j may be identically zero on $C(j)$, in which event $C(j)$ must be closed.) The feasible set defined by the constraints is a "general convex polyhedron with some of its faces possibly missing" (if the intervals $C(j)$ are not all closed). Let

$$\mathscr{C} = \left\{ x \in R^J \mid \ \exists z \in R^n \quad \text{with } x(j) = l_j(z) \quad \text{for all } \ j \in J \right\}. \quad (13)$$

Then $\mathscr{C}$ is a subspace of R^J, and the problem takes the form 11A(9). Since any subspace of R^J can easily be expressed as in (13) for some collection of linear functions (e.g., using a Tucker representation), it is clear that the abstract model (12) is equivalent to 11A(9) and hence to our standardized primal problem (or for that matter, to our standardized dual problem).

If we call a convex function F on R^n *preseparable* when it can be expressed as a sum of linear functions composed with convex functions of a single variable as in (12), the situation can be summarized as follows. *A monotropic programming problem is an optimization problem where a preseparable convex function is minimized over a convex polyhedron.* This description does not fully capture the essence of monotropic programming, however. The particular formulation of the linear system of variables, and the combinatorial properties that result from that formulation, are of direct importance to the subject too, especially in connection with pivoting algorithms, as will be seen later.

11C.   DESCENT BY ELEMENTARY VECTORS

Monotropic programming is distinguished from other areas of optimization by the special way that feasible solutions can be improved. In generating a direction for line search, one does not have to operate with a continuum of possibilities. If improvement can be made at all, it can be made in an "elementary direction," of which there are only finitely many. Such directions are given by elementary vectors of the spaces $\mathscr{C}$ or $\mathscr{D}$, as defined in Section 10B.

We now explain this role of elementary vectors and survey some properties related to line search that will provide background for the theorems in Section 11D as well as the algorithms in Sections 11H and 11I.

Let x be any feasible solution to the primal problem, and for any z belonging to the subspace $\mathscr{C}$, consider the function

$$f(t) = \Phi(x + tz) = \sum_{j \in J} f_j(x(j) + tz(j)).$$

Since each f_j is convex and continuous relative to the closure $[c^-(j), c^+(j)]$ of its finiteness interval $C(j)$, the same holds for f. Thus f is a closed proper convex function of $t \in R$ with $f(0) < \infty$, and in particular, it possesses right and left derivatives $f'_+(t)$ and $f'_-(t)$. Since $z \in \mathscr{C}$ (i.e., $Ez = 0$), the vector $x + tz$ satisfies $E(x + tz) = Ex = b$ for all t and therefore is a feasible solution to the primal problem as long as t is such that $f(t) < \infty$ (i.e., as long as $x(j) + tz(j) \in C(j)$ for all $j \in J$). Information about $f'_+(t)$ and $f'_-(t)$ can be useful in determining whether x can be improved by passing $x + tz$ for some t, and if so, what value of t to choose.

We define a *descent vector* at x for the primal problem to be a vector $z \in \mathscr{C}$ such that $f'_+(0) < 0$. For such a z there is of course a $\tau > 0$ such that for all $t \in (0, \tau)$, $x + tz$ is a feasible solution better than x in the sense that $\Phi(x + tz) < \Phi(x)$. Of prime interest will be the case where z is an elementary vector of $\mathscr{C}$, or more specifically (since positive scalar multiples of descent vectors are descent vectors, only the direction matters) where $z = e_P$ for some elementary primal support P. We then say that P gives an *elementary direction of descent* for Φ at x.

The derivatives of f can be expressed in terms of those of the f_j's. For $z = e_P$ we have

$$f(t) = \Phi(x + te_P) = \sum_{j \in P^+} f_j(x(j) + te_P(j)) + \sum_{j \in P^-} f_j(x(j) + te_P(j)),$$

$$(1)$$

where $e_P(j) > 0$ in the first sum and $e_P(j) < 0$ in the second, and the formulas work out to

$$f'_+(t) = \sum_{j \in P^+} f'_{j+}(x(j) + te_P(j))e_P(j) + \sum_{j \in P^-} f'_{j-}(x(j) + te_P(j))e_P(j),$$

$$f'_-(t) = \sum_{j \in P^+} f'_{j-}(x(j) + te_P(j))e_P(j) + \sum_{j \in P^-} f'_{j+}(x(j) + te_P(j))e_P(j).$$

$$(2)$$

This is easy to calculate from the definitions of the derivatives as one-sided limits of difference quotients (Section 8B), as long as t is such that $x + te_P$ is *regularly* feasible, that is, has $x(j) + te_P(j) \in \tilde{C}(j)$ for all j; then $f'_{j+}(x(j) + te_P(j)) > -\infty$ and $f'_{j-}(x(j) + te_P(j)) < \infty$, so the question of $\infty - \infty$ does not arise in the sums in (2). In fact (2) is valid for all $t \in R$ if one interprets $\infty - \infty$ as ∞ in the first formula and $-\infty$ in the second. For t such $x + te_P$ is feasible but not regularly feasible, this is seen from the same limit argument, whereas for other values of t it follows from the conventions in Section 8A about the values of the right and left derivatives of f_j outside the interval $C(j)$ (Exercise 11.14).

From (2) we have that *P gives an elementary direction of descent at x if and only if*

$$\sum_{j \in P^+} f'_{j+}(x(j))e_P(j) + \sum_{j \in P^-} f'_{j-}(x(j))e_P(j) < 0. \qquad (3)$$

Primal Optimality Theorem. *A feasible solution x to the primal problem is optimal if and only if there is no elementary direction of descent for* Φ *at x. Indeed, if x is feasible but not optimal, and if* $\varepsilon > 0$ *is any number such that there is no feasible solution u to the dual problem satisfying* $\Phi(x) - \Psi(u) \leq \varepsilon$, *then there must exist an elementary primal support P and a value* $t > 0$ *such that* $x + te_P$ *is feasible with*

$$\Phi(x + te_P) < \Phi(x) - \varepsilon/|J|.$$

The proof of this result will be given in Section 11I. Procedures for finding elementary directions of descent will be discussed in both Sections 11H and 11I.

Once an elementary direction of descent has been determined, there is still the question of minimizing the function (1) in t. The minimum occurs at t if and only if $f'_-(t) \leq 0 \leq f'_+(t)$ (see Exercise 8.11), and since $f'_+(0) < 0$, such a t (if it exists) must satisfy $t > 0$; it can fail to exist only if f is a decreasing function on R (see Exercise 8.12). Recall that f is said to be *decreasing* if $f(t_1) > f(t_2)$ when $t_1 < t_2$, unless $f(t_1) = \infty = f(t_2)$. It is *nonincreasing* if $f(t_1) \geq f(t_2)$ when $t_1 > t_2$. It is *strictly decreasing* if there is a $\lambda > 0$ such that $f(t_1) - \lambda(t_2 - t_1) > f(t_2)$ when $t_1 > t_2$, unless $f(t_1) = \infty = f(t_2)$.

A characterization of the monotonicity properties just mentioned will be helpful later in connection with the existence of optimal solutions (Section 11D). For this purpose we need to look again at the interval

$$D(P) = \sum_{j \in J} e_P(j)D(j) = \left\{ \sum_{j \in J} e_P(j)\eta_j \, \middle|\, \eta_j \in D(j) \right\}, \qquad (4)$$

already encountered in the feasibility theory of Section 10I, and at its possibly smaller companion

$$\tilde{D}(P) = \sum_{j \in J} e_P(j)\tilde{D}(j) = \left\{ \sum_{j \in J} e_P(j)\eta_j \, \middle|\, \eta_j \in \tilde{D}(j) \right\}. \qquad (5)$$

Both intervals have the same closure $[d^-(P), d^+(P)]$, where

$$d^+(P) = \sum_{j \in P^+} e_P(j)d^+(j) + \sum_{j \in P^-} e_P(j)d^-(j),$$

$$d^-(P) = \sum_{j \in P^+} e_P(j)d^-(j) + \sum_{j \in P^-} e_P(j)d^+(j), \qquad (6)$$

because $D(j)$ and $\tilde{D}(j)$ have the same closure $[d^-(j), d^+(j)]$ (see Section 8E). The bounds (6) have the alternative expression

$$d^+(P) = \lim_{t \uparrow \infty} f'_-(t) = \lim_{t \uparrow \infty} f'_+(t)$$

$$d^-(P) = \lim_{t \downarrow -\infty} f'_-(t) = \lim_{t \downarrow -\infty} f'_+(t), \qquad (7)$$

by virtue of (2) and the limits in 8E(9). Obviously,

$$\tilde{D}(P) = D(P) = [d^-(P), d^+(P)] \quad \text{if} \quad \tilde{D}(j) = D(j) = [d^-(j), d^+(j)]$$

$$\text{for all } j.$$

Recalling that in general

$$\eta_j \in \tilde{D}(j) \Leftrightarrow \begin{cases} f'_{j+}(\xi_j) \geq \eta_j & \text{when } \xi_j \text{ is sufficiently high,} \\ f'_{j-}(\xi_j) \leq \eta_j & \text{when } \xi_j \text{ is sufficiently low,} \end{cases}$$

(see 8E(8) and 8E(7)), we deduce from the derivative formulas (2) that

$$\eta \in \tilde{D}(P) \Leftrightarrow \begin{cases} f'_+(t) \geq \eta & \text{when } t \text{ is sufficiently high,} \\ f'_-(t) \leq \eta & \text{when } t \text{ is sufficiently low} \end{cases} \tag{8}$$

(Exercise 11.15), or in other words,

$$\tilde{D}(P) = \begin{bmatrix} \text{projection of the maximal monotone relation } \Gamma \\ \text{corresponding to } f \text{ onto the vertical axis} \end{bmatrix}. \tag{9}$$

Extending the terminology in Section 8J, we define an elementary primal support P to be *unbalanced* if $\tilde{D}(P) \subset (-\infty, 0)$. It follows then from (8) that

$$P \text{ is unbalanced} \Leftrightarrow f'_+(t) < 0 \quad \text{for all } t \in R. \tag{10}$$

In comparison, since $d^+(P)$ is the upper endpoint of $\tilde{D}(P)$, we have from (8) that

$$d^+(P) \leq 0 \Leftrightarrow f'_+(t) \leq 0 \quad \text{for all } t \in R \tag{11}$$

$$d^+(P) < 0 \Leftrightarrow \exists \lambda > 0 \quad \text{with} \quad f'_+(t) \leq -\lambda \quad \text{for all } t \in R. \tag{12}$$

Since the properties on the right in (10), (11), and (12) characterize the three monotonicity properties we intended to analyze (see Exercise 8.10), the following theorem has been obtained.

Primal Descent Theorem. Let P be any elementary primal support, and let $f(t) = \Phi(x + te_P)$, where x is any feasible solution to the primal optimization problem. Then:

    1. f is nonincreasing $\Leftrightarrow d^+(P) \leq 0$.

    2. f is decreasing $\Leftrightarrow P$ is unbalanced.

    3. f is strongly decreasing $\Leftrightarrow d^+(P) < 0$.

Corollary. If P is such that the function $f(t) = (x + te_P)$ is nonincreasing, decreasing, or strongly decreasing for one choice of the feasible solution x, then it has the same property for every feasible x.

Corresponding results can be obtained for the dual optimization problem. In that problem a certain function Ψ is maximized, but it is convenient here and in the development of algorithms to speak instead of minimizing $-\Psi$. That way facts about convex functions of a single variable, as presented in Chapter 8, can be applied without translating them first into "concave form."

Let u be any feasible solution to the dual problem. For arbitrary $w \in R^I$, the function $f(t) = -\Psi(u + tw)$ is a closed proper convex function of $t \in R$ with $f(0) < \infty$ (Exercise 11.16). We say w is a descent vector at u if $f'_+(0) < 0$.

It has been noted in Sections 10I and 11A that the expression $u \cdot b$ depends only on the vector $v = -uE$ (inasmuch as b belongs to $\mathscr{B}$, the column space of E), so that the value $\Psi(u)$ depends only on v. Likewise the value $f(t) = -\Psi(u + tw)$ depends only on $v + tz$, where $z = -wE$. We shall concentrate in our analysis on the case where the vector $z \in \mathscr{D}$ is actually an elementary vector of $\mathscr{D}$, or more specifically, where

$$-wE = e_Q \quad \text{for some elementary dual support } Q. \tag{13}$$

We say then that Q gives an elementary direction of descent for $-\Psi$ at u (or v).

In the case of (13) we have

$$f(t) = -\Psi(u + tw) = u \cdot b - tb_Q + \sum_{j \in J} g_j(v(j) + te_Q(j)), \tag{14}$$

where b_Q is the quantity that corresponds to Q in the sense of Section 10I, namely, $b_Q = -w \cdot b$. The derivative formulas are then

$$f'_+(t) = -b_Q + \sum_{j \in Q^+} g'_{j+}(v(j) + te_Q(j))e_Q(j)$$

$$+ \sum_{j \in Q^-} g'_{j-}(v(j) + te_Q(j))e_Q(j),$$

$$f'_-(t) = -b_Q + \sum_{j \in Q^+} g'_{j-}(v(j) + te_Q(j))e_Q(j)$$

$$+ \sum_{j \in Q^-} g'_{j+}(v(j) + te_Q(j))e_Q(j), \tag{15}$$

with the convention that $\infty - \infty$ is interpreted as ∞ in the extended arithmetic of the first formula and as $-\infty$ in the second (Exercise 11.17). In particular, Q gives an elementary direction of descent if and only if

$$\sum_{j\in Q^+} g'_{j+}(v(j))e_Q(j) + \sum_{j\in Q^-} g'_{j-}(v(j))e_Q(j) < b_Q. \qquad (16)$$

Dual Optimality Theorem.   *A feasible solution u to the dual problem is optimal if and only if there is no elementary direction of descent for* $-\Psi$ *at u. Indeed, if u is feasible but not optimal, and if* $\varepsilon > 0$ *is any number such that there is no feasible solution x to the primal problem satisfying* $\Phi(x) - \Psi(u) \leq \varepsilon$, *then there must exist an elementary dual support Q and a value* $t > 0$ *such that, for any w satisfying* $-wE = e_Q$, *the vector* $u + tw$ *is feasible with*

$$\Psi(u + tw) > \Psi(u) + \varepsilon/|J|.$$

This could be derived from the primal optimality theorem (by reformulating the dual problem in primal form), but like the latter it will receive a direct algorithmic proof in Section 11I.

Consider now the intervals

$$C(Q) = \sum_{j\in J} e_Q(j)C(j) = \left\{ \sum_{j\in J} e_P(j)\xi_j \,\middle|\, \xi_j \in C(j) \right\}, \qquad (17)$$

$$\tilde{C}(Q) = \sum_{j\in J} e_Q(j)\tilde{C}(j) = \left\{ \sum_{j\in J} e_P(j)\xi_j \,\middle|\, \xi_j \in \tilde{C}(j) \right\}. \qquad (18)$$

These both have $[c^-(Q), c^+(Q)]$ as their closure, where

$$c^+(Q) = \sum_{j\in Q^+} e_Q(j)c^+(j) + \sum_{j\in Q^-} e_Q(j)c^-(j),$$

$$c^-(Q) = \sum_{j\in Q^+} e_Q(j)c^-(j) + \sum_{j\in Q^-} e_Q(j)c^+(j), \qquad (19)$$

or alternatively (by 8E(10))

$$c^+(Q) = \lim_{t\uparrow\infty} f'_-(t) = \lim_{t\uparrow\infty} f'_+(t),$$

$$c^-(Q) = \lim_{t\downarrow-\infty} f'_-(t) = \lim_{t\downarrow-\infty} f'_+(t). \qquad (20)$$

Note that

$$\tilde{C}(Q) = C(Q) = [c^-(Q), c^+(Q)] \quad \text{if} \quad \tilde{C}(j) = C(j) = [c^-(j), c^+(j)]$$

$$\text{for all } j.$$

Let us define an elementary dual support Q to be *unbalanced* if $\tilde{C}(Q) \subset (-\infty, b_Q)$. (This extends the terminology in Section 8K.)

Dual Descent Theorem. *Let Q be any elementary dual support, and let $f(t) = -\Psi(u + tw)$, where u is any feasible solution to the dual optimization problem and w is any vector such that $-wE = e_Q$. Then:*
1. *f is nonincreasing $\Leftrightarrow c^+(Q) \le b_Q$.*
2. *f is decreasing $\Leftrightarrow Q$ is unbalanced.*
3. *f is strongly decreasing $\Leftrightarrow c^+(Q) < b_Q$.*

This can be derived by arguments parallel to those for the primal descent theorem (Exercise 11.18).

11D. DUALITY AND THE EXISTENCE OF SOLUTIONS

The main theorems of monotropic programming will now be stated. They are close generalizations of the network programming results in Chapter 8, requiring at most an expansion in notation and terminology. First on the agenda is a pair of theorems sharpening the assertion of 11A(7).

Duality Theorem. *Except in the case where the infimum in the primal problem is ∞ (i.e., no primal feasible solution exists) and the supremum in the dual problem is $-\infty$ (i.e., no dual feasible solution exists), one has*

$$\begin{bmatrix} \text{inf in primal} \\ \text{problem} \end{bmatrix} = \begin{bmatrix} \text{sup in dual} \\ \text{problem} \end{bmatrix}.$$

Proof. From 11A(7) it is evident that [inf] $\ge$ [sup] always. The primal optimality theorem excludes the possibility that $\infty > $ [inf] $> $ [sup]: the existence of a number ε satisfying $0 < \varepsilon < $ [inf] $-$ [sup] would imply by the second assertion of that theorem that *every* primal feasible solution could be improved by at least $\varepsilon/|J|$, so that [inf] $= \infty$ (a contradiction). Similarly, the dual optimality theorem excludes the possibility that [inf] $> $ [sup] $> -\infty$. Therefore strict inequality cannot hold unless [inf] $= \infty$ and [sup] $= -\infty$.

For an example where [inf] $= \infty$ and [sup] $= -\infty$; see Exercise 8.38(a).

Equilibrium Theorem. *For x and u to solve the equilibrium problem, it is necessary and sufficient that x be an optimal solution to the primal problem and u be an optimal solution to the dual problem.*

Proof. We know from 11A(7) that x and u solve the equilibrium problem if and only if x satisfies $Ex = b$ and $\Phi(x) = \Psi(u)$. Since Φ never takes on the value $-\infty$ and Ψ never takes on ∞, it is impossible to have $\Phi(x) = \Psi(u)$ unless $\Phi(x)$ and $\Psi(u)$ are both finite, in which case x and u must actually be feasible (as long as $Ex = b$ too). Invoking the duality theorem, we conclude that x and u solve the equilibrium problem if and only if x and u are feasible solutions such that $\Phi(x)$ equals the infimum in the primal problem and $\Psi(u)$ equals the supremum in the dual problem. This is what we wanted to prove.

Existence of Feasible and Regularly Feasible Solutions

As already pointed out in Section 11A, the question of whether a feasible solution exists to the primal or dual problem is answered by the theorems in Section 10I. Thus the primal problem has a feasible solution if and only if

$$b_Q \in C(Q) \quad \text{for all elementary dual supports } Q \tag{1}$$

where b_Q is the number defined in 10I(4) and $C(Q)$ is the interval in 10I(6) (see also 11C̃(17)). The dual problem has a feasible solution if and only if

$$0 \in D(P) \quad \text{for all elementary primal supports } P, \tag{2}$$

where $D(P)$ is the interval in 10I(11) (see also 11C(4)).

Corresponding results about regularly feasible solutions are obtained simply by replacing $C(j)$ and $D(j)$ by $\tilde{C}(j)$ and $\tilde{D}(j)$ in the theorems of Section 10I; then $C(Q)$ and $D(P)$ are replaced by the intervals $\tilde{C}(Q)$ and $\tilde{D}(P)$ in 11C(18) and 11C(8), respectively. Thus the primal problem has a regularly feasible solution if and only if

$$b_Q \in \tilde{C}(Q) \quad \text{for all elementary dual supports } Q, \tag{3}$$

whereas the dual problem has a regularly feasible solution if and only if

$$0 \in \tilde{D}(P) \quad \text{for all elementary primal supports } P. \tag{4}$$

Of course (3) fails if and only if there exists an elementary dual support Q that is *unbalanced* as defined in Section 11C, that is, in the sense that $\tilde{C}(Q) \subset (-\infty, b_Q)$. Likewise, (4) fails if and only if there exists an elementary primal support that is unbalanced in the sense that $\tilde{D}(P) \subset (-\infty, 0)$. These conditions on Q and P reduce to $c^+(Q) < b_Q$ and $d^+(P) < 0$ for problems with $\tilde{C}(j) = C(j) = [c^-(j), c^+(j)]$ and $\tilde{D}(j) = D(j) = [d^-(j), d^+(j)]$.

Existence of Optimal Solutions

Unbalanced supports are important in connection with regularly feasible solutions as just noted, but we have also seen in the theorems of Section 11C that they give elementary directions of descent in which the descent goes on forever, that is, no finite stepsize t exhausts the potential for improvement of the objective. We look now at the role of unbalanced supports in characterizing the existence of optimal solutions.

Primal Existence Theorem. *Suppose the primal problem has a feasible solution. Then the following are equivalent:*

1. *The primal problem has an optimal solution.*
2. *The dual problem has a regularly feasible solution.*
3. *No elementary primal support P is unbalanced.*
4. *For every feasible solution x to the primal problem and every elementary primal support P, the function $f(t) = \Phi(x + te_P)$ attains its minimum at some $t \in R$.*

Dual Existence Theorem. *Suppose the dual problem has a feasible solution. Then the following are equivalent:*

1. *The dual problem has an optimal solution.*
2. *The primal problem has a regularly feasible solution.*
3. *No elementary dual support Q is unbalanced.*
4. *For every feasible solution u to the dual problem and every elementary dual support Q and vector w satisfying $-wE = e_Q$, the function $f(t) = -\Psi(u + tw)$ attains its minimum at some $t \in R$.*

These results, which extend the network theorems of Section 8L, will be proved in Section 11E.

Corollary. *The following are equivalent:*

1. *The equilibrium problem has a solution.*
2. *The primal and dual problems both have regularly feasible solutions.*
3. *No elementary primal or dual support is unbalanced.*

The corollary is obtained by combining the two existence theorems with the equilibrium theorem stated earlier in this section.

Primal Characterization Theorem. *Suppose the primal problem has a regularly feasible solution. Then every optimal solution to the primal problem is regularly feasible, and indeed, the optimal solutions to the primal problem (if any) are the vectors x such that, for some u, x and u solve the equilibrium problem.*

Proof. The existence of a feasible solution to the primal problem ensures by the duality theorem that

$$[\text{sup in dual}] = [\text{inf in primal}] < \infty.$$

Thus either the infimum in the primal is $-\infty$ (and no primal optimal solution exists) or the supremum in the dual is finite. In the latter case the dual has a feasible solution, and since Condition 2 of the dual existence theorem is satisfied, the dual actually has an optimal solution. Then from the equilibrium theorem we know that x is an optimal solution to the primal if and only if x and some u solve the equilibrium problem. The conditions of the equilibrium problem imply the regular feasibility of x because $\tilde{C}(j)$ is the projection of Γ_j on the horizontal axis: if $(x(j), v(j)) \in \Gamma_j$, then $x(j) \in \tilde{C}(j)$.

Dual Characterization Theorem. *Suppose the dual problem has a regularly feasible solution. Then every optimal solution to the dual problem is regularly feasible, and indeed, the optimal solutions to the dual problem (if any) are the vectors u such that, for some x, x, and u solve the equilibrium problem.*

Proof. The argument is the same as for the preceding theorem with only an interchange of the roles of x and u.

Uniqueness of Optimal Solutions

Simple conditions developed in Exercises 11.26 through 11.30 identify situations where the primal or dual problem can have at most one optimal solution.

11E.* BOUNDEDNESS PROPERTY

The question of boundedness of optimizing sequences to the primal and dual problems is tied in with the existence of optimal solutions as well as the behavior of algorithms. In this section we develop boundedness results and use them to prove the existence theorems in Section 11D.

An observation about the primal objective function Φ is helpful at the start. This function is continuous relative to the closed set

$$\{x \in R^J \mid \ c^-(j) \leqq x(j) \leqq c^+(j) \quad \text{for all} \ \ j \in J\}, \tag{1}$$

because each of the functions f_j is continuous relative to the closure $[c^-(j), c^+(j)]$ of $C(j)$ (see Section 8A). Every set of the form

$$\{x \in R^J \mid \ x \ \ \text{feasible with} \ \ \Phi(x) \leqq \alpha\} \quad \text{for} \ \ \alpha \in R \tag{2}$$

is therefore closed.

We say that *the primal problem has the boundedness property* if the sets (2) are all bounded. It can be shown (Exercise 11.37) that this holds if and only if every minimizing sequence for the primal problem is bounded. (A minimizing sequence is a sequence of feasible solutions x^k such that $\Phi(x^k)$ tends to the infimum in the problem.) When every minimizing sequence is bounded, then of course every minimizing sequence has cluster points, and because of the continuity of Φ on the set (1) these cluster points must all be optimal solutions to the primal problem.

In the dual problem the objective function Ψ is continuous relative to the set

$$\{u \in R^I \mid d^-(j) \leq v(j) \leq d^+(j) \quad \text{for all } j, \text{where } v = -uE\},$$

so that every set of the form

$$\{u \in R^I \mid u \text{ is feasible with } \Psi(u) \geq \beta\} \quad \text{for } \beta \in R \qquad (3)$$

is closed. However, no set of the latter form can be bounded unless E has full row rank; that is, there is a unique u corresponding to each $v \in \mathcal{D}$. We therefore look at the sets

$$\{v \in R^J \mid \exists \text{ feasible } u \text{ with } -uE = v \text{ and } \Psi(u) \geq \beta\} \quad \text{for } \beta \in R, \qquad (4)$$

which likewise are closed (recall that $\Psi(u)$ actually depends only on v; see Section 11A).

We say that the *dual problem has the boundedness property* if the sets (4) are all bounded. This implies that for every maximizing sequence of feasible solutions u^k to the dual, the corresponding sequence of vectors $v^k \in \mathcal{D}$ is bounded and therefore has cluster points. Every u corresponding to any such cluster point v is optimal.

In the two theorems we now state, the quantities $d^+(P)$ and $c^+(Q)$ are the ones defined in 10I(12) and 10I(7), respectively (also 11C(6) and 11C(19)).

Primal Boundedness Theorem. *Suppose the primal problem has at least one feasible solution. Then the following properties are equivalent:*

1. *The primal problem has the boundedness property.*
2. $d^+(P) > 0$ *for all elementary primal supports P.*
3. *The index set $F_d = \{j \in J \mid d^-(j) = d^+(j)\}$ contains no elementary primal supports, and there is a feasible solution u to the dual problem such that $d^-(j) < v(j) < d^+(j)$ for all $j \notin F_d$.*

Dual Boundedness Theorem. *Suppose the dual problem has at least one solution. Then the following properties are equivalent:*

1. *The dual problem has the boundedness property.*
2. $c^+(Q) > b_Q$ *for all elementary dual supports Q.*
3. *The index set $F_c' = \{j \in J \mid c^-(j) = c^+(j)\}$ contains no elementary dual supports, and there is a feasible solution x to the primal problem such that $c^-(j) < x(j) < c^+(j)$ for all $j \notin F_c'$.*

Note that in the terminology of Section 10C, F_d is an *independent* index set in the first theorem, whereas $J \setminus F_c'$ is a *spanning* index set in the second. For a combined version of the two theorems stated in terms of a basis F instead of the sets F_d and F_c', see Exercise 11.39.

Proof of the Primal Boundedness Theorem. The argument will be a copy of the one for the network case in Section 8M, except for terminology.

Condition 1 $\Rightarrow$ Condition 2. If there were an elementary primal support P with $d^+(P) \leq 0$, then by the primal descent theorem in Section 11C no nonempty set of the form (2) could be bounded.

Condition 2 $\Rightarrow$ Condition 3. For each $j \notin F_d$, let $\hat{D}(j)$ be the open interval $(d^-(j), d^+(j))$, whereas for $j \in F_d$, let $\hat{D}(j)$ be the degenerate interval consisting of the point $d^-(j) = d^+(j)$. Then for an elementary primal support P the interval

$$\hat{D}(P) = \sum_{j \in J} e_P(j)\hat{D}(j) = \left\{ \sum_{j \in J} e_P(j)\eta_j \mid \eta_j \in \hat{D}(j) \right\}$$

reduces to $(d^-(P), d^+(P))$ if $P \not\subset F_d$, but to the single number $D^-(P) = d^+(P)$ if $P \subset F_d$. The reverse support P' of course has $d^+(P') = -d^-(P)$. Therefore Condition 2 is equivalent to the condition that for every elementary primal support P one has $P \not\subset F_d$ and $0 \in \hat{D}(P)$. But if $0 \in \hat{D}(P)$ for every elementary primal support P, there exists by the dual feasibility theorem in Section 10I a $u \in R^I$ such that the vector $v = -uE$ has $v(j) \in \hat{D}(j)$ for all $j \in J$. According to the choice of $\hat{D}(j)$, this means that Condition 3 holds.

Condition 3 $\Rightarrow$ Condition 1. Let u be such that $d^-(j) < v(j) < d^+(j)$ for all $j \notin F_d$. Then as explained in the proof of the boundedness theorem for flows in Section 8M (the argument for "Condition 3 $\Rightarrow$ Condition 1"; nothing needs modification), there exist $\varepsilon > 0$ and $\mu \in R$ such that

$$|x(j)| \leq (\alpha + \mu)/\varepsilon \quad \text{for all} \quad j \notin F_d \quad \text{when} \quad x \quad \text{belongs to (2).} \tag{5}$$

Fix any $\bar{x}$ in (2). For any other x in (2) the vector $x - \bar{x}$ belongs to $\mathscr{C}$, because $E(x - \bar{x}) = b - b = 0$, and we have

$$|x(j) - \bar{x}(j)| \leq 2(\alpha + \mu)/\varepsilon \quad \text{for all} \quad j \notin F_d \tag{6}$$

by (5). Since F_d contains no elementary primal supports by hypothesis, it is an independent index set and therefore included in some basis F. The corresponding Tucker representation of $\mathscr{C}$ gives us for certain coefficients $d(j, k)$ an expression

$$x(j) - \bar{x}(j) = \sum_{k \notin F} a(j, k)[x(k) - \bar{x}(k)] \quad \text{for all} \quad j \in F$$

which is valid in particular whenever x belongs to the set (2). Since indexes $k \notin F$ are among those covered by (6), we obtain

$$|x(j) - \bar{x}(j)| \leq 2\left[\sum_{k \notin F} |a(j, k)| \right](\alpha + \mu)/\varepsilon \quad \text{for all} \quad j \in F_d.$$

This and (6) being satisfied for every x in (2), we conclude that the set (2) is bounded, regardless of the choice of α.

Proof of the Dual Boundedness Theorem. This result can be obtained by reformulating the dual problem in primal form (Exercise 11.38), or it can be derived by arguments parallel to those for the primal problem.

Proof of the Primal Existence Theorem. The equivalence of Conditions 2 and 3 in the primal existence theorem in Section 11D follows from the primal feasibility theorem in Section 10I as applied to the intervals $\tilde{D}(j)$ in place of $D(j)$. This was already noted in the discussion preceding the statement of the theorem. The equivalence of Conditions 3 and 4 is maintained by the primal descent theorem in 11C (see also Exercise 8.12). The corollary of that result shows too that Condition 1 implies Condition 4, for if Condition 4 were false, there would exist an elementary primal support P such that $\Phi(x + te_P)$ is a decreasing function of t for *every* feasible x and then no x could be optimal.

We know therefore that Conditions 2, 3, and 4 are equivalent and implied by Condition 1. The remaining task is to demonstrate that Condition 1 is implied by Conditions 2, 3, and 4. The argument is virtually identical to the one given in Section 8M in the proof of the existence theorem for flows: one need only make a few trivial changes in terminology, such as substituting "elementary primal support" for "elementary circuit."

Proof of the Dual Existence Theorem. Again one could derive this by putting the dual problem in primal form or by tracing arguments parallel to the primal case.

11F.* DECOMPOSITION

The conditions in the equilibrium problem provide a characterization of optimal solutions to the primal and dual problems of monotropic programming, as we have observed in Section 11D. This characterization has many interesting and useful consequences. We explore these now in the case where extra relations or variables have been added to a linear system that would otherwise be comparatively simple, perhaps because of a network being involved. A sort of decomposition turns out to be possible in which the simplicity of the linear system is restored by incorporating new cost terms in the objective. The coefficients in these terms can be determined by solving an auxiliary optimization problem.

Figure 11.1 provides the general setting for a discussion of decomposition. Here the index set I is partitioned into I_0 and and I_1. A primal problem

$$\text{minimize} \quad \sum_{j \in J} f_j(x(j)) \quad \text{subject to} \quad [x(j) \in C(j)] \quad \text{and}$$

$$\sum_{j \in J} e(i, j)x(j) = b(i) \quad \text{for} \quad i \in I_0$$

(1)

x

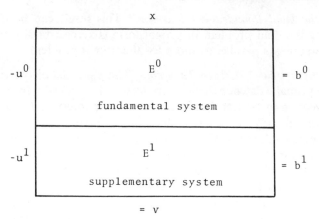

Figure 11.1

has been made more complicated by additional constraints

$$\sum_{j \in J} e(i, j)x(j) = b(i) \quad \text{for} \quad i \in I_1. \tag{2}$$

Traffic problems and multicommodity transportation problems as described in terms of the linear systems in Section 10F fit this picture, for instance.

In harmony with the notation in Figure 11.1 (where u^0 is the element of R^{I_0} obtained by restricting u to I_0, etc.) let us write

$$v = v^0 + v^1 \quad \text{with} \quad v^0 = -u^0 E^0, v^1 = -u^1 E^1.$$

The equilibrium conditions associated with problem (1) (2) consist of the equations in (1) (2) on x plus the existence of $u^0 \in R^{I_0}$ and $u^1 \in R^{I_1}$ such that

$$\left(x(j), v^0(j) + v^1(j)\right) \in \Gamma_j \quad \text{for all} \quad j \in J. \tag{3}$$

Consider now, in terms of an element $u^1 \in R^{I_1}$, the functions

$$\hat{f}_j(x(j)) = f_j(x(j)) + \left[\sum_{i \in I_1} u^1(i)e(i, j)\right]x(j)$$

$$= f_j(x(j)) - v^1(j)x(j) \quad \text{for} \quad j \in J. \tag{4}$$

Obviously, $\hat{f}_j$ is a closed proper convex function on R having the same interval of finiteness as f_j, namely, $C(j)$, and its characteristic curve $\hat{\Gamma}_j$ is the same as Γ_j except for being shifted downward by the amount $v^1(j)$. Thus the relations (3) can be written as

$$\left(x(j), v^0(j)\right) \in \hat{\Gamma}_j \quad \text{for all} \quad j \in J.$$

But these relations, together with the equations in (1), are the equilibrium conditions for the problem

$$\text{minimize} \quad \sum_{j \in J} f_j(x(j)) = \sum_{j \in J} f_j(x(j)) + \sum_{i \in I_1} u^1(i) \left[\sum_{j \in J} e(i, j) x(j) \right]$$

$$\text{subject to} \quad \sum_{j \in J} e(i, j) x(j) = b(i) \quad \text{for all} \quad i \in I_0. \tag{5}$$

The equilibrium conditions for problem (1) (2) can be identified therefore with the equations (2) plus the equilibrium conditions for problem (5) for some u^1 in (4). Note that (5) is a monotropic programming problem whose linear system of variables omits the extra relations that had made the original problem more complicated.

It will be demonstrated that the vectors u^1 which bring about this reduction are the solutions to an auxiliary optimization problem in the space R^{I_1}, namely,

$$\text{maximize} \quad p(u^1) - b^1 \cdot u^1 \quad \text{over all} \quad u^1 \in R^{I_1}, \text{ where}$$

$$p(u^1) = [\text{inf in problem (5)}]. \tag{6}$$

This is *not* a monotropic programming problem, although it actually does amount to maximizing a continuous concave function over a certain convex polyhedron, the closure of the set of points u^1 where $p(u^1) > -\infty$ (see Exercise 11.40).

Decomposition Theorem. *Suppose that problem (1) (2) has a regularly feasible solution. Then*

$$[\text{inf in (1) (2)}] = [\text{max in (6)}] \quad (\text{attained}). \tag{7}$$

Moreover the following conditions on a pair of vectors x and u^1 are equivalent:

1. *x is an optimal solution to (1) (2), and u^1 is an optimal solution to (6).*
2. *u^1 is such that x is an optimal solution to the corresponding problem (5) and x also happens to satisfy (2).*

Proof.   Any regularly feasible solution to (1) (2) is also a regularly feasible solution to (5) for arbitrary u^1. The existence of such regularly feasible solutions implies by the primal characterization theorem in Section 11D that the optimal solutions to these problems are characterized by corresponding equilibrium conditions. The analysis carried out earlier therefore applies: x is an optimal solution to (1) (2) if and only if Condition 2 holds for some u^1. In fact the possible vectors u^1 such that Condition 2 holds for a given x are precisely the ones such that, for some u^0, the pair (u^0, u^1) satisfies with x the equilibrium conditions associated with (1) (2) such pairs (u^0, u^1) being the optimal solutions to the dual of problem (1) (2) according to the equilibrium theorem in Section 11D. Likewise, for such a vector u^1 the vectors u^0 with the latter property are precisely the ones such that u^0 satisfies with x the equilibrium conditions associated with (5), such vectors u^0 being the optimal solutions to the dual of problem (5).

We need to tie these observations in with the auxiliary problem (6). Specifically, we must show that (7) is true, and that u^1 is an optimal solution to (6) if and only if there is a u^0 such that (u^0, u^1) is optimal for the dual of (1) (2).

The dual of problem (1) (2) can be expressed as

$$\text{maximize} \quad \Psi(u^0, u^1) = -b^0 \cdot u^1 - \sum_{j \in J} g_j(v^0(j) + v^1(j))$$

$$\text{over all} \quad u^0 \in R^{I_0} \quad \text{and} \quad u^1 \in R^{I_1}. \tag{8}$$

On the other hand, the dual of problem (5) (which depends on u^1) is

$$\text{maximize} \quad \hat{\Psi}(u^0) = -b^0 \cdot u^0 - \sum_{j \in J} \hat{g}_j(v^0(j))$$

$$\text{over all} \quad u^0 \in R^{I_0}, \tag{9}$$

where $\hat{g}_j$ is conjugate to the function $\hat{f}_j$ in (4). One has

$$\hat{g}_j(v^0(j)) = g_j(v^0(j) + v^1(j)),$$

so that

$$\hat{\Psi}(u^0) = b^1 \cdot u^1 + \Psi(u^0, u^1). \tag{10}$$

Our feasibility assumption ensures by the duality and existence theorems in Section 11D that

$$[\text{inf in (1) (2)}] = [\text{max in (8)}] \qquad (\text{attained by some } u^0, u^1),$$

$$p(u^1) = [\text{max in (9)}] \qquad (\text{attained by some } u^0),$$

(where attainment is trivial when both sides of the equation are $-\infty$). Referring to (10), we see that

$$p(u^1) - b^1 \cdot u^1 = \max_{u^0 \in R^{I_0}} \Psi(u^0, u^1)$$

and consequently

$$[\text{max in (8)}] = [\text{max in (6)}] \qquad (\text{attained by some } u^1).$$

Hence (7) is valid, and the optimal solutions to (6) are indeed the u^1 components of the optimal solutions to (8).

Corollary. *Suppose that problem* (1) (2) *has a regularly feasible solution and at least one optimal solution. Let u^1 be an optimal solution to problem* (6) *and suppose that the auxiliary problem* (5) *corresponding to u^1 has a unique optimal solution x. Then x is the unique optimal solution to* (1) (2).

The uniqueness property in the corollary is certainly present if enough of the functions f_j are strictly convex as in Exercise 11.29.

The theorem decomposes a given problem (1) (2) into three stages, at least in principle:

1. Solve problem (6) for u^1.
2. Next solve the corresponding "simple" problem (5) for x.
3. Then find among such possible x's one that also satisfies Equations (2).

Here the third stage could be troublesome, suggesting as it does that not just any solution to (5) will be acceptable. However, we have seen in the corollary that this stage is often superfluous. In practice, the first stage too can present difficulties because problem (6) may be solvable only in an approximate sense. Furthermore the objective function in (6) is itself defined in terms of optimization and may not be so easy to work with.

One approach to handling the objective function in the auxiliary problem (6) is to note that it has the form

$$q(u^1) = \max_{x \in X} \left\{ [E^1x - b^1] \cdot u^1 + \Phi(x) \right\}, \tag{11}$$

where

$$\Phi(x) = \sum_{j \in J} f_j(x(j)),$$

$$X = \left\{ x \in R^J \mid x(j) \in C(j) \quad \text{and} \quad E^1x = b^1 \right\}.$$

$$= [\text{set of feasible solutions to (5)}].$$

This function q can be viewed as the pointwise infimum of a collection of affine functions

$$l_x(u^1) = [E^1x - b^1] \cdot u^1 + \Phi(x)$$

indexed by the elements x of X; see Figure 11.2. The x's which give the infimum in (11) for a given u^1 are the optimal solutions to the problem (5)

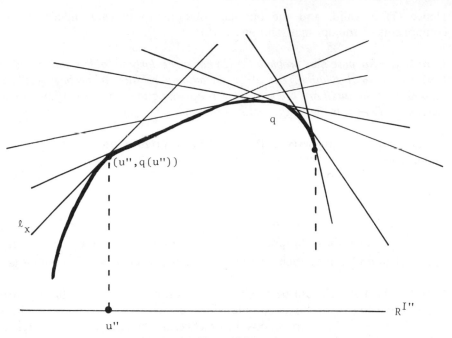

Figure 11.2

corresponding to u^1. Determining such an x amounts to determining an affine function l_x whose graph (a certain hyperplane in $R^{I_1} \times R$) touches the graph of q at the point $(u^1, q(u^1))$.

Algorithms can be devised that make use of such structure. A sequence of u^1 vectors is generated along with corresponding x vectors and affine functions l_x. The functions l_x generated up to a certain stage provide information about q that helps in choosing the next value of u^1. The u^1 sequence tends to the maximum of q, whereas the x sequence, or another constructed from it, yields in the limit an optimal solution to the original problem (1) (2). We can say no more here because it would take us too far into the general theory of convex programming.

For purposes of computation the factorization theorem in Section 10F provides an alternative way of taking advantage of special structure in the subsystem $E^0 x = b^0$.

11G.*   APPLICATION TO TRAFFIC EQUILIBRIUM

A fine example of how the decomposition idea in the preceding section may be applied occurs in the theory of traffic equilibrium. This concerns the optimal sharing of a single network G_* by several kinds of flow at the same time.

Let A_* and N_* denote the arc set and node set for G_*. The problem we wish to consider involves r separate vectors $x_k \in R^{A_*}$ and their sum z:

$$\text{minimize} \quad \sum_{j \in A_*} f_j(z(j)) \quad \text{subject to}$$

$$z(j) \in C(j) \quad \text{and} \quad x_k(j) \in C_k(j) \quad \text{for all} \quad j \in A_*, \quad \text{where} \quad z = \sum_{k=1}^{r} x_k,$$

$$\text{and} \quad y_k(k) = b_k(i) \quad \text{for all} \quad i \in N_*, \quad \text{where} \quad y_k = \operatorname{div} x_k. \tag{1}$$

Here x_k for $k = 1,\ldots,r$ is interpreted as the flow in G_* corresponding to the kth "commodity" or kind of traffic. Each such flow has its own distribution constraints specified by *closed* capacity intervals $C_k(j)$ (possibly $(-\infty, \infty)$) on the arcs and supplies $b_k(i)$ (possibly 0) at the nodes. However, the costs incurred in the arcs depend only on the *total flow* z, at least as we have expressed them (see Exercise 11.4 for a generalization). Adjustments must therefore be made between the different x_k's in minimizing overall cost.

Each f_j is of course a closed proper convex function on R with $C(j)$ as its interval of finiteness. The usual conventions will be followed in our analysis: g_j is the conjugate function, Γ_j the characteristic curve, $\tilde{C}(j)$ the projection of Γ_j on the horizontal axis, and so forth.

For present purposes, it is best to think of problem (1) as associated with the linear system in Figure 11.3, where we wish to have $y_k = b_k$ and $z' = 0$. (The column index set for this tableau consists really of $r + 1$ copies of A_*, say $A_*^0,\ldots,A_*^r$, but we shall continue to write $x_k(j)$ rather than $x(j^k)$, with j^k the copy of j in A_*^k. Likewise, the row index set can be identified with $N_*^1 \cup \cdots \cup N_*^r \cup A_*$, where each N_*^k is a copy of N_*.) Introducing the functions

$$f_{kj}(x_k(j)) = \begin{cases} 0 & \text{if } x_k(j) \in C_k(j), \\ \infty & \text{if } x_k(j) \notin C_k(j), \end{cases}$$

and recalling that $f_j(z(j)) = \infty$ when $z(j) \in C(j)$, we can write the problem equivalently in these terms as

$$\text{minimize} \quad \Phi(z, x_1,\ldots,x_r) = \sum_{j \in A_*} f_j(z(j)) + \sum_{k=1}^{r} \sum_{j \in A_*} f_{kj}(x_k(j))$$

$$\tag{2}$$

$$\text{subject to} \quad y_k = b_k \quad \text{for} \quad k = 1,\ldots,r \quad \text{and} \quad z' = 0.$$

We are ready then to invoke the decomposition scheme with (in the notation of Section 11F) $y^0 = (y_1,\ldots,y_r)$, $y^1 = z'$, $u^0 = (u_1,\ldots,u_r)$, $u^1 = w$. Note that

$$v_k = v_k^* - w, \quad \text{where} \quad v_k^* = \Delta u_k.$$

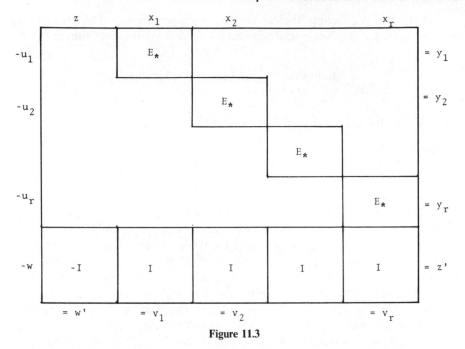

Figure 11.3

For each choice of the vector w, we consider the modified problem

$$\text{minimize} \quad \sum_{j \in A_*} f_j(z(j)) + \sum_{k=1}^{r} \sum_{j \in A_*} f_{kj}(x_k(j))$$

$$+ \sum_{j \in A_*} w(j) \left[-z(j) + \sum_{k=1}^{r} x_k(j) \right]$$

over all $z \in R^{A_*}$ and $x_k \in R^{A_*}$ such that $y_k = b_k$ for $k = 1, \ldots, r$,

(3)

where z can be chosen independently of the x_k's. We also look at the auxiliary problem

$$\text{maximize} \quad p(w) \quad \text{over all} \quad w \in R^{A_*}, \quad \text{where} \quad p(w) = [\inf \text{ in } (3)]. \quad (4)$$

We know from the decomposition theorem in Section 11F that if problem (2) is regularly feasible, then

$$[\inf \text{ in } (2)] = [\max \text{ in } (4)] \qquad \text{(attained)}.$$

Furthermore the vectors $(z, x_1, \ldots, x_r)$ and w such that $(z, x_1, \ldots, x_r)$ is an optimal solution to (2) and w an optimal solution to (4) are precisely the ones such that $(z, x_1, \ldots, x_r)$ also happens to satisfy $z = x_1 + \cdots + x_r$.

Our aim is to translate all of this back into the context of the original problem (1). First we make the crucial observation that the objective function in the modified problem (3) can be written as

$$\Phi_0^w(z) + \Phi_1^w(x_1) + \cdots + \Phi_r^w(x_r),$$

where

$$\Phi_0^w(z) = \sum_{j \in A_*} \left[f_j(z(j)) - w(j)z(j) \right],$$

$$\Phi_k^w(x_k) = \sum_{j \in A_*} \left[f_{kj}(x_k(j)) + w(j)x_k(j) \right]$$

$$= \begin{cases} \sum_{j \in A_*} w(j)x_k(j) & \text{if } x_k(j) \in C_k(j) \text{ for all } j \in A_*, \\ \infty & \text{otherwise.} \end{cases}$$

Correspondingly, (3) breaks down into $r + 1$ separate problems: minimize $\Phi_0^w(z)$ overall $z \in R^{A_*}$, and for $k = 1, \ldots, r$, minimize $\Phi_k^w(x_k)$ subject to $y_k = b_k$. The first of these breaks down even further: for each $j \in A_*$, minimize $f_j(z(j)) - w(j)z(j)$ over all $z(j) \in R$. Of course the infimum for the latter is $-g_j(w(j))$, and it is attained if and only if $(z(j), w(j)) \in \Gamma_j$. Minimizing $\Phi_k^w(x_k)$ subject to $y_k = b_k$, on the other hand, is a *linear optimal distribution problem* in G_* with cost coefficients $w(j)$:

$$\text{minimize} \quad \sum_{j \in A_*} w(j)x_k(j) \quad \text{subject to}$$

$$x_k(j) \in C_k(j) \quad \text{for all} \quad j \in A_* \quad \text{and} \quad y_k(i) = b_k(i) \quad \text{for all} \quad i \in N_*.$$

$$(5)$$

In particular then, the auxiliary problem (4) reduces to

$$\text{maximize} \quad - \sum_{j \in A_*} g_j(w(j)) + \sum_{k=1}^r p_k(w) \quad \text{over all } w \in R^{A_*},$$

$$\text{where} \quad p_k(w) = [\inf \text{ in } (5)], \quad \text{for } k = 1, \ldots, r.$$

The result stated for problems (2) and (4) now turns into the following.

Traffic Theorem.   *Suppose problem (1) is regularly feasible, that is, there exist flows* $\tilde{x}_1, \ldots, \tilde{x}_r$ *satisfying the constraints in (1), such that for* $\tilde{z} = \tilde{x}_1 + \cdots + \tilde{x}_r$

one actually has $\tilde{z}(j) \in \tilde{C}(j)$ for all $j \in A_*$. Then

$$[inf \ in \ (1)] = [max \ in \ (6)] \qquad (attained).$$

Moreover the following conditions on a set of flows $x_1, \ldots, x_r$ and vector w are equivalent:

  1. *$(x_1, \ldots, x_r)$ is an optimal solution to (1), and w is an optimal solution to (6).*

  2. *w is such that x_k is an optimal solution to the corresponding problem (5) for $k = 1, \ldots, r$, and $z = x_1 + \cdots + x_r$ satisfies $(z(j), w(j)) \in \Gamma_j$ for all $j \in A_*$.*

The interpretation is appealing: $w(j)$ is the marginal cost for additional flow in the arc j, and it depends only on the total $z(j)$ already flowing in j. The breakdown of $z(j)$ into the different kinds of flow $x_k(j)$ depends in turn only on these marginal costs. Each x_k is the solution to a separate linear optimal distribution problem with the $w(j)$'s as cost coefficients. The cost vector w itself solves an auxiliary optimization problem.

Condition (2) is said to define a generalized *Wardrop equilibrium* for traffic, relative to the curves Γ_j. (Here, if desired, the optimality of x_k could be replaced by the assertion that a potential u_k exists satisfying with x_k the equilibrium conditions for the corresponding problem (5).) Note that any maximal monotone relations Γ_j could serve in this role, at least mathematically. One could always construct f_j's with these as their characteristic curves (see Section 8B). What the theorem does is to identify every Wardrop equilibrium with a pair of variational principles in $(x_1, \ldots, x_r)$ and w.

11H.   BASIC DESCENT ALGORITHMS

The optimal distribution algorithm in Section 9A and optimal differential algorithm in Section 9C are just the network versions of conceptual algorithms that can be applied to any monotropic programming problem and its dual. These general methods will now be explained in the notation and terminology of Section 11A.

Primal Descent Algorithm

Given any regularly feasible solution x to the primal optimization problem, consider the dual feasibility problem of Section 10I with respect to the intervals

$$D_x(j) = [d_x^-(j), d_x^+(j)] = [f'_{j-}(x(j)), f'_{j+}(x(j))]$$

$$= \{\eta \in R | (x(j), \eta) \in \Gamma_j\}. \tag{1}$$

(Regular feasibility of x ensures that these are all nonempty.) Invoke any subroutine that resolves the corresponding alternatives in the dual feasibility theorem, or in other words, constructs either a $u \in R^I$ satisfying $d_x^-(j) \le v(j)$ $\le d_x^+(j)$ for all $j \in J$ or, if none exists, an elementary primal support P with $d_x^+(P) < 0$. (For instance, the primal rectification algorithm in Section 10J could accomplish this; it could be executed with refinements as described in Sections 10K and 10L.)

If a vector u is obtained, terminate; x solves the primal optimization problem, and u solves the dual optimization problem.

If a support P is obtained, then e_P is an elementary descent vector at x in the sense of Section 11C: for $f(t) = \Phi(x + te_P)$ one has $f_+'(0) = d_x(P) < 0$. Let

$$\alpha = \min\{t > 0 | f_+'(t) \ge 0\}. \tag{2}$$

Then $\alpha > 0$. If $\alpha = \infty$, that is, if $f_+'(t) < 0$ for all $t > 0$, terminate; P is unbalanced. The primal optimization problem has no optimal solution, and the dual optimization problem has no regularly feasible solution.

If $\alpha < \infty$, then α is the lowest of the values of t at which f attains its minimum. Let $x' = x + \alpha e_P$. Then x' is another regularly feasible solution to the primal optimization problem, and $\Phi(x') < \Phi(x)$. Repeat the procedure.

Justification

If a vector u is obtained in the subroutine, the corresponding v has $(x(j), v(j))$ $\in \Gamma_j$ for all $j \in J$ by the definition of the intervals (1). Since x, being feasible, satisfies $Ex = b$, we conclude that x and v solve the equilibrium problem. Hence by the equilibrium theorem in Section 11D, x and u are optimal solutions to the primal and dual problems.

If an elementary primal support P is obtained instead, one has $f_+'(0) = d_x^+(P) < 0$ by formula 11C(3) and the definition (1) of $d_x^+(j)$ and $d_x^-(j)$. If it happens that $f_+'(t) < 0$ for all $t > 0$, then since f_+' is a nondecreasing function (cf. Exercise 8.15) it must be true that $f_+'(t) < 0$ for all $t \le 0$ too, so that f itself is a decreasing function on R. Then by the primal descent theorem in Section 11C, P must be unbalanced. It follows in this case from the primal existence theorem in Section 11D that the primal optimization problem has no optimal solution, and the dual optimization problem has no regularly feasible solution.

On the other hand, if $f_+'(t) \ge 0$ for some $t > 0$, then $0 < \alpha < \infty$ in (2) by the right continuity of f_+'. In fact the limit relations between f_+' and f_-' in 8A(8) tell us that α is the lowest value of t such that $f_-'(t) \le 0 \le f_+'(t)$. The latter inequalities characterize the values of t where f attains its minimum (see Exercise 8.11). Obviously then, $f(\alpha) < f(0)$, that is, $\Phi(x') < \Phi(x)$; furthermore $f_+'(\alpha) > -\infty$ and $f_-'(\alpha) < \infty$, so that x' is again regularly feasible.

Step-Size Calculation

The exact calculation of α by (2) is easy for piecewise linear or quadratic monotropic programming problems. This is explained in Section 9A for the special case which is the optimal distribution algorithm. That explanation carries over to the present case under the observation that the breakpoints for $f(t) = \Phi(x + te_P)$ are the values of t such that $x(j) + te_P(j)$ is a breakpoint of f_j for some j.

Obviously, it is not necessary in all of this actually to calculate the (normalized) vector e_P, at each iteration. Any positive multiple of e_P, such as might be read from a Tucker tableau, would serve just as well and involve merely a change in the t scale.

Moving all the way to an exact minimum of $f(t)$ in the line search, as required by formula (2), is not really necessary to the notion of a descent algorithm. One can imagine other step-size rules that might work effectively, as long as they lead to a decrease of the objective function value $\Phi(x)$ from one iteration to the next. For instance, in piecewise linear programming, one could descend merely to the first breakpoint of f, beyond $t = 0$, before looking around again.

Termination

Even the optimal distribution algorithm can fail without extra assumptions and precautions, so one cannot expect more of the primal descent algorithm. The latter could, in general, produce an infinite sequence of regularly feasible solutions x^k such that $\Phi(x^k)$ decreases to a value μ which is not the infimum in the problem. This could even happen in the piecewise linear case. (The max flow example in Exercise 3.12 can be portrayed as an application of the primal descent algorithm where such misbehavior occurs.)

For optimal flow and optimal potential problems with piecewise linear costs, commensurability arguments were available for a proof of termination under minor restrictions. What we can claim, in general, is the following. *For monotropic programming problems that are piecewise linear, if the primal descent algorithm is implemented with*

1. *the dual rectification algorithm invoked as the subroutine in each iteration (the vector u used to initiate the subroutine each time being the u left over from the preceding application; the very first u can be chosen arbitrarily), and with*

2. *the dual rectification algorithm itself implemented in terms of the painted index algorithm with red-green preprocessing and OK priorities as described in the termination discussion in Section 10J (the Tucker tableau used to initiate this procedure each time being the one left over from the*

> *preceding round; the very first Tucker tableau can be chosen arbitrarily),*
> *and if*

3. *the step-size formula (2) is replaced by a certain more complicated rule*
    that generally gives smaller α values (i.e., 11K(3), where the intervals are
    the ones in formula (3) which follows),

then the algorithm must terminate in finitely many iterations.

Termination means of course that we either get optimal solutions x and u to
the primal and dual problems or an unbalanced elementary primal support P
indicating that

$$[\text{inf in primal}] = [\text{sup in dual}] = -\infty$$

(no other situations being possible in the piecewise linear case). Proof of this
result will be given in Section 11K: the implementation just described reduces
the primal descent algorithm to a special case of the general out-of-kilter
algorithm that will be described there (see Exercise 11.63).

Finite termination of the primal descent algorithm is also assured under a
different implementation, the one presented as the *primal simplex method* in
Section 11J. Then, however, the initial x must be chosen in a special way, and
precautions must be taken against a certain kind of degeneracy (see Section
11J).

Dual Descent Algorithm

Given any regularly feasible solution u to the dual optimization problem and
corresponding vector v, consider the primal feasibility problem of Section 10I
with respect to b and the intervals

$$C_u(j) = \left[c_u^-(j), c_u^+(j) \right] = \left[g'_{j^-}(v(j)), g'_{j^+}(v(j)) \right]$$

$$= \left\{ \xi \in R | (\xi, v(j)) \in \Gamma_j \right\}. \tag{3}$$

(Regularly feasibility of u ensures that these are all nonempty.) Invoke any
subroutine that resolves the corresponding alternatives in the primal feasibility
theorem, or in other words, constructs either an $x \in R^J$ satisfying $c_u^-(j) \leq$
$x(j) \leq c_u^+(j)$ for all $j \in J$, $Ex = b$, or, if none exists, an elementary dual
support Q with $c_u^+(Q) < b_Q$. (The dual rectification algorithm in Section 10J
could be of service here.)

If a vector x is obtained, terminate; x solves the primal optimization
problem, and u solves the dual optimization problem.

If a support Q is obtained, then any w satisfying $-wE = e_Q$ is a descent
vector for $-\Psi$ in the sense of Section 11C: for $f(t) = -\Psi(u + tw)$ one has

$f'_+(0) = c_u^+(Q) - b_Q < 0$. Let

$$\alpha = \min\{t > 0 | f'_+(t) \geq 0\}. \tag{4}$$

Then $\alpha > 0$. If $\alpha = \infty$, that is, if $f'_+(t) < 0$ for all $t > 0$, terminate; Q is unbalanced, the dual optimization problem has no optimal solution, and the primal optimization problem has no regularly feasible solution.

If $\alpha < \infty$, then α is the lowest of the values of t at which f attains its minimum. Let $u' = u + \alpha w$ (and $v' = v + \alpha e_Q$). Then u' is another regularly feasible solution to the dual optimization problem, and $\Psi(u') > \Psi(u)$. Repeat the procedure.

Justification

The arguments on which this algorithm is based are just like the ones given earlier for the primal descent algorithm (Exercise 11.44).

Stepsize Calculation

Again there are elementary procedures for obtaining the exact α in (4) in the case of piecewise linear or quadratic problems. The description given in Section 9A is valid with the breakpoints of f being the values of t such that $v(j) + te_Q(j)$ is a breakpoint of g_j for some j. Further e_Q could be replaced by any of its positive multiples. Formula (4) might profitably be weakened in some situations to allow for partial descent, not necessarily all the way to a minimizing value $t = \alpha$ for $f(t)$.

Termination

We have the same situation as for the primal descent algorithm. Termination is not assured without extra precautions and assumptions, although any sequence $\{u^k\}$ generated by the procedure does consist of regularly feasible solutions with ever-higher objective values $\Psi(u^k)$. The following can at least be demonstrated, however. *For monotropic programming problems that are piecewise linear, if the dual descent algorithm is implemented with*

1. *the primal rectification algorithm invoked as the subroutine in each iteration (the vector x used to initiate the subroutine each time being the x left over from the preceding application; the very first x can be chosen arbitrarily, subject only to $Ex = b$), and with*
2. *the primal rectification algorithm itself implemented in terms of the painted index algorithm with red-green preprocessing and OK priorities as described in the termination discussion in Section 10K (the Tucker tableau used to initiate this procedure being the one left over from the previous round; the initial Tucker tableau can be chosen arbitrarily), and if*

3.   *the step-size formula* (4) *is replaced by a certain more complicated rule that generally gives smaller α values (i.e., 11K(3)),*

then the algorithm must terminate in finitely many iterations. Thus in finitely many steps one will either get optimal solutions x and u to the primal and dual optimization problems or an unbalanced elementary dual support Q indicating that

$$[\text{inf in primal}] = [\text{sup in dual}] = \infty.$$

We shall see in Section 11K that this implementation of the dual descent algorithm reduces it to a form of the general out-of-kilter algorithm, and the termination results for the latter then yield the fact just stated.

Implementation as the dual simplex algorithm in Section 11J offers another approach to finite termination of the dual descent algorithm when applied to piecewise linear problems.

11I.   FORTIFIED AND DISCRETIZED DESCENT

The procedures and results in Sections 9H and 9I can be extended to general monotropic programming problems with hardly any alteration. We merely need to state the main facts in updated terminology. We observe, among other things, that the optimality theorems formulated in Section 11C and used to prove the existence and duality theorems in Section 11D are thereby established.

Fortified Primal Descent Algorithm

Fix any $\varepsilon > 0$, and let $\delta = \varepsilon/|J|$. Follow the same procedure as the primal descent algorithm, but only require x to be feasible (not regularly feasible), and take the intervals instead to be

$$D_x(j) = [d_x^-(j), d_x^+(j)] = \left[f_{j-}^{\prime\delta}(x(j)), f_{j+}^{\prime\delta}(x(j)) \right],$$

as defined in formula 9H(2) (see Figure 9.13 and Exercise 9.19). If the dual feasibility subroutine yields a vector u, then x and u satisfy $\Phi(x) - \Psi(u) \leq \varepsilon$ and are ε-optimal solutions to the primal and dual optimization problems. (They are feasible solutions that furnish values of the objective no more than ε from the optimal value.) If it terminates rather with an elementary primal support P giving an elementary direction of descent, then

$$\inf_{t>0} \Phi(x + te_P) < \Phi(x) - \delta.$$

In particular, if $\alpha < \infty$, one obtains $\Phi(x') < \Phi(x) - \delta$.

Fortified Dual Descent Algorithm

Again fix any $\varepsilon > 0$, and let $\delta = \varepsilon/|J|$. Follow the same procedure as in the dual descent algorithm, but only require u to be feasible (not regularly feasible), and take the intervals instead to be

$$C_u(j) = \left[c_u^-(j), c_u^+(j) \right] = \left[g_j'^{-\delta}(v(j)), g_j'^{\delta}(v(j)) \right],$$

as defined in formula 9H(7). If the primal feasibility subroutine terminates with a vector x, then x and u satisfy $\Phi(x) - \Psi(u) \le \varepsilon$ and are ε-optimal solutions to the primal and dual optimization problems. If it terminates instead with an elementary dual support Q and the corresponding w with $-wE = e_Q$, then

$$\sup_{t>0} \Psi(u + tw) > \Psi(u) + \delta.$$

In particular, if $\alpha < \infty$, one obtains $\Psi(u') > \Psi(u) + \delta$.

Remarks

The extension of the arguments of Section 9H to those algorithms is just a matter of expressing them more broadly; see Exercise 11.45. Note that the primal algorithm must terminate in finitely many iterations if the infimum in the primal problem is not $-\infty$, since an improvement of at least δ is made each time in the value of Φ. Likewise, the dual algorithm must terminate in finitely many iterations if the supremum in the dual problem is not $+\infty$.

Proof of the Optimality Theorems

The primal and dual optimality theorems in Section 11C are established by the properties of the respective algorithms just discussed. If there do not exist x and u satisfying $\Phi(x) - \Psi(u) \le \varepsilon$ (and $Ex = b$), then whichever algorithm is being used, the subroutine must yield an elementary support making possible an improvement by more than δ in the corresponding objective value.

Primal Discretized Descent Algorithm

This procedure depends on a fixed $\delta > 0$ and is the same as the primal descent algorithm, but x is required only to be feasible (not regularly feasible), and the intervals are instead taken to be

$$D_x(j) = \left[d_x^-(j), d_x^+(j) \right] = \left[\Delta_-^{\delta} f_j(x(j)), \Delta_+^{\delta} f_j(x(j)) \right]$$

as defined by formula 9I(1) (forward and backward difference quotients). If the dual feasibility subroutine yields a vector u, then for the value $\varepsilon = w(x, \delta)$ defined in 9I(5) (see also 9I(4) and the comments that follow it) one has $\Phi(x) - \Psi(u) \le \varepsilon$, and x and u are ε-optimal solutions to the primal and dual optimization problems. Furthermore x is δ-quasi-optimal in the sense that $\Phi(x + te_P) \ge \Phi(x)$ for all elementary primal supports P and all $t \ge \delta$.

If, on the other hand, the dual feasibility subroutine yields an elementary primal support P such that $d_x^+(P) < 0$, a discretized line search is executed as follows. Let n be the smallest integer such that

$$\Phi(x + (n + 1)\delta e_P) \ge \Phi(x + n\delta e_P).$$

Unless P is unbalanced (in which event the procedure terminates with the information that the primal optimization problem has no optimal solution and the dual optimization problem has no regularly feasible solution) such an n must exist, and $n \ge 1$. Then $x' = x + n\delta e_P$ is a feasible solution with $\Phi(x') < \Phi(x)$, and the procedure can be repeated.

Dual Discretized Descent Algorithm

Again the procedure depends on a fixed $\delta > 0$ and is the same as the dual descent algorithm, but u is required only to be

$$C_u(j) = [c_u^-(j), c_u^+(j)] = [\Delta_-^\delta g_j(v(j)), \Delta_+^\delta g_j(v(j))]$$

as defined by formula 9I(15). If the primal feasibility subroutine yields a vector x, then for the value $\varepsilon = z(u, \delta)$ defined in 9I(17) (see also 9I(16)) one has $\Phi(x) - \Psi(u) \le \varepsilon$, and x and u are ε-optimal solutions to the primal and dual problems. Furthermore u is δ-quasi-optimal in the sense that $\Psi(u + tw) \le \Psi(u)$ for all $t \ge \delta$ when w is a vector such that $-wE = e_Q$ for some elementary dual support Q.

If, on the other hand, the primal feasibility subroutine yields an elementary dual support Q such that $c_u^+(Q) < b_Q$, a discretized line search is executed as follows. Let n be the smallest integer such that

$$\Psi(u + (n + 1)\delta w) \ge \Psi(u + n\delta w),$$

where w is a vector such that $-wE = e_Q$. Unless Q is unbalanced (in which event the procedure terminates with the information that the dual optimization problem has no optimal solution and the primal optimization problem has no regularly feasible solution), such an n must exist, and $n \ge 1$. Then $u' = u + n\delta w$ is a feasible solution with $\Psi(u') > \Psi(u)$, and the procedure can be repeated.

Remarks

Justification follows the lines in Section 9I exactly. Termination with x and u having the properties described is guaranteed in finitely many iterations of the primal algorithm if the primal problem has the boundedness property in

Section 11E, and in finitely many iterations of the dual algorithm if the dual problem has the boundedness property.

The results in Sections 9K and 9L on calculating ε-optimal solutions for a preassigned ε do *not* transfer automatically to the general case of monotropic programming, because they are based on commensurability arguments that are unavailable outside of networks, or more exactly, the context of linear systems which are unimodular in the sense of Exercise 10.2.

11J.   SIMPLEX METHODS

The classical simplex method and its variants are the techniques most commonly used for solving a general linear programming problem, but in broadened formulation they can also be applied to any monotropic programming problem. Methods of this type can be viewed as realizations of the descent algorithms in Section 11H which are characterized by employing Tucker representations that simplify as far as possible the search for an elementary direction of descent at each iteration. They have the feature, however, that unless "degeneracy" is assumed away or addressed by supplementary procedures, a sequence of iterations could be produced in which the step size α is 0 and the current feasible solution and objective value remain the same.

Recall that a number $\xi \in R$ is said to be a *first-order breakpoint* of f_j if $f_j'^-(\xi) < f_j'^+(\xi)$. Any such number belongs of course to the interval $\tilde{C}(j)$. If f_j is linear on $[c^-(j), c^+(j)]$, the only first-order breakpoints are $c^-(j)$ and $c^+(j)$ (when finite). The case where f_j is *piecewise differentiable* can be identified with the one where f_j has finitely many first-order breakpoints.

For a regularly feasible solution x to the primal optimization problem, the set

$$F_x = \left\{ j \in J \,|\, x(j) \text{ is not a first-order breakpoint of } f_j \right\} \qquad (1)$$

will be important. A basis $F \subset J$ will be said to be *of primal simplex type with respect to* x if the number of indexes in $F_x \cap F$ cannot be increased by pivoting, in other words, if in the corresponding Tucker tableau there is no entry $a(\bar{j}, \bar{k}) \neq 0$ with $\bar{j} \in F_x, \bar{k} \notin F_x$. Any basis not already of this type can be made into one by pivoting on pairs $(\bar{j}, \bar{k})$ as just described. Each such pivot transformation puts another index of F_x into row position without taking any out of row position, so a finite number of steps will suffice.

Later on, for a regularly feasible solution u to the dual optimization problem and corresponding $v = -uE$, we shall be concerned with the set

$$F_v' = \left\{ j \in J \,|\, v(j) \text{ is not a first-order breakpoint of } g_j \right\}. \qquad (2)$$

A basis $F \subset J$ will be said to be *of dual simplex type with respect to* u if the number of indexes in $F_v' \cap F'$, where $F' = J \setminus F$, cannot be increased by pivoting, in other words, if in the corresponding Tucker tableau there is no entry $a(\bar{j}, \bar{k}) \neq 0$ with $\bar{j} \notin F_v'$, $\bar{k} \in F_v'$. Again, by pivoting on pairs $(\bar{j}, \bar{k})$ with the latter property as long as they are present, we can transform an arbitrary basis into one of this type.

Primal Simplex Method

Given any regularly feasible solution x to the primal optimization problem along with a basis F and corresponding Tucker tableau (with entries $a(j, k)$ for $j \in F$ and $k \in F'$), first perform pivot steps as just described, if necessary, so as to ensure that the basis is of primal simplex type with respect to x. Then for the intervals

$$D_x(j) = \left[d_x^-(j), d_x^+(j) \right] = \left[f_{j}'^-(x(j)), f_{j}'^+(x(j)) \right]$$

choose numbers $v(j) \in D_x(j)$ for $j \in F$ and get the corresponding numbers $v(k) = -\sum_{j \in F} v(j) a(j, k)$ for $k \in F' = J \setminus F$. If it happens that $v(k) \in D_x(k)$ for all $k \in F'$, terminate; x is an optimal solution to the primal problem, and v (or rather, any u with $-uE = v$) is an optimal solution to the dual problem.

Otherwise there must be an index $\bar{k} \in F'$ such that either $v(\bar{k}) > d_x^+(\bar{k})$ or $v(\bar{k}) < d_x^-(\bar{k})$. In the first of these cases let $P = P_{\bar{k}}$ (where $P_{\bar{k}}$ is the elementary primal support associated with $\bar{k}$ and F; this corresponds to the $\bar{k}$ column of the Tucker tableau; see Section 10C), but in the second case let P be the reverse of $P_{\bar{k}}$. Next do a line search in the direction of the elementary vector e_P: for $f(t) = \Phi(x + t e_P)$ let

$$\alpha = \min\left\{ t \geq 0 \,|\, f_+'(t) \geq 0 \right\} \in [0, \infty].$$

If $\alpha = \infty$, that is, if $f_+'(t) < 0$ for all $t \geq 0$, terminate; P is unbalanced. The primal optimization problem has no optimal solution, and the dual optimization problem has no regularly feasible solution. If $0 < \alpha < \infty$, replace x by $x + \alpha e_P$, and return to the beginning. (The basis F may no longer be of primal simplex type with respect to the new x, but this property will be restored by the pivoting rule at the beginning of the algorithm.)

The case where $\alpha = 0$ is termed the *degenerate* case; then $d_x^+(P) = f_+'(0) \geq 0$, so P does not give an elementary direction of descent at x. In this event there must be at least one $\bar{j} \in F \setminus F_x$ with $a(\bar{j}, \bar{k}) \neq 0$. Pivot on any such $(\bar{j}, \bar{k})$, and return to the beginning with the same x but this modified basis. (An alternative approach to the degenerate case is described later in this section.)

Rule of Thumb (optional)

If more than one index is eligible to be chosen as $\bar{k}$, choose one for which the distance of $v(\bar{k})$ from $D_x(\bar{k})$ is maximal.

Justification and Comments

Everything follows the pattern of the primal descent algorithm in Section 11H and is justified accordingly, except for the assertion in the degenerate case about the existence of $\bar{j}$. Before verifying the latter, it will be good to look more generally at the nature of the line search and see why bases that are of primal simplex type offer an advantage.

Remember that P is either $P_{\bar{k}}$ or its reverse, depending on whether $v(\bar{k}) > d_x^+(\bar{k})$ or $v(\bar{k}) < d_x^-(\bar{k})$. Therefore e_P or its negative can be constructed from the entries $a(j, \bar{k})$ in the $\bar{k}$ column of the Tucker tableau corresponding to F as in the basis theorem in Section 10C. This construction involves a normalization factor $\lambda(\bar{k}) > 0$ that can be ignored for present purposes. Thus the line search can just as well be carried out in terms of

$$h(s) = \Phi\left(x + s\lambda(\bar{k})e_{P_{\bar{k}}}\right)$$

$$= f_{\bar{k}}\left(x(\bar{k}) + s\right) + \sum_{j \in F} f_j\left(\bar{x}(j) + sa(j, \bar{k})\right), \tag{3}$$

with minimization subject to $s \geq 0$ when $d_x^+(\bar{k}) - v(\bar{k}) < 0$, but $s \leq 0$ when $v(\bar{k}) - d_x^-(\bar{k}) < 0$. The quantity $\lambda(\bar{k})d_x^+(P)$ equals $h_+'(0)$ in the first case and $-h_-'(0)$ in the second. Observe now that since $d_x^-(j) \leq v(j) \leq d_x^+(j)$ for $j \in F$, one has

$$h_+'(0) = d_x^+(\bar{k}) + \sum_{\substack{j \in F \\ a(j, \bar{k}) > 0}} d_x^+(j)a(j, \bar{k}) + \sum_{\substack{j \in F \\ a(j, \bar{k}) < 0}} d_x^-(j)a(j, \bar{k})$$

$$\geq d_x^+(\bar{k}) + \sum_{j \in F} v(j)a(j, \bar{k}) = d_x^+(\bar{k}) - v(\bar{k}),$$

$$-h_-'(0) = -d_x^-(\bar{k}) - \sum_{\substack{j \in F \\ a(j, \bar{k}) > 0}} d_x^-(j)a(j, \bar{k}) - \sum_{\substack{j \in F \\ a(j, \bar{k}) < 0}} d_x^+(j)a(j, \bar{k})$$

$$\geq -d_x^-(\bar{k}) - \sum_{j \in F} v(j)a(j, \bar{k}) = v(\bar{k}) - d_x^-(\bar{k}),$$

where equality holds in both formulas when

$$d_x^-(j) = v(j) = d_x^+(j) \quad \text{for every } j \in F \text{ with } a(j, \bar{k}) \neq 0,$$

or in other words, when

$$a(j, \bar{k}) = 0 \quad \text{for all } j \in F \setminus F_x \text{ (if any)}. \tag{4}$$

Therefore

$$d_x^+(P) < 0 \quad \text{when (4) holds (in particular, when } F \subset F_x). \tag{5}$$

This last fact is the key to understanding the simplex method. By keeping the set $F \setminus F_x$ minimal (as required by the concept of a basis of "simplex type"), we not onl̲   ̲ ̲minate some of the ambiguity in the choice of $v(j) \in D_x(j)$ for $j \in F$ but also increase the chance that (4) holds. Thus we increase the chance that the simple detection of a $\bar{k}$ with $v(\bar{k}) > d_x^+(\bar{k})$ or $v(\bar{k}) < d_x^-(\bar{k})$ will immediately furnish an elementary direction of descent.

Sometimes of course one may have $d_x^+(P) \geq 0$ and be unable to make any improvement in x. (The line search as described here in terms of $h(s)$ terminates then with $s = 0$.) Then, however, there must exist by (5) some $\bar{j}$ with $a(\bar{j}, \bar{k}) \neq 0$, as claimed in the algorithm. The purpose of pivoting on $(\bar{j}, \bar{k})$ in this case is to alter the basis so as to provide a fresh opportunity on the next round. There is no guarantee that anything is really accomplished by such pivoting, since it does not involve any change in x or $\Phi(x)$.

Note that when x is replaced by $x + \alpha e_P$ this means in terms of the optimal value $s = \beta$ in the line search of $h(s)$ in (3) that $x(j)$ is replaced by $x(j) + \beta a(j, \bar{k})$ for $j \in F$ and $x(\bar{k})$ by $x(\bar{k}) + \beta$. The possible effect on the index set F_x is that one or more of the indexes $j \in F$ with $a(j, \bar{k}) \neq 0$, or $\bar{k}$ itself, could enter or leave F_x, while all other indexes keep their status. If $\bar{k}$ enters F_x, it will be necessary, in restoring the basis to primal simplex type, to see whether there is a $\bar{j} \in F \setminus F_x$ with $a(\bar{j}, \bar{k}) \neq 0$; pivoting on such a $(\bar{j}, \bar{k})$ will increase the number of elements of F_x that are in row position. Typically this is the most that ever happens in the restoration process, but sometimes there might be other indexes k that already belonged to $F_x \cap F'$ and previously could not be pivoted into row position but now can be, because the status of certain indexes $j \in F$ has been altered.

Piecewise Linear Case

When the problem is piecewise linear, that is, every f_j is (linear or) piecewise linear relative to the interval $C(j)$, it makes sense to say that the feasible solution x is *quasi extreme* if the index set F_x is independent in the sense of Section 10C (i.e., if every basis F of primal simplex type relative to x has $F \supset F_x$). Any extreme feasible solution as defined in Section 10M is in particular quasi extreme, and the two concepts coincide when every f_j is actually linear relative to $C(j)$. There are at most finitely many feasible solutions of this special kind under our assumption of piecewise linearity (Exercise 11.46). A feasible solution that is quasi extreme can readily be constructed (see Exercise 11.47).

We claim that *if the primal simplex method is applied to a (linear or) piecewise linear programming problem and initiated with a feasible solution that is quasi extreme, then every feasible solution produced by the algorithm will be quasi extreme*. Indeed, in this case when it comes to restoring the basis to primal simplex type after x has been improved, there can be at most one index of F_x in column position and consequently targeted for transfer to row position, namely, $\bar{k}$. Moreover in this case $\bar{k}$ must just have entered F_x due to the improvement, since all the elements of F_x were in row position before the improvement. It follows that $\alpha \neq 0$ in the line search; in fact the function f in the line search is piecewise linear, and since α is by definition the lowest value of t that minimizes t, α must be a breakpoint of f. This means that *for some* $j \in J$ *such that* $x(j)$ *is altered by the improvement, the new* $x(j)$ *must be a breakpoint of* f_j (see Exercise 11.50). But only $x(\bar{k})$ and the $x(j)$'s with $a(j, \bar{k}) \neq 0$ are affected. Hence if $\bar{k}$ has entered F_x (i.e., $x(\bar{k})$ has ceased to be a breakpoint of $f_{\bar{k}}$), there must be some $\bar{j} \in F$ with $a(\bar{j}, \bar{k}) \neq 0$ such that $\bar{j}$ does not belong to the updated F_x. Pivoting on $(\bar{j}, \bar{k})$ is then possible and results in all the elements of F_x being in row position, namely, $F_x \subset F$. This single step not only restores the basis to primal simplex type, it shows that the improved x is again quasi extreme.

Termination

Let us call a feasible solution x *nondegenerate* if F_x is a spanning set of indexes in the sense of Section 10C. Then for F to be a basis of simplex type with respect to x, one must have $F \subset F_x$. In that case when P is determined, one will necessarily have $d_x^+(P) < 0$ by (5), so that an improvement in x will be possible. It follows that *if every feasible solution in the sequence produced by the algorithm is nondegenerate, then the corresponding sequence of objective values will be decreasing, so that no feasible solution can ever be repeated.*

Combining this with the facts observed previously in the piecewise linear case, we get a simple criterion for finite termination. *Suppose the primal simplex method is applied to a piecewise linear programming problem with the property that every quasi extreme feasible solution is nondegenerate. If the initial feasible solution is quasi extreme, then the algorithm is sure to terminate in finitely many iterations.* Namely, every iteration must lead to a quasi extreme feasible solution different from all previous ones, so an infinite number of iterations is impossible.

Circumventing Degeneracy

One way of limiting the threat of degenerate iterations, which unfortunately could far outnumber the good iterations in some applications of the simplex method (especially to network problems), is to realize the method even more thoroughly as an instance of the primal descent algorithm in Section 11H. Specifically, apply the dual rectification algorithm as a subroutine relative to the intervals $D_x(j)$, taking as initial v a vector constructed in the manner described, namely, by choosing arbitrary $v(j) \in D_x(j)$ for $j \in F$ and then setting $v(k) = -\sum_{j \in F} v(j)a(j, k)$ for $k \notin F$. For such a v the dual rectification algorithm will immediately yield the same P as before, if $d_x^+(P) < 0$, but otherwise it will continue searching until a true direction of descent is found. As a matter of fact, if the dual rectification algorithm is implemented in terms of the painted index algorithm (with red-green preprocessing and the OK priority rule to ensure termination, as described in Section 10J), the Tucker tableau at hand serves for a start, and l can be chosen to be an index $\overline{k}$ as before. When the procedure is finished, either we will have determined a v such that $v(j) \in D_x(j)$ for all $j \in J$ (leading to the conclusion that x and v are optimal), or we will have passed to a tableau that yields from a certain column an elementary direction of descent that can be utilized just as before.

In this mode the simplex method has finite termination when applied to any piecewise linear programming problem and initial feasible solution that is quasi extreme. No assumption of nondegeneracy is needed. Termination follows simply from the fact that a sequence of quasi-extreme feasible solutions is generated in which the values of the objective function Φ are strictly decreasing. Since there are only finite many quasi-extreme feasible solutions to the problem altogether (see Exercise 11.46), the sequence cannot be infinite.

Dual Simplex Method

Given any regularly feasible solution u to the dual optimization problem and corresponding $v = -uE$, along with a basis F and its associated Tucker tableau, first perform pivot steps as described earlier (just before the statement of the primal simplex method), if necessary, so as to ensure that the basis is of dual simplex type with respect to u. Then for the intervals

$$C_u(j) = \left[c_u^-(j), c_u^+(j)\right] = \left[g_j^-(v(j)), g_j^+(v(j))\right]$$

choose numbers $x(k) \in C_u(k)$ for $k \in F' = J \setminus F$, and determine the unique corresponding numbers $x(j)$ for $j \in F$ such that $Ex = b$ (see remarks that follow). Next check whether $x(j) \in C_u(j)$ for every $j \in F$. If so, terminate; x is an optimal solution to the primal problem, and v is an optimal solution to the dual problem.

Otherwise there must be an index $\bar{j} \in F$ such that either $x(j) > c_u^+(j)$ or $x(j) < c_u^-(j)$. In the first of these cases let $Q = Q_{\bar{j}}$ (where $Q_{\bar{j}}$ is the elementary dual support associated with $\bar{j}$ and F; this corresponds to the $\bar{j}$ row of the Tucker tableau; see Section 10C), but in the second case let Q be the reverse of $Q_{\bar{j}}$. Next do a line search in the direction of the elementary vector e_Q: setting $f(t) = -\Psi(u + tw)$ for any $w \in R^I$ with $-wE = e_Q$, let

$$\alpha = \min\{t \geq 0 | f_+'(t) \geq 0\} \in [0, \infty].$$

If $\alpha = \infty$, that is, if $f_+'(t) < 0$ for all $t \geq 0$, terminate; Q is unbalanced. The dual optimization problem has no optimal solution, and the primal optimization problem has no regularly feasible solution. If $0 < \alpha < \infty$, replace u by

$u + \alpha w$, and return to the beginning. (The basis F may no longer be of dual simplex type with respect to the new u, but this property will be restored by the pivoting rule at the beginning of the algorithm.)

The case where $\alpha = 0$ is termed the *degenerate* case; then $c_u^+(Q) - b_Q \geq 0$, so Q does not provide an elementary direction of descent at u. Then there must exist at least one $\bar{k} \in F' \setminus F_v'$ with $a(\bar{j}, \bar{k}) \neq 0$. Pivot on any such $(\bar{j}, \bar{k})$, and return to the beginning with the same u but this modified basis.

Remarks

The assertion that once the values $x(k)$ for $k \in F'$ have been chosen, there exist unique values $x(j)$ for $j \in F$ such that $Ex = b$ is valid because the columns of E corresponding to indexes $j \in F$ form a basis for the column space of E (by the definition of "basis"). In fact these values can be determined from equations of the form

$$x(j) = \sum_{k \in F'} a(j, k) x(k) + r(k) \quad \text{for } j \in F, \tag{6}$$

where the $a(j, k)$'s are still the entries in the Tucker tableau associated with F and the $r(k)$'s are unique numbers depending only on F and b (Exercise 11.51); the pivoting rules for the $a(j, k)$'s can readily be extended to the $r(k)$'s. (Of course one could also arrange that $b = 0$, namely, by reformulating the problems in terms of the extended spaces $\mathscr{C}$ and $\mathscr{D}$; see Section 11A.)

The various properties of the primal simplex method hold true for the dual simplex method with slight changes of notation. These are laid out in Exercises 11.56 through 11.58.

11K.   GENERAL OUT-OF-KILTER ALGORITHM

The out-of-kilter algorithm in Section 9F can be extended to all monotropic programming problems that are piecewise linear. The two rectification algorithms in Section 10J turn out then to be special cases, as do the primal and dual descent algorithms of Section 11H in a certain implementation. The most remarkable feature is that the method is symmetric with respect to the primal and dual optimization problems and can be initiated with any pair of vectors x and u, not necessarily feasible. This makes it especially attractive in situations where a change in data requires a process of reoptimization of an x or u that used to be optimal.

We are supposing now that f_j and g_j are piecewise linear for every $j \in J$, so

$$C(j) = \tilde{C}(j) = [c^-(j), c^+(j)] \quad \text{and} \quad D(j) = \tilde{D}(j) = [d^-(j), d^+(j)].$$

An unbalanced primal support P is therefore one that satisfies $d^+(P) < 0$ (cf.

10I(12)), whereas an unbalanced dual support Q is one that satisfies $c^+(Q) < b_Q$ (see 10I(4)(7)). The characteristic curves Γ_j are of "staircase" type. For $x \in R^J$ and $u \in R^I, v = -uE$, we define as usual

$$[d_x^-(j), d_x^+(j)] = [f_{j^-}'(x(j)), f_{j^+}'(x(j))],$$

$$[c_u^-(j), c_u^+(j)] = [g_{j^-}'(v(j)), g_{j^+}'(v(j))], \tag{1}$$

so that

$$(x(j), v(j)) \in \Gamma_j \Leftrightarrow d_x^-(j) \le v(j) \le d_x^+(j) \Leftrightarrow c_u^-(j) \le x(j) \le c_u^+(j). \tag{2}$$

(When $x(j) > c^+(j)$, both $d_x^-(j)$ and $d_x^+(j)$ are regarded as $+\infty$, and when $x(j) < c^-(j)$ they are $-\infty$; similarly for $c_u^-(j)$ and $c_u^+(j)$.)

Out-of-Kilter Algorithm

Given a piecewise linear pair of monotropic programming problems, begin with any $x \in R^J$ satisfying $Ex = b$ and any $u \in R^I$ and $v = -uE$. Call an index $j \in J$ *out of kilter* if $(x(j), v(j)) \notin \Gamma_j$ (otherwise *in kilter*). If there are no out-of-kilter arcs, terminate: x and u solve the primal and dual problems, respectively.

Otherwise paint the indexes $j \in J$ according to the location of the point $(x(j), v(j))$ relative to Γ_j, just as in the network version of the algorithm; see Figure 11.4. Select any out-of-kilter index l, which will necessarily be white or black, and apply the painted index algorithm (or any other procedure for deciding between the alternatives of the painted index theorem).

CASE 1.   There is an elementary primal support P as in Alternative 1 of the painted index theorem. Calculate

$$\alpha = \min \begin{cases} [c_u^+(j) - x(j)]/e_P(j) & \text{for } j \in P^+, \\ [c_u^-(j) - x(j)]/e_P(j) & \text{for } j \in P^-, \\ \bar{\alpha}, \end{cases} \tag{3}$$

where

$$\bar{\alpha} = \begin{cases} [c_u^-(l) - x(l)]/e_P(l) & \text{if } l \text{ is white } (l \in P^+), \\ [c_u^+(l) - x(l)]/e_P(l) & \text{if } l \text{ is black } (l \in P^-). \end{cases}$$

Then $\alpha > 0$. If $\alpha = \infty$, terminate; P is unbalanced. The dual problem has no feasible solution. If the primal problem has a feasible solution at all, it has

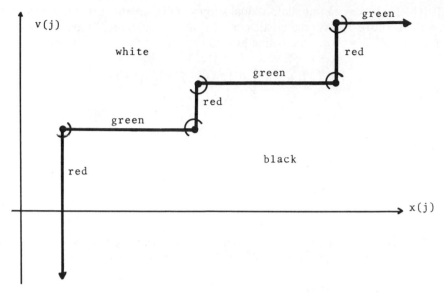

Figure 11.4

[inf] $= -\infty$. Otherwise replace x by $x + \alpha e_P$, and return to the beginning with this new x and the same u.

CASE 2.   There is an elementary dual support Q as in (b) of the painted index algorithm. Calculate

$$\alpha = \min \begin{cases} [d_x^+(j) - v(j)]/e_Q(j) & \text{for } j \in Q^+, \\ [d_x^-(j) - v(j)]/e_Q(j) & \text{for } j \in Q^-, \\ \bar{\alpha}, \end{cases} \tag{4}$$

where

$$\bar{\alpha} = \begin{cases} [d_x^-(l) - v(l)]/e_Q(l) & \text{if } l \text{ is white } (l \in Q^+), \\ [d_x^+(l) - v(l)]/e_Q(l) & \text{if } l \text{ is black } (l \in Q^-). \end{cases}$$

Then $\alpha > 0$. If $\alpha = \infty$, terminate; Q is unbalanced. The primal problem has no feasible solution. If the dual problem has a feasible solution at all, it has [sup] $= \infty$. Otherwise replace v by $v + \alpha e_Q$ (and u by $u + \alpha w$ where w is any vector satisfying $-wE = e_Q$), and return to the beginning with the same x and this new u.

Improvement Property

The distances of $x(j)$ from $[c_u^-(j), c_u^+(j)]$ and of $v(j)$ from $[d_x^-(j), d_x^+(j)]$ are never increased under this procedure for any $j \in J$. In the case of the out-of-kilter index l in a given iteration, one distance or the other is definitely decreased, unless the interval in question degenerates to the empty set (because both of the bounds in question are the same infinity). In particular, *once an index is in kilter, it stays forever in kilter*.

This fact can easily be verified from the formulas, along with the other claims made in the statement of the algorithm about the terminal situations, just as in the network case.

Relation to the Rectification Algorithms

Any sequence of iterations of the out-of-kilter algorithm in which x stays fixed and only u is improved can be identified with a sequence of iterations of the dual rectification algorithm in Section 10J relative to the intervals $[d_x^-(j),$ $d_x^+(j)]$. Indeed, the color code in Figure 11.4 designates j as:

$$\begin{aligned}
&\text{red} &&\text{if } d_x^-(j) < v(j) < d_x^+(j)\\
&\text{black} &&\text{if } v(j) \le d_x^-(j),\, v(j) < d_x^+(j)\\
&\text{white} &&\text{if } v(j) \ge d_x^+(j),\, v(j) > d_x^-(j)\\
&\text{green} &&\text{if } d_x^-(j) = v(j) = d_x^+(j)
\end{aligned}$$

This is identical to the painting in Section 10J for the intervals in question. In the same way any sequence of iterations in which u stays fixed and only x is improved can be identified with a sequence of iterations of the primal rectification algorithm in Section 10J relative to the intervals $[c_u^-(j), c_u^+(j)]$. There is a slight generalization, however. Previously only nonempty intervals were treated, whereas now there is the possibility (if x is not feasible) that $d_x^-(j) = d_x^+(j) = \infty$ or $d_x^-(j) = d_x^+(j) = -\infty$. Similarly, one might (if u is not feasible) have $c_u^-(j) = c_u^+(j) = \infty$ or $c_u^-(j) = c_u^+(j) = -\infty$.

In particular, if the out-of-kilter algorithm is applied to the relations Γ_j in Figure 11.5a with $u = 0$ ($v = 0$) initially, it reduces to the primal rectification

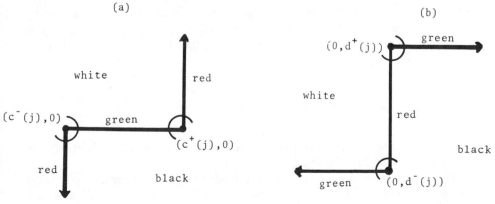

Figure 11.5

algorithm as stated in Section 10J (see Exercise 11.61). The relations in Figure 11.5b with $x = 0$, $b = 0$, give the dual rectification algorithm of Section 10J (Exercise 11.62).

Relation to the Descent Algorithms

With the color code of Figure 11.4 interpreted as in (5), it becomes clear that if x is feasible, each iteration of the out-of-kilter algorithm that modifies x in terms of a primal support P has $d_x'(P) < 0$ and corresponds therefore to an iteration of the primal descent algorithm of Section 11H. From this and the preceding observation about iterations which modify v corresponding to iterations of the dual rectification algorithm, we see that *the out-of-kilter algorithm is identical to the version of the primal descent algorithm in which the u left over from one application of the dual rectification subroutine is used to initiate the next,* except that the step-size α can be smaller and x need not be feasible at the start (although $Ex = b$). For $f(t) = \Phi(x + te_P)$ and the α in (3), it is true that $f_-'(\alpha) \leq 0$ but not necessarily that $f_+'(\alpha) \geq 0$, although α is a breakpoint of f, and $f(\alpha) < f(0)$ (Exercise 11.63).

At the same time, for parallel reasons, *the out-of-kilter algorithm is identical to the version of the dual descent algorithm in which the x left over from one application of the primal rectification subroutine is used to initiate the next,* except that the step-size α given by (4) could fall short of the one specified by the line search in Section 11H, and u need not be feasible at the start.

These observations, combined with the termination result for the out-of-kilter algorithm furnished in the next section, give proof of the assertions in Section 11H about finite termination of the primal and dual descent algorithms when applied in certain ways to piecewise linear programming.

Suboptimality Property

As just noted, x-improving iterations in which x is feasible bring about a decrease in $\Phi(x)$, whereas u-improving iterations in which u is feasible bring about an increase in $\Psi(u)$. Hence feasibility of x or u is preserved, once it is attained, and when x and u are both feasible, the difference $\Phi(x) - \Psi(u)$ is decreased in every iteration. Since

$$\Phi(x) \geq [\text{min in primal}] = [\text{max in dual}] \geq \Psi(u),$$

we have in the latter case that x and u are ε-optimal solutions to the respective problems for $\varepsilon = \Phi(x) - \Psi(u)$.

The out-of-kilter algorithm thus has the advantage, when applied to feasible x and u, of providing a measure of how far x and u are from optimality, in case it might be necessary to stop computations before optimality has been achieved.

11L.*   OTHER FORMATS AND TERMINATION

The out-of-kilter algorithm for piecewise linear programming problems can be applied in special modes where the nature of the out-of-kilter variables is automatically much simpler. As a matter of fact one can usually arrange that only one index is out of kilter. Before investigating such possibilities, we must deal with the question of termination of the algorithm in its general form.

Termination

It will be demonstrated now that regardless of the initial choice of x and u, which need not be feasible (aside from having $Ex = b$), *the out-of-kilter algorithm is sure to terminate in finitely many iterations if it is implemented in terms of the painted index subroutine with red-green preprocessing* (Section 10H) *and OK priorities* (Section 10J) (The Tucker tableau left over from one round of the subroutine is used to initiate the next.) The OK priority rule referred to here is the extension of Bland's rule already invoked in Section 10J in establishing finite termination of the rectification algorithms. Under this rule indexes are assigned priorities in such a way that out-of-kilter indexes are preferred over in-kilter indexes, and such priorities are followed in resolving ties that may arise in the selection of the lever index l or the pivot row or column (except during red-green preprocessing, where attention to priorities is unnecessary). As far as the selection of l is concerned, the desired effect of the rule is to keep the same index as l until it is finally brought into kilter.

The arguments already given for the rectification algorithms in Section 10J, along with our observations above about how the latter correspond to different kinds of iterations of the out-of-kilter algorithm, serve to exclude the possibility of an infinite sequence of iterations in which only x is improved, or only v is improved. Thus we need only eliminate the possibility of the out-of-kilter algorithm going through infinitely many iterations in which l is always the same and improvements continue to be made in both $x(l)$ and $v(l)$.

Altogether there are only finitely many elementary primal supports P, finitely many possible paintings, and finitely many subsets J_0 of J. Hence in an infinite sequence of iterations of the sort just mentioned, there would have to be infinitely many that involve the same out-of-kilter index set J_0, the same painting, and the same x-improving P compatible with this painting. Then since $e_P \in \mathscr{C}$ and $v \in \mathscr{D}$, the equation $e_P \cdot v = 0$ would always hold, and this can be written

$$\sum_{j \in J_0 \cap P^+} e_P(j)v(j) + \sum_{j \in J_0 \cap P^-} e_P(j)v(j) = - \sum_{j \in P \setminus J_0} e_P(j)v(j), \quad (1)$$

where the index l appears in one of the sums on the left.

The indexes on the left of (6) are all out of kilter. Those in the first sum are white, whereas those in the second are black. For white out-of-kilter indexes j,

$v(j)$ can only decrease, whereas for black ones it can only increase. The total on the left side of (1) must therefore be lower each time we return to the same painting after an improvement in $v(l)$, which must happen infinitely often. This is in conflict with a property of the right side of (1), however. All the indexes on the right are in kilter, and none is red. For each such index, $v(j)$ must be equal to one of the finitely many values corresponding to the horizontal levels in the relation Γ_j according to the nature of the painting (see Figure 11.4). The right side of (1) therefore has only finitely many possible values in this situation, although the left side must take on a decreasing infinite sequence of values.

The contradiction proves that there cannot be an infinite sequence of iterations; the out-of-kilter algorithm must terminate eventually as claimed.

Augmented Format

A given primal-dual pair of piecewise linear programming problems can always be reformulated with the variables $y(i)$ and $u(i)$ treated on a par with $x(j)$ and $v(j)$. Specifically, the equilibrium conditions

$$(x(j), v(j)) \in \Gamma_j \quad \text{for all } j \in J, \qquad y(i) = b(i) \quad \text{for all } i \in I,$$

can be written as

$$(x(j), v(j)) \in \Gamma_j \quad \text{for all } j \in J, \qquad (y(i), u(i)) \in \Gamma_i \quad \text{for all } i \in I, \quad (2)$$

where the characteristic curve Γ_i is the vertical line in $R \times R$ consisting of all the points whose first coordinate is $b(i)$. This amounts to translating the given optimization and equilibrium problems into the context of the subspaces $\mathscr{C}$ and $\overline{\mathscr{D}}$ in $R^{I \cup J}$ (see 11A(13)(14)(15)).

The virtue of such a formulation is that it gives one control over the sort of characteristic curves to be dealt with in connection with out-of-kilter indexes. One can arrange that these all be vertical lines. To accomplish this, it is merely necessary to have an initial feasible solution to the dual problem, in other words, any u and $v = -uE$ such that, for each $j \in J$, $v(j)$ belongs to the interval $D(j) = [d^-(j), d^+(j)]$ (which in the present piecewise linear case is precisely the projection of Γ_j on the vertical axis of $R \times R$); recall that this condition is trivial in particular when $C(j)$ is bounded, because $D(j)$ is then all of R. Since $v(j) \in D(j)$, it is possible to choose for each j a value $x(j)$ satisfying $(x(j), v(j)) \in \Gamma_j$. The vector x obtained this way is unlikely to satisfy $Ex = b$, but no matter. We have freed ourselves from this requirement by reformulating the optimality conditions in terms of (2). We simply regard indexes $i \in I$ with $y(i) \neq b(i)$ as out of kilter. All other indexes, including all those in J, are in kilter by this choice of x and u.

In the special case corresponding to the primal feasibility problem, where all the curves Γ_j have the form in Figure 11.5a, we can start with $u = 0$, $v = 0$.

The out-of-kilter algorithm in this mode then becomes the primal rectification algorithm in the augmented format described in Section 10L.

Aggregated Format

Building on the choices made so far, we now add a trick that aggregates all the "out-of-kilterness" into a single index. This is a new "artificial" index l that labels a new row of the augmented tableau, as shown in Figure 10.20. The coefficients in this row are obtained as follows.

Depending on the *initial* $y = Ex$, which by assumption is not identical to b (or all the optimality conditions would already be fulfilled), choose a weighting vector $w \in R^I$ that satisfies

$$w(i) > 0 \quad \text{for all } i \in I \text{ with } y(i) > b(i),$$

$$w(i) < 0 \quad \text{for all } i \in I \text{ with } y(i) < b(i), \tag{3}$$

and define

$$e(l, j) = \sum_{i \in I} w(i)e(i, j) \quad \text{for all } j \in J,$$

$$b(l) = \sum_{i \in I} w(i)b(i).$$

The variable

$$y(l) = \sum_{j \in J} e(l, j)x(j)$$

corresponding to the new row then satisfies

$$y(l) = \sum_{i \in I} w(i)y(i)$$

and consequently

$$y(l) - b(l) = \sum_{i \in I} w(i)[y(i) - b(i)] > 0. \tag{4}$$

Thus l is out of kilter, as are the indexes mentioned in (3), but no others. (The dual variable $u(l)$ corresponding to the new row is assigned 0 as its initial value.)

The out-of-kilter algorithm can now be applied with l taken as the lever index in the subroutine. The important feature is that when this index l is brought into kilter, every index will be in kilter. The improvement property of the out-of-kilter algorithm yields this result: the indexes $j \in J$ stay in kilter, and for the indexes $i \in I$ the quantities $y(i) - b(i)$ never move farther from zero or change sign. Thus (3) is preserved, and (4) holds as long as any index $i \in I$ is out of kilter. When $y(l) = b(l)$, an optimal solution x to the given primal problem will have been determined.

An optimal solution to the given dual will also have been determined, but it needs to be recovered from the altered notation. The addition of a new l row to the matrix E introduced a new dual variable $u(l)$ (see Figure 10.20) which must be eliminated. The vector v on hand at the termination of the algorithm needs to be re-expressed in terms of the original E. As a matter of fact one has only to set

$$\tilde{u}(i) = u(i) + u(l)w(i) \quad \text{for all } i \in I$$

to get a vector $u \in R^I$ with $-\tilde{u}E = v$. Then $\tilde{u}$ and v solve the dual.

11M.*   PARAMETRIC PROGRAMMING

In many situations one cannot be content with solving just a single monotropic programming problem and its dual. The problem may depend on parameters, and one may need to know how the minimum (or maximum) value and the optimal solutions can be expressed as functions of these parameters.

In this section we study a prototypical case where a single parameter is involved. The optimal value function and its derivatives are characterized. We see that for piecewise linear problems they can be calculated by an adapted form of the out-of-kilter algorithm, and that this yields at the same time a description of how optimal solutions to such a problem and its dual change as the parameter varies.

Basic Model

Consider a pair of monotropic programming problems in which the vector b depends on a parameter $\xi \in R$:

$$b(i) = b_0(i) + \xi h(i) \text{ for } i \in I.$$

The problems thus have the form

$$\begin{aligned}
\text{minimize } & \sum_{j \in J} f_j(x(j)) \quad \text{subject to} \quad x(j) \in C(j) \quad \text{for all } j \in J \\
& \text{and} \quad y(i) = b_0(i) + \xi h(i) \quad \text{for} \quad i \in I,
\end{aligned} \tag{1}$$

$$\begin{aligned}
\text{maximize } & -\sum_{i \in I} u(i)[b_0(i) + \xi h(i)] - \sum_{j \in J} g_j(v(j)) \\
& \text{subject to} \quad v(j) \in D(j) \quad \text{for} \quad j \in J.
\end{aligned} \tag{2}$$

For the present we do not impose any special conditions like piecewise linearity on the f_j's. Suppose, however, that *for $\xi = 0$ we know a primal optimal solution x_0 and a dual optimal solution u_0.* How can these vectors be modified to maintain optimality as ξ moves away from zero?

Further questions concern the infimum and supremum in the two problems. Let us observe that no matter what the choice of ξ, the dual problem (2) has at least one regularly feasible solution. (Namely, x_0 and u_0 are regularly feasible for $\xi = 0$ by the existence theorems in Section 11D, and in (2) the feasible and regularly feasible solutions are the same for all ξ.) Hence by the duality theorem the infimum in (1) and the supremum in (2) always coincide (and are not $-\infty$), and by the primal existence theorem *the infimum in (1) is attained whenever it is not ∞.* In particular, the function

$$f(\xi) = [\inf \text{ in } (1)] = [\sup \text{ in } (2)] \quad \text{for all } \xi \in R \tag{3}$$

is well defined and satisfies

$$f(\xi) > -\infty \quad \text{for all} \quad \xi, f(0) < \infty. \tag{4}$$

What are the properties of f, and how can it and its possible derivatives be determined systematically?

The following result gives a foundation for answering these questions.

Proposition 1. *The function f defined in (3) (under our assumption that optimal solutions exist for (1) and (2) when $\xi = 0$) is a closed proper convex function on R whose conjugate is*

$$g(\eta) = -\sup\left\{ -\sum_{i \in I} u(i)b(i) - \sum_{j \in J} g_j(v(j))| - \sum_{i \in I} u(i)h(i) = \eta \right\}, \tag{5}$$

and whose characteristic curve is

$$\Gamma = \left\{ (\xi, \eta) \in R^2 | \exists \text{ optimal } u \text{ to } (2) \text{ for } \xi \text{ with } - \sum_{i \in I} u(i)h(i) = \eta \right\}. \tag{6}$$

Moreover the supremum in (5) is attained whenever it is not $-\infty$.

The proof of these assertions is based on a reformulation in terms of slightly broader problems, and this reformulation will also be the key later to a computational procedure. We introduce a new index l corresponding to a new column adjoined to the matrix E as in Figure 11.6. Whatever function f_l might be specified (with conjugate g_l and characteristic curve Γ_l), the choice being left

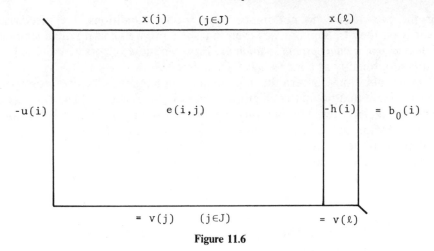

Figure 11.6

open until later, we get primal and dual problems as follows:

minimize $\displaystyle\sum_{j\in J} f_j(x(j)) + f_l(x(l))$ subject to

$$x(j) \in C(j) \quad \text{for } j \in J, \quad x(l) \in C(l), \quad \text{and}$$

$$\sum_{j\in J} e(i, j)x(j) - h(i)x(l) = b_0(i) \quad \text{for } i \in I, \tag{7}$$

maximize $\displaystyle - \sum_{i\in I} u(i)b(i) - \sum_{j\in J} g_j(v(j)) - g_l(v(l))$

subject to $v(j) \in D(j)$ for $j \in J$,   (8)

$$v(l) = \sum_{i\in I} u(i)h(i).$$

The corresponding equilibrium conditions consist of the relation $(x(l), v(l)) \in \Gamma_l$, yet to be specified, plus

$$x(j), v(j)) \in \Gamma_j \quad \text{for } j \in J,$$

$$\sum_{j\in J} e(i, j)x(j) - h(i)x(l) = b_0(i) \quad \text{for } i \in I,$$

$$v(j) = - \sum_{i\in I} u(i)e(i, j) \quad \text{for } j \in J,$$

$$v(l) = \sum_{i\in I} u(i)h(i). \tag{9}$$

We can reduce (7) and (8) to (1) and (2) by taking

$$f_l(x(l)) = \begin{cases} 0 & \text{if } x(l) = \xi, \\ \infty & \text{if } x(l) \neq \xi, \end{cases}$$

$$g_l(v(l)) = \xi v(l), \tag{10}$$

so that Γ_l is the vertical line in R^2 with horizontal coordinate ξ. In fact the supremum in (8) in this case can also be viewed as the supremum of $\xi \eta - g(\eta)$ over all ξ, so the equation in the definition (3) of f tells us that

$$f(\xi) = \sup_{\eta \in R} \{\xi \eta - g(\eta)\} \quad \text{for all } \xi \in R. \tag{11}$$

This expression for f implies in conjunction with (4) that f is a closed proper convex function (see Exercise 11.70).

If instead of (10) we take

$$f_l(x(l)) = -\eta x(l),$$

$$g_l(v(l)) = \begin{cases} 0 & \text{if } v(l) = -\eta, \\ \infty & \text{if } v(l) \neq -\eta, \end{cases}$$

so that Γ_l is a horizontal line in R^2 with vertical coordinate $-\eta$, the infimum in problem (7) is the infimum of $f(\xi) - \eta \xi$ over all $\xi \in R$, whereas the supremum in (8) is $-g(\eta)$ (see (5)). Furthermore (7) has at least one regularly feasible solution (the x_0 noted earlier). Hence by the dual existence theorem the supremum defining $-g(\eta)$ is attained whenever it is not $-\infty$, as claimed, and by the duality theorem

$$\inf_{\xi \in R} \{f(\xi) - \eta \xi\} = -g(\eta), \text{ or } g(\eta) = \sup_{\xi \in R} \{\eta \xi - f(\xi)\}.$$

Thus g is the function conjugate to f (see Section 8E).

It follows now from 8E(9) that the characteristic curve Γ consists of the pairs (ξ, η) such that η achieves the supremum in (11). Recalling that this supremum can be identified with the one in (8) under the choice (10), that is, the supremum in (2), and using the fact that the supremum in (5) is attained by some u whenever it is finite, we conclude that Γ is expressible by (6).

Corollary.   *Any pair of vectors x and u satisfying (9) yields, for $\xi = x(l)$ and $\eta = -v(l)$, optimal solutions to problems (1) and (2) as well as a point $(\xi, \eta) \in \Gamma$. In fact every point of Γ can be obtained in this way.*

The corollary is obtained from (6) and the equilibrium theorem in Section 11D by identifying the optimal solutions to (1) and (2) with those of (7) and (8) under (10). The equilibrium conditions for the latter consist of (9), and the equation $x(l) = \xi$. (Remember that (1) has an optimal solution whenever the value in (3) is finite, which is certainly true when (2) has an optimal solution.)

Consequences for Computation

The postulated optimal solutions x_0 and u_0 to (1) and (2) for $\xi = 0$ furnish an initial point

$$(0, \eta_0) \in \Gamma, \quad \text{where } \eta_0 = -v_0(l) = -\sum_{i \in I} u_0(i)h(i).$$

Since Γ is a maximal monotone relation (see Section 8B), it is a continuous curve which in principle can be traced in either direction from $(0, \eta_0)$. If the tracing can be accomplished in terms of continuous modification of x_0 and u_0 into other vectors x and u which satisfy (9), in such a way that $(x(l), -v(l))$ moves continuously along Γ, then a parametric representation of optimal solutions to (1) and (2) is achieved at the same time.

Knowing Γ, we know all the left and right derivatives of f and g and can construct these functions by integration from the known value $f(0) = -g(\eta_0)$.

Parametric Computations in the Piecewise Linear Case

For piecewise linear programming problems (1) and (2), where the Γ_j's are staircase relations, the out-of-kilter algorithm can be used to trace Γ in the manner just described. Starting with x_0 and u_0 as vectors known to satisfy (9) with $(x_0(l), -v_0(l)) = (0, \eta_0)$, we proceed as if a maximal monotone relation Γ_l has been assigned to the index l, but in such a way that l is out of kilter and white (the point $(x_0(l), v_0(l))$ lies "northwest" of Γ_l). No actual choice of Γ_l needs to be made. We simply imagine the curve to be sufficiently remote that the iterations we perform do not bring l into kilter.

We apply the out-of-kilter algorithm. Since every index $j \in J$ is in kilter from the start (see (9)), but l is out of kilter and white, each iteration will preserve these properties and either modify x to increase $\xi = x(l)$ (leaving u the same) or modify u to increase $\eta = -v(l)$ (leaving x the same). These modifications can be realized in continuous fashion: instead of passing all at once from x to $x + \alpha e_P$ in the first case, one can observe that the pair $(x + te_P, u)$ satisfies (9) for all $t \in [0, \alpha]$. The situation is similar in the second case, where u is modified to $u + \alpha w$ for certain w with $-wE = e_Q$; the pair $(x, u + tw)$ actually satisfies (0) for $0 \leq t \leq \alpha$. The same treatment can be given when $\alpha = \infty$.

Iterations of the out-of-kilter algorithm that modify x thus serve to trace part of a horizontal segment of Γ toward the right, whereas iterations that modify u trace part of a vertical segment of Γ upward. Moreover such tracing is done with solutions to (9) maintained in a continuous fashion. This yields an optimal solution x to (1) as a continuous function of ξ. (The u vector cannot be viewed as a continuous function of ξ, since its changes take place while ξ is fixed.)

Following the same procedure but with l painted black (i.e., the point $(x_0(l), v_0(l)) = (0, -\eta_0)$ imagined as "southeast" of Γ_l), we trace Γ in the opposite direction. Either way an iteration with $\alpha = \infty$ identifies a final infinite segment of Γ, whether horizontal or vertical.

Termination

If the out-of-kilter algorithm is applied in this mode using the painted index subroutine with red-green preprocessing and Bland's priority rule (the added features of the OK priority rule being superfluous, inasmuch as only one index is ever out of kilter), *it must in each direction eventually terminate with an infinite segment of Γ as just described; thus it will trace all of Γ in finitely many steps.* This follows from the properties of the algorithm that have already been observed in Section 11K. The termination argument in that section did not depend really on fixing a particular Γ_l. It simply showed that the procedure could not make more than a finite number of improvements. Thus we only need imagine Γ_l far enough away from the initial point $(x_0(l), v_0(l))$ to see that the algorithm must sooner or later terminate with $\alpha = \infty$, termination with $(x(l), v(l)) \in \Gamma_l$ being impossible.

Proposition 2. *In the case of piecewise linear programming, the functions f and g in Proposition 1 are piecewise linear, and Γ is of staircase form.*

This result is immediate from our termination argument and the observation that each iteration of the algorithm traces a horizontal or a vertical portion of Γ.

Application to Solving a Particular Problem

Suppose what we are really interested in is solving problem (1) or (2) for a particular value of ξ, say $\xi = 1$, but all we know is an optimal pair x_0, u_0 for $\xi = 0$. We can apply the out-of-kilter algorithm in just the manner described, except that Γ_l is specifically chosen as the vertical line in R^2 with ξ-coordinate 1 (see (10) with $\xi = 1$). The procedure will trace Γ from $(0, \eta_0)$ only as far as a point $(1, \eta)$ and then terminate with the desired optimal solution. (If there is no such point on Γ, the targeted problem is unsolvable, and the algorithm terminates with an infinite vertical segment of Γ before ξ reaches 1.)

This solution procedure is very similar to the aggregated format for the out-of-kilter algorithm in Section 11L.

11N.*   EXERCISES

11.1.   (*Problem Formulation*). Express the following as a monotropic programming problem in primal form:

$$\text{minimize } \sum_{j=1}^{n} |x(j) - d(j)| \text{ subject to } x(j) \geq 0$$

$$\text{and } \sum_{j=1}^{i} x(j) = b(i) \text{ for } i = 2,\ldots,m,$$

where $d(j) > 0$ and $b(i) > 0$ are given. Then determine the corresponding dual problem and equilibrium problem.

11.2.   (*Problem Formulation*). Express the following as a monotropic programming problem in primal form:

$$\text{minimize } \frac{1}{p} \sum_{i=1}^{m} \left| b(i) - \sum_{j=1}^{n} e(i, j) x(j) \right|^{p}$$

$$\text{subject to } |x(j)| \leq 1 \text{ for } j = 1,\ldots,n,$$

where $b(i)$ and $e(i, j)$ are given, and $1 < p < \infty$. Then determine the corresponding dual problem and equilibrium problem.

(*Hint.*   You will need to use the augmented format in 11A(13) (14) (15).)

11.3.   (*Problem Formulation*). Express the following as a monotropic programming problem in *dual* form:

$$\text{maximize } \sum_{i=1}^{m} \log[u(i) - u(i - 1)]$$

$$\text{subject to } 0 \leq u(0) \leq u(1) \leq \cdots \leq u(m) \leq 1$$

$$\text{(with the convention } \log 0 = -\infty\text{)}.$$

Then determine the corresponding primal problem and equilibrium problem.

(*Hint.*   Introduce variables $v(j)$ which are linear functions of the $u(i)$'s such that everything can be expressed in terms of costs $g_j(v(j))$ and constraints $v(j) \in D(j)$ on the $v(j)$'s individually.)

11.4. (*Problem Formulation*). For the equilibrium problem

$$\text{find} \quad x(j) \quad \text{and} \quad u(i) \quad \text{such that for} \quad v(j) = -\sum_{i=1}^{m} u(i)e(i, j)$$

$$\text{one has} \quad x(j) = v(j)^3 \quad \text{for } j = s + 1,\ldots,r$$
$$v(j) = \tan[x(j)] \quad \text{for } j = r + 1,\ldots,s,$$
$$v(j) = x(j) \quad \text{for } j = s + 1,\ldots,n,$$

$$1 + \sum_{j=1}^{n} e(i, j)x(j) = 0 \quad \text{for } i = 1,\ldots,m,$$

what are the corresponding primal and dual monotropic programming problems?

(*Hint.* Here $e(i, j)$'s are supposed as given, but you must specify $b(i)$'s and curves Γ_j. The functions f_j and g_j can then be determined by integration as in Section 8B.)

11.5. (*Feasibility*). Give an example of a monotropic programming problem which has feasible solutions but no regularly feasible solutions.

11.6. (*Feasibility*). Prove that the set of all feasible solutions to a mono-tropic programming problem in primal form is a *convex* subset of R^J (i.e., if x' and x'' are in this set, then so is $x = (1 - \lambda)x' + \lambda x''$ for arbitrary $\lambda \in (0, 1)$). Do the same also for the set of all regularly feasible solutions.

11.7. (*Optimality*). Prove that the objective function Φ in the primal prob-lem is convex, in the sense that for all x', x'', one has

$$\Phi\big((1 - \lambda)x' + \lambda x''\big) \leq (1 - \lambda)\Phi(x') + \lambda\Phi(x'') \quad \text{when } \lambda \in (0, 1).$$

Use this to show that the set of all optimal solutions to the problem is convex (in the sense used in the preceding exercise).

11.8. (*Linear Programming*). Demonstrate that every piecewise linear pro-gramming problem is equivalent to some linear programming problem in the general sense.

(*Hint.* Each j with a piecewise linear cost term that is not actually linear must be replaced by two or more elements much in the manner of Section 7I.)

11.9. (*Linear Programming*). Show that every linear programming problem in the general sense is equivalent to some "elementary" problem, as defined in Example 1 in Section 11B.

(*Hint.* Follow a pattern like the one used in the network case in Section 7I.)

11.10. (*Linear Programming*). In the case where the primal monotropic programming problem is a linear programming problem in the general sense, show that one does not have $-\infty < d^-(j) < d^+(j) < \infty$ for any j; moreover for the corresponding dual monotropic programming problem to be linear too, it is then necessary and sufficient that one does not have $-\infty < c^-(j) < c^+(j) < \infty$ for any j.

11.11. (*Quadratic Programming*). Formulate the least-squares problem

$$\text{minimize} \quad \tfrac{1}{2} \sum_{j=1}^{n} \left[a(j) - \sum_{i=1}^{m} u(i)e(i, j) \right]^2 \quad \text{over all } u \in R^m$$

as a monotropic programming problem in dual form, and then determine the corresponding primal problem and equilibrium problem.

11.12. (*Quadratic Programming*). Formulate the quadratic programming problem

$$\text{minimize} \quad \tfrac{1}{2}|Sx|^2 \quad \text{subject to } x \geq 0, Wx \geq w$$

(where $x \in R^n$, $S \in R^{p \times n}$, $W \in R^{q \times n}$, $w \in R^q$) as a monotropic programming problem in primal form, and then determine the corresponding dual problem and equilibrium problem.

(*Hint.* Work with the augmented format in 11A(13) (14) (15).)

11.13. (*Piecewise Quadratic Programming*). Demonstrate that when the quadratic programming problem

$$\text{minimize} \quad \frac{1}{2} \sum_{j=1}^{n} a(j)x(j)^2$$

$$\text{subject to} \quad |x(j)| \leq 1 \text{ for } j = 1,\dots,n,$$

$$\text{and} \quad \sum_{j=1}^{n} e(i, j)x(j) = b(i) \quad \text{for } i = 1,\dots,m$$

(with $a(j) > 0$) is expressed as a monotropic programming problem in primal form, the corresponding dual problem is only piecewise quadratic.

11.14. (*Line Search*). For the function f in 11C(1), verify the full formulas in 11C(2) for f'_+ and f'_- (in terms of the conventions explained after the statement of these formulas).

11.15. (*Line Search*). Verify the description given in 11C(8) for the interval $\check{D}(P)$ defined in 11C(5) (where P is an elementary primal support).

11.16. (*Line Search*). Prove that for any feasible solution u to the dual problem and any $w \in R^I$, the function $f(t) = -\Psi(u + tw)$ is a closed proper convex function on R, as defined in Section 8A.

11.17. (*Line Search*). For the function f in 11C(14) and w satisfying 11C(13), derive the formulas given in 11C(15) for f'_+ and f'_-. (Here $b_Q = -w \cdot b$. Watch out for the convention explained following 11C(15).)

11.18. (*Line Search*). Prove the dual descent theorem at the end of Section 11C. (Follow arguments like those for the primal descent theorem.)

11.19. (*Line Search*). For the function $f(t) = \Phi(x + te_P)$ and the interval $D(P)$ in 11C(4), where x is feasible, prove that $D(P) \subset (-\infty, 0)$ if and only if $\inf\{f(t) | t > 0\} = -\infty$.

(*Hint.* Verify that $D(P)$ is the interval of finiteness for the closed proper convex function g which is conjugate to f. Help is provided by 11C(9) and the fact that g is continuous relative to the closure of its interval of finiteness; recall that $D(P)$ and $\tilde{D}(P)$ have the same closure $[d^-(P), d^+(P)]$.)

(*Remark.* A similar result holds for the dual problem.)

11.20. (*Elementary Descent*). Consider the problem of minimizing $F(z) = \sum_{j \in J} f_j(a_j \cdot z)$ over all $z \in R^n$, where each f_j is a closed proper convex function on R (with finiteness interval $C(j)$). Recall that by setting $x_j = a_j \cdot z$, one gets a correspondence between vectors $z \in R^n$ and vectors x forming a certain subspace $\mathscr{C}$ of R^J. One then has a monotropic programming problem in the sense of Section 11A: minimize $\Phi(x) = \sum_{j \in J} f_j(x_j)$ over all $x \in \mathscr{C}$ (see Example 3 in Section 11B). Suppose now that the a_j's span R^n, so that the correspondence $z \leftrightarrow x$ is one to one. Then every elementary vector of $\mathscr{C}$ corresponds to a uniquely determined vector $w \in R^n$. Prove that w is of this type if and only if $w \neq 0$ but there exist $n - 1$ linearly independent a_j's with $a_j \cdot w = 0$.

(*Interpretation.* An elementary direction of descent, viewed in terms of F at a given z, is a direction of a ray from z which is perpendicular to $n - 1$ linearly independent vectors chosen from the a_j's, and along which F is initially decreasing.)

11.21. (*Elementary Linear Programming*). Specialize each of the theorems in Section 11D to the "elementary" case in Example 1 of Section 11B, namely, where the primal problem is

$$\text{minimize} \quad d \cdot x \quad \text{subject to} \quad x \geq c, \, Ex = b,$$

the dual problem is

$$\text{maximize} \quad -u \cdot [b + Ec] \quad \text{subject to} \quad -uE \leq d,$$

and the equilibrium conditions on $x \in R^J$ and $u \in R^I$ reduce to

$$x \geq c, \quad -uE \geq d, \quad [x - c] \cdot [uE + d] = 0.$$

(Include a description of just what it means in this case for an elementary primal or dual support to be "unbalanced.")

11.22. (*Piecewise Quadratic Programming*). Prove that for piecewise quadratic problems in the sense of Example 2 in Section 11B, the following are equivalent:

(a) The primal problem has an optimal solution.

(b) The infimum in the primal problem is finite.

(c) The dual problem has an optimal solution.

(d) The supremum in the dual problem is finite.

(e) The primal and dual problems both have feasible solutions.

(*Remark.* This is true then for linear or piecewise linear problems in particular. In fact it holds for any class of monotropic programming problems having $\tilde{C}(j) = C(j) = [c^-(j), c^+(j)]$ and $\tilde{D}(j) = D(j) = [d^-(j), d^+(j)]$ for all $j \in J$.)

11.23. (*Generalized Duality*). Prove that for an arbitrary subspace $\mathscr{C}$ of R^J and arbitrary closed proper convex functions f_j, one has

$$\inf_{x \in \mathscr{C}} \left\{ \sum_{j \in J} f_j(x_j) \right\} = \sup_{v \in \mathscr{D}} \left\{ - \sum_{j \in J} g_j(v(j)) \right\}$$

unless the "inf" is $+\infty$ and the "sup" is $-\infty$; here g_j is conjugate to f_j, and $\mathscr{D} = \mathscr{C}^\perp$.

(*Hint.* This result corresponds to the model problems (10) and (11) in Section 11A. Reformulate these, and apply the duality theorem in Section 11D.)

11.24. (*Generalized Duality*). Prove that for an arbitrary matrix $E \in R^{I \times J}$ and closed proper convex functions f_i for $i \in I$ and f_j for $j \in J$, one has

$$\inf_{x \in R^J} \left\{ \sum_{j \in J} f_j(x(j)) + \sum_{i \in I} f_i\left(\sum_{j \in J} e(i, j) x(j) \right) \right\}$$

$$= \sup_{u \in R^I} \left\{ - \sum_{i \in I} g_i(u(i)) - \sum_{j \in J} g_j\left(- \sum_{i \in I} u(i) e(i, j) \right) \right\}$$

unless the "inf" is $+\infty$ and the "sup" is $-\infty$; here g_i and g_j are conjugate to f_i and f_j.

(*Hint.* Apply the duality theorem of Section 11D to a reformulated version of the model problems (13) and (14) in Section 11A.)

11.25. (*Equilibrium Problem*). Suppose that the pair (x, u) solves the equilibrium problem and also the pair (x', u'). Show that the pairs (x', u') and (x', u) then solve it too.

(*Hint.* Apply the equilibrium theorem in Section 11D.)

11.26. (*Nonunique Optimal Solutions*). Suppose the primal and dual optimization problems have optimal solutions. Prove that:

    (a) If x and x' are both primal optimal, then for each $j \in J$ such that
        $x(j) \neq x'(j)$, the function f_j must be affine (linear-plus-a-con-
        stant) on the interval between $x(j)$ and $x'(j)$.

    (b) If u and u' are both dual optimal, then for each $j \in J$ such that
        $v(j) \neq v'(j)$, the function g_j must be affine on the interval
        between $v(j)$ and $v'(j)$.

    (*Hint.*   A function is affine relative to an open interval if its right and
    left derivatives have a certain constant value on that interval. Make
    use of the equilibrium conditions as a characterization of optimality;
    see Section 11D.)

11.27. (*Nonunique Optimal Solutions*). Suppose the primal and dual optimization problems have optimal solutions. Let

$$J_1 = \{\, j \in J | \exists \text{ primal optimal } x, x', \text{ with } x(j) \neq x'(j)\},$$

$$J_2 = \{\, j \in J | \exists \text{ dual optimal } u, u', \text{ with } v(j) \neq v'(j)\}.$$

Demonstrate that $J_1 \cap J_2 = \varnothing$. (Thus for every $j \in J$, either the $x(j)$
value is uniquely determined for all the primal optimal solutions, or
the $v(j)$ value is uniquely determined for all the dual optimal solutions, or both.)

(*Hint.*   Make use of the equilibrium conditions as a characterization
of optimality; see Section 11D.)

11.28. (*Uniqueness Theorem*). Prove the following:

    (a) The primal problem has at most one optimal solution if the dual
        problem has an optimal solution u such that for

$$F_v' = \{\, j \in J | g_{j-}'(v(j)) = g_{j+}'(v(j))\}$$

$$= \{\, j \in J | v(j) \text{ is not a first-order breakpoint for } g_j\}$$

        the set $J \setminus F_v'$ is independent in the sense of Section 10C (i.e.,
        there exists a basis F with $F_v' \supset F' := J \setminus F$).

    (b) The dual problem has at most one optimal solution as far as the
        vector v is concerned (although different u's could give the same
        v, unless $uE = 0$ implies $u = 0$), if the primal problem has an
        optimal solution x such that the set

$$F_x = \{\, j \in J | f_{j-}'(x(j)) = f_{j+}'(x(j))\}$$

$$= \{\, j \in J | x(j) \text{ is not a first-order breakpoint for } f_j\}$$

spans in the sense of Section 10C (i.e., there exists a basis F with $F_x \supset F$).

(*Hint.* Using the equilibrium conditions, show in (a) that F_v' must be disjoint from the index set J_1 defined in Exercise 11.27, so that J_1 is included in some basis F. Then use the fact that the values $x(j)$ for $j \in F$ are uniquely determined by the values $x(k)$ for $k \in F'$. A similar argument works for (b).)

11.29. (*A Priori Uniqueness Criterion*). Prove that the primal problem cannot have more than one optimal solution if the following condition is fulfilled: the primal problem has a regularly feasible solution, and there is a basis F such that for every $j \notin F$ the relation Γ_j includes no horizontal segments. Likewise, the dual problem cannot have more than one optimal solution as far as the vector v is concerned (although different u's might still correspond to the same v) if the following condition is fulfilled: the dual problem has a regularly feasible solution, and there is a basis F such that for every $j \in F$ the relation Γ_j includes no vertical segments.

(*Hint.* Apply the theorem stated in the preceding exercise.)

(*Remark.* This criterion is sufficient but far from necessary. It cannot be applied to piecewise linear programming at all. In contrast to the result given in the preceding exercise, however, it does not require advance knowledge about any particular optimal solution to the primal or dual problem. The condition that Γ_j contain no horizontal segment means that f_j is strictly convex on $C(j)$, or equivalently that $\tilde{D}(j)$ is open and g_j is differentiable throughout $\tilde{D}(j)$. Dually, the condition that Γ_j contain no vertical segments means that g_j is strictly convex on $D(j)$, or equivalently that $\tilde{C}(j)$ is open and f_j is differentiable throughout $\tilde{C}(j)$.)

11.30. (*Unique Equilibrium*). State and prove a sufficient condition in terms of the relations Γ_j for the equilibrium problem to have a unique solution.

(*Hint.* Apply the result in the preceding exercise. The condition obtained in this way is far from being *necessary* of course, but it is easily verified in many situations.)

11.31. (*Optimal Solution Sets*). Prove that the set of all optimal solutions to the primal problem is actually a convex polyhedron, and so too is the set of all optimal solutions to the dual problem. (A set is a convex polyhedron if it can be represented as the set of all vectors satisfying some system of finitely many linear equations or weak linear inequalities.)

(*Hint.* First prove the assertion in the case where the primal problem has a regularly feasible solution, making use of the primal characterization theorem and the nature of the equilibrium conditions. Next demonstrate that for any primal problem one can obtain by modification of the f_j's another primal problem having the same optimal solution set but also possessing at least one regularly feasible solution.)

11.32. (*Infinite Optimal Values*). Suppose that the primal problem has inf = $-\infty$. Prove that there must exist an elementary primal support P such that for every feasible solution x to the primal, the function $f(t) = \Phi(x + te_P)$ is decreasing with $\lim_{t \to \infty} f(t) = -\infty$.

(*Hint*. Demonstrate by means of results in Section 11D that the dual optimization problem cannot have a feasible solution. Plug this into the result stated in Exercise 11.19.)

(*Remark*. This can be dualized.)

11.33. (*Saddlepoint Optimality Criterion*). Show that for x and u to be optimal solutions to the primal and dual problems, respectively, it is necessary and sufficient that (x, u) be a *saddlepoint* of the function

$$L(x, u) = \sum_{j \in J} f_j(x(j)) - \sum_{i \in I} u(i)b(i) + \sum_{i \in I, j \in J} u(i)e(i, j)x(j),$$

in the sense that $L(x', u) \geq L(x, u) \geq L(x, u')$ for all $x \in R^J$ and $u' \in R^I$.

11.34. (*Saddlepoint Optimality Criterion*). Demonstrate that the result in the preceding exercise carries over to the extended formulation of the primal and dual problems in 11A(13) (14) if one takes

$$L(x, u) = \sum_{j \subset J} f_j(x(j)) - \sum_{i \in I} g_i(u(i)) - \sum_{i \in I, j \in J} u(i)e(i, j)x(j)$$

with the convention that $\infty - \infty = \infty$. (The convention $\infty - \infty = -\infty$ would work just as well. Alternatively, one could restrict attention to vectors x and u such that $x(j) \in C(j)$ for all j and $u(i) \in D(i)$ for all $i \in I$; then the formula for L does not involve $+\infty$ or $-\infty$.)

11.35. (*Saddlepoint Optimality Criterion*). Consider the primal and dual problems in the formulation of 11A(10) (11) (i.e., with $b = 0$). Let F be any basis, and let $F' = J \setminus F$. Let $a(j, k)$ be the coefficients in the corresponding Tucker representation of the spaces $\mathscr{C}$ and $\mathscr{D}$, so that vectors $x \in \mathscr{C}$ and $v \in \mathscr{D}$ are characterized by

$$x(j) = \sum_{k \in F'} a(j, k)x(k) \quad \text{for all } j \in F,$$

$$v(k) = - \sum_{j \in F} v(j)a(j, k) \quad \text{for all } k \in F'.$$

Denoting by $x^{F'}$ and v^F the projections of x on $R^{F'}$ and v on R^F, define

$$L_F(x^{F'}, v^F) = \sum_{k \in F'} f_k(x(k)) - \sum_{j \in F} g_j(x(j))$$

$$- \sum_{k \in F', j \in F} v(j)a(j,k)x(k)$$

with the convention that $\infty - \infty = \infty$. Prove that x and v are optimal for the respective problems if and only if $(x^{F'}, v^F)$ is a saddlepoint of L_F on $R^{F'} \times R^F$, that is,

$$L_F(\xi, v^F) \geq L_F(x^{F'}, v^F) \geq L_F(x^{F'}, \eta)$$

for all $\xi \in R^{F'}$ and $\eta \in R^F$.

(*Hint*. Show that this property holds if and only if the equilibrium conditions in 11A(12) are fulfilled.)

11.36. (*Strict Complementary Slackness*). In the case of piecewise linear programming, where every Γ_j is of staircase form (see Example 1 in Section 11B), prove that the equilibrium problem, if it can be solved at all, has a solution (x, u) such that for each $j \in J$ the point $(x(j), v(j))$ is either interior to some horizontal segment of Γ_j or interior to some vertical segment of Γ_j.

(*Hint*. Verify that the following procedure, starting with an arbitrary solution (x, u), modifies it in finitely many iterations into one having the property described. Paint the indexes $j \in J$ as indicated in Figure 11.4; remember that each $(x(j), v(j))$ actually lies on Γ_j. If there are no white or black indexes, then (x, u) has the desired property. Otherwise select any white or black index l, and apply the painted index algorithm. If this produces an elementary primal support P, then by setting $x' = x + \varepsilon e_P$ for $\varepsilon > 0$ sufficiently small, one gets a solution (x', u) having fewer white or black indexes than did (x, u). Likewise, if the painted index algorithm produces an elementary dual support Q, then by setting $u' = u + \varepsilon w$ for $\varepsilon > 0$ sufficiently small, where $-wE = e_Q$ (and consequently $v' = v + \varepsilon e_Q$ for $v' = -u'E$), one gets a solution (x, u') having fewer white or black indexes than did (x, u). Either way, keep repeating the procedure until there are no more white or black indexes.)

11.37. (*Boundedness Property*). Verify that the primal problem has the boundedness property in Section 11E if and only if every minimizing sequence $\{x^k\}$ to the primal problem is bounded. (Similarly, the dual problem has the boundedness property if and only if for every maximizing sequence $\{u^k\}$ to the dual problem the corresponding sequence $\{v^k\}$ is bounded.)

(*Hint*. The "only if" part is easy; to prove the "if" part, use the fact established in the primal boundedness theorem that if the boundedness property did not hold, there would have to exist an elementary primal support P with $d^+(P) \geq 0$. Then apply the primal descent theorem.)

11.38. (*Boundedness Property*). Prove the dual boundedness theorem in Section 11E by identifying the dual problem with a certain problem of primal form.

11.39. (*Boundedness Property*). Suppose both the primal and dual optimization problems have feasible solutions. Show that for both problems to have the boundedness property, it is necessary and sufficient that they have feasible solutions x and u which for some basis $F \subset J$ satisfy:

(a) $c^-(j) < x(j) < c^+(j)$ for every $j \in F$,

(b) $d^-(j) < v(j) < d^+(j)$ for every $j \notin F$.

(*Hint*. Establish first that the sets F_d and F_c' in the primal and dual boundedness theorems of Section 11E are disjoint; use the fact that $[c^-(j), c^+(j)]$ and $[d^-(j), d^+(j)]$ are the closures of the projections of Γ_j on the horizontal and vertical axes. Next verify the existence of a basis F satisfying $F_d \subset F \subset F \setminus F_c'$, and show by means of the associated Tucker representations that any primal feasible solution x that satisfies (a) can be perturbed into one that actually has $c^-(j) < x(j) < c^+(j)$ for all $j \in J$ with $c^-(j) < c^+(j)$; similarly for u satisfying (b).)

11.40. (*Decomposition*). Demonstrate that the function p giving the infimum in problem 11F(5) as a function of the parameter vector $u^1 \in R^{I_1}$ is in fact a *concave* function such that the set $\{u^1 | p(u^1) > -\infty\}$ consists of all the vectors satisfying a certain system of finitely many linear inequalities, weak or strict (i.e., is the intersection of a finite collection of closed or open half-spaces in R^{I_1}, a "convex polyhedron with some faces possibly missing").

(*Hint*. For the concavity, argue that p is the pointwise infimum of a collection of affine functions of u^1, one for each feasible solution x to problem 11F(5). Next apply the duality theorem to the problem in question, and verify that

$$p(u^1) = \sup\left\{-b^0 \cdot u^0 - \sum_{j \in J} g_j(v^0(j) + v^1(j)) | u^0 \in R^{I_0}\right\},$$

so that $p(u^1) > -\infty$ if and only if

$$\exists u^0 \quad \text{with} \quad v^0(j) \in D^0(j) \quad \text{for all } j \in J,$$

where $v^0 = -u^0 E^0$ and $D^0(j) = D(j) - v^1(j)$.

Applying the dual feasibility theorem of Section 10I; prove that the latter holds if and only if u^1 has $e_p \cdot v^1 \in D(P)$ for every elementary primal support P relative to the matrix E^0.)

(*Remark.* The function p is also continuous relative to the convex polyhedron which is the closure of the set $\{u^1 | p(u^1) > -\infty\}$. The proof of this involves continuity properties of concave functions beyond what can be developed here, however.)

11.41. (*Generalized Traffic Theorem*). Consider the generalized traffic problem where in 11G(1) there is for each $j \in J$ and $k = 1, \ldots, r$ a closed proper convex function f_j^k with interval of finiteness $C_k(j)$ (not necessarily closed), and the expression to be minimized is

$$\sum_{j \in A_*} \left[f_j(z(j)) + \sum_{k=1}^{r} f_j^k(x_k(j)) \right].$$

Show that the traffic theorem remains true if the expression to be minimized in the subproblem 11E(5) is taken to be

$$\sum_{j \in A_*} \left[w(j) x_k(j) + f_j^k(x_k(j)) \right].$$

(Of course the regular feasibility described in the theorem would in this case also require $x_k(j) \in \tilde{C}_k(j)$ for all $j \in J$ and $k = 1, \ldots, r$.)

11.42. (*Decomposition*). Consider the problem

$$\text{minimize} \quad \sum_{l \in L} f_l(z(l)) + \sum_{k=1}^{r} \sum_{j \in J_k} f_j^k(x_k(j)) \quad \text{subject to}$$

$$z(l) \in C(l) \quad \text{for} \quad l \in L, \qquad x_k(j) \in C_k(j) \quad \text{for} \quad j \in J_k,$$

$$y_k(i) = b_k(i) \quad \text{for} \quad i \in I_k,$$

where

$$z(l) = \sum_{k=1}^{r} \sum_{j \in J_k} e_k^1(l, j) x_k(j) \quad \text{for} \quad l \in L,$$

$$y_k(i) = \sum_{j \in J_k} e_k^0(l, j) x_k(j) \quad \text{for} \quad i \in I_k \quad \text{and} \quad k = 1, \ldots, r,$$

and f and f_j^k are closed proper convex functions on R with finiteness intervals $C(l)$ and $C_k(l)$. This corresponds to the matrix in Figure 11.7 and generalizes the traffic problem in Section 11E in replacing the submatrices E_* and I in Figure 11.3 by arbitrary matrices not neces-

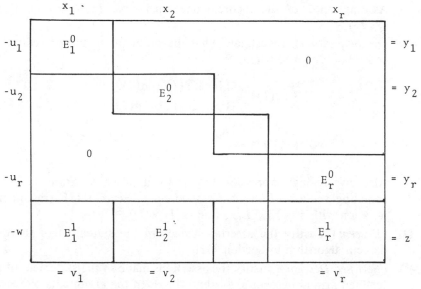

Figure 11.7

sarily related to any network; it generalizes at the same time in the direction of the preceding exercise. Show by arguments resembling those in Section 11E that the traffic theorem remains valid when the subproblem 11E(5) is generalized correspondingly to

$$\text{minimize} \quad \sum_{j \in J_k} \left[f_j^k(x(j)) + \left(\sum_{l \in L} w(l) e_k^1(l, j) \right) x(j) \right]$$

$$\text{subject to} \quad x_k(j) \in C_k(j) \quad \text{for } j \in J_k \quad \text{and}$$

$$y_k(i) = b_k(i) \quad \text{for } i \in I_k,$$

and (with $p_k(w)$ denoting the infimum in the latter as a function of the parameter vector w) the auxiliary problem 11E(6) is written as

$$\text{maximize} \quad -\sum_{l \in L} g_l(w(l)) + \sum_{k=1}^{r} p_k(w) \quad \text{over all } w \in R^L.$$

(Notation in the theorem must be updated to accommodate the new situation. In particular, the assumption of regular feasibility now requires the existence of vectors $\tilde{x}_k \in R^{J_k}$ satisfying $\tilde{x}_k(j) \in \tilde{C}_k(j)$ and $E_k^0 \tilde{x}_k = b_k$, such that for $\tilde{z} = E_1^1 \tilde{x}_1 + \cdots + E_r^1 \tilde{x}_r$, one has $\tilde{z}(l) \in \tilde{C}(l)$. Here $\tilde{C}(l)$ and $\tilde{C}_k(j)$ are the projections on the horizontal axis of the characteristic curves Γ_l and Γ_j^k associated with f_l and f_j^k. In

Assumption 2 of the theorem one must write $(z(l), w(l)) \in \Gamma_l$ for $l \in L$.)

11.43. (*Decomposition*). Investigate what the problems in the preceding exercise boil down to when

$$f_l(z(l)) = \begin{cases} 0 & \text{if } z(l) = a(l), \\ \infty & \text{if } z(l) \neq a(l), \end{cases}$$

$$f_j^k(x_k(j)) = \begin{cases} d_k(j)x_k(j) & \text{if } x_k(j) \geq 0, \\ \infty & \text{if } x_k(j) < 0. \end{cases}$$

Also give a dual expression for $p_k(w)$ in this case. State the corresponding version of the traffic theorem in Section 11E (in updated notation with $z = \sum_{k=1}^{r} E_k^1 x_k$ and $(z(l), w(l)) \in \Gamma_l$ for $l \in L$).

11.44. (*Descent*). Justify the assertions made in the statement of the dual descent algorithm in Section 11H.

11.45. (*Fortified Descent*). Justify the assertions made in the statement of the fortified primal descent algorithm. (Broaden the arguments of Section 9H in notation and terminology.)

11.46. (*Quasi-Extreme Solutions*). A feasible solution x to the primal optimization problem is quasi extreme, as defined in Section 11J, if the index set $F_x = \{ j \in J | f_j'(x(j)) = f_{j+}'(x(j)) \}$ is independent, in other words, contained in some basis F. Show that in the case of a piecewise linear problem the set of all such vectors x is finite and includes all feasible solutions that are extreme in the sense of Section 10M.

11.47. (*Quasi-Extreme Solutions*). Let x_0 be any feasible solution to a given piecewise linear programming problem in primal form. Develop a procedure that in finitely many applications of the painted index algorithm constructs from x_0 a feasible solution x that is quasi extreme as in the preceding exercise, provided the problem satisfies the following condition: there does not exist any elementary primal support P such that for all $j \in P$, the function f_j is without any breakpoints (i.e., is linear-plus-a-constant on $C(j) = (-\infty, \infty)$).

(*Hint.* For each j such that $x_0(j)$ is not a breakpoint of f_j, let $C_0(j)$ be the "linearity interval" for f_j which contains $x(j)$, whereas for j such that $x_0(j)$ is a breakpoint, let $C_0(j)$ consist solely of the value $x_0(j)$. The idea is to construct an extreme solution to the primal feasibility problem given by these intervals $C_0(j)$ and the same vector b as in the original problem; see Section 10M. Verify that this will accomplish the task.)

11.48. (*Quasi-Extreme Solutions*). Consider a piecewise linear programming problem that satisfies the condition in the preceding exercise. Prove that if the problem has an optimal solution at all, then it has an optimal solution which is quasi extreme (see Exercise 11.46).

(*Hint.* Give a constructive argument based on the simplex method and its termination properties when used with a rectification subroutine so as to avoid degeneracy; see Section 11J. Invoke the facts in Exercises 11.46 and 11.47.)

11.49. (*Linear Simplex Method*). Specialize the statement of the simplex method in Section 11J to the case of a general linear programming problem, that is, a primal monotropic programming problem in which

$$f_j(x(j)) = \begin{cases} d(j)x(j) & \text{if } c^-(j) \leq x(j) \leq c^+(j), \\ \infty & \text{otherwise.} \end{cases}$$

Assume that the method begins with an extreme feasible solution in the sense of Section 10M, and work out what will actually take place each iteration in restoring the basis to primal simplex form. What simplifications will there be in all this if $[c^-(j), c^+(j)] = [0, \infty)$ for all j? (The latter case is the standard one treated in linear programming texts.)

11.50. (*Breakpoints*). For $f(t) = \Phi(x + te_p)$ as in Section 11C, suppose that f has a first-order breakpoint at $t = \alpha$ (i.e., $f'_-(\alpha) < f'_+(\alpha)$). Show that for $x' = x + \alpha e_p$ at least one of the functions f_j must have a first-order breakpoint at $x'(j)$.

(*Hint.* Make use of the derivative formulas in Section 11C.)

11.51. (*Extended Pivoting*). Consider the system of equations

$$\sum_{j \in J} e(i, j)x(j) = b(i) \text{ for } i \in I.$$

Show that for any basis $F \subset J$ the system can be written equivalently in the form

$$x(j) = \sum_{k \in F} a(j, k)x(k) + r(j) \quad \text{for } j \in F,$$

where the $r(j)$'s are uniquely determined constants and the $a(j, k)$'s are the coefficients in the Tucker tableau corresponding to F; in the latter equations the values of the $x(k)$'s can be chosen freely. Demonstrate further that in passing from F to an adjacent basis by pivoting on an index pair $(\bar{j}, \bar{k})$, the $r(j)$'s can be updated by the same formulas used for updating the $a(j, k)$'s: simply extend the rules to the bordered matrix shown in Figure 11.8.

11.52. (*Simplified Linear Programming*). Demonstrate that every elementary linear programming problem in the sense of Example 1 in Section 11B is actually equivalent to a linear programming problem of the following still more special type. There is a subspace $\mathscr{C}$ of R^J, and for a

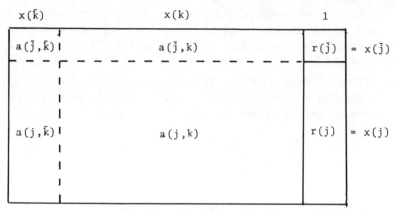

Figure 11.8

particular pair of indexes j_0 and k_0 one wants to

$$\text{minimize} \quad x(j_0) \quad \text{over all} \quad x \in \mathscr{C} \quad \text{satisfying} \quad x(k_0) = 1 \quad \text{and}$$

$$x(j) \geq 0 \quad \text{for all other indexes.}$$

(*Hint.* First reduce the elementary problem to the case where $c(j) = 0$. Then introduce two new variables $x(j_0)$ and $x(k_0)$; set $x(j_0) = \sum_{j \in J} d(j)x(j)$, and in each of the equations $\sum_{j \in J} e(i, j)x(j) = b(i)$ replace $b(i)$ by $b(i)x(k_0)$. The new index set is $\bar{J} = J \cup \{j_0, k_0\}$.)

11.53. (*Simplified Linear Programming*). Show that the dual of the special type of linear programming problem in the preceding exercise is

$$\text{maximize} \quad -v(k_0) \quad \text{over all} \quad v \in \mathscr{D} \quad \text{satisfying} \quad v(j_0) = 1$$

$$\text{and} \quad v(j) \leq 0 \quad \text{for all other indexes } j,$$

where $\mathscr{D} = \mathscr{C}^{\perp}$. (Investigate what happens to the problems in Section 11A when $b = 0$, $f_{j_0}(\xi) = \xi$, $f_{k_0}(\xi)$ is 0 if $\xi = 1$ but ∞ if $\xi \neq 1$, and for all other j the function f_j is identically 0 on $C(j) = [0, \infty)$.)

11.54. (*Simplified Linear Programming*). Consider a primal and dual pair of linear programming problems of the special type in Exercises 11.52 and 11.53. For a basis F with $j_0 \in F$ and $k_0 \in F' = J \setminus F$, the corresponding Tucker tableau takes the form indicated in Figure 11.9. Show that if $a(j, k_0) \geq 0$ for all $j \in F \setminus \{j_0\}$, then an extreme feasible solution to the primal problem is obtained by setting $x(k_0) = 1$

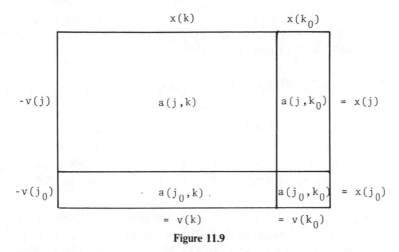

Figure 11.9

and

$$x(j) = a(j, k_0) \quad \text{for all } j \in F,$$
$$x(k) = 0 \quad \text{for all } k \in F'\backslash\{k_0\}.$$

Likewise, if $a(j_0, k) \geq 0$ for all $k \in F'\backslash\{k_0\}$, then an extreme feasible solution to the dual problem is obtained by setting $v(j_0) = 1$ and

$$v(k) = -a(j_0, k) \quad \text{for all } k \in F',$$
$$v(j) = 0 \quad \text{for all } j \in F\backslash\{j_0\}.$$

Finally, show that if both conditions are satisfied, that is, $a(j, k_0) \geq 0$ for all $j \in F\backslash\{j_0\}$ and $a(j_0, k) \geq 0$ for all $k \in F'\backslash\{k_0\}$, then the x and v just described are in fact optimal solutions to the two problems.

(*Remark.*   From this observation it is clear that the two problems can be solved by finding a sequence of pivot steps that transforms the tableau in Figure 11.9 into one satisfying $a(j, k_0) \geq 0$ for $j \neq j_0$ and $a(j_0, k) \geq 0$ for $k \neq k_0$.)

11.55.   (*Simplified Linear Programming*). In the context of the problems in Exercises 11.52 and 11.53 as represented by tableaus in Figure 11.9, suppose the primal simplex method is applied to the extreme feasible solution x corresponding to such a tableau with $a(j, k_0) \geq 0$ for all $j \neq j_0$ (see Exercise 11.54). Show that it turns out to be the following procedure (the *classical simplex method* in this notation):

(a)   If $a(j_0, k) \geq 0$ for all $k \neq k_0$, terminate; the extreme feasible solutions x and v corresponding to the tableau are optimal (see Exercise 11.54). Otherwise select $\bar{k}$ such that $a(j_0, \bar{k}) < 0$.

(b)   If $a(j, \bar{k}) \geq 0$ for all $j \neq j_0$, terminate; the primal problem has inf $= -\infty$, whereas the dual problem has no feasible solution. Otherwise calculate

$$\alpha = \min\{-a(j, k_0)/a(j, \bar{k}) | j \in F \setminus \{j_0\}, a(j, \bar{k}) < 0\}$$

and let $\bar{j}$ be an index for which this min is achieved.

(c)   Pivot on $(\bar{i}, \bar{j})$, and return to (a).

(*Remark.* This classical form of the simplex method can also be viewed as a special case of the out-of-kilter algorithm. See Exercise 11.65.)

11.56.   (*Dual Simplex Method*). A feasible solution u to the dual problem is said to be *quasi extreme* if for the index set F_v' in 11J(2), $J \setminus F_v'$ is a spanning set in the sense of Section 10C. It is said to be *nondegenerate* if $J \setminus F_v'$ is an independent set in the sense of Section 10C. Prove that if the dual simplex method is applied to a piecewise linear programming problem and initiated with a dual feasible solution u that is quasi extreme, then every dual feasible solution in the sequence generated by the algorithm will be quasi extreme. Moreover, if these feasible solutions are all nondegenerate, the corresponding sequence of values of the dual objective function Ψ will be increasing, so that no feasible solution can ever be repeated. (Develop arguments parallel to those in the primal case in Section 11J.)

11.57.   (*Dual Simplex Method*). Show that in the case of piecewise linear programming, the vectors v corresponding to quasi-extreme feasible solutions u to the dual problem form a finite set. Use this to prove that when the dual simplex method is applied to such a problem, starting with a quasi extreme feasible solution, the method must terminate in finitely many iterations unless it comes to a degenerate dual feasible solution u which it stays with forever. (Use the facts in the preceding exercise.)

11.58.   (*Dual Simplex Method*). Indicate how degeneracy can be circumvented in applying the dual simplex method to piecewise linear problems by using the primal rectification algorithm of Section 10J as a subroutine (see the primal simplex case in Section 11J). Demonstrate that in this implementation the dual simplex method is sure to terminate in finitely many iterations if initiated with a quasi-extreme feasible solution u to the dual. (Use the facts in Exercises 11.56 and 11.57.)

11.59.   (*Dual Simplex Method*). Work out the details of the dual algorithm corresponding to the primal one in Exercise 11.54, namely, the procedure obtained when the dual simplex method is applied to the special pair of problems in Exercises 11.52 and 11.53.

11.60. (*Out-of-Kilter Algorithm*). Prove that in the sequence of pairs (x, u) generated by the out-of-kilter algorithm, feasibility is preserved: once x is feasible, it stays feasible, and similarly for u.

11.61. (*Out-of-Kilter Algorithm*). Suppose the out-of-kilter algorithm is applied to the special type of primal optimization problem in which all the functions have the form

$$f_j(x(j)) = \begin{cases} 0 & \text{if } x(j) \in C(j) = [c^-(j), c^+(j)], \\ \infty & \text{otherwise} \end{cases}$$

(The corresponding characteristic curves Γ_j are displayed in Figure 11.5a.) Demonstrate that when the algorithm is initiated with arbitrary x but $u = 0$ ($v = 0$), it reduces to the primal rectification algorithm as stated in Section 10J. (For the optimization problem in question, optimality is the same as feasibility.)

11.62. (*Out-of-Kilter Algorithm*). Suppose the out-of-kilter algorithm is applied to the special type of dual optimization problem in which $b = 0$ and all the functions have the form

$$g_j(v(j)) = \begin{cases} 0 & \text{if } v(j) \in D(j) = [d^-(j), d^+(j)], \\ \infty & \text{otherwise.} \end{cases}$$

(The corresponding characteristic curves Γ_j are displayed in Figure 11.5b.) Demonstrate that when the algorithm is initiated with arbitrary u but $x = 0$, it reduces to the dual rectification algorithm as stated in Section 10J.

11.63. (*Out-of-Kilter Algorithm*). Prove that for $f(t) = \Phi(x + te_P)$ and the α in the step-size rule 11K(3) for out-of-kilter algorithm, it is true that $f'_-(\alpha) \le 0$, although not necessarily that $f'_+(\alpha) \ge 0$ (in contrast to the step-size rule 11H(3) for the basic form of the primal descent algorithm). Nonetheless, α is a breakpoint value for f (which is piecewise linear in this context), and $f(\alpha) < f(0)$.

(*Hint.* Work with the formulas for f'_+ and f'_- in Section 11C.)

11.64. (*Out-of-Kilter Algorithm*). Restate the out-of-kilter algorithm in terms of the coefficients $a(j, k)$ in the current Tucker tableau rather than the normalized elementary vectors e_P and e_Q. (See the basis theorem in Section 10C and the formulation of the painted index algorithm in Section 10G for the notational connections.)

11.65. (*Simplex versus Out-of-Kilter*). Suppose that the out-of-kilter algorithm is applied to a primal linear programming problem of the form described in Exercise 11.52 (and its dual in Exercise 11.53), using red-green preprocessing and the painted index subroutine (the Tucker tableau left over from one iteration being the one used to start the next). Show that if the algorithm is initiated with a Tucker tableau of the kind in Figure 11.9 with the initial x taken to be the extreme primal feasible solution corresponding to this tableau as in Exercise 11.54, but *with the initial v taken to be the zero vector $(v(j) = 0$ for all j, including $j = j_0, j = k_0)$*, then the algorithm reduces to the classical simplex method in Exercise 11.55. (The only out-of-kilter index is j_0.) (*Remark.* It follows from this observation and what was proved in Section 11K about termination of the out-of-kilter algorithm that the classical simplex method in Exercise 11.55 is sure to terminate if Bland's priority rule is used to resolve ties in the selection of the pivot column and pivot row. Of course this could also be proved directly.)

11.66. (*Simplex versus Out-of-Kilter*). Consider a primal linear programming problem of the following kind: there is a subspace $\mathscr{C}$ of R^J, and for a particular index j_0 one wants to

$$\text{minimize} \quad x(j_0) \quad \text{over all} \quad x \quad \text{satisfying}$$

$$x(j) \in C(j) = [c^-(j), c^+(j)] \quad \text{for all } j \neq j_0,$$

where $x(j_0)$ itself is unrestricted $(C(j_0) = (-\infty, \infty))$. (Every linear programming problem in the general sense of Example 1 in Section 11B can be put in this form by the means suggested in Exercise 11.52. The characteristic curve Γ_j has the form in Figure 11.5a for $j \neq j_0$, whereas Γ_{j_0} is a horizontal line at level 1.)

Show that when the primal simplex method of Section 11J is applied to this problem, starting with any feasible solution x, the resulting procedure is the same as when the out-of-kilter algorithm with red-green preprocessing is applied, starting with the same x but with $v = 0$. (The painted index subroutine is to be used in the out-of-kilter algorithm, and the initial basis F is to be chosen to contain j_0, which will be the only out-of-kilter index.)

11.67. (*Simplex Approach to Feasibility*). Consider a primal feasibility problem in the augmented format with aggregated infeasibility as in Section 10L. (For an initial $\bar{x}$ satisfying $c^-(j) \leq \bar{x}(j) \leq c^+(j)$ for all j, one has introduced weights $w(i)$ which satisfy 10L(3) for the initial $\bar{y} = E\bar{x}$, and these have been used to define a new l row for the tableau; see Figure 10.20 and 10L(4) (5).) Partitioning the given index set I into

$$I_+ = \{i \mid \bar{y}(i) > b(i)\}, \qquad I_0 = \{i \mid \bar{y}(i) = b(i)\},$$

$$I_- = \{i \mid \bar{y}(i) < b(i)\},$$

show that a solution to the primal feasibility problem, if one exists, can be obtained by solving the linear programming problem:

$$\text{minimize} \quad y(l) \quad \text{subject to} \quad c^-(j) \leq x(j) \leq c^+(j) \quad \text{for all } j \in J,$$

$$y(i) \geq b(i) \quad \text{for } i \in I_+, \qquad y(i) = b(i) \quad \text{for } i \in I_0,$$

$$y(i) \leq b(i) \quad \text{for } i \in I_-.$$

(This is a problem in terms of the spaces $\overline{\mathscr{C}}$ and $\overline{\mathscr{D}}$ as in 11A(13) with all functions f_j, f_i, and f_l linear relative to their domains of finiteness.)

Specifically, demonstrate that this problem does have at least one optimal solution, and if the minimum is $b(l)$ the optimal solutions are the solutions to the primal feasibility problem. If, on the other hand, the minimum is not $b(l)$, any optimal solution to the corresponding dual problem (see 11A(14)) satisfies 10L(6). (This dual solution therefore yields by 10L(7) (8) a vector pair $(\tilde{u}, \tilde{v}) \in R^I \times R^J$ such that 10L(9) holds. Such a pair substitutes in some situations for having an elementary dual support Q such that $c^+(Q) < b_Q$; see the beginning of Section 10L.)

11.68. (*Simplex Approach to Feasibility*). Suppose that in applying the simplex method to the problem in the preceding exercise, an index i in I_+ or I_- is transferred to I_0 as soon as $y(i) = b(i)$ (so that this equality is maintained forever more). Show that with this added feature the method still works and as a matter of fact is equivalent to applying the primal rectification algorithm in the aggregated feasibility mode in Section 10L using the painted index subroutine with red-green preprocessing and the following slightly different painting: out-of-kilter indexes $i \in I$ (not l) are painted green (rather than white or black) until brought into kilter (when they will naturally become red).

11.69. (*Primal-Dual Algorithm*). Demonstrate that when the out-of-kilter algorithm is applied in the aggregated format in Section 11L, using the painted index subroutine with red-green preprocessing, it still works (has the same termination properties) if the out-of-kilter indexes $i \in I$ (not l) are painted green (rather than white or black) until brought into kilter (when they will naturally become red).

(*Remark*. The possible advantage of the modification lies in removing the requirement that the distance between $y(i)$ and $b(i)$ can never be increased, thereby making it easier to scan a pivot column for an appropriate pivot row. Out-of-kilter indexes will be kept in row position, and except for l, the rows they label will not enter the pivot selection process.

This procedure amounts to applying the dual descent algorithm to a pair of piecewise linear programming problems with the simplex feasibility routine in Exercise 11.68 used in place of the primal rectification algorithm. (The x left over from one round is used to initiate the next.) The step-size rule, however, is slightly different: instead of performing the line search all the way to a minimum, one just descends to a lower breakpoint.

11.70. (*Convex Functions*). Suppose f is a function on R that has the form

$$f(\xi) = \sup_{\eta \in R} \{\xi\eta - h(\eta)\},$$

where h is some other function on R (possibly with infinite values. Prove that if $f(\xi_0)$ is finite for some ξ_0, then f is a closed proper convex function on R as defined in Section 8A.

(*Hint.* First show that $h(\eta)$ cannot be ∞ for all η, and consequently $f(\xi) > -\infty$ for all ξ. Then show convexity:

$$f((1-\lambda)\xi + \lambda\xi') \leq (1-\lambda)f(\xi) + \lambda f(\xi')$$

holds for all ξ', ξ'' and all $\lambda \in (0, 1)$. Next verify that the sets $\{\xi \in R \mid f(\xi) \leq \alpha\}$ for $\alpha \in R$ are closed and this implies

$$\liminf_{\zeta \to \xi} f(\zeta) \geq f(\xi) \quad \text{for all } \xi.$$

Finally, for $D = \{\xi \in R \mid f(\xi) < \infty\}$ show that

$$\limsup_{\substack{\zeta \to \xi \\ \zeta \in D}} f(\zeta) \leq f(\xi) \quad \text{for all } \xi \in \text{cl } D.$$

Do this by representing ζ as $(1-\lambda)\xi + \lambda\xi'$ for $\xi \in D$ and taking the limit in the convexity inequality as $\lambda \downarrow 0$.)

11P.* COMMENTS AND REFERENCES

The word "monotropic" refers to something that turns or changes in one direction only. Its use here is intended to reflect both the convexity properties of the functions f_j and g_j and the monotonicity of the curves Γ_j. What we call monotropic programming problems have also been called "separable convex programming problems with linear constraints," but that designation, besides being unwieldy (there are also separable convex programming problems with separable convex constraints), puts too much emphasis on the "separability" and does not capture the full scope of the class of problems being treated. As seen in Section 11B, the class includes all quadratic convex programming problems and more generally all linearly constrained problems where the objective function is a sum of linear functions composed with convex functions of a single variable. Such objective functions are not separable in themselves, although they can be reduced to separable form.

The duality, equilibrium, and existence theorems in Section 11D were first proved by Rockafellar [1963, 1967] through arguments that implicitly contained the boundedness theorem of Section 11E. The original statements made no mention of elementary vectors or supports; everything concerning existence and boundedness was expressed directly in terms of feasibility with respect to the intervals $C(j)$ and $D(j)$, or $\tilde{C}(j)$ and $\tilde{D}(j)$, or $\hat{C}(j)$ and $\hat{D}(j)$. To get the present statements, it is merely necessary to plug in the elementary vector characterizations of feasibility in Section 10I (see Section 10P for the history of the latter). Note that the case of nonclosed intervals plays a crucial role here. The proof of the duality and equilibrium theorems by way of the results in Section 10C and the concept of elementary directions of descent come from Rockafellar [1981]. The uniqueness criteria laid out in Exercises 10.26 through 10.29 have not previously been published.

In the context of linear programming the existence, duality, and equilibrium theorems in Section 11D reduce to celebrated results of Gale, Kuhn, and Tucker [1951]. In contrast to the latter, however, they are directly applicable to linear programs in the general sense (with the possibility of both upper and lower bounds on the variables) and even to piecewise linear programs. They do not require that a problem first be formulated in one of the standard linear programming schemes, which has been a nuisance in the past and an obstacle to clarity of interpretation. The theorems are also applicable to (convex) quadratic programs, as explained in Section 11B, and they then yield results of Dorn [1960] and Cottle [1963]. As for the piecewise quadratic programs which they cover (see Section 11B), no one until now has singled out that class of problems for special treatment. (It should be noted that convex objective functions of the form

$$F(z) = \max_{k=1,\ldots,r} \left\{ \alpha_k + d_k \cdot z + \tfrac{1}{2} z \cdot D_k z \right\}$$

with each matrix D_k positive semidefinite are not necessarily "preseparable" as defined in Section 11B and hence cannot always be treated by the methods of monotropic programming, even though they might be termed "piecewise quadratic" in a broader sense.)

The strict complementary slackness result for piecewise linear programming problems in Exercise 11.36 is a new extension of a classical linear programming theorem of Tucker [1956].

Monotropic programming problems are not the only kind of optimization problems to exhibit duality, although they do so in a superlative manner. Convexity is the key to a general theory of dual problems having any comparable degree of symmetry and richness. See Rockafellar [1970, 1974] for the outlines of a theory that is capable of application to both finite and infinite-dimensional models and furnishes a thorough interpretation of the optimal solutions to either problem as Lagrange multiplier vectors for the other. Such vectors are "subgradients" of certain optimal value functions and help in expressing the sensitivity of the problems to perturbation. We have not gone into this aspect of duality here, although Exercises 11.33 and 11.34 hint at a related game-theoretic interpretation of optimality.

For an exploration of duality in the neighboring case of separable convex programming problems with separable convex (not necessarily linear) constraints, see Rockafellar [1970′].

Decomposition in the sense of Section 11F dates back to Dantzig and Wolfe [1960] for linear programming and has become a familiar technique in convex programming. The notable feature of the freshly developed version here for monotropic programming is that the results can be stated so sharply and yet so flexibly. The corresponding new version of the traffic equilibrium theorem (Section 11G) avoids many of the technical and conceptual limitations of previous results and lends itself as a model for many variants and extensions, such as the ones suggested in Exercises 11.41 and 11.42. It is easy in the same way to handle multicommodity or traffic problems with resource constraints or costs, where quantities other than arc capacity are jointly consumed in sustaining a flow. For instance, the transport of goods or people in some models may depend on how a limited number of vehicles or drivers are allocated among the various transportation links (see Wollmer [1972]). See Potts and Oliver [1972] and Florian [1976] for more on traffic equilibrium.

The basic descent algorithms in Section 11H serve as the framework for a large class of procedures, some of which, like the linear network programming algorithms in Chapter 7 and the primal, dual, and primal-dual simplex methods in general linear programming, are widely familiar. Our emphasis in presenting these algorithms is on their essential form and nature, unobscured by the details of any particular implementation. We also wish to stress the fact that either the primal or the dual approach leads to the simultaneous solution of both the primal and dual problems. This latter feature is an important consequence of the duality theory and, although well known in linear programming, has not heretofore been perceived as such a direct and practical possibility for any broader class of problems.

One should keep in mind that the subroutine to be invoked at each iteration in testing which of the alternatives of the primal or dual feasibility theorem holds does *not* have to be the corresponding rectification algorithm in Section 10J (or one of its variants in Section 10K or 10L). Thus in network problems there are combinatorial subroutines that can be used instead, as we already know from Chapter 9 where the basic descent algorithms appear in the form of the optimal distribution algorithm and the optimal differential algorithm. For networks with gains and monotropic programming problems over unimodular linear systems, similar specializations are possible.

In implementing the descent algorithms it is not only the exact choice of the direction-finding subroutine that needs to be worked out but also the numerical procedure to be used in the line search, namely, the minimization of $f(t)$

over $t \geq 0$. Exact minimization, as specified in our simple statement of the algorithms, may be hard to carry out, but it may not actually be necessary or even desirable under all circumstances. Thus in the piecewise linear case, where there is no difficulty in finding the exact minimum, one may prefer to descend only to some lower breakpoint of f. The general out-of-kilter algorithm can be interpreted as such a realization of partial descent, according to observations in Section 11K.

There is a large literature on line search for convex functions; see, for example, the references of Avriel [1976]. For a recent method that seems especially suited to the present application, where f could well be piecewise twice differentiable, but with breakpoints that might be a chore to sort out completely in advance, see Mifflin [1983].

A gap in the theory of the basic descent algorithms, whether exact minimization is insisted upon or not, is the absence of any really general criterion for convergence to an optimal solution. With no commensurability properties at hand like those in the network case, the best that can be said is that convergence is assured for piecewise linear problems in special implementations that rely on pivoting rules and the fact that there are only finitely many possible Tucker representations (compare with the piecewise linear simplex methods and the out-of-kilter algorithm). Better results along these lines may be discovered in the future.

The fortified and discretized algorithms in Section 11I are in some measure an attempt to address this troublesome issue of convergence. The fortified algorithms (Rockafellar [1981]) choose the elementary direction of descent with extra care. In this way they are able to avoid the kind of "jamming" that in a naive implementation could lead to a sequence of feasible solutions with corresponding objective values that decrease in each iteration but do not tend to the optimum. (Recall the jamming example of Ford and Fulkerson in the special case of the max flow algorithm; see Exercise 3.12.) For feasible solutions which are not yet ε-optimal (for a prescribed $\varepsilon > 0$), an improvement of at least $\varepsilon/|J|$ in the objective value is guaranteed in each iteration. This is an unusual property for an optimization algorithm, but general circumstances under which it is possible in convex programming were elucidated some time ago by Bertsekas and Mitter [1973]. No previous algorithms have been able to take advantage of these circumstances in a truly practical manner, however.

The chief drawback to the fortified algorithms, as opposed to the basic ones, is that the interval bounds used in the test for ε-optimality are harder to determine and in some cases can only be approximated numerically through a line search. On the other hand, once these bounds have been determined and a corresponding elementary direction of descent has been found, it is possible actually to proceed without a line search at all. The step-size

$$\bar{\alpha} = \min \begin{cases} t^+(j)/e_P(j) & \text{for } j \in P^+, \\ t^-(j)/e_P(j) & \text{for } j \in P^-, \end{cases}$$

where $t^+(j)$ and $t^-(j)$ are numbers yielding the minimum in the special expression for $d_x^+(j)$ and $d_x^-(j)$ in this case (possibly infinite under the right interpretation), is sure to improve the objective value by at least $\varepsilon/|J|$; see Rockafellar [1982]. (Of course one can continue to move in the same direction by increments of this same amount until there is no more improvement.)

The discretized descent algorithms get around the convergence issue in a different way. They use interval bounds that are merely difference quotients and therefore relatively easy to determine, as well as line search with a preassigned step size, but they terminate only with quasi optimality. They are capable, nevertheless, of generating optimizing sequences, by repeated restarts with tighter termination criteria (progressively smaller values of δ).

It is worthwhile underlining again that by virtue of duality, each of these variants of basic descent provides *two* ways of solving a given monotropic programming problem. This is true also of the simplex methods in Section 11J, even in the nonlinear case.

The original simplex method of Dantzig [1951] was formulated for primal linear programming problems in "standard form," that is, with all intervals of type $C(j) = [0, \infty)$. The corresponding method for the dual problem was introduced by Lemke [1954]. Many variants and refinements have since been developed; see the book of Dantzig [1963] for an overview. These algorithms have proved themselves to be remarkably robust and efficient over the years in myriad applications. For theoreticians this good behavior has been a challenge to explain, because worst-case analysis shows that simplex methods can fail miserably on certain artificially constructed examples. Attention has focused on trying to establish good behavior in some probabilistic sense, and recently there has been notable progress toward that end; see Smale [1983].

This state of affairs needs to be borne in mind in making any assessment of either the generalized simplex methods in Section 11J or the closely related out-of-kilter methods in Sections 11K, 11L, and 11M. These are likely to behave much better than theory can at present account for, but the particular mode of implementation could have a big influence too. Judging from the intense effort that has gone into the current implementations of the original simplex method, not to mention the shortcuts that have been discovered for cases of special structure, important refinements in the generalized methods are possible and offer a worthy goal for research. Until such research has been carried out, definitive comparisons between the various approaches can hardly be attempted.

Simplex methods are popularly viewed as being almost synonymous with pivoting algorithms that descend from one "extreme point" to another, but they are really more special than that in some respects and more general in others. They are not the only optimization procedures to employ pivoting (the out-of-kilter algorithm does too, when implemented in certain ways), and they are not necessarily restricted to dealing with "extreme points" of some kind (the convex simplex method of Zangwill [1967, 1969] being an example). Of course the pivoting that is involved need not be executed in a matrix format, either. In the network setting in Chapters 7 and 9 there is a direct combinatorial procedure for pivoting, and that sort of approach is extendable to networks with gains and to unimodular systems, for example.

The formulation of a "simplex method" in Section 11J is more general than any given before. In our view such a method in primal form, say, is an implementation of the primal descent algorithm which in generating a sequence of bases and associated Tucker tableaus

1. Tries to keep out of the basis, as far as possible, indexes j such that the current $x(j)$ is a breakpoint for f_j.

2. Initiates the optimality test in each iteration with dual variables chosen so that indexes in the basis are all in kilter.

Such a method can be started with any regularly feasible solution x and, without any assumptions of linearity or piecewise linearity, will produce a sequence of feasible solutions with decreasing objective values (at least if nondegeneracy holds, or if degeneracy is circumvented by a more careful pivot selection procedure corresponding to internal iterations of the dual rectification algorithm, as suggested in Section 11J).

Although the simplex method was extended to problems with nonlinear objectives (not necessarily "preseparable," incidentally) by Zangwill [1967, 1969], that extension, in contrast to the present one, was confined to differentiable functions and, in particular, did not cover piecewise linear programming. A piecewise linear simplex method has more recently been sketched by Fourer [1981], however. Zangwill's work in the differentiable nonlinear case raises some challenges for the general simplex method that indicate the need for further research. He proves *global convergence* to an optimal solution in that case under the assumption of nondegeneracy (and differentiability) and using a supplementary criterion on the basis. A rough statement of this criterion in our context is the following: in choosing a basis F from the set F_x of nonbreakpoint indexes (which is a spanning set by nondegeneracy), do this so that the indexes $j \in F$ have $x(j)$ as far as possible from being a breakpoint of f_j. (More specifically, for each $j \in F_x$ let δ_j denote the distance of $x(j)$ from the nearest breakpoint of f_j; choose F to maximize $\delta(F) = \min\{\delta_j | j \in F\}$ over all bases included in F_x.) This criterion is no doubt harder to work with in our general context than in Zangwill's. At any rate a global convergence result for the general algorithms is still lacking for problems that are not piecewise linear.

Zangwill [1969] also shows how the convex simplex method can be coupled with a conjugate direction method to ensure finite termination in quadratic programming and presumably a better local rate of convergence for other nonlinear situations. This is surely an important area still to be explored in monotropic programming. How can second-order techniques be combined with the methods in this chapter and applied to problems where the cost functions are piecewise of class $\mathscr{C}^2$, say? The idea is to take advantage of the underlying combinatorial structure as reflected in the mechanisms for generating elementary directions of descent, at least relative to *some* of the variables, where breakpoints (and interval bounds) are active, but at the same time to use second-order techniques to accelerate convergence with respect to other variables that locally are behaving as if they are unconstrained.

In the same connection it should be pointed out that pivoting algorithms with finite termination properties already exist for quadratic programming problems; see the article of Cottle and Dantzig [1968] and its bibliography. Such methods are not sensitive to the combinatorial structure of the primal and dual linear systems of variables, however. Thus when applied to network programming problems, for example, they cannot be reduced to graphical subroutines. Also their extension to piecewise quadratic functions, or other, nonquadratic functions, is not apparent. The descent methods of monotropic programming do make use of combinatorial structure and can be applied very generally, yet for quadratic programming problems they do not enjoy finite termination. Surely some marriage of these two approaches will be possible.

For the general algorithms of convex and nonconvex programming, see the books of Avriel [1976] and McCormick [1983].

The general out-of-kilter algorithm presented in Section 11K is essentially new, although it has a predecessor in the case of linear network programming in the algorithm of Fulkerson [1961]; see Chapter 7. As implemented in the aggregated format in Section 11L, it resembles the primal-dual algorithm of Dantzig, Ford, and Fulkerson [1956]. The latter was formulated only for linear problems, however, and corresponds to using the linear simplex method in a certain way instead of one of the rectification algorithms in Section 10J as a feasibility subroutine (see Exercises 11.67 and 11.68). In the parametric mode in Section 11M the out-of-kilter algorithm reduces for linear problems to something close to the well-known parametric methods in linear programming (see Dantzig [1963] and Murty [1976]). However, it is applicable more generally to problems that are merely piecewise linear, and it displays greater flexibility relative to the nature and number of indexes that can be out of kilter. (The classical simplex method itself is contained as a special case; see Exercise 11.65). Furthermore the out-of-kilter algorithm is formulated in such a way that combinatorial subroutines can readily be inserted when available (in place of the painted index algorithm as realized in tableau form).

Among the nice features of the out-of-kilter algorithm is the fact that it can be applied to *arbitrary* initial x and u, and when these are feasible, it maintains that property and thereby provides a test of ε-optimality at all times. This could be advantageous in large-scale problems where one is willing to forgo exact optimality. Primal and dual simplex methods do not provide such a test and do not even yield feasible solutions to both problems if stopped short of optimality.

For problems in networks with gains, specializations of all these algorithms are possible in terms of combinatorial subroutines for generating elementary primal and dual supports (see the characterizations in Sections 10D and 10E).

The original method of Jewell [1962], for instance, corresponds roughly to such a realization of the linear case of the out-of-kilter algorithm in parametric form. (Jewell overlooked the fact that without the integrality properties present in the ordinary network case, however, a sequence of improving feasible solutions can fail to be an optimizing sequence.)

See Section 10P for further references to computational procedures for networks with gains and also for multicommodity networks and traffic problems. Weintraub and Gonzalez [1980] give additional update of the latter. Factorization as in Section 10F and decomposition as in Sections 11F and 11G can be very helpful in such applications. In solving the master problem in the decomposition scheme, direct methods can be used, but techniques of subgradient optimization of nondifferentiable functions like those used in integer programming by Held, Wolfe, and Crowder [1974] can also be tried; see Kennington and Shalaby [1977].

BIBLIOGRAPHY

H. A. Aashtriani and T. L. Magnanti. 1976. "Implementing primal-dual network flow algorithms." Working Paper OR-055-76. Operations Research Center, Massachetts Institute of Technology.

G. M. Adelson-Velsky, E. A. Dinic, and A. V. Karzanov. 1975. *Flow Algorithms*. Nauka (Moscow, in Russian).

D. L. Adolphson. 1980. "A nondegenerate network simplex algorithm." Preprint.

A. V. Aho, J. E. Hopcroft, and J. D. Ullman. 1974. *The Design and Analysis of Computer Algorithms*. Addison-Wesley.

A. Ali, R. V. Helgason, J. L. Kennington, and H. S. Lall. 1977. "Primal simplex techniques: state-of-the-art implementation technology." *Networks* 8:315–340.

A. Ali, R. V. Helgason, J. L. Kennington, and H. S. Lall. 1980. "Computational comparison among three multicommodity network flow algorithms." *Operations Res.* 28:995–1000.

A. A. Assad. 1978. "Multicommodity network flows: a survey." *Networks* 8:37–92.

M. Avriel. 1976. *Nonlinear Programming: Analysis and Methods*. Prentice-Hall.

A. Bachem and B. Korte. 1978. "An algorithm for quadratic optimization over transportation polytopes." *Z. Angew. Math. Mech.* 58:T459–T461.

M. L. Balinski. 1970. "On a selection problem." *Manag. Sci.* 17:230–231.

M. L. Balinski and R. E. Gomory. 1963. "A mutual primal-dual simplex method." In *Recent Advances in Math. Programming*. R. L. Graves and P. Wolfe, eds. McGraw-Hill, pp. 17–26.

Z. Baranyai. 1973. "On the factorization of the complete uniform hypergraph." *Colloq. Math. Soc. J. Bolyai* 10:91–108.

R. S. Barr, F. Glover, and D. Klingman. 1974. "An improved version of the out-of-kilter method and a comparative study of computer codes." *Math. Prog.* 7:60–86.

R. S. Barr, F. Glover, and D. Klingman. 1977. "The alternating basis algorithm for assignment problems." *Math. Prog.* 13:1–13.

R. S. Barr, F. Glover, and D. Klingman. 1978. "The generalized alternating path algorithm for transportation problems." *European J. Operations Res.* 2:137–144.

M. S. Bazaraa and J. J. Jarvis. 1977. *Linear Programming and Network Flows*. Wiley.

C. Berge. 1962. *The Theory of Graphs and its Applications*. Wiley.

C. Berge and A. Ghouila-Houri. 1962. *Programmation, Jeux et Réseaux de Transport*. Dunod. (English translation. 1965. *Programming, Games and Transportation Networks*. Wiley.)

D. P. Bertsekas. 1976. "A new algorithm for solution of restrictive networks involving diodes." *IEEE Trans. Circuits Sys.* CAS-23:599–608.

D. P. Bertsekas. 1979. "Algorithms for nonlinear multicommodity network flow problems." *International Symp. on Systems Optimization and Analysis*. A. Bensoussan and J. L. Lions, eds. Springer-Verlag, pp. 210–224.

D. B. Bertsekas and S. K. Mitter. 1973. "Descent numerical methods for optimization problems with nondifferentiable cost functionals." *SIAM J. Control* 11:637–652.

G. Birkhoff. 1946. "Three observations on linear algebra." *Rev. Univ. Tucumán*, A5:147–151.

G. Birkhoff and J. B. Diaz. 1956. "Nonlinear network problems." *Quart. Appl. Math.* 13:431–444.

R. E. Bixby and W. H. Cunningham. 1980. "Converting linear programs to network problems." *Math. Operations Res.* 5:321–357.

R. G. Bland. 1974. *Complementary Orthogonal Subspaces of R^n and Orientability of Matroids*. Thesis. Cornell University.

R. G. Bland. 1977. "A combinatorial abstraction of linear programming." *J. Combinatorial Theory*, B 23:33–57.

R. G. Bland. 1978. "New finite pivoting rules for the simplex methods." *Math. Operations Res.* 3:103–107.

R. G. Bland. 1979. "Generalizations of Hoffman's existence theorem for circulations." *Networks* 11:243–254.

R. G. Bland and M. Las Vergnas. 1978. "Orentiability of matroids." *J. Combinatorial Theory*, B 24:94–123.

R. G. Bland and M. Las Vergnas. 1979. "Minty colorings and orientations of matroids." *Ann. New York Acad. Sci.* 319:86–92.

J. A. Bondy and U. S. R. Murty. 1976. *Graph Theory with Applications*. American Elsevier.

G. H. Bradley, G. G. Brown, and G. W. Graves. 1977. "Design and implementation of large-scale primal transshipment problems." *Manag. Sci.* 24.

J. Bruno and L. Weinberg. 1976. "Generalized networks: networks embedded on a matroid," *Networks* 6:53–94 (part I); 231–272 (part II).

R. E. Burkard and H. Hamacher. 1981. "Minimal cost flows in regular matroids." *Math. Programm. Study* 14:32–47.

R. G. Busacker and P. J. Gowen. 1961. "A procedure for determining a family of minimal-cost network flow patterns." O.R.O. Technical Paper 15.

R. G. Busacker and T. L. Saaty. 1965. *Finite Graphs and Networks: An Introduction with Applications*. McGraw-Hill.

P. Camion. 1968. "Modules unimodulaires." *J. Combinatorial Theory* 4:301–362.

A. Charnes and W. W. Cooper. 1958. "Nonlinear network flows and convex programming over incidence matrices." *Naval Res. Logist. Quart.* 5:321–340.

W. K. Chen. 1971. *Applied Graph Theory*. North-Holland.

N. Christophides. 1975. *Graph Theory: An Algorithmic Approach*. Academic Press.

M. Collins, L. Cooper, R. Helgason, J. Kennington, and L. LeBlanc. 1978. "Solving the pipe network analysis problem using optimization techniques." *Manag. Sci.* 24:747–760.

R. W. Cottle. 1963. "Symmetric dual quadratic programs." *Quart. Appl. Math.* 21:237–243.

R. W. Cottle. 1983. "A triple decomposition algorithm for the constrained matrix problem," forthcoming.

R. W. Cottle and G. B. Dantzig. 1968. "Complementary pivot theory of mathematical programming." *Lin. Alg. Appl.* 1:103–125.

W. H. Cunningham. 1976. "A network simplex method." *Math. Prog.* 11:105–116.

W. H. Cunningham. 1979. "Theoretical properties of the network simplex method." *Math. Operations Res.* 4:196–208.

G. B. Dantzig. 1948. "Programming in a linear structure." Comptroller, United States Air Force. Washington, D.C.

G. B. Dantzig. 1951. "Application of the simplex method to a transportation problem." In *Activity Analysis of Production and Allocation*. T. C. Koopmans, ed. Wiley, pp. 359–373.

G. B. Dantzig. 1951. "Maximization of a linear function of variables subject to linear inequalities." In *Activity Analysis of Production and Allocation*. T. C. Koopmans, ed. Wiley, pp. 339–347.

G. B. Dantzig. 1963. *Linear Programming and Extensions*. Princeton University Press.

G. B. Dantzig, L. R. Ford, Jr., and D. R. Fulkerson. 1956. "A primal-dual algorithm for linear problems." In *Linear Inequalities and Related Systems* (H. W. Kuhn and A. W. Tucker, eds.), *Ann. of Math. Studies* 38:171–182.

G. B. Dantzig and D. R. Fulkerson. 1954. "Minimizing the number of tankers to meet a fixed schedule." *Naval Res. Logist. Quart.* 1:217–222.

G. B. Dantzig and D. R. Fulkerson. 1956. "On the max-flow min-cut theorem of networks." In *Linear Inequalities and Related Systems* (H. W. Kuhn and A. W. Tucker, eds.), *Ann. of Math. Studies* 38:215–221.

G. B. Dantzig and P. Wolfe. 1960. "The decomposition algorithm for linear programming." *Operations Res.* 101–111.

R. S. Dembo and J. G. Klincewicz. 1981. "A scaled reduced-gradient algorithm for convex network flow problems." *Math. Programm. Study* 15:125–147.

J. B. Dennis. 1959. *Mathematical Programming and Electrical Networks*. Wiley.

N. Deo. 1974. *Graph Theory with Applications to Engineering and Computer Science*. Prentice-Hall.

E. W. Dijkstra. 1959. "A note on two problems in connexion with graphs." *Num. Math.* 1:269–271.

R. Dial, F. Glover, D. Karney, and D. Klingman. 1979. "A computational analysis of alternative algorithms and labeling techniques for finding shortest path trees." *Networks* 9:215–248.

R. P. Dilworth. 1959. "A decomposition theorem for partially ordered sets." *Ann. of Math.* 51:161–166.

E. A. Dinic. 1970. "Algorithm for solution of a problem of maximum flow in a network with power estimation." *Soviet Math. Dokl.* 11:1277–1280.

V. Dolezal. 1979. *Monotone Operators and Applications in Control and Network Theory*. American Elsevier.

W. S. Dorn. 1960. "Duality in quadratic programming." *Quart. Appl. Math.* 18:155–162.

S. E. Dreyfus. 1969. "An appraisal of some shortest-path algorithms." *Operations Res.* 17:395–412.

S. E. Dreyfus 1977. *The Art and Theory of Dynamic Programming*. Academic Press.

R. J. Duffin. 1947. "Nonlinear networks. IIa." *Bull. Amer. Math. Soc.* 53:963–971.

J. Edmonds and D. R. Fulkerson. 1970. "Bottleneck extrema." *J. Combinatorial Theory* 8:299–306.

J. Edmonds and R. Giles. 1977. "A min-max relation for submodular functions on graphs. *Ann. Discrete Math.* 1:185–204.

J. Edmonds and R. M. Karp. 1972. "Theoretical improvements in algorithmic efficiency for max flow problems." *J. ACM* 19:248–264.

J. Egerváry. 1931. "Matrixok kombinatoricus tulajonsagairol." *Mat. és Fiz. Lapok* 38:16–28. Translation by H. W. Kuhn. 1955. "On combinatorial properties of matrices." George Washington University Logistics Papers 11.

P. Elias, A. Feinstein, and C. E. Shannon. 1956. "Note on maximum flow through a network." *IRE Trans. Inf. Theory* IT-2:117–119.

J. R. Evans. 1976. "Max flow in probabilistic graphs." *Networks* 6:161–183.

J. R. Evans, J. J. Jarvis, and R. A. Duke. 1977. "Graphic matroids and the multicommodity transportation problem." *Math. Prog.* 323–328.

S. Even. 1976. "The max flow algorithm of Dinic and Karzanov." Report MIT/LCS/TM-80. Laboratory of Computer Science, Massachusetts Institute of Technology.

K. Fan. 1956. "On systems of linear inequalities." In *Linear Inequalities and Related Systems* (H. W. Kuhn and A. W. Tucker, eds.), *Ann. of Math. Study* 38:99–156.

J. Farkas. 1902. "Über die Theorie der einfachen Ungleichungen." *J. Reine Angew. Math.* 124:1–24.

W. Fenchel. 1949. "On conjugate convex functions." *Canad. J. Math.* 1:73–77.

M. Florian. 1976. *Proceedings of the International Symposium on Traffic Equilibrium Methods.* Lecture Notes in Econ. and Math. Systems No. 118. Springer-Verlag.

J. Folkman and J. Lawrence. 1978. "Oriented matroids." *J. Combinatorial Theory*, B 25:199–236.

J. J. H. Forrest and J. A. Tomlin. 1972. "Updated triangular factors on the basis to maintain sparsity in the product form simplex method." *Math. Prog.* 2:263–278.

L. R. Ford, Jr., and D. R. Fulkerson. 1956. "Maximal flow through a network." *Canad. J. Math.* 8:339–404.

L. R. Ford, Jr., and D. R. Fulkerson. 1957. "A simple algorithm for finding maximal network flows and application to the Hitchcock problem." *Canad. J. Math.* 9:210–218.

L. R. Ford, Jr., and D. R. Fulkerson. 1957. "A primal-dual algorithm for the capacitated Hitchcock problem." *Naval Res. Logist. Quart.* 4:47–54.

L. R. Ford and D. R. Fulkerson. 1962. *Flows in Networks.* Princeton University Press.

R. Fourer. 1981. "Notes on 'semi-linear' programming." Draft report, Dept. of Industrial Engineering and Management Sciences, Northwestern University.

H. Frank and I. T. Frisch. 1971. *Communication, Transmission and Transportation Networks.* Addison-Wesley.

M. Frank and P. Wolfe. 1956. "An algorithm for quadratic programming." *Naval Res. Logist. Quart.* 3:95–110.

A. M. Frieze. 1976. "Shortest path algorithms for knapsack type problems." *Math. Prog.* 11:150–157.

G. Frobenius. 1912. "Über Matrizen mit nicht negativen Elementen." *S.-B. Berlin. Akad.* 23:456–477.

D. R. Fulkerson. 1961. "An out-of-kilter method for minimal cost flow problems." *SIAM J.* 9:18–27.

D. R. Fulkerson. 1966. "Flow networks and combinatorial operations research." *Amer. Math. Monthly* 73:115–138.

D. Gale. 1957. "A theorem of flows in networks." *Pacific J. Math.* 7:1073–1082.

D. Gale, H. W. Kuhn, and A. W. Tucker. 1951. "Linear programming and the theory of games." In *Activity Analysis of Production and Allocation.* Cowles Commission Monograph No. 13, pp. 317–328.

Z. Galil and A. Naamad. 1979. "Network flow and generalized path compression." *ACM Sym. Theory Comput.* 11:13–26.

G. Gallo and C. Sodini. 1979. "Adjacent extreme flows and application to min concave cost flow problems." *Networks* 9:95–121.

B. Gavish and P. Schweitzer. 1974. "An algorithm for combining truck trips." *Transportation Sci.* 8:13–23.

B. Gavish, P. Schweitzer, and E. Shlifer. 1977. "The zero pivot phenomenon in transportation problems and its computational implications." *Math. Prog.* 12:226–240.

A. M. Geoffrion. 1977. "Objective function approximations in mathematical programming." *Math. Prog.* 13:23–37.

P. E. Gill, W. Murray, M. A. Saunders, and M. H. Wright. 1982. "Procedures for optimization problems with a mixture of bounds and general linear constraints." Technical Report SOL 82-6. Systems Optimization Laboratory, Stanford University.

F. Glover, J. Elam, and D. Klingman. 1974. "A strongly convergent primal algorithm for generalized networks." *Math. Operations Res.* 4:39–59.

F. Glover, D. Karney, and D. Klingman. 1974. "Implementation and computational comparisons of primal, dual, and primal-dual computer codes for minimum cost network flow problems." *Networks* 4:191–212.

F. Glover, D. Karney, D. Klingman, and A. Napier. 1974. "A computational study on start procedures: basic change criteria and solution algorithms for transportation problems." *Manag. Sci.* 20:793–813.

F. Glover and D. Klingman. 1973. "On the equivalence of some generalized network problems to pure network problems." *Math. Prog.* 4:269–278.

F. Glover and D. Klingman. 1975. "Real world applications of network related problems and breakthroughs in solving them efficiently." *ACM Trans. on Math. Software* 1:47–55.

F. Glover, D. Klingman, and J. Stutz. 1973. "Extensions of the augmented index method to generalized network problems." *Transportation Sci.* 7:377–384.

B. Golden. 1976. "Shortest-path algorithms: a comparison." *Operations Res.* 24:1164–1168.

B. L. Golden and T. L. Magnanti. 1977. "Deterministic network optimization: a bibliography." *Networks* 7:149–183.

P. Gordan. 1873. "Über die Auflösung linearer Gleichungen mit reelen Coefficienten." *Math. Ann.* 6:23–28.

R. C. Grinold. 1973. "Calculating maximal flows in a network with positive gains." *Operations Res.* 21:528–541.

O. Gross. 1959. "The bottleneck assignment problem." Rand Corporation Paper p-1630.

M. Hall, Jr. 1956. "An algorithm for distinct representatives." *Amer. Math. Monthly* 63:716–717.

M. A. Hall. 1976. "Hydraulic network analysis using (generalized) geometric programming." *Networks* 6:105–130.

M. A. Hall. 1977. "Civil engineering applications of (generalized) geometric programming." *J. of Engineering Opt.* 2.

P. Hall. 1935. "On representatives of subsets." *J. London Math. Soc.* 10:26–30.

J. Halpern and I. Priess. 1974. "Shortest paths with time constraints on movement and parking." *Networks* 4:241–253.

H. Hamacher. 1979. "Numerical investigations on the maximal flow algorithm of Karzanov." *Computing* 22:17–29.

H. Hamacher. 1980. "Algebraic flows in regular matroids." *Discrete Appl. Math.* 2:27–38.

H. Hamacher. 1981. *Flows in Regular Matroids.* Oelgeschlager, Gunn and Hain, Inc.

G. Y. Handler and P. Mirchandani. 1979. *Location on Networks: Theory and Algorithms.* MIT Press.

G. Y. Handler and I. Zang. 1980. "A dual algorithm for the constrained shortest path problem." *Networks* 10:293–310.

F. Harary. 1969. *Graph Theory.* Addison-Wesley.

R. Hassin. 1981. "Generalizations of Hoffman's existence theorem for circulations." *Networks* 11:243–254.

M. Held, P. Wolfe, and H. Crowder. 1974. "Validation of subgradient optimization." *Math. Prog.* 6:62–88.

R. V. Helgason and J. L. Kennington. 1977. "An efficient specialization of the convex simplex method for nonlinear network flows problems." Technical Report IEOR 99017. Southern Methodist University.

E. Hellerman and D. Rarick. 1971. "Reinversion with the preassigned pivot procedure." *Math. Prog.* 1:195–216.

F. L. Hitchcock, 1941. "The distribution of a product from several sources to numerous locations." *J. Math. Phys.* 20:224–230.

A. J. Hoffman. 1960. "Some recent applications of the theory of linear inequalities to extremal combinatorial analysis." *Proc. Sym. Appl. Math.* 10:113–128.

A. J. Hoffman and J. B. Kruskal. 1956. "Integral boundary points of convex polyhedra." *Linear Inequalities and Related Systems* (H. W. Kuhn and A. W. Tucker, eds.). *Ann. of Math. Studies* 38:233–246.

A. J. Hoffman and H. W. Kuhn. 1956. "Systems of distinct representatives and linear programming." *Amer. Math. Monthly* 63:455–460.

J. Hopcroft and R. Karp. 1973. "An $n^{5/2}$ algorithm for maximal matchings in bipartite graphs." *SIAM J. Comput.* 2:225–231.

T. C. Hu. 1961. "The maximum capacity route problem." *Operations Res.* 9:898–900.

T. C. Hu. 1969. *Integer Programming and Network Flows.* Addison-Wesley.

M. Iri. 1960. "A new method of solving transportation network problems." *J. Operations Res. Soc. Japan* 3:27–87.

M. Iri. 1969. *Network Flow, Transportation and Scheduling.* Academic Press.

W. S. Jewell. 1961. "Optimal flow through networks." Interim Technical Report No. 8. Massachusetts Institute of Technology.

W. S. Jewell. 1962. "Optimal flow through networks with gains." *Operations Res.* 10:476–499.

E. L. Johnson. 1966. "Networks and basic solution." *Operations Res.* 14:619–624.

L. Kantorovitch, 1942. "On the translocation of masses." *Compt. Rend. (Doklady) Acad. Sci. de l'URSS* 37:199–201.

C. Y. Kao and R. R. Meyer. 1979. "Secant approximation methods for convex approximation." Computer Sciences Dept. Tech. Rep. No. 382, University of Wisconsin, Madison.

R. M. Karp. 1980. "An algorithm to solve the $m \times n$ assignment problem in expected time $O(m^2n)$." *Networks* 10:143–152.

A. V. Karzanov. 1974. "Determining the maximal flow in a network by the method of preflows." *Soviet Math. Dokl.* 15:434–437.

A. Kaufmann and G. Desbazeille. 1969. *The Critical Path Method.* Gordon and Breach.

J. L. Kennington. 1978. "A survey of linear cost multicommodity network flow problems." *Operations Res.* 26:209–236.

J. L. Kennington and R. V. Helgason. 1980. *Algorithms for Network Programming.* Wiley-Interscience.

J. L. Kennington and M. Shalaby. 1977. "An effective subgradient procedure for minimal cost multicommodity flow problems." *Manag. Sci.* 23:994–1004.

G. Kirchhoff. 1847. *Annalen der Physik and Chemie* 72:497.

V. L. Klee. 1979. "Combinatorial optimization: what is the state of the art?" In *Information Linkage between Applied Mathematics and Industry.* Academic Press, pp. 71–36.

M. Klein, 1967. "A primal method for minimal cost flows with applications to assignment and transportation problems." *Manag. Sci.* 14:205–220.

R. W. Klessig. 1974. "An algorithm for nonlinear multicommodity flow problems." *Networks* 4:343–355.

D. König. 1931. "Graphok és matrixok." *Mat. és Fiz. Lapok* 38:116–119.

D. König. 1933. "Über trennende Knotenpunkte in Graphen (nebst Anwendungen auf Determinanten und Matrizen)." *Acta Litterarum ac Scientarum (Sectio Scientarum Mathematicarum) Szeged* 6:155–179.

D. König. 1936. *Theorie der Endlischen und Unendlichen Graphen.* Akademische Verlagsgesellschaft, Leipzig. Reprint 1950. Chelsea.

J. B. Kruskal, Jr. 1956. "On the shortest spanning subtree of a graph and the traveling salesman problem." *Proc. Amer. Math. Soc.* 7:48–50.

H. W. Kuhn. 1955. "The Hungarian method for the assignment problem." *Naval Res. Logist. Quart.* 2:83–97.

H. W. Kuhn. 1956. "Solvability and consistency for linear equations and inequalities." *Amer. Math. Monthly* 63:217–232.

E. L. Lawler. 1976. *Combinatorial Optimization: Networks and Matroids.* Holt, Rinehart and Winston.

J. Lawrence. 1982. "Oriented matroids and multiply ordered sets." *Lin. Alg. Appl.* 48:1–12.

J. Lawrence and L. Weinberg. 1980. "Unions of oriented matroids." *Ann. Discrete Math.* 8:29–34.

J. Lawrence and L. Weinberg. 1981. "Unions of oriented matroids." *Lin. Alg. Appl.* 41:183–199.

C. E. Lemke. 1954: "The dual method of solving the linear programming problem." *Naval Res. Logist. Quart.* 1:48–54.

D. G. Luenberger. 1973. *Introduction to Linear and Nonlinear Programming,* Addison-Wesley.

H. B. Mann and H. J. Ryser. 1953. "Systems of distinct representatives." *Amer. Math. Monthly* 60:397–401.

J. F. Maurras. 1972. "Optimization of the flow through networks with gains." *Math. Prog.* 3:135–144.

W. Mayeda. 1972. *Graph Theory.* Wiley-Interscience.

G. P. McCormick. 1983. *Nonlinear Programming: Theory, Algorithms, Applications.* Wiley.

N. S. Mendelssohn and A. L. Dulmage. 1958. "Some generalizations of distinct representatives." *Canad. J. Math.* 10:230–241.

K. Menger. 1927. "Zur algemeninen Kurventheorie." *Fund. Math.* 10:96–115.

K. Menger. 1932. *Kurventheorie.* Akademische Verlaggesellschaft, Leipzig.

R. R. Meyer. 1977. "A class of nonlinear integer programs solvable by a single linear program." *SIAM J. Control Opt.* 15:935–946.

R. R. Meyer. 1979. "Two-segment separable programming." *Manag. Sci.* 25:285–295.

R. R. Meyer and M. L. Smith. 1978. "Algorithms for a class of 'convex' nonlinear integer programs." In *Computers and Math. Programming.* W. W. White, ed. National Bureau of Standards Publication 502.

R. Mifflin. 1983. "A superlinearly convergent algorithm for one-dimensional constrained minimization problems with convex functions." *Math. Operations Res.* 8:185–195.

W. Miller. 1951. "On some general theorems for nonlinear systems possessing resistance." *Philos. Mag.* 42:1150–1177.

C. E. Miller. 1963. "The simplex method for local separable programming." In *Recent Advances in Mathematical Programming.* R. L. Graves and P. Wolfe, eds. McGraw-Hill, pp. 89–100.

R. Miller. 1963. *Schedule, Cost and Profit Control with PERT.* McGraw-Hill.

E. Minieka. 1978. *Optimization Algorithms for Networks and Graphs.* Marcel Dekker.

G. J. Minty. 1957. "A comment on the shortest route problem." *Operations Res.* 5:724.

G. J. Minty. 1960. "Monotone networks." *Proc. Roy. Soc. London* A 257:194–212.

G. J. Minty. 1961. "Solving steady-state nonlinear networks of 'monotone' elements." *IRE Trans. Circuit Theory* CT-8:99–104.

G. J. Minty. 1966. "On the axiomatic foundations of the theories of directed linear graphs, electrical networks, and network programming." *J. Math. Mech.* 15:420–485.

G. J. Minty. 1974. "A 'from scratch' proof of a theorem of Rockafellar and Fulkerson." *Math. Prog.* 7:368–375.

T. S. Motzkin. 1933. "Beiträge zur Theorie der Linearen Ungleichungen." Dissertation. Basel.

H. Müller-Merbach. 1970. "Die Method der direkten Koeffizientenanpassung (η-Form) des Separablen Programming." *Unternehmungsforschung* 14:198–214.

K. Murtz. 1976. *Linear and Combinatorial Programming.* Wiley.

J. Muth and G. Thompson. 1963. *Industrial Scheduling*. Prentice-Hall.

O. Ore. 1956. "Studies on directed graphs I." *Ann. of Math*. 63:383–406.

C. H. Papadimitriou and K. Steiglitz. 1982. *Combinatorial Optimization: Algorithms and Complexity*. Prentice-Hall.

U. Pape. 1974. "Implementation and efficiency of Moore-algorithms for the shortest route problems." *Math. Prog*. 7:212–222.

T. D. Parsons. 1970. "A combinatorial approach to convex quadratic programming." *Lin. Alg. App*. 3:359–378.

J. C. Picard and H. D. Ratliff. 1975. "Minimum cuts and related problems." *Networks* 5:357–370.

H. Pollack. 1961. "The maximum capacity route through a network." *Operations Res*. 9:722–736.

R. B. Potts and R. M. Oliver. 1972. *Flows in Transportation Networks*. Academic Press.

W. G. Prager. 1965. "Mathematical programming and the theory of structures." *J. Soc. Indus. Appl. Math*. 13:312–332.

J. Rhys. 1970. "Shared fixed cost and networks flows." *Manag. Sci*. 17:200–207.

R. T. Rockafellar. 1963. *Convex Functions and Dual Extremum Problems*. Thesis. Department of Mathematics, Harvard University.

R. T. Rockafellar. 1967. "Convex programming and systems of elementary monotonic relations." *J. Math. Anal. Appl*. 19:167–187.

R. T. Rockafellar. 1969. "The elementary vectors of a subspace of R^n." In *Combinatorial Math. and its Applications*. Proc. of the Chapel Hill Conference 1967. R. C. Bose and T. A. Dowling, eds. University of North Carolina Press, pp. 104–127.

R. T. Rockafellar. 1970. *Convex Analysis*. Princeton University Press.

R. T. Rockafellar. 1970. "Some convex programs whose duals are linearly constrained." In *Nonlinear Programming*. J. B. Rosen et al, eds. Academic Press, pp. 293–322.

R. T. Rockafellar. 1974. *Conjugate Duality and Optimization*. Regional Conference Series in Applied Mathematics 16. SIAM.

R. T. Rockafellar. 1978. "Duality in optimal control," In *Mathematical Control Theory*. W. A. Coppel, ed. Lecture Notes in Math. No. 680. Springer-Verlag, pp. 219–257.

R. T. Rockafellar. 1981. "Monotropic programming: descent algorithms and duality." In *Nonlinear Programming* 4. O. L. Mangasarian et al., eds., Academic Press, pp. 327–366.

R. T. Rockafellar. 1983. "Automatic step sizes for the fortified descent algorithms in monotropic programming." In *Mathematical Programming*, R. V. Cottle, M. Kelmanson, and B. Korte, eds. North-Holland.

D. R. Ryan and S. Chan. 1981. "A comparison of three algorithms for finding fundamental cycles in a directed graph." *Networks* 11:1–12.

H. J. Ryser. 1957. "Combinatorial properties of matrices of zeros and ones." *Canad. J. Math*. 9:371–377.

R. Saigal. 1967. *Multicommodity Flows in Directed Networks*. Thesis. Stanford University.

P. D. Seymour. 1977. "The matroids with the max-flow min-cut property." *J. Combinatorial Theory*, B 23:189–222.

P. D. Seymour. 1980. "Decomposition of regular matroidsatorial" *J. Combinatorial Theory Series B* 28:305–359.

J. Shapiro. 1977. "A note on the primal-dual and out-of-kilter algorithms for network optimization problems." *Networks* 7:81–88.

D. R. Shier. 1979. "On algorithms for finding the k shortest paths in a network." *Networks* 9:195–214.

D. R. Shier and C. Witzgalls. 1980. "Arc tolerances in shortest path and network flow problems." *Networks* 10:277–291.

A. W. Shogan. 1977. "Bouding distributions for a stochastic PERT network." *Networks* 7:359–381.

P. Slepian. 1968. *Mathematical Foundations of Network Analysis*. Springer-Verlag.

S. Smale. 1983. "The problem of the average speed of the simplex method." In *Mathematical Programming: The State of the Art*. A. Bachem et al, eds. Springer–Verlag, pp. 530–539.

V. Srinivasan and G. L. Thompson. 1973. "Benefit-cost analysis of coding techniques for the primal transportation algorithm." *J. ACM* 20:194–213.

E. Stiemke. 1915. "Über positive Lösungen homogener linearer Gleichungen." *Math. Ann.* 76:340–342.

N. Tomizawa. 1972. "On some techniques useful for solution of transportation network problems." *Networks* 1:173–194.

J. A. Tomlin. 1972. "Modifying triangular factors of the basis in the simplex method." In *Sparse Matrices and their Applications*. D. J. Rose and R. Willoughby, eds. Plenum.

A. W. Tucker. 1956. "Dual systems of homogeneous relations," in *Linear Inequalities and Related Systems* (H. W. Kuhn and A. W. Tucker, eds.). *Ann. of Math. Studies* 38:3–18.

A. W. Tucker. 1960. "A combinatorial equivalence of matrices." In *Combinatorial Analysis. Amer. Math. Soc*. Proc. Symposia in Appl. Math. (R. Bellman and M. Hall, Jr., eds.) 10:129–140.

A. W. Tucker. 1963. "Combinatorial theory underlying linear programs." In *Recent Advances in Math Programming*. R. L. Graves and P. Wolfe, eds. McGraw-Hill, pp. 1–16.

W. T. Tutte. 1971. *Introduction to the Theory of Matroids*. American Elsevier.

J. G. Wardrop. 1952. "Some theoretical aspects of road traffic research." *Proc. Inst. Civil Engineers*, part 2, 1(2):325–378.

A. Weintraub and J. Gonzalez. 1980. "An algorithm for the traffic assignment problem." *Networks* 10:197–209.

D. J. A. Welsh. 1976. *Matroid Theory*. Academic Press.

H. Whitney. 1935. "On the abstract properties of linear dependence." *Amer. J. Math.* 57:509–533.

A. W. Willson, Jr. 1974. *Nonlinear Networks: Theory and Analysis*. IEEE Press.

R. J. Wilson. 1972. *Introduction to Graph Theory*. Academic Press.

R. D. Wollmer. 1972. "Multicommodity networks with resource constraints." *Networks* 1:245–263.

N. Zadeh. 1973a. "A bad network problem for the simplex method and other minimum cost flow problems." *Math. Prog.* 5:255–266.

N. Zadeh. 1973b. "More pathological examples for network flow problems." *Math. Prog.* 5:217–224.

N. Zadeh. 1979. "A simple alternative to the out-of-kilter algorithm." Technical Report 25. Department of Operations Research, Stanford University.

N. Zadeh. 1980. "What is the worst case behavior of the simplex algorithm." Technical Report 37. Department of Operations Research, Stanford University.

W. I. Zangwill. 1967. "The convex simplex method." *Manag. Sci.* 3:221–238.

W. I. Zangwill. 1969. *Nonlinear Programming: A Unified Approach*. Prentice-Hall.

INDEX